AF708880

ENVIRONMENTAL BIOTECHNOLOGY

Reducing Risks from Environmental Chemicals through Biotechnology

BASIC LIFE SCIENCES

Alexander Hollaender, Founding Editor

Recent volumes in the series:

Volume 28 GENETIC CONTROL OF ENVIRONMENTAL POLLUTANTS
Edited by Gilbert S. Omenn and Alexander Hollaender

Volume 29 SISTER CHROMATID EXCHANGES: 25 Years of Experimental Research
Part A: The Nature of SCEs
Part B: Genetic Toxicology and Human Studies
Edited by Raymond R. Tice and Alexander Hollaender

Volume 30 PLASMIDS IN BACTERIA
Edited by Donald R. Helinski, Stanley N. Cohen, Don B. Clewell, David A. Jackson, and Alexander Hollaender

Volume 31 GENETIC CONSEQUENCES OF NUCLEOTIDE POOL IMBALANCE
Edited by Frederick J. de Serres

Volume 32 TISSUE CULTURE IN FORESTRY AND AGRICULTURE
Edited by Randolph R. Henke, Karen W. Hughes, Milton J. Constantin, and Alexander Hollaender

Volume 33 ASSESSMENT OF RISK FROM LOW-LEVEL EXPOSURE TO RADIATION AND CHEMICALS: A Critical Overview
Edited by Avril D. Woodhead, Claire J. Shellabarger, Virginia Pond, and Alexander Hollaender

Volume 34 BASIC AND APPLIED MUTAGENESIS: With Special Reference to Agricultural Chemicals in Developing Countries
Edited by Amir Muhammed and R. C. von Borstel

Volume 35 MOLECULAR BIOLOGY OF AGING
Edited by Avril D. Woodhead, Anthony D. Blackett, and Alexander Hollaender

Volume 36 ANEUPLOIDY: Etiology and Mechanisms
Edited by Vicki L. Dellarco, Peter E. Voytek, and Alexander Hollaender

Volume 37 GENETIC ENGINEERING OF ANIMALS: An Agricultural Perspective
Edited by J. Warren Evans and Alexander Hollaender

Volume 38 MECHANISMS OF DNA DAMAGE AND REPAIR: Implications for Carcinogenesis and Risk Assessment
Edited by Michael G. Simic, Lawrence Grossman, and Arthur C. Upton

Volume 39 ANTIMUTAGENESIS AND ANTICARCINOGENESIS MECHANISMS
Edited by Delbert M. Shankel, Philip E. Hartman, Tsuneo Kada, and Alexander Hollaender

Volume 40 EXTRACHROMOSOMAL ELEMENTS IN LOWER EUKARYOTES
Edited by Reed B. Wickner, Alan Hinnebusch, Alan M. Lambowitz, I. C. Gunsalus, and Alexander Hollaender

Volume 41 TAILORING GENES FOR CROP IMPROVEMENT: An Agricultural Perspective
Edited by George Bruening, John Harada, Tsune Kosuge, and Alexander Hollaender

Volume 42 EVOLUTION OF LONGEVITY IN ANIMALS: A Comparative Approach
Edited by Avril D. Woodhead and Keith H. Thompson

Volume 43 PHENOTYPIC VARIATION IN POPULATIONS: Relevance to Risk Assessment
Edited by Avril D. Woodhead, Michael A Bender, and Robin C. Leonard

Volume 44 GENETIC MANIPULATION OF WOODY PLANTS
Edited by James W. Hanover and Daniel E. Keathley

Volume 45 ENVIRONMENTAL BIOTECHNOLOGY: Reducing Risks from Environmental Chemicals through Biotechnology
Edited by Gilbert S. Omenn

A Continuation Order Plan is available for this series. A continuation order will bring delivery of each new volume immediately upon publication. Volumes are billed only upon actual shipment. For further information please contact the publisher.

ENVIRONMENTAL BIOTECHNOLOGY

Reducing Risks from Environmental Chemicals through Biotechnology

Edited by

Gilbert S. Omenn

University of Washington
Seattle, Washington

Associate Editors

Rita Colwell, A. M. Chakrabarty,
Morris Levin, and Perry McCarty

Technical Editors

Claire M. Wilson and
Gregory Kuny

Council for Research Planning in Biological Sciences, Inc.
Washington, D.C.

PLENUM PRESS • NEW YORK AND LONDON

Library of Congress Cataloging in Publication Data

Conference on Reducing Risks from Environmental Chemicals through Biotechnology (1987: Seattle, Wash.)
Environmental biotechnology.

(Basic life sciences; v. 45)
"Proceedings of a Conference on Reducing Risks from Environmental Chemicals through Biotechnology, held July 19–23, 1987, in Seattle, Washington"—T.p. verso.
Includes bibliographical references and index.
1. Sewage—Purification—Biological treatment—Congresses. 2. Chemicals—Environmental aspects—Congresses. 3. Pollution—Environmental aspects—Congresses. 4. Microbial ecology—Congresses. I. Omenn, Gilbert S. II. Colwell, Rita R., 1934– . III. Title. IV. Series.
TD755.C67 1987 628 88-17886
ISBN 0-306-42984-5

Proceedings of a conference on Reducing Risks from Environmental Chemicals through Biotechnology, held July 19–23, 1987, in Seattle, Washington

This book was copyedited and entirely retyped by the staff of the Council for Research Planning in Biological Sciences, Inc., located on the premises of Associated Universities, Inc., of which the Council is a guest.

The opinions expressed herein reflect the views of the authors, and mention of any trade names or commercial products does not necessarily constitute endorsement by the funding sources.

© 1988 Plenum Press, New York
A Division of Plenum Publishing Corporation
233 Spring Street, New York, N.Y. 10013

All rights reserved

No part of this book may be reproduced, stored in a retrieval system, or transmitted in any form or by any means, electronic, mechanical, photocopying, microfilming, recording, or otherwise, without written permission from the Publisher

Printed in the United States of America

DEDICATION

Alexander Hollaender

(1898-1986)

Pioneering scientist, inspirational mentor, discerning art collector. Dr. Hollaender was the father of radiation biology, longtime director of the Biology Division at Oak Ridge National Laboratory, a founder of the Radiation Research Society and the Environmental Mutagen Society, member of the National Academy of Sciences (1957), recipient of the Enrico Fermi Award (1983). He had a remarkable capacity to look beyond the well-recognized problems of the day, even at age 87, as this Conference was organized.

The Editors

CONTENTS

CORPORATE PANEL

PUBLIC PARTICIPATION

WELCOME AND INTRODUCTION

Gilbert S. Omenn

Dean, School of Public Health and
Community Medicine
University of Washington
Seattle, Washington 98195

On behalf of the University of Washington, the City of Seattle, the Steering Committee, and the sponsoring agencies, corporations, and organizations, I welcome you. We all expect this Conference to stimulate further what is becoming an important application of biotechnology in an area in which our society experiences considerable frustration and gloom: the management of hazardous wastes.

It is an all-too-frequent refrain that technology has its benefits and its risks. To many--in the lay public, at least--the damaging notion has taken hold that we are capable of creating problems but are less capable of finding solutions. Chemical streams from industry, agriculture, municipal operations, and household operations have contaminated groundwater, drinking water, and soils, and have undermined the productivity of agriculture and the quality of life. In the meantime, however, we have improved our quality of life in immeasurable ways through some related developments. The challenge is to continue the enhancements while modifying or preventing the damage.

With regard to wastes, the basic strategies are (1) to minimize the amount of waste generated, (2) to reduce the toxicity of wastes that are generated, and (3) to find more satisfactory ways of disposing of and degrading wastes which have accumulated. It is no secret that the "bury or burn" practice is being exhausted. Landfills are filling up and leaking, and it is harder and harder to find sites for new landfills. While there are some exciting developments about incineration, it is nearly as hard to find acceptable sites for incinerators. It is certain that the demonstration of techniques to degrade and detoxify current wastes and to decontaminate environmental compartments could go a long way toward addressing public frustration.

The 1984 Amendments to the Resource Conservation and Recovery Act (RCRA) declared reliance on land disposal of untreated hazardous wastes to be no longer acceptable. A series of land disposal restrictions was established in those Hazardous Solid Waste Amendments which require the U.S. Environmental Protection Agency (EPA) to ban land disposal of entire classes of hazardous wastes or define predisposal treatment standards for them. The Superfund Amendment and Reauthorization Act of 1986 (SARA) revised waste management practices under the Comprehensive Environmental Response, Compensation and Liability Act (CERCLA). Permanent on-site remedies are now explicitly preferred; "front-end" management and treatment have priority over off-site clean-up remedies. In addition, the SARA capacity certification requirement--that states identify, by October of 1989, sufficient treatment, storage, or disposal capacity to handle all hazardous wastes expected to be generated within the state over the next 20 years (a very unrealistic mandate)--stimulates states to consider ways to reduce that burden of wastes.

Our Conference and this resulting book were organized as follows. Part I updates and stimulates us with advances in the scientific underpinnings for approaches to degradation and detoxification of hazardous wastes. This work, from microbial ecology, molecular biology, and a host of interdisciplinary efforts, brings together ecology, physiology, and genetics to select organisms and--in the future--to design organisms that have the least likelihood of risk and the most likelihood of functioning in the environmental tasks to which they are applied. There should be no doubt that in this work the aim is to introduce microorganisms into the environment. We are not talking about inadvertent releases to the environment. Thus, the development of strategies to satisfy the public, the regulators, and the scientific community that the approaches taken will have the least likelihood of risk--based upon good scientific knowledge, tight engineering controls, and stepwise field site demonstrations--is crucial to these applications.

Part II is about applications. We here record dramatic advances since our 1983 Conference and 1984 book, Genetic Control of Environmental Pollutants (G.S. Omenn and A. Hollaender, eds., Plenum Press, New York). Most of the applications are real field demonstrations in important groundwater and soil sites--with bioreactors taken into the field and with biodegradation and major engineering capabilities in combination. As a general theme, it is not microbiology alone, not engineering alone, but the capacity to marry the two that is generating good results and much excitement. Companies, academic groups, and federal agencies are joining forces to identify good opportunities and to accomplish practical field applications. The plenary sessions are supplemented by more than 30 poster presentations, abstracts of which are published herein.

Part III addresses policy aspects: the role of biotechnology in pollution control, and the role of genetic engineering in biotechnology; regulatory approaches; corporate strategies; and public participation.

We may take our 1983 conference conclusions as a point of departure:

(1) Nature offers remarkable biodegradative capabilities in certain organisms.

(2) Ecosystems present severe challenges for survival and effectiveness of genetically engineered or selected organisms.

(3) The best strategy for efficacy and safety is to isolate organisms from environments of interest (sediments, sludges, dumps, soils).

(4) The most promising targets of opportunity are:

a. point sources of industrial effluents with high concentrations of specific chemicals;

b. spills of particular chemicals in contained areas; and

c. dump sites being prepared for encapsulation or excavation.

(5) The marketplace will be a tough test for genetic approaches.

(6) There was considerable support for guidelines or regulations to inform and protect industrial and academic applications of genetic engineering in pollution control.

With regard to item 6, there has been a concerted effort of the Reagan administration, led by the Office of Science and Technology Policy--and, lately, chaired by Dr. David Kingsbury of the National Science Foundation--to analyze the existing regulatory framework and orchestrate a reasonable array of intra-agency and interagency guidelines and regulatory processes. Our Conference syllabus contains a copy of the relevant Federal Register notice (26 June 1986).

Several reports have since appeared from scientific and technical meetings. A particularly noteworthy recent publication, given its expression of a mandate by EPA, is the "The Proceedings of the U.S. EPA Workshop on Biotechnology and Pollution Control" (November, 1986). Its conclusions are the following:

(1) The nation needs alternative approaches to complement present "burn or bury" approaches to chemical pollutants.

(2) Improvements are needed within the microbiological treatment arena, some of which might draw upon genetic engineering methods.

(3) EPA's regulatory scheme and stimulation of innovation should encourage stepwise testing in the laboratory and in field sites.

(4) Federally designated field sites are essential.

(5) Cooperation between EPA and developers of the technology, and communication with the public, can be fostered in several specific ways.

As chair of that workshop, I remain committed to stimulating federal agencies, especially EPA, to recognize their mandate not only to regulate but also to foster development and application of better technologies. Item (4) is particularly needed. It should not be left to the academics or companies to try to identify sites that are acceptable to the regulatory agencies. There are plenty of sites on the National Priority List and many less complicated sites that would be suitable for field tests of newly emerging technologies and combinations of technologies.

I want to thank all of our sponsors. The National Science Foundation, the U.S. Environmental Protection Agency, the U.S. Air Force, the National Institute for Environmental Health Sciences, and the Department of Energy have all contributed generously to this meeting, both with funding and with suggestions for program development. The American Institute of Chemical Engineers and the Water Pollution Control Federation have been helpful in publicizing the Conference and in the development of the program. We're very pleased to have corporate sponsorship from the Ecova Corporation (Redmond, WA), the Electric Power Research Institute, Procter and Gamble, DuPont, Amgen, Amoco, Monsanto, Standard Oil of Ohio, Chevron, and Texaco. Finally, the Council for Research Planning in Biological Sciences, the late Dr. Alexander Hollaender's organization, and the University of Washington have been involved from the start. To all these sponsors, our thanks and our pleasure that they are represented here in our scientific, engineering, and policy discussions.

This Conference and its book join a continuing series, stimulated by Dr. Hollaender and published by Plenum Press, about important applications of genetic engineering. I am proud to dedicate the Conference and the book to Dr. Hollaender's memory.

Next, I thank our Steering Committee: Martin Alexander, Cornell University; A.M. Chakrabarty, University of Illinois; Rita Colwell, University of Maryland; Arnold Demain, Massachusetts Institute of Technology; William Farland, Morris Levin, and Mark Segal of the U.S. Environmental Protection Agency; Alexander Hollaender, Council for Research Planning; James Spain, U.S. Air Force Engineering Center (Pensacola, FL); William St. John, Ecova Corporation; James Tiedje, Michigan State University; Michael Torpy, then Argonne National Laboratory and now Ecova Corporation; and Michael Weishaar, Monsanto Company.

Finally, I want to thank Sharon Morris, Linda Crew, Marge Falkner, and Virginia Daigle of the University of Washington, and Claire Wilson, Greg Kuny, and Cosette Saslaw of the Council for Research Planning, for excellent staff work on the conference and the book.

In the Appendix to this Volume, we have provided several well-selected lists of chemicals which have been given priority status in the water, air, and hazardous waste programs of the EPA, and under Proposition 65 in the State of California. These lists surely indicate areas of need and targets of opportunity for applications of biotechnology.

I trust that the 400 participants in this meeting will find the program productive, the contacts useful, and Seattle enjoyable. We all hope that the book will give this field another boost.

MICROBIAL ECOLOGY

THE ECOLOGY OF AN ANAEROBIC DECHLORINATING CONSORTIUM

James M. Tiedje and Todd O. Stevens

Departments of Microbiology and Public Health
and of Crop and Soil Sciences
Michigan State University
East Lansing, Michigan 48824

INTRODUCTION

The mechanisms used by anaerobic bacterial communities to degrade organic molecules are of interest not only to increase our understanding of biogeochemical cycling, but also to explore the potential of microorganisms to degrade xenobiotic molecules in anaerobic environments (12). This has been of increasing concern in recent years, as many man-made chemicals are found to be toxic or mutagenic. New methods must be found to dispose of these compounds when they are no longer of use and to clean up contaminated areas. Digestion by anaerobic bacteria is an attractive possibility for disposal of synthetic chemicals, because it can require less energy than other methods of waste disposal; construction costs for anaerobic treatment facilities are lower; and anaerobic digestion offers the possibility of recovering useful products, such as intermediary metabolites or methane gas (30). Studying these processes also aids in predicting the fate of xenobiotics that have been introduced into natural anaerobic environments.

ANAEROBIC COMMUNITIES AND XENOBIOTIC DEGRADATION

A given anaerobic microorganism can rarely take up a complex organic molecule and mineralize it to carbon dioxide. Rather, a succession of specialized organisms modify the molecule in turn, each deriving a small amount of carbon and/or energy from the reaction (14,17,24). For example, a typical organic polymer may be degraded into monomers by one organism. The monomers may then be fermented by one or more other organisms into volatile fatty acids (VFAs), and these may then be fermented to CO_2 and acetate which can be used by other organisms to produce methane or CO_2 (15,32). Complex trophic webs can arise from the interactions among these organisms. In addition, reactions that may be thermodynamically unfavorable under in situ conditions--for example, the oxidation of benzoate--are able to proceed when the H_2 partial pressure is reduced by a syntrophic hydrogen-consuming organism. Some of these interactions are so tightly coupled that associations of two organisms were classified as a single species for years before they were discovered to be mixed cultures (6,20).

Because anaerobic bacterial communities consist of a large number of different specialized organisms, such a community might be successfully selected to degrade a compound more readily than will a given single organism. When incubated with slowly increasing concentrations of a compound, bacteria that are able to derive carbon or energy from the compound, or its metabolites, multiply. The population then becomes enriched with higher numbers of these species. New associations of specialized anaerobes can join into syntrophic reactions to degrade the compound without as many major genetic mutations and rearrangements as would be required if a single species were to degrade the compound.

Such enrichment experiments have been used to explore the degradation of a number of halogenated aromatic compounds. Halogenated aromatic compounds are common constituents of pesticides, are found in industrial waste streams, and are common components of hazardous waste sites. They include such serious contaminants as dioxins and polychlorinated biphenyls (PCBs). Environments such as sewage treatment plants have experienced selection of haloaromatic-degrading organisms. It is not entirely unreasonable to expect to find natural enzymes that can degrade these compounds, since some chlorinated aromatic compounds are found in nature, although most are probably of marine origin (13,16,21). Enzyme systems that have evolved to degrade nonchlorinated compounds may also be nonspecific enough to degrade their chlorinated homologs.

Anaerobic dechlorination of several classes of chlorinated aromatic compounds has been demonstrated in the laboratories of Steven Boyd and myself. Examples from the various chemical classes are summarized in Fig. 1. These conversions are carried out by anaerobic microbial consortia that have been successfully enriched on the chlorinated compound as substrate, or by consortia that dechlorinate directly in samples from the natural habitat, albeit at a slower rate. Substrates that have yielded enriched consortia include 3-chlorobenzoate (29), 2,4-dichlorophenoxyacetic acid (2,4-D), 2,4,5-trichlorophenoxyacetic acid (2,4,5-T) (18,28), chlororesorcinol (11), and various chlorinated phenols (5), including pentachlorophenol (19). Hexachlorobenzene (12) and at least some PCBs (31) were dechlorinated by sewage sludge and river sediments, respectively. The dechlorination of these nonpolar compounds is particularly important since they represent the more persistent problem chemicals in contamination sites. These compounds are not completely dechlorinated, however, and do not readily result in an enriched dechlorinating population. Thus, it is difficult to accelerate the rate of degradation and to isolate the organisms resposible for these useful transformations. For this reason, we have used consortia that enrich and, in one case, yielded the isolate DCB-1 (23), to study the ecology of the consortia members as well as the mechanism of the reaction.

Because of the interactions necessary among consortia members, they are, not surprisingly, often found in clumps. Figure 2 shows a clump of bacteria from a chlorophenol-degrading enrichment. The close proximity of the organisms likely aids in interspecies transfer of hydrogen and other metabolites, and allows the group of organisms to function as a single unit.

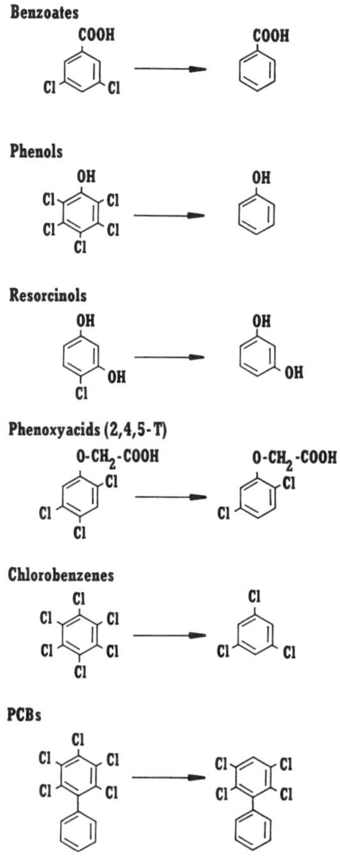

Fig. 1. Examples of reductive dechlorination reactions by anaerobic microbial communities. The identity of the PCB product shown was indicated by the gas chromatographic retention time, but the meta dechlorination product also remains a possible identity for this dechlorination product.

THE 3-CHLOROBENZOATE MINERALIZING CONSORTIUM

One model system for studying the ecology of microorganism-xenobiotic interactions is a sludge community that has been enriched on chlorobenzoates for a number of years (22). Eight anaerobic bacteria have been isolated from this enrichment that can grow on 3-chlorobenzoate as their sole carbon and energy source (23). We have studied these organisms both in axenic cultures and in more "realistic" mixed cultures, to gain an understanding of their interactions. After physiological characterization of the isolates, it appeared that at least three were needed to grow on the 3-chlorobenzoate substrate.

The dechlorinating bacterium, designated strain DCB-1, is so far unique in its morphology and metabolism. At present, this is the only obligately anaerobic bacterium isolated that is known to carry out reductive

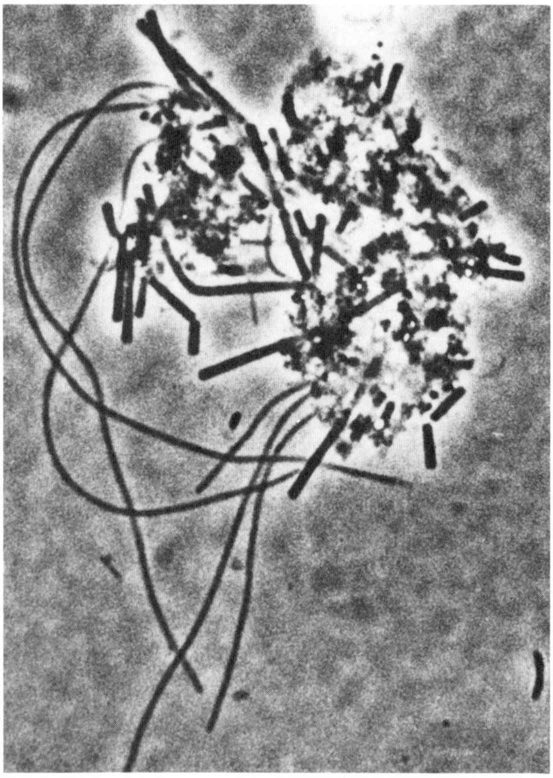

Fig. 2. A typical clump of bacteria in an anaerobic chlorophenol-degrading enrichment. Note the physical contact and diversity of bacterial types. Magnification, 4,800X. (Culture courtesy of M. Mikesell and S. Boyd.)

dechlorination. In this syntrophic association, it carries out the following reaction (9,23):

$$\text{3-Cl-Benzoate} + H_2 \rightarrow \text{benzoate} + Cl^- + H^+ \tag{1}$$

It does not grow on benzoate or chlorobenzoate, and presumably receives some carbon from later stages in the oxidation of benzoate. Strain DCB-1 is a long, rod-shaped, nonmotile bacterium with a distinctive "collar" structure near one end (23). It has a very specialized substrate range; only pyruvate can be fermented in mineral medium, although a few other substrates can be used mixotrophically (25). Carbon dioxide forms a large part of the cellular material. DCB-1 can also conserve energy by reducing sulfidogenic electron acceptors, allowing some growth on acetate plus H_2 (26). This bacterium can also reductively demethoxylate 3-methoxybenzoate, but this activity has been found to be unrelated to its dechlorinating ability (8).

The benzoate oxidizer, designated strain BZ-2, has been grown only in co-culture with a hydrogen-consuming bacterium, such as a methanogen or a sulfidogen (23). Benzoate is the only substrate known to support its growth, and is oxidized stoichiometrically to acetate by the reaction:

$$\text{Benzoate} + 7H_2O \rightarrow \text{3-acetate} + 3H_2 + HCO_3^- + 3H^+ \qquad (2)$$

The methanogen, designated strain PM-1, has been identified as a *Methanospirillum* species, and is required to keep the hydrogen levels low to enable benzoate oxidation to take place (23). The only known substrate for *Methanospirillum* is formate or H_2/CO_2 (3). In this association, it evidently carries out the reaction:

$$4H_2 + CO_2 \rightarrow CH_4 + 2H_2O \qquad (3)$$

Another methanogen, a *Methanobacterium* sp., designated strain PM-2, was also isolated from the consortium.

In addition, the consortium contained a sulfidogen, designated as *Desulfovibrio* strain PS-1, and two butyrate-oxidizing bacteria. One of the butyrate oxidizers, a spore former, was designated strain SF-1, and the other, a nonspore former, was designated strain NSF-2. These organisms were not required for mineralization of 3-chlorobenzoate, and their place in the food web was not established (23,30).

Integrating the above three reactions suggests that the organisms responsible for these reactions, when combined, might be able to grow on 3-chlorobenzoate and produce methane, acetate, and chloride. Indeed, this was the case: the reconstructed consortium, consisting of DCB-1, BZ-2, and PM-1, grew as fast on 3-chlorobenzoate as did the original enrichment (9). This successful result raised three questions about how these organisms interact, especially how DCB-1 grows by simply removing chlorine from the aromatic ring. The questions are: (i) where does DCB-1 get its energy?; (ii) where does it get its reductant?; and (iii) where does it get its carbon for cell material? Figure 3 summarizes the transformations by each organism and previews the answers to some of these questions.

Recent experiments using the reconstructed consortium have shown that the consortium apparently derives energy from the reductive dechlorination, since the cell yield is greater when the consortium is grown on 3-chlorobenzoate than when it is grown on benzoate (10). Other experiments have shown that the association is a true trophic "web," rather than a "chain," since one third of the hydrogen from benzoate oxidation is returned to the dechlorinator for reduction of 3-chlorobenzoate (9). Presumably, since strain DCB-1 is incapable of growing on benzoate or on 3-chlorobenzoate, some form of carbon is also returned to it from benzoate oxidation. Under microscopic observation, the benzoate-oxidizing bacterium is often seen attached to the dechlorinating bacterium, as shown in Fig. 4, suggesting that intimate contact may facilitate transfer of some intermediate in 3-chlorobenzoate mineralization.

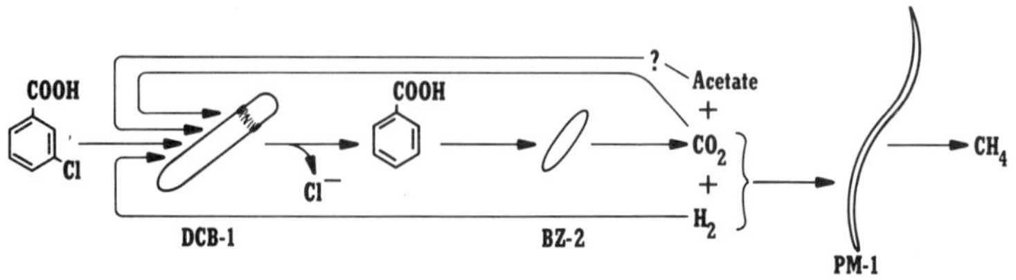

Fig. 3. Members of the reconstituted consortium and the reactions they carry out. Strain DCB-1 is the dechlorinating sulfidogen; BZ-2 is probably a strain of Syntrophus buswellii; and PM-1 is a strain of Methanospirillum. DCB-1 apparently competes effectively for H_2 and CO_2 with PM-1 to provide the reductant and cell carbon for the dechlorinator.

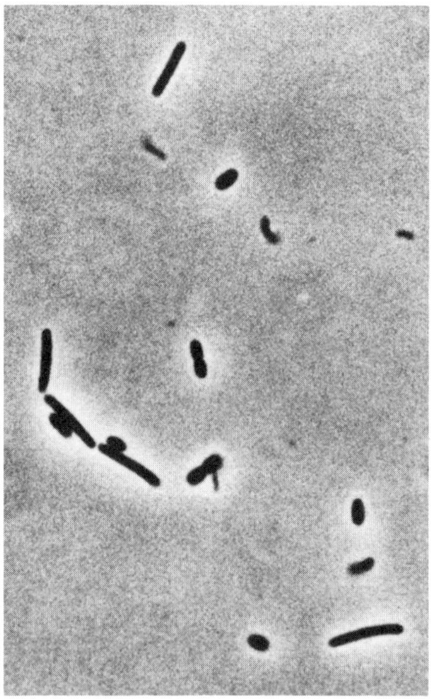

Fig. 4. Bacteria of the 3-chlorobenzoate-degrading consortium. Note the close physical association of the shorter bacterium, strain BZ-2, with the longer bacterium, strain DCB-1. Magnification, 4,800X.

In recent experiments, we investigated the flow of ^{14}C-labeled substrates through the consortium and found that CO_2 appears to be the major source of DCB-1 cell carbon (T.O. Stevens, J. Dolfing, and J.M. Tiedje, unpubl. data). But, to quantitatively assess how much carbon is derived from various intermediates in the food chain, a means is needed to separate DCB-1 from the other two members. We have used two approaches: (i) separation of the cells by their buoyant densities using a Percoll gradient (4); and (ii) separation of the members by a membrane during growth. Although separation of the three species was sometimes possible, as shown in Fig. 5, the buoyant density was variable, and satisfactory separation could not be achieved with aged cultures.

For separation of consortia members by a membrane, the members were grown in a sealed spinner flask and separated by a 0.2-micron pore size membrane. Strain DCB-1, the dechlorinating organism, was on one side, and the benzoate degrader and a methanogen were on the other. 3-Chlorobenzoate was degraded, as shown in Fig. 6, although at a much slower rate than occurred when the organisms were intimately associated. If the dividing membrane was ruptured, the combined culture grew and dechlorinated at the faster rate seen for the bottle-grown consortium. This observation indicates that the medium and conditions in this vessel were adequate for normal growth, and that the physical separation of the members of the consortium was apparently responsible for the slow growth and dechlorination. Because of the slow growth and projected low molecular weight of the "exchanged" substrates, slow diffusion through the membrane is not a likely explanation for the limited growth. As evidence of this, 3-chlorobenzoate concentrations were the same on both sides of the membrane (Fig. 6). Perhaps physical contact is required for, or stimulates, this syntrophic interaction.

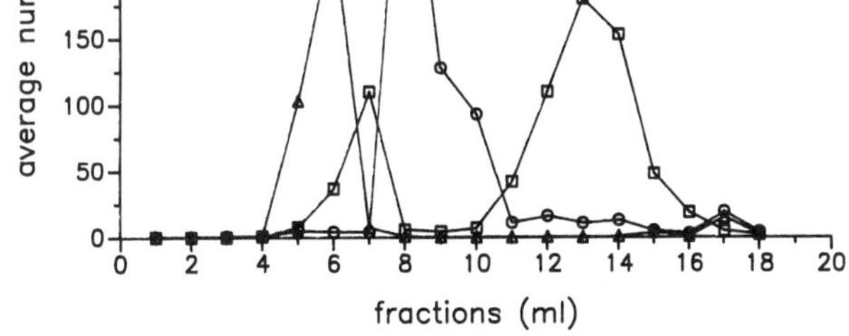

Fig. 5. Distribution of 3-chlorobenzoate-degrading consortium members in a Percoll gradient.

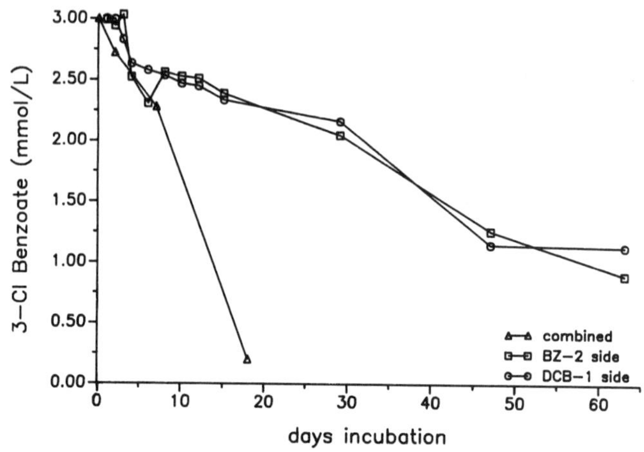

Fig. 6. Degradation of 3-chlorobenzoate by the membrane-divided consortium and by the combined consortium.

STIMULATION OF DECHLORINATING ACTIVITY

When originally isolated, the organisms in the chlorobenzoate consortium grew very slowly, to only low densities, and had low halobenzoate-degrading activity (23). Obviously, the laboratory conditions that were used to grow such organisms are only an approximation of the conditions in their natural environment and may not be the best suited for their growth. In addition, the reaction of interest may be dependent on, or regulated by, some other metabolic reaction or condition which remains unknown and is, therefore, not considered in determining the culture conditions.

A number of improvements in culture techniques have been worked out for strain DCB-1, the dehalogenating organism. Since its isolation, the growth rate, maximum density, and dechlorination rate have all increased dramatically.

Medium Reductant

Strict anaerobes require chemical reduction of the growth medium, which is typically done with sodium sulfide or cysteine-HCl. This was the method used in initial cultures of the organisms in this consortium (23). Because of the toxicity of sulfide, we investigated the use of other reducing agents. The cytochrome-containing membrane fraction of Escherichia coli has been suggested to be an efficient, nontoxic reductant for use in isolating and propagating anaerobic bacteria (7). We compared the rate of dechlorination by strain DCB-1 in medium reduced by 500 μM sulfide, the E. coli membrane fraction, and 500 μM sodium dithionite. As seen in Fig. 7, reducing the medium with either dithionite or the membrane fraction resulted in much greater rates of dechlorination. Growth of the bacterium was also stimulated. Because of its greater ease of preparation, we now use sodium dithionite routinely in culturing strain DCB-1.

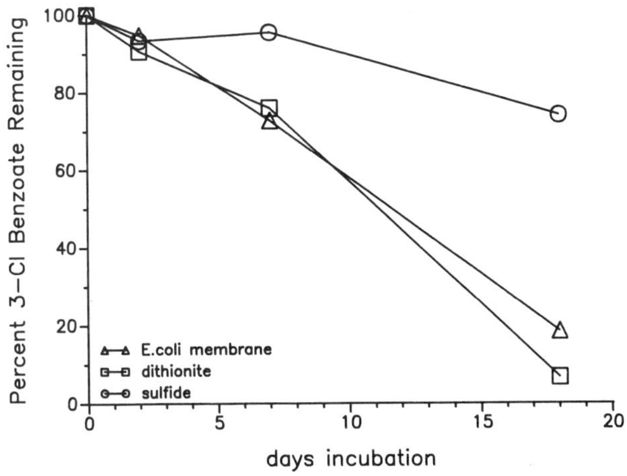

Fig. 7. Dechlorination of 3-chlorobenzoate by strain DCB-1 in media with different reducing agents.

Media Supplements

Clarified rumen fluid (RF) is a complex substance rich in organic acids and many other compounds, and is commonly used as a media supplement for anaerobic bacteria. Rumen fluid was originally used in the isolation of strain DCB-1 and was, in fact, the sole carbon source (23). Even though DCB-1 can now be grown without RF, in the absence of sulfide (26), the addition of RF results in increased growth and dechlorination. Although the nature of the stimulatory compounds in RF is unknown, at least some of these compounds appear to be volatile, since their effectiveness is dependent upon the method of clarification. This is done before addition to the medium, usually by a combination of filtration, centrifugation, and autoclaving. We have found that autoclaving RF in vented containers, while a prudent safety precaution, reduces its effectiveness. Adding the RF prior to autoclaving the medium in sealed serum bottles results in an optical density of over 1.0 O.D. units. Also, sterile RF stored at 4°C loses its effectiveness over a period of several months, while it remains active if frozen at -20°C.

Rumen fluid was originally believed to be required for dechlorination, although relatively very slow dechlorination is now observed in mineral medium without RF. Various volatile aromatic compounds that might be found in RF were added to mineral medium in an attempt to stimulate dechlorination. Tryptophan appeared to stimulate dechlorination somewhat, while indole and skatole did not (Tab. 1). Interestingly, although in pure culture DCB-1 stoichiometrically converted 3-chlorobenzoate to benzoate in the presence of RF, some benzoate appears to have been consumed during dechlorination in mineral medium.

When strain DCB-1 was first isolated, its apparent K for dechlorination was 67 µM (27). Current cultures exhibit apparent K values less than 1 µM (8). Initial growth of DCB-1 rarely reached densities of 0.1

Tab. 1. Dehalogenation of 1,000 μM 3-chlorobenzoate by strain DCB-1 in mineral medium, with and without addition of 500 μM of other aromatic substances, after 35 days.

Addition (500 μM)	3-Chlorobenzoate (μM)	Benzoate (μM)
None	382	480
Tryptophan	203	754
Indole	496	363
Skatole	566	300

Note: Concentrations are means of triplicate cultures.

O.D. units in an incubation period of a month, but cultures now routinely attain densities of 1.0 O.D. units in two weeks. Although we have gradually altered our culture conditions and methods to better stimulate growth and dechlorination by this unique organism, it is worth noting that the reverse is also probably true--the organism has doubtlessly been adapting to our culture conditions. Experiments conducted with such laboratory-adapted organisms can reveal much about biochemical processes and community interactions, as well as provide organisms for industrial applications. However, since these are no longer precisely the same organisms that were present in the original environment, any conclusions about in situ activities or the original communities must be tested by autoecological methods.

ACKNOWLEDGEMENTS

The research summarized here was supported in part by the United States Air Force, the Monsanto Corporation, the United States Environmental Protection Agency, the United States Department of Agriculture, and the Michigan REED Program.

REFERENCES

1. Alexander, M. (1973) Nonbiodegradable and other recalcitrant molecules. Biotechnol. Bioengineer. 15:611-647.
2. Alexander, M.A (1979) Recalcitrant molecules, fallible microorganisms. In Microbial Ecology: A Conceptual Approach, J.M. Lynch and N.J. Poole, eds. Blackwell Scientific Publications, Oxford.
3. Balch, W.E., G.E. Fox, L.J. Magrum, C.R. Woese, and R.S. Wolfe (1979) Methanogens: Reevaluation of a unique biological group. Microbiol. Rev. 43:160-296.
4. Beaty, P.S., N.Q. Wofford, and M.J. McInerney (1987) Separation of Syntrophomonas wolfei from Methanospirillum hungatei in syntrophic cocultures using Percoll gradients. Appl. Environ. Microbiol. 53:1183-1185.

5. Boyd, S.A., and D.R. Shelton (1984) Anaerobic biodegradation of chlorophenols in fresh and acclimated sludge. Appl. Environ. Microbiol. 47:272-277.
6. Bryant, M.P., E.A. Wolin, M.J. Wolin, and R.W. Wolfe (1967) Methanobacillus omelianskii, a symbiotic association of two species of bacteria. Archiv für Mikrobiol. 59:20-31.
7. Crow, W.D., R. Machanoff, and H.I. Adler (1985) Isolation of anaerobes using an oxygen reducing membrane fraction: Experiments with acetone butanol producing organisms. J. Microbiol. Meth. 4:133-139.
8. DeWeerd, K.A., J.M. Suflita, T.G. Linkfield, J.M. Tiedje, and P.H. Pritchard (1986) The relationship between reductive dehalogenation and other aryl substituent removal reactions catalyzed by anaerobes. FEMS Microbiol. Ecol. 38:331-339.
9. Dolfing, J., and J.M. Tiedje (1986) Hydrogen cycling in a three-tiered food web growing on the methanogenic conversion of 3-chlorobenzoate. FEMS Microbiol. Ecol. 38:293-298.
10. Dolfing, J., and J.M. Tiedje (1987) Growth yield increase linked to reductive dechlorination in a defined 3-chlorobenzoate degrading methanogenic coculture. Arch. Microbiol. (in press).
11. Fathepure, B.Z., J.M. Tiedje, and S.A. Boyd (1987) Reductive dechlorination of 4-chlororesorcinol by anaerobic microorganisms. Environ. Toxicol. Chem. Vol. 6 (in press).
12. Fathepure, B.Z., J.M Tiedje, and S.A. Boyd (1988) Reductive dechlorination of hexachlorobenzene to tri- and dichlorobenzenes in anaerobic sewage sludge. Appl. Environ. Microbiol. (in press).
13. Gschwend, P.M., J.K. MacFarlane, and K.A. Newman (1985) Volatile halogenated organic compounds released to seawater from temperate marine macroalgae. Science 227:1033-1035.
14. Hamilton, W.A. (1979) Microbial energetics and metabolism. In Microbial Ecology: A Conceptual Approach, J.M. Lynch and N.J. Poole, eds. Blackwell Scientific Publications, Oxford.
15. Hobson, P.N., S. Bousfield, and R. Summers (1974) Anaerobic digestion of organic matter. Crit. Rev. Environ. Control 4:131-191.
16. King, G.M. (1986) Inhibition of microbial activity in marine sediments by a bromophenol from a hemichordate. Nature 323:257-259.
17. Lovely, D.R., and M.J. Klug (1982) Intermediary metabolism of organic matter in the sediments of a eutrophic lake. Appl. Environ. Microbiol. 43:552-560.
18. Mikesell, M.D., and S.A. Boyd (1985) Reductive dechlorination of the pesticides 2,4-D, 2,4,5-T and pentachlorophenol in anaerobic sludges. J. Environ. Qual. 14:337-340.
19. Mikesell, M.D., and S.A. Boyd (1986) Complete reductive dechlorination and mineralization of pentachlorophenol by anaerobic microorganisms. Appl. Environ. Microbiol. 52:861-865.
20. Postgate, J.R. (1979) The Sulphate Reducing Bacteria, Cambridge University Press, Cambridge.
21. Sato, T., M. Mukaida, Y. Ose, H. Nagase, and T. Ishikawa (1985) Mutagenicity of chlorinated products from soil humic substances. The Science of the Total Environment 46:229-241.
22. Shelton, D.R., and J.M. Tiedje (1984) General method for determining anaerobic biodegradation potential. Appl. Environ. Microbiol. 47:850-857.
23. Shelton, D.R., and J.M. Tiedje (1984) Isolation and partial characterization of bacteria in an anaerobic consortium that mineralizes 3-chlorobenzoic acid. Appl. Environ. Microbiol. 48:840-848.

24. Sleat, R., and J.P. Robinson (1984) The bacteriology of anaerobic degradation of aromatic compounds. J. Appl. Bacteriol. 57:381-394.
25. Stevens, T.O., and J.M. Tiedje (1988) Carbon dioxide fixation and mixotrophic metabolism by strain DCB-1, a unique sulfidogenic dehalogenating organism. Appl. Environ. Microbiol. (submitted for publication).
26. Stevens, T.O., T.G. Linkfield, and J.M. Tiedje (1988) Physiological characterization of strain DCB-1, a unique dehalogenating sulfidogenic bacterium. Appl. Environ. Microbiol. (submitted for publication).
27. Suflita, J.M., J. Robinson, and J.M. Tiedje (1983) Kinetics of microbial dehalogenation of haloaromatic substrates in methanogenic environments. Appl. Environ. Microbiol. 45:1466-1473.
28. Suflita, J M., J. Stout, and J.M. Tiedje (1984) Dechlorination of (2,4,5-trichlorophenoxy) acetic acid by anaerobic microorganisms. J. Agric. Food Chem. 32:218-221.
29. Suflita, J.M., A. Horowitz, D.R. Shelton, and J.M. Tiedje (1982) Dehalogenation: A novel pathway for the anaerobic biodegradation of haloaromatic compounds. Science 218:1115-1117.
30. Tiedje, J.M., S.A. Boyd, and B.Z. Fathepure (1987) Anaerobic degradation of chlorinated aromatic hydrocarbons. Dev. Indus. Microbiol. 27:117-127.
31. Tiedje, J.M., S.A. Boyd, and J.F. Quensen III (1987) Reductive dechlorination of PCBs in anaerobic microbial communities. In Research and Development Program for the Destruction of PCBs, H.L. Finkbeiner and S.B. Hamilton, eds. General Electric Company Corporate Research and Development, Sixth Progress Report, Schenectady, New York, pp. 29-33.
32. Wolin, M.J. (1979) The rumen fermentation: A model for microbial interactions in anaerobic ecosystems. Adv. Microb. Ecol. 3:49-77.

PERFORMANCE OF BIODEGRADATIVE MICROORGANISMS IN SOIL: XENOBIOTIC CHEMICALS AS UNEXPLOITED METABOLIC NICHES

Dennis D. Focht

Department of Soil and Environmental Sciences
University of California
Riverside, California 92521

NATURAL SELECTION AND SURVIVAL OF MICROORGANISMS

Microorganisms are where they are because there is a food source or a metabolic niche to exploit. The word "niche," as derived from the old French *nicher* (i.e., "to nest"), has been synonymous in macroecology to location or habitat. In microbial ecology, niche refers to function rather than location. This is particularly significant with respect to bacteria, since the basic principle of nutritional selection has been used for isolating them from soil since the time of Beijerinck and Winogradsky a century ago. Reduced to basic simplicity, nutrition is the single most powerful selection force in microbial ecology in the absence of extreme environments. The microorganisms that are the best fitted for exploiting the metabolic niches of the environment are the ones that comprise the microbial community in accordance with Darwinian principles of natural selection. Thus, cellulitic fungi and proteolytic bacteria co-exist side by side spatially, while exploiting different niches, and can be isolated exclusively from each other on different selective media. The vast diversity of metabolic niches ranges in simplicity from ammonium to large, uncharacterized macromolecules.

Inoculation of soil with alien organisms or with those obtained from that same soil has usually been an exercise in futility, as recognized by the early soil microbiologists (31). Alien microorganisms are generally unable to compete with the indigenous microflora for the existing resources, except in the case of some rhizobial inoculants or in the case of xenobiotic compounds. Indigenous microorganisms are unable to be maintained at the high artificial population densities under which they are cultured in vitro; consequently, they decline to levels commensurate with the available resources. Since the time of Winogradsky and Beijerinck, many schemes and dreams have arisen to inoculate soil for improving crop yields or for some other noble effort. They have been largely unsuccessful because survival of the microorganisms is dependent upon existing concentrations of the available resources. No amount of wishful thinking will make *Azotobacter*, for example, fix more than its meager share of N_2 without a concomitant input of energy into the system.

Why, then, should specialized bacterial inoculants not also fail in the cleanup of contaminated soil? The answer is consistent with the same principles of natural selection and competition. Xenobiotic chemicals are potential energy sources, yet they are not always exploited as metabolic niches by the indigenous microflora. If indigenous microorganisms are not capable of metabolizing a xenobiotic chemical, then it follows that microorganisms that are isolated from a more dynamic environment (e.g., activated sludge) and that can metabolize the specific xenobiotic chemical should be successful, since competition for that resource does not exist. Several specific examples where inoculation with nonindigenous bacteria has been successful in promoting biodegradation of the xenobiotic compound are presented in Tab. 1.

When indigenous microorganisms are enriched for biodegradation by addition of xenobiotics or when inocula are added to soil, their survival is dependent on the xenobiotic compound. When it is depleted, their numbers will eventually decrease to the level at which they were originally present (in the case of indigenous microorganisms) or to undetectable levels (in the case of inoculants). Thus, when biphenyl was added to soil (Fig. 1) for enhancement of polychlorinated biphenyl (PCB) metabolism, the numbers of biphenyl-utilizers exhibited a characteristic growth and death curve in response to the presence of the specific niche. Similar results were noted with the constructed strain of *Pseudomonas cepacia* ATCC 1100 (10): when 2,4,5-trichlorophenoxyacetic acid (2,4,5-T) was depleted, the number of viable cells declined exponentially. These findings should be reassuring to those concerned about the fate of genetically modified bacteria used for decontamination of toxic organics in soil.

THE SOIL ENVIRONMENT

The principles that hold for the metabolism of natural products are equally applicable to the metabolism of xenobiotic compounds. Soil organic matter is more abundant in humid, cool climates than in dry, hot climates because of greater plant productivity and slower microbial breakdown of humus in the former. Therefore, the 30-year persistence of DDT in forest

Tab. 1. Enhancement of xenobiotic biodegradation in soil with bacterial inoculants.

Compound	Bacterial species	Reference(s)
Pentachlorophenol	*Flavobacterium* sp.	11
Parathion	Nonaxenic consortia	4
2,4,5-Trichlorophenoxyacetic acid	*Pseudomonas cepacia* ATCC 1100	10
3-Chlorobenzoate	*Pseudomonas alcaligenes* C-0	19
Polychlorinated biphenyls	*Acinetobacter* sp. P6	9, 18

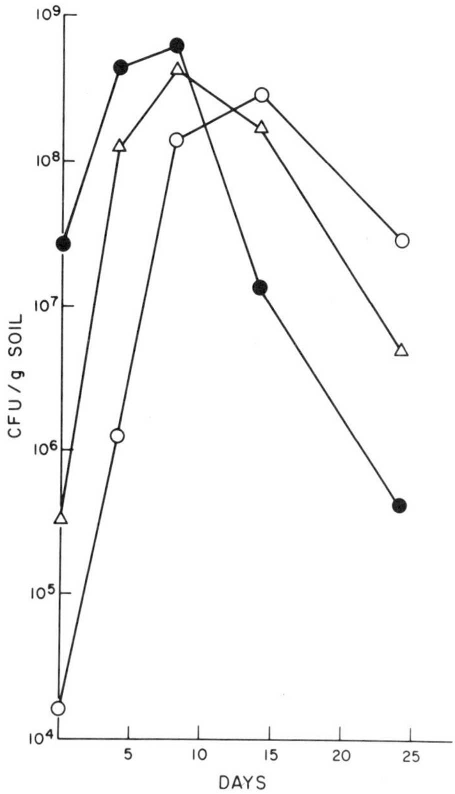

Fig. 1. Growth and death of biphenyl-utilizing bacteria in an Altamont sandy loam treated with biphenyl to enrich for biodegradation of PCBs. Treatments represent inoculation with Acinetobacter sp. P6 at a high (●) and low (△) inoculum density and without inoculation (○). From Focht and Brunner (18).

soils of Maine (14) compared with a 38% disappearance over four weeks in agricultural soils of Colorado (23) is consistent with basic principles of organic matter decomposition in soil.

Temperature

Unlike enteric bacteria, many soil bacteria do not grow optimally--if at all--at 37°C. Common cultural procedures for growth of pseudomonad bacteria are usually optimal between 25° and 30°C. To quantify the relationship of temperature to rate process, the Q_{10} coefficient has been used to evaluate the increase in the rate over a 10° interval. Most biological systems have Q_{10} values of 1.5 to 3. Thus, a rate process that is optimal at 30°C will be 4.5- to 9-fold less at 10°C. Some bacteria may not survive temperatures above 30°C. At the other end of the temperature spectrum (below 10°C), rate processes generally do not adhere to Arrhenius kinetics, and the effect of temperature (Q_{10}) is proportionally much greater. These conditions must be considered in the in situ treatment of xenobiotic compounds with inocula.

pH

Most saprophytic bacteria grow best at neutral to slightly alkaline pH and grow poorly, if at all, at pH 5 or lower. Where high concentrations of chlorinated aromatic compounds are present and where soils have a low base saturation (e.g., sandy soils), liming may be necessary because of the production of HCl that accompanies dehalogenation.

It should not go unnoticed that far more pure culture work on xenobiotic metabolism has been done with bacteria than with fungi. This is because enrichment cultures are established at neutral or slightly alkaline conditions, which favor selection of bacteria over fungi.

Aeration

Aeration is critical for the metabolism of aromatic compounds that are oxidized by dioxygenases and for the growth of the inoculants, which are generally obilgate aerobes. Flooding is to be avoided, since liquid diffusion of O_2 is about 10^{-5} slower through water than through air. Gaseous diffusion essentially ceases when the fraction of the air-filled pore space in soil is below 0.12 (44).

There is some indication that chlorinated aromatic hydrocarbons (CAHs) that are refractory to aerobic attack may be partially metabolized by one or more reductive dehalogenations. DDT, for example, has been demonstrated to be metabolized by a series of reductive dehalogenations of the aliphatic moiety in culture by *Aerobacter aerogenes* (42,43) and in soil (23). Further extensive degradation (namely, ring fission) of these anaerobic metabolites occurs by the separate actions of a pseudomonad bacterium (17) and a dehalogenating fungus (15) separately in vitro. The same pseudomonad bacterium does not metabolize DDT aerobically, but reduces it anaerobically to 4,4'-dichlorobiphenylmethane, which it then can oxidize aerobically to 4-chlorophenylacetic acid. This latter compound is then mineralized by an *Arthrobacter* (33). The production of $^{14}CO_2$ from ^{14}C-labeled DDT in sewage cultures apparently comes from the combined efforts of anaerobic-aerobic consortia (33). Similarly, greater mineralization of methoxychlor occurs when a series of anaerobic and aerobic incubations is employed as opposed to either alone (20). The more refractory chlorinated congeners (hexachloro- and higher) of PCBs, which are not metabolized by aerobic bacteria, are reduced to mono- and dichlorinated products in sediments (8).

Anaerobic biodegradation is much slower in comparison to the aerobic mode. Only recently has the metabolism of mixed anaerobic microbial consortia been elucidated (Tiedje and Stevens, this Volume). Less yet is known about how aerobes and anaerobes specifically interact in nature to metabolize refractory CAHs.

Nutrients

One of the major problems in the treatment of petroleum wastes in soil and in marine systems is the low availability of the two macro elements N and P. Addition of these two elements is necessary when concentrations of xenobiotic chemicals are high because the inoculum may not be as effective in competing with indigenous microflora if these elements are limiting for metabolism of other carbonaceous compounds. Moreover, the high proportion of C to N and P will affect assimilation of available N and P from soil. Finally, micronutrient deficiency may affect the survival and maintenance of the inoculum, although there has been no indication of this being a problem in previous studies (Tab. 1).

Depth

Microbial activity is greatest near the surface, where the organic matter and nutrient concentrations are greatest. Thus, land-farming of contaminated surface soils presents no engineering obstacle. The greatest physical impediment is the treatment of subsurface soil. Microbial numbers generally decline exponentially in subsoils (39,40), where they probably reach a base level of about 10^3/g as reported for groundwaters (7). Since microbial activity is proportional to population density, it is not surprising that the biodegradation of xenobiotic compounds in subsurface soils and in aquifers is a very slow process by comparison to topsoil. (See McCarty, this Volume, for discussion on the methodology and problems encountered with in situ treatment of subsurface soils and groundwater.)

Diffusion

Diffusion of substrate to cells appears to be much slower in soil than in liquid culture. Consequently, some concern may be raised about the optimal density of the inoculum. An example with 3-chlorobenzoate (Fig. 2) shows that low inoculum densities can be satisfactorily used in uniformly mixed soil under laboratory conditions. Although the growth rate (ψ_{max}) in both soil and culture was the same (19), the saturation constant as determined by Monod kinetics was 30 times greater in soil, and the fit to Monod kinetics became poorer as the inoculum density decreased. Thus, the use of formal kinetics derived from well-mixed liquid systems has questionable application to soil.

Spatial Variability

The static and semiporous solid matrix of soil contributes to more formidable predictive problems than are encountered in waste streams, treatment plants, or liquid cultures. Concentrations of soluble, very mobile compounds such as nitrate are log-normally, rather than arithmetically, distributed (32). Consequently, the success of in situ treatment is far more difficult to evaluate than in laboratory studies because of the large number of samples that must be taken in the former. Spatial variability can also have adverse effects if the xenobiotic compound is toxic at high concentrations. Thus, treatment of pentachlorophenol-contaminated soil is not done by land-farming but rather by entire physical removal of the soil to a biological waste-treatment reactor (Pflug and Burton, this Volume). Since contaminated soil in general is still physically removed and hauled off to Class I dumpsites, the economics and engineering practices for reactor treatment of soil may be more practical than land-farming.

CO-METABOLISM AND ANALOG ENRICHMENT

Many microorganisms are capable of effecting one or more reaction steps in a catabolic pathway, yet are unable to further transform the

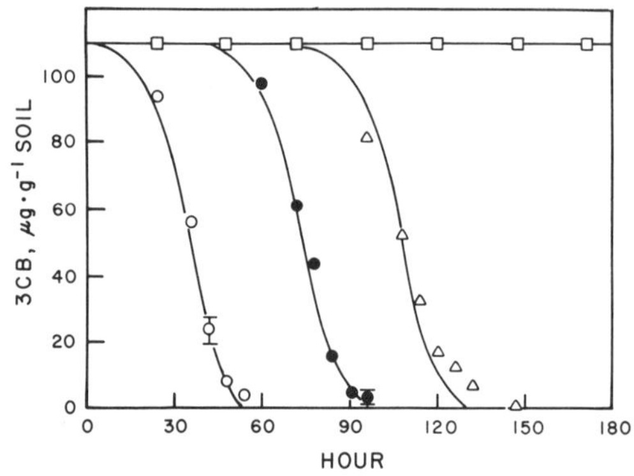

Fig. 2. Effect of different inoculum densities of Pseudomonas alcaligenes C-O on biodegradation of 3-chlorobenzoate (3CB) in a Greenfield soil: none (□), 1.9 x 10^{-1} (○), 1.9 x 10^{-3} (●), 1.9 x 10^{-5} (△), μg cells (dry mass) . g^{-1} soil. Theoretical curves are fit to Monod kinetics using the same parameter estimates for the growth rate constant (ψ_{max} = 0.32 h^{-1}), saturation constant (K_s = 1.1 μmol 3CB · g^{-1} soil), yield coefficient (Y = 34 μg cells · μmol^{-1} 3CB), and initial substrate concentration (S_0 = 0.696 μmol 3CB · g^{-1} soil). From Focht and Shelton (19).

product for growth and energy. This phenomenon has been referred to as co-metabolism (24,25). It can be demonstrated with growing cells in the presence of an energy source or with washed dense-cell suspensions in the absence of an energy source. The mechanism for this seemingly peculiar action is best described by Dagley (12) as being a case of fortuitous metabolism, in which the enzymes that evolved for the catabolism of aromatic hydrocarbons effect the same reaction with the chlorinated analog. Thus, these enzymes have low substrate specificity. However, at some point downstream in the catabolic pathway, enzymes of high substrate specificity are involved, and they cannot metabolize the chlorinated degradation product.

Without exception, all bacteria that grow on aromatic hydrocarbons utilize the type of pathway shown in Fig. 3. The key features of this pathway are the oxidation of the ring by a dioxygenase and the fission of the catechol structure by another dioxygenase, adjacent to one of the hydroxyl-bearing carbons. Despite considerable transformation of the CAH, none of the steps in the pathway involves respiratory electron transport; thus, the organism receives no return for its effort. Even with biphenyl as the growth substrate, respiratory electron transport begins only when the products of ring fission are converted to pyruvate, acetaldehyde, and formate. In the case of CAHs, the same is true, except that the co-metabolizing organism is unable to dehalogenate the ring-fission product, which accumulates in pure culture. With PCB degradation in soil, the situation is similar, except that chlorinated benzoates and other degradation products are metabolized by the indigenous microflora (9,18). Thus, any useful energy gained from the mineralization of PCBs and CAHs is obtained not by the co-metabolizers, but by the commensals that dehalogenate the chlorinated products of co-metabolism and utilize them for growth.

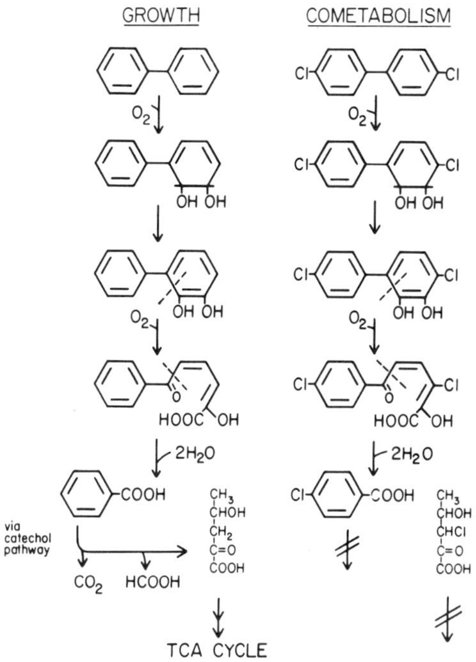

Fig. 3. Fortuitous catabolic pathways of biphenyl and 4,4'-dichlorobiphenyl in pure culture (1,21). In soil, the chlorinated degradation products are mineralized by the indigenous microflora (9, 18). From Focht (16).

If the metabolism of CAH compounds were a syntrophic relationship, by which both members would derive benefits by the action of the other, then CAHs would not be persistent in the environment. In a commensal relationship, the commensal does not return the favor it derived from the co-metabolizer. Since the co-metabolizer cannot make a living off the CAH, a nonchlorinated analog must be added to support the growth; otherwise, the density and activity of co-metabolizers would be too low to effect significant transformation of the CAH. This is apparently why the biodegradation of PCBs in soil is so slow. Focht and Brunner (18) found that indigenous bacteria that utilize biphenyl were below 10^5/g, whereas quantities of 10^9/g were required to effect significant mineralization of PCBs. The addition of growth-supporting nonchlorinated analog has been referred to as "analog enrichment," since it enriches for indigenous co-metabolizing bacteria or supports the maintenance of inoculants. Analog enrichment has also been useful in enrichment and isolation of bacteria that co-metabolize CAHs.

Although co-metabolism has taken on considerable relevance to the catabolism of CAHs, the actual phenomenon was first noticed many years ago by Leadbetter and Foster (30), who used the term "co-oxidation" to describe growth on methane of methylotrophic bacteria, which simultaneously oxidized ethane to ethanol as a final product. It has since become well-established that ATP is not generated from the initial oxidation of hydrocarbons, but rather from the oxidation of alcohols and aldehydes. Thus, the inability of methylotrophs to further metabolize ethanol would explain why ethane oxidation does not support growth. Beam and Perry (5) described a co-metabolic-commensalistic relationship, in which *Mycobacterium*, growing on propane, co-metabolized cyclohexane to cyclohexanone. The product, in contrast to the substrate, was utilized for growth by *Pseudomonas*. There are many other examples that have been given in reviews of the subject (3,24).

Soil contaminated by industrial wastes frequently contains more than one type or class of compounds, so that compounds that support growth of co-metabolizers may induce for co-metabolism of a broad range of structurally similar compounds. Thus, multicomponent organic wastes may be less problematic for treatment if substrate specificity is broad. Clearly, the role that hydrocarbons and xenobiotic compounds play as fortuitous inducers in biodegradative pathways is pertinent to the in situ biological treatment of xenobiotic wastes. This topic has been recently addressed by Pflug and Burton (this Volume).

EVOLUTIONARY ASPECTS OF MICROBIAL INFALLIBILITY

Gale (22) raised the concept that microorganisms were essentially infallible in catabolizing any conceivable organic compound in existence. Noting that the introduction of synthetic compounds to the environment has been a recent event in contrast to the period of microbial evolution, Alexander (2) asked if it was reasonable to expect that microorganisms with novel biodegradative abilities would evolve in our lifetime. At the time of his classic review, very little was known about the catabolism of xenobiotic compounds. (The word "xenobiotic" was not used until 15 years later.) Scarcely a half-dozen years had passed before considerable catabolic diversity had been observed with respect to chlorinated aromatic compounds in Alexander's laboratory and in others. This prompted Horvath (24) to ask if the fallibility to demonstrate biodegradation of recalcitrant compounds resided not with microbial species but with the species *Homo sapiens*! In looking back to the questions raised by Alexander and Horvath over 15 years ago, the answer appears to be yes in both cases. Were it otherwise (at least negating Horvath's question), the conference upon which the present Volume is based would not have been held.

The question of evolution is an intriguing one. Its relation to the lack of technological innovation is addressed in the last section of this chapter. There has been a recent flurry of independent reports on the isolation of bacteria capable of growth on dichlorobenzenes (13,36,38) from long-term (six to ten months) enrichment cultures of polluted environments. This prolonged period of adaptation cannot be explained kinetically, because the growth rates of the novel bacteria are sufficiently rapid that a single cell should produce a turbid enrichment culture in a few weeks or less from when the enrichment culture was started. Moreover, all of these bacteria use the *ortho*-pyrocatechase pathway of ring fission between the two hydroxyls instead of the *meta*-pyrocatechase pathway adjacent to one of the hydroxyls, as illustrated in Fig. 4. The former pathway is normally used by bacteria that grow on chlorinated aromatic acids, while the latter is used by bacteria that grow on nonchlorinated aromatic hydrocarbons. Thus, the specter of genetic exchange among two or more organisms that contain the complementary genes for the entire catabolic pathway must be considered.

Another potential example of genetic exchange and subsequent natural selection may arise from the rapid (one to two weeks) breakdown of many methyl carbamate insecticides and thiocarbamate herbicides that has been observed recently in random locations throughout the United States (26). Normally, these pesticides persist during the length of a growing season, and it has not been possible to isolate bacteria from enrichment culture. In contrast, bacteria that utilize the pesticide as a sole carbon source can be obtained readily from those soils in which degradation is rapid. The inference is clearly that the slower breakdown of the pesticide in the "normal soil" is due to the presence of two or more species, neither of which can utilize it as a growth substrate, while the rapid breakdown in the "problem soils" is due to the genetic mobilization of the entire catabolic pathway into a single organism, which when faced with a continual selection pressure (i.e., yearly application of the same compound) is established as a new component of the indigenous microflora.

Since the short generation time of bacteria is unparalleled in biology, evolutionary changes in bacteria are relatively common occurrences. The development of antibiotic-resistant bacteria was noted within several years after the widespread medicinal use of these drugs. Although spontaneous mutation and rapid growth rates were recognized as normal events in the natural selection and evolution of these bacteria, multiple resistance to so many antibiotics by a single genotype was an unpredicted occurrence. That antibiotic-resistant genes were coded by plasmids and readily transferred among different strains was finally established by Watanabe (41). Not surprisingly, many--if not most--of the genes coding for aromatic catabolism are plasmid-coded. With no intent of being frivolous, it is possible that, if we are patient, organisms will evolve in 50 to 100 years that may grow on DDT or PCBs as sole carbon sources. Nevertheless, a more realistic scheme to achieve these results much faster and more reproducibly is presented in the following section.

STRAIN CONSTRUCTION BY PROGENITIVE MANIFESTATION

Considerable advances have been made in the area of molecular biology of aromatic compounds. Some of the exciting and promising molecular techniques for combining the genes encoding complementary catabolic pathways from two or more bacteria into a single bacterial strain are reported elsewhere in this Volume (Chapman; Cohen et al.; Timmis et al., this Volume). Kellogg et al. (27) introduced the term "molecular breeding" to describe the natural growth, mating, and selection of novel genotypes in a chemostat. They reasoned that such recombinant strains would be more stable and less problematic with respect to gene regulation and expression than those made by splicing and cloning of DNA segments. Using this concept, Kilbane et al. (28) added a strain of *P. cepacia*, which grew on 2,4-dichlorophenoxyacetic acid (2,4-D) to a chemostat containing a nonaxenic soil inoculum. Selection pressure with 2,4,5-T as the sole carbon source was gradually increased until they eventually obtained *P. cepacia* ATCC 1100, which grew on 2,4,5-T as the sole carbon source. The successful application and decontamination of 2,4,5-T in soil with this strain has been mentioned previously. A somewhat similar approach was used by Reineke and Knackmuss (34) to construct a chlorobenzene-degrading strain from a benzene-degrader over a period of ten months by gradually increasing the selections pressure by replacement of benzene with chlorobenzene.

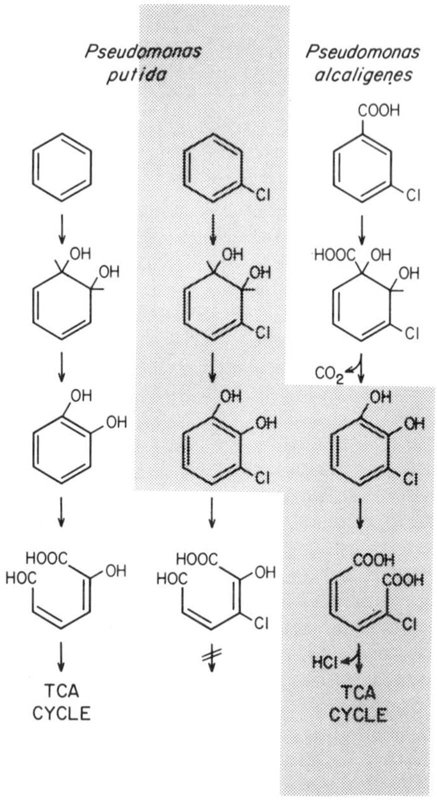

Fig. 4. Theoretical construction of a chlorobenzene-utilizing recombinant bacterium from two parental strains having parts of the complete pathway for metabolism of chlorobenzene. Note that P. putida and P. alcaligenes both produce 3-chlorocatechol from their respective chlorinated substrates, but the former uses the nonproductive meta fission route in contrast to the productive ortho fission route used by the latter. The strategy, as advanced by Reineke and Knackmuss (35), is to route metabolism of chlorobenzene through the ortho fission pathway by combining all the genes in both parents to effect conversion through the shaded pathway.

The transfer of genetic material in the chemostat has been discussed by Slater and Lovatt (37), along with strategies to be used in mobilizing all the catabolic genes from all the organisms involved in the desired pathway into a single recombinant. The major problem with the chemostat is the inability to maintain physiological conditions suitable for two or more organisms that may compete for the same substrate or that may have vastly different growth rates.

The frequency at which genetic material is transferred across species is uncertain. Recombination frequencies among Hfr strains of Escherichia coli are rather high at 10^{-6} (45), yet the frequency in nature is surely much lower. This means that if a high dilution rate (fast doubling time) is used, the recombinant is likely to be washed out. Thus, low dilution rates are normally employed to guard against this possibility. The trade-off with a low growth rate is that the effects of evolution and selection of novel genotypes take much longer to be manifested.

To solve all the problems listed above, we (29) have developed a novel system which uses a multichemostat approach. The two different strains, which between them contain the complementary enzymes for the entire degradation pathway (Fig. 4), are grown in separate chemostats with their respective substrates (Fig. 5), because neither is capable of utilizing chlorobenzene as a growth substrate. The chemostat can be set to ψ_{max} for maximum growth, but this is generally not good practice, since the outflow will have a high residual substrate concentration that may interfere with the selection pressure from the desired substrate, chlorobenzene. We have found it best to run the two chemostats at about one third of ψ_{max}. Continuous amalgamation of new progeny contributes to the cumulative frequency of a rare event, and the recombinant that is formed is eventually manifested by growth and natural selection. The major advantage is the ability to grow the two organisms independently without concern for substrate competition, product inhibition, or vastly different growth rates. The results and reproducibility of this method have been discussed in more detail elsewhere (29).

Continuous growth is necessary to achieve rapid progenitive manifestation. If the probability of two strains exchanging genetic material to form a recombinant is very low (e.g., 10^{-80}), then this event will not be observed in a stationary culture that contains 10^9 cell/ml in a 100-ml system. However, if sufficient time is allowed for creation of new parental progeny, the cumulative probability of a rare event approaches unity (i.e., certainty). One might think of the analogy that a genius of the stature of Einstein or Newton comes but once for every one billion births; hence, what was once not very likely to be manifested in a lifetime 100 years ago, is today manifested every 10 years. A simple exercise in algebra will illustrate this point with bacterial growth in the chemostat. If two strains have doubling times (t') of 4 hr (1/6 day), the time required for one cell to produce 10^{80} new cells is 44 days by simple exponential growth kinetics ($10^{80} = 2^{t/t'}$). Progenitive manifestation should be much faster, since recombination events are not likely to be as rare as the example given above. In fact, this was indeed the case with the construction of P. putida CB1-9: the event was manifested in one week, and the strain was purified by growth on chlorobenzene in one more week.

The rapid, reproducible, and simple features of this method should make it possible now to construct bacteria that can grow on one or more congeners of PCBs. We (18) and others (Ref. 6; Unterman et al., this Volume) already have strains that are capable of extensive co-metabolism of the congeners of commercial PCB mixtures, including penta- and hexa-chlorobiphenyls. Many bacteria that grow on chlorinated benzoates (the products of PCB co-metabolism) have been isolated and characterized (35). Co-culture metabolism of 4,4'-dichlorobiphenyl with Acinetobacter sp. P6 (co-metabolizer) and Acinetobacter sp. 4CB-1 (commensal that grows on 4-chlorobenzoate) has been demonstrated (P. Adriaens and D.D. Focht, poster abstracts, this Volume). Thus the two requisite gene pools exist for the creation of a recombinant capable of growing on 4,4'-dichlorobiphenyl as a sole carbon source by the progenitive manifestation method.

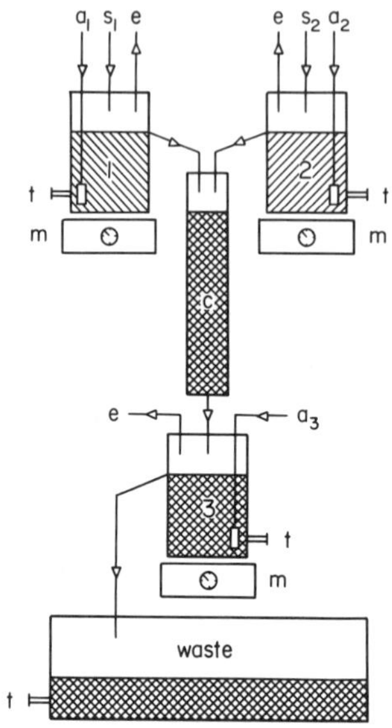

Fig. 5. Construction of a chlorobenzene-utilizing bacterium, P. putida CB1-9, with a dual three-stage chemostat. Pseudomonas putida R5-3 is grown in the left chemostat, which is supplied with a constant nutrient feed (S_1) and a stream of toluene-saturated air (a_2), while P. alcaligenes is grown in the right chemostat, which is supplied with a constant concentration of benzoate in the nutrient solution feed (S_2) and a constant aeration rate (a_2). The cultures are mixed together continually on ceramic beads contained in the reactor column (c), which receives a constant flow of chlorobenzene vapors by unsaturated mass flow from the third stage reactor (a_3). All three chemostats are mixed continually by magnetic stirrers (m). Exhaust gases (e) and sampling ports (t) enable constant monitoring if desired. Manifestation of genetic exchange is noted by the appearance of inorganic chloride (by addition of acidic $AgNO_3$) in the bottom chemostat. The waste effluent is autoclaved before discharge. From Krochel and Focht (29), with the convention changed from left to right to match the same order of the two respective bacteria depicted in Fig. 4.

The advantages of using a recombinant capable of metabolizing PCBs in contrast to the use of a co-culture system are summarized below. The recombinant overcomes all of the following problems of co-culture:

(a) A substrate analog (biphenyl) is required for growth and enzyme induction.

(b) There may be competition for the growth substrate (biphenyl) from indigenous bacteria.

(c) There are likely to be diffusion limitations of PCB co-metabolic products to commensals.

(d) The growth rates of the co-cultures may be very different and difficult to manage.

Clearly, the method of progenitive manifestation is far easier and more rapid than current recombinant DNA technology. It contains the desirable features discussed by Kellog et al. (27) with respect to genetic stability. More than a year has passed since the construction and isolation of P. putida CB1-9, and the organism remains phenotypically and genotypically stable. Progenitive manifestation may also be useful in exploring the quantitative dynamics of genetic exchange among different species and genera as well as in construction of desirable bacterial strains for biodegradative performance. Whether or not accelerated evolution falls under the recent Federal Register Guidelines for genetic manipulation of microorganisms is a topic open to prudent debate and reasonable interpretation.

ACKNOWLEDGEMENTS

I am grateful to Occidental Chemical Corporation and to the National Science Foundation (ECE-84191315) for their financial support of my research.

REFERENCES

1. Ahmed, M., and D.D. Focht (1973) Degradation of polychlorinated biophenyls by two species of Achromobacter. Can. J. Microbiol. 19:47-52.
2. Alexander, M. (1965) Biodegradation: Problems of molecular recalcitrance and microbial fallibility. Adv. Appl. Microbiol. 7:35-76.
3. Alexander, M. (1981) Biodegradation of chemicals of environmental concern. Science 211:132-138.
4. Barles, R.W., C.G. Daughton, and D.P.H. Hsieh (1979) Accelerated parathion degradation on soil inoculated with acclimated bacteria under field conditions. Arch. Environ. Contam. Toxicol. 8:647-660.
5. Beam, H.W., and J.J. Perry (1974) Microbial degradation of cycloparaffine hydrocarbons via cometabolism and commensalism. J. Gen. Microbiol. 83:163-169.
6. Bedard, D.L., R.E. Wagner, M.J. Brennan, M.L. Haberl, and J.F. Brown, Jr. (1987) Extensive degradation of Aroclors and environmentally transformed polychlorinated biphenyls by Alcaligenes eutrophus H850. Appl. Environ. Microbiol. 53:1094-1102.
7. Bitton, G., and C.P. Gerba (1984) Groundwater pollution microbiology: The emerging issue. In Groundwater Pollution Microbiology, G. Bitton and C.P. Gerba, eds. John Wiley and Sons, New York, pp. 1-8.
8. Brown, Jr., J.F., R.E. Wagner, D.L. Bedard, M.J. Brennan, J.C. Carnahan, R.J. May, and T.J. Tofflemire (1984) PCB transformations in upper Hudson sediments. Northeast Environ. Sci. 3:167-179.
9. Brunner, W., F.H. Sutherland, and D.D. Focht (1985) Enhanced biodegradation of polychlorinated biphenyls in soil by analog enrichment. J. Environ. Qual. 14:324-328.
10. Chatterjee, D.K., J.J. Kilbane, and A.M. Chakrabarty (1982) Biodegradation of 2,4,5-trichlorophenoxyacetic acid in soil by a pure culture of Pseudomonas cepacia. Appl. Environ. Microbiol. 44:514-516.

11. Crawford, R.L., and W.W. Mohn (1985) Microbiological removal of pentachlorophenol from soil using a Flavobacterium. Enzyme Microb. Technol. 7:617-620.
12. Dagley, S. (1984) Introduction. In Microbial Degradation of Organic Compounds, D.T. Gibson, ed. Marcel Dekker, Inc., New York, pp. 1-11.
13. DeBont, J.A.M., M.J.A.W. Vorage, S. Hartmans, and W.J.J. van den Tweel (1986) Microbial degradation of 1,3-dichlorobenzene. Appl. Environ. Microbiol. 52:677-680.
14. Dimond, J.B., G.Y. Belyea, R.E. Kadunce, A.S. Getchell, and J.A. Blease (1970) DDT residues in robins and earthworms associated with contaminated forest soils. Canadian Entomologist 102:1122-1130.
15. Focht, D.D. (1972) Microbial degradation of DDT metabolites to carbon dioxide, water, and chloridae. Bull. Environ. Contam. Toxicol. 7:52-56.
16. Focht, D.D. (1987) Ecological and evolutionary considerations on the metabolism of xenobiotic chemicals in soil. In Future Developments in Soil Science Research, American Society of Agronomy, Madison, Wisconsin, pp. 157-167.
17. Focht, D.D., and M. Alexander (1971) Aerobic cometabolism of DDT analogues by Hydrogenomonas sp. J. Agric. Food Chem. 19:20-22.
18. Focht, D.D., and W. Brunner (1985) Kinetics of biphenyl and polychlorinated biphenyl metabolism in soil. Appl. Environ. Microbiol. 50:1058-1063.
19. Focht, D.D., and D. Shelton (1987) Growth kinetics of Pseudomonas alcaligenes C-0 relative to inoculation and 3-chlorobenzoate metabolism in soil. Appl. Environ. Microbiol. (in press).
20. Fogel, S., R.L. Lancione, and A.E. Sewall (1982) Enhanced biodegradation of methoxychlor in soil under sequential environmental conditions. Appl. Environ. Microbiol. 44:113-120.
21. Furukawa, K.N., N. Tomizuka, and A. Kamibayashi (1979) Effect of chlorine substitution on the bacterial metabolism of various polychlorinated biphenyls. Appl. Environ. Microbiol. 38:301-310.
22. Gale, E.F. (1952) The Chemical Activities of Bacteria, Academic Press, London.
23. Guenzi, W.D., and W.E. Beard (1967) Anaerobic biodegradation of DDT to DDD in soil. Science 156:1116-1117.
24. Horvath, R.S. (1972) Microbial cometabolism and the degradation of organic compounds in nature. Bacteriol. Rev. 36:146-155.
25. Jensen, H.L. (1963) Carbon nutrition of some microorganisms decomposing halogen-substituted aliphatic acids. Acta Agr. Scand. 13:404-412.
26. Kaufman, D. (1985) Microbial adaptation and metabolism of pesticides. In Agricultural Chemicals of the Future (BARC Symposium 8), J.L. Hilton, ed. Rowman and Allanheld, Totawa, New Jersey, pp. 437-451.
27. Kellogg, S.T., D.K. Chatterjee, and A.M. Chakrabarty (1982) Plasmid-assisted molecular breeding: New technique for enhanced biodegradation of persistent toxic chemicals. Science 214:1133-1135.
28. Kilbane, J.J., D.K. Chatterjee, J.S. Karns, S.T. Kellogg, and A.M. Chakrabarty (1983) Biodegradation of 2,4,5-trichlorophenoxyacetic acid by a pure culture of Pseudomonas cepacia. Appl. Environ. Microbiol. 33:72-78.
29. Kröckel, L., and D.D. Focht (1987) Construction of chlorobenzene-utilizing recombinants by progenitive manifestation of a rare event. Appl. Environ. Microbiol. (in press).

30. Leadbetter, E.R., and J.W. Foster (1959) Oxidation products formed from gaseous alkanes by the bacterium Pseudomonas methanica. Arch. Biochem. Biophys. 82:491-492.
31. Lipman, J.G. (1908) Bacteria in Relation to Country Life, Macmillan, New York.
32. Nielsen, D.R., J.W. Biggar, and K.T. Erh (1973) Spatial variability of field-measured soil-water properties. Hilgardia 43:215-259.
33. Pfaender, F.K., and M. Alexander (1972) Extensive microbial degradation of DDT in vitro and DDT metabolism by natural communities. J. Agric. Food Chem. 20:842-846.
34. Reineke, W. (1984) Microbial degradation of halogenated aromatic compounds. In Microbial Degradation of Organic Compounds, D.T. Gibson, ed. Marcel Dekker, Inc., New York, pp. 319-360.
35. Reineke, W., and H.-J. Knackmuss (1984) Microbial metabolism of haloaromatics: Isolation and properties of a chlorobenzene-degrading bacterium. Appl. Environ. Microbiol. 47:395-402.
36. Schraa, G., M.L. Boone, M.S.M. Jetten, A.R.W. vanNeerven, P.J. Colberg, and A.J.B. Zehnder (1986) Degradation of 1,4-dichlorobenzene by Alcaligenes sp. strain A175. Appl. Environ. Microbiol. 52:1374-1381.
37. Slater, J.H., and D. Lovatt (1984) Biodegradation and the significance of microbial communities. In Microbial Degradation of Organic Compounds, D.T. Gibson, ed. Marcel Dekker, Inc., New York, pp. 439-485.
38. Spain, J.C., and S.F. Nishino (1987) Degradation of 1,4-dichlorobenzene by a Pseudomonas sp. Appl. Environ. Microbiol. 53:1010-1019.
39. Starc, S.A. (1942) Mikrobiologische Untersuchungen einiger podsoliger Boden Kroatiens. Arch. Mikrobiol. 12:329-352.
40. Waksman, S.A. (1916) Bacteria numbers in soils at different depths, and in different seasons of the year. Soil Sci. 1:363-380.
41. Watanabe, T. (1963) Infective heredity of multiple drug resistance in bacteria. Bacteriol. Rev. 27:87-115.
42. Wedemeyer, G. (1967) Biodegradation of dichlorodiphenyltrichloroethane: Intermediates in dichlorophenylacetic acid metabolism by Aerobacter aerogenes. Appl. Microbiol. 15:1494-1495.
43. Wedemeyer, G. (1967) Dechlorination of 1,1,1-trichloro-2,2-bis (p-chlorophenyl) ethane by Aerobacter aerogenes. Appl. Microbiol. 15:569-574.
44. Wesseling, J., and W.R. van Wijk (1957) Land drainage in relation to soils and crops: I. Soil physical conditions in relation to drain depth. In Drainage of Agricultural Lands, L.N. Luthin, ed. American Society of Agronomy, Madison, Wisconsin, pp. 461-504.
45. Wollman, E.L., F. Jacob, and W. Hayes (1956) Conjugation and genetic recombination in Escherichia coli K-12. In Papers on Bacterial Genetics, E.A. Adelberg, ed. Little, Brown and Company, Boston, pp. 300-333.

TRACKING MICROORGANISMS AND GENES IN THE ENVIRONMENT

Ronald M. Atlas[1] and Gary S. Sayler[2]

[1]Department of Biology
University of Louisville
Louisville, Kentucky 40292

[2]Graduate Ecology Program
University of Tennessee
Knoxville, Tennessee 37916

INTRODUCTION

Many concerns have been expressed about the possible hazards resulting from the release into natural environments of genetically engineered microorganisms (GEMs) derived from recombinant DNA (rDNA) technology (7,15,42). These concerns center around the possibilities that GEMs, or host cells into which rDNA might move, could demonstrate pathogenicity, causing disease in plants or animals, or that GEMs or subsequent hosts for the rDNA could have detrimental effects on natural ecosystems. Many different disaster scenarios have been postulated, ranging from a deadly "Andromeda strain" pathogen produced by rDNA technology causing massive mortalities, to the release of a GEM altering the world climate. Many similar concerns were raised in the early 1970s when the technology for producing GEMs was first developed, leading to the Asilomar Conference in February 1975, and the development of guidelines and oversight by the Recombinant Advisory Committee (RAC) of the National Institutes of Health (NIH) for the safe handling of GEMs.

The thrust of the guidelines for handling GEMs administered by the RAC is to use methods that will prevent accidental environmental contamination, that is, "Don't let the GEM escape from the laboratory." A different set of guidelines that took into account ecological concerns, as well as human health concerns, became necessary in the early 1980s when GEMs were developed for intentional release into the environment. The deliberate release of GEMs has the potential of increasing agricultural productivity and of reducing environmental levels of chemical contaminants, such as polychlorinated biphenyls (PCBs); but introducing novel organisms, particularly microbes, has the potential for altering various processes--such as biogeochemical cycling--in which microorganisms play critical roles (8,15, 18).

Many aspects of deliberate release for desired effects, such as prediction of inoculum needed and sites for inoculation, also depend on quantitative data about persistence, growth, metabolic function, degradative function, and fate of the organism and rDNA (genes) in time and space.

In dealing with the potential adverse and desired effects of deliberately released GEMs, it is necessary to be able to track the introduced microbes and their genes in the environment. In other words, it is necessary to determine the fate of the rDNA and the host cells in which it resides so that the survival and movement of the rDNA in the environment can be monitored and correlated with any effects that are detected. Many factors influence rates of survival and gene transfer in the environment (26,45,46,49). The development of techniques for tracking GEMs in the environment is a necessary part of the process of risk assessment, whereby the advantages and disadvantages of the deliberate release of a GEM can be assessed (24,38). Because the need for practical monitoring methods is a prerequisite to the biotechnological use of engineered microorganisms in the environment, the development of sensitive, specific, reliable, and convenient monitoring techniques is very important (28). Monitoring can be accomplished using several different methodologies, each of which has significant limitations when dealing with diverse environments and organisms (13,20,23).

Tracking GEMs and their genes is analogous to determining the fate in the environment of xenobiotic (man-made) chemicals, but there are some critical differences. When released into the environment, xenobiotics can be degraded biologically or abiotically, or they may be recalcitrant (not subject to degradation), in which case the absolute amount of the chemical will remain the same. Although xenobiotics may become concentrated in biological tissues and hence be biomagnified, they cannot increase in absolute amounts. Genetically engineered microorganisms, on the other hand, are living organisms and have the potential for proliferation in the environment. Furthermore, the rDNA genes of the GEM can have an entirely different fate than the GEM itself (Tab. 1). Hence, one must examine any method that is proposed for tracking GEMs in terms of its ability to monitor the rDNA of the introduced organism, even if it moves into another organism, as well as the introduced organism itself.

As listed by Omenn (33), the desirable characteristics of a detection method for monitoring the environmental fate of GEMs and their genes are:

(a) Applicable under a wide variety of environmental conditions.

(b) Technically simple to apply.

(c) Able to detect, identify, and enumerate GEMs.

(d) Sensitive and specific enough to detect small populations.

(e) Capable of differentiating specific GEMs from other organisms in the environment.

(f) Able to discriminate GEMs from other strains of the same species.

(g) Efficient, cost-effective, and time-economic.

Tab. 1. Possible environmental fates of genetically engineered microorganisms and their genes.

Fate of microorganism	Fate of genome	Explanation
Constant	Constant	The organism and the novel genes that it contains persist in the environment. There may be oscillations in the levels of both, but overall neither increases or is eliminated.
Constant	Decrease	The organism persists in the environment, but the novel genes are eliminated. This can occur when an organism loses a plasmid containing the novel genes.
Decrease	Constant	The introduced organism dies out and is eliminated. However, before this occurs the novel genes are transferred to another organism that persists along with the acquired genes.
Increase	Increase	The organism and the novel genes that it contains proliferate in the environment.
Increase	Decrease	The organism proliferates in the environment but the novel genes are eliminated. This can occur when an organism loses a plasmid containing the novel genes even though the organism proliferates.
Decrease	Increase	The introduced organism dies out and is eliminated. However, before this occurs the novel genes are transferred to another organism that proliferates along with the acquired genes.

In this chapter we will examine some of the conventional and newly developed methodologies that can be used to detect and monitor GEMs and their genes. We will also examine the limitations of each technique, particularly with respect to sensitivity and specificity.

CONVENTIONAL SELECTIVE PLATING AND ENRICHMENT TECHNIQUES

Plate Count Methods

The conventional plate count technique for the enumeration of microorganisms has been a standard method in microbiology for almost a century. This technique has been, and continues to be, useful for the detection and enumeration of specific bacterial populations from environmental samples. The principal application of conventional plating cultivation techniques for tracking GEMs is to detect the rDNA host microorganism (i.e., the GEM itself) rather than the specific rDNA sequences (i.e., the rDNA genes). In plating methods, cultivation media are usually formulated to take advantage of specific traits of the GEM, such as nutritional capabilities and/or resistances to specific antibiotics, so that growth of the GEM is favored over other, nonrecombinant strains. Generally, when using conventional selective culture techniques for detecting GEMs, selection for those strains harboring the specific rDNA sequences is indirect, indicating that the selection is based upon determinants such as antibiotic resistance associated with the host organism's chromosome, cloning vector, or subcloned genes, rather than the specific rDNA gene sequence of the GEM. As a consequence, conventional cultivation techniques are limited in terms of specificity for rDNA sequences and must be considered only to be presumptive tests, requiring confirmation of rDNA presence.

Aside from the problem of specificity for detecting rDNA-containing organisms, the conventional plate count procedure has serious limitations because it requires growth of the organisms under the specific conditions used in the particular plate count procedure and thus is subject to criticism for underestimating populations in samples (1). Conventional culture techniques are constrained by the lack of a universal medium that permits the growth of all potential host organisms within the sample, by the requirement for growth of target organisms, and, as already noted, by the fact that the fate of the recombinant genes may be independent of that of the original host organism (19). Problems encountered with plate counts include: (a) poor or slow growth of the population(s); (b) no growth due to the need for a co-metabolic substrate or auxotrophic nutrient requirement(s); (c) the need for another population to contribute to cross-feeding; (d) lack of sensitivity to a population(s) representing a small proportion of the total population; (e) colony-colony inhibition; and (f) the failure to detect poorly selectable phenotypes or genotypes that are not expressed on the specific media (40). Additionally, some microorganisms have been found to be viable, but nonculturable in environmental samples. It is now known that plate counts underestimate or produce false-negative results if the organisms are sufficiently stressed at the immediate time of sampling and analysis.

Despite these limitations, it is likely that viable culture procedures will have a necessary role in tracking GEMs because they allow selection of specific populations and amplify low-density populations that may exist in various environments. Such selective plating methods were employed to track the first deliberate experimental release of a genetically engineered microbe, an ice-minus strain of *Pseudomonas syringae*.

To improve the detection capabilities of viable plating methods, it is possible to include specific markers in the GEMs that can be easily detected (9). Modern genetic recombination techniques make possible the labeling of organisms in unique ways. The use of marker genes in GEMs may prove to be a sensitive and relatively inexpensive technology to follow the fate of released organisms in the environment. To be suitable for use as a marker: (a) the gene should be expressed in the host; (b) the gene product should be stable in the host; (c) the gene product should be transported to its final location (e.g., cell wall, outer membrane); and (d) the gene product should be readily detectable by a standard, reliable assay, or provide a selection technique, or both. As an example, the inclusion of antibiotic resistance genes permits plating on media containing the particular antibiotics to select for those organisms that are antibiotic-resistant. Similarly, the inclusion of genes with readily observable gene products, such as the lux genes that code for luminescence, permits the specific detection of the GEMs amid numerous other organisms.

Potentially, several types of markers can be used, including chromogenic markers, markers of metabolic traits, antibiotic resistance markers, and heavy metal resistance markers. Selective or enrichment media can easily be based on specific genetic expression, such as the resistance markers, and media can be selective and differential for various metabolic traits. The choice of one or more antibiotics or heavy metals as resistance markers upon which selection can be based should be such that the growth of susceptible organisms (non-GEMs) is suppressed while the GEM is able to grow from a small initial population to a population size that can be readily detected. Because resistance to clinically useful antibiotics is a major public health problem, if antibiotic resistance markers are employed, it is important to use ones that are not commonly used for treating diseases of humans or other animals. It is also desirable to introduce at least two markers into the GEM(s) as selective traits to ensure a high degree of specificity in the recovery process.

A good example of genetic engineering for the creation of a marker strain was the introduction of the lacZ and lacY genes of Escherichia coli (encoding β-galactosidase and lactose permease, respectively) into a Pseudomonas fluorescens strain, thereby creating a phenotypically Lac$^+$ Pseudomonas (9). This genetically engineered pseudomonad strain can be selected from the wild as a pseudomonad, and then screened for the particular strain by growth on lactose-minimal agar. Alternatively, primary selection can be done on a medium that contains the chromogenic substrate 5-bromo-4-chloro-3-indolyl-β-D-galactoside (X-gal). The recombinant pseudomonad will be able to hydrolyze the X-gal, producing blue colonies, whereas the natural pseudomonads will produce white colonies, providing a quick means of following competition among the strains for survival. Monsanto has applied to the United States Environmental Protection Agency for permission to conduct a small-scale field test to examine the sensitivity and reliability of using the lacZY genes to track GEMs in soil.

The inclusion within GEMs of marker traits directly linked to the rDNA sequences of interest permits the detection by selective cultivation techniques of GEMs and even their genes. The inclusion of such genetically engineered traits will have practical applications for direct rDNA monitoring. Accordingly, conventional cultivation methods can be used in conjunction with molecular biological techniques to derive suitable approaches for tracking GEMs in the environment. This topic will be discussed later in this chapter when considering gene probes and colony hybridization methods.

Detection by the Most-Probable-Number (MPN) Method

Like the conventional plate count procedure, the most-probable-number (MPN) procedure has been widely used for enumeration of specific microbial populations, particularly in sanitary and food microbiology. The MPN procedure has many of the advantages and limitations of plate count procedures. Paramount among these is the need to establish the appropriate conditions for culturing the microorganisms of interest. The media used in MPN procedures permit enrichment (growth) of specific microbial populations, thereby amplifying the numbers of microbes of interest (i.e., increasing the sensitivity for the detection of specific microbes). Unlike the plate count procedure, MPN techniques generally use liquid media, making selection for populations that can metabolize a single carbon source easier than when using agar-based solid media. As with the plate count procedure, the best applications of MPN techniques for tracking GEMs are likely to occur when the sensitivity of the MPN method is combined with a molecular biological method, such as DNA-DNA hybridization, that provides the specificity needed for monitoring GEMs and their genes (12).

DIRECT MICROSCOPIC DETECTION

Another approach for enumerating GEMs in the environment is to use fluorescence microscopy with fluorescent antibodies that permit the direct visualization of specific populations. This approach does not require culturing and hence is not adversely affected by the physiological state of the organisms. Fluorescent antibody techniques have been successfully used for the identification and enumeration of specific microorganisms in natural habitats, and for the exploration of ecological questions in mixed culture environments (6). The main problem with this approach is establishing the appropriate degree of specificity to the particular GEM. In some cases there is nonspecific binding of the fluorescent antibody to materials or organisms other than the target; in other cases, other organisms in the natural sample have the same target as the novel GEM for binding of the fluorescent antibody. Thus, erroneous results can occur with this method. In the best case, the fluorescent antibody is designed to react specifically with a gene product made exclusively by the GEM. This thereby provides the needed specificity, assuming that there is expression of that gene so that there is the right antigenic product to react with the fluorescent antibody.

GENE PROBES

In the area of molecular biology, nucleic acid hybridization probes have become indispensable for the detection of specific, complementary nucleic acid sequences (25). Hybridization techniques have been modified to identify, select, and monitor specific rDNA sequences in bacterial colonies (14,16). A gene probe specific for the complementary DNA base sequences present in a GEM provides a reliable, powerful, and specific detection and monitoring tool (31). Gene probes for GEMs and rDNA sequences can detect GEMs in various environmental samples, detect functional groups of microorganisms having identical functional genes, reveal gene flow in natural microbial communities, and detect nonculturable organisms (2).

Probe technology is based on the reannealing of two complementary denatured DNA strands. If one DNA strand is labeled either radioactively (for example, by ^{32}P) or nonisotopically, the presence and identity of another strand of unknown origin (the target) can be demonstrated by its ability to bind (i.e., hybridize) to the labeled probe (29).

In a typical hybridization procedure, the DNA of interest is attached in single-stranded form to a filter; the filter with the bound target DNA is incubated with the probe DNA (in single-stranded form and with incorporated radioactive nucleotides); excess probe is washed from the filter; and an X-ray film is laid over the filter. A spot will show up wherever hybridization occurred and wherever the radioactive nucleotides exposed the film. For detection purposes, nucleic acid probes are most commonly labeled with ^{32}P as [^{32}P]deoxynucleoside triphosphate, most often by the process of nick translation (37). There are a number of alternative procedures for labeling specific nucleic acid probes nonisotopically, e.g., biotinylated probes (22) or probes that can be detected immunologically (47). The use of nonradioactive probes overcomes the disadvantage of handling radioactive probes (11,30,35,36). However, there may be a number of interfering factors that restrict the use of nonradioactively labeled probes with environmental samples.

There are a variety of factors that affect the sensitivity of nucleic acid hybridization as well as specific detection limits (2,4,40), including:

(a) The length of target and probe nucleic acid sequences.

(b) The abundance of target sequence host organisms.

(c) The specific activity of isotopically labeled probe sequences.

(d) The type of nucleic acid label.

(e) The stringency of the hybridization procedure.

The length of the probe can vary from a short oligonucleotide to thousands of base pairs. Short sequences of nucleotides have greater specificity than longer ones. However, longer probes prepared by nick translation can incorporate greater amounts of radiolabel and therefore achieve greater detectability (sensitivity) than short probes (4). Given the right probe and hybridization conditions, it is possible to achieve sensitivities of sub-picogram levels of target DNA, which would equate to the detection of a single copy of an rDNA gene sequence. Analogous methods may be used to detect ribosomal RNA or mitochondrial DNA.

Colony Hybridization

Colony hybridization is one method that employs gene probes and nucleic acid hybridization to detect and monitor specific populations (10,39,40,41). This method involves the cultivation of microorganisms on a suitable solid medium. Colonies that develop are transferred to nitrocellulose or nylon filters and lysed so that the DNA from the cells is released and can bind directly to the filters. The filter is then incubated with a specific segment of labeled DNA or RNA which acts as a gene probe to detect the specific sequence of nucleotides that characteristically identifies the GEM or its novel rDNA genes.

It is essential in these procedures that the DNA probe reflect the DNA sequences that are unique to the GEMs. This can be achieved by either preparing the probe to reflect a specific plasmid present in a GEM or incorporating a small segment of rDNA sequence into a plasmid or chromosome of the GEM. The probe then can be labeled, with ^{32}P for example, and hybridized with DNA from colonies of the GEMs that have been transferred to the filter surface (such as nitrocellulose or Biotrans^{R}). Following autoradiography, GEM colonies can be located on the original filter surface. Thus, colony hybridization allows one to scan many cells from a sample to pick out those with the recombinant DNA sequences, thereby permitting the rapid screening of samples in a relatively inexpensive fashion.

The colony hybridization method has been demonstrated to detect various specific microbes and genes (3,10,34,40). Colony hybridization has been applied to detect and monitor the critical populations recovered from the environment. Examples include the enumeration of naphthalene and toluene degradative populations in environmental microcosms contaminated with synthetic oils (40) and the rapid isolation and identification of microorganisms capable of degrading polychlorinated biphenyls (34) and mercury resistance genes in environmental isolates (3). Yates et al. (51) were able to monitor individual bacterial strains and species in an actual metal biomining operation.

Sayler et al. (40) detected as few as one colony containing target genes in 10^6 colonies from the environment when plasmids TOL and NAH were used as probe DNAs. The maintenance of a genetically modified strain of *Pseudomonas putida* (carrying plasmids TOL and RK2) inoculated in groundwater aquifer microcosms has been successfully monitored over an eight-week period by the use of specific gene probes (20). Also, the utility, significance, and sensitivity of a naphthalene-degrading DNA probe (NAH7 plasmid) to monitor the naphthalene-degrading population in a continuous complex biological wastewater system (a completely mixed aerobic reactor) have been demonstrated (5). Using colony hybridization, the number of naphthalene-degrading organisms was four to five times higher than those determined by traditional plate analysis when naphthalene was the sole carbon source.

While relatively sensitive and specific--potentially detecting as few as one colony containing target genes in one million background colonies from the environment (40)--the colony hybridization method still suffers from the fact that the organisms must be cultured as part of the procedure. As such, it has the same limitations as the viable plating method discussed earlier. In general, the sensitivity of the colony hybridization procedure is only as good (or as bad) as the original isolation/enumeration techniques used.

Most-Probable-Number Gene Probe Methods

A gene probe method has also been developed for use with MPN procedures (12,50). In this procedure, multiple serial dilutions of a sample are placed into microtiter plates and a suitable enrichment medium added. After incubation to allow for microbial growth (amplification of rDNA sequences), the microbes are lysed and the released DNA transferred to a nitrocellulose filter. The target DNA bound to the filter is then incubated with a labeled gene probe. Hybridization, if any occurs (e.g., as evidenced by black spots on an autoradiogram if the gene probe is labeled with ^{32}P), is scored as a positive; the pattern of positives and negatives at the various dilutions is used to calculate the most probable number of microorganisms in the sample.

Colony ELISA Blot

Another new method, called "Colony ELISA Blot," has been developed for the detection and indirect screening of recombinant genes or rDNA sequences (27). In this method the cells of interest are filtered from an environmental sample onto a hydrophobic grid membrane, and colonies develop. A replica is made onto nitrocellulose paper and the cells are lysed with chloroform, immobilizing the proteins to the paper. A rabbit antibody preparation made against a particular antigen is used to detect the rDNA sequences. For example, antibodies can be made against the gene products of a recombinant DNA sequence, such as an enzyme involved in the initial metabolism of the pesticide 2,4,5-trichlorophenoxyacetic acid, in order to detect indirectly the presence of the gene coding for that protein. The major limitations of this method are cost, the requirement that organisms be able to grow on the grid membrane and the nitrocellulose paper, and the fact that direct quantitation of organisms may not be possible.

Direct Probing of DNA

An alternative approach to using probes in conjunction with plate count or MPN procedures is to directly probe DNA extracted from environmental samples. This involves the direct quantitative recovery of the microbial community DNA from an environmental sample (48). The DNA from environmental samples can be recovered by (a) lysis of cells followed by extraction of DNA (32), or (b) differential centrifugation to separate microbial cells from other particles, lysis of cells to release DNA, and purification of the recovered DNA (21,43,44,48). The advantage of the direct DNA extraction approach is that no cultivation is required. Thus, the method is entirely independent of the ability to cultivate the members of the population, and persistence of genes in environmental samples can be accurately followed. The gene(s) of interest can be detected, monitored, and quantified by using hybridization to a specific probe. For example, direct extraction of DNA and filter hybridization have been used to detect and quantify DNA sequences homologous to specific probe DNA on the NAH7 plasmid in total community DNA extracted from activated sludge (41).

The disadvantage of the direct extraction and gene probe approach is that sensitivity is limited by the efficiency of DNA recovery from the total microbial community of complex environmental samples and by the limited amount of DNA that can be adsorbed onto a filter for hybridization. The typical detection limit with filter hybridization is 10^4 organisms/g sample. Sensitivity can be improved by using labeled RNA or single-stranded DNA probes and solution hybridization to achieve a detection limit of about one organism/g sample (43).

As part of a study on the persistence of GEMs in the environment, we have examined the capabilities of several probe methods for tracking the GEMs in freshwater microcosms. The microcosms consisted of glass carboys containing 19 liters of water collected from Fort Loudoun Reservoir in Knoxville, Tennessee. The water was supplemented with a glucose-mineral salts solution and agitated to permit limited microbial growth. The microcosms were inoculated with two different GEMs with and without the xenobiotic chemicals that they are capable of degrading (Tab. 2). The microcosms were incubated at 25°C and samples were collected periodically from each microcosm for analysis. Analytical procedures that were run included selective plating based on antibiotic resistance markers, plating on a general growth medium coupled with colony hybridization, MPN enrichment coupled with filter hybridization, and direct community DNA extraction coupled with dot blot hybridization.

Tab. 2. Descriptions of microcosm treatments.

Microcosm set	Inoculation organism	Chemical addition*
1	None	None
2	*Alcaligenes* sp. A5	None
3	*Alcaligenes* sp. A5	4CBP
4	*Pseudomonas cepacia* DC102	None
5	*Pseudomonas cepacia* DC102	2,4,5-T

*4CBP, 4-chlorobiphenyl; 2,4,5-T, 2,4,5-trichlorophenoxy-acetic acid.

Note: *Alcaligenes* sp. A5 degrades biphenyl and naphthalene. *Pseudomonas cepacia* DC102 degrades 2,4,5-T. 4-Chlorobiphenyl was added to give a final concentration of 25 mg/l; 2,4,5-T was added to give a final concentration of 100 mg/l.

With regard to *Alcaligenes* A5, the colony hybridization failed to detect the organism, indicating loss of the organism from the microcosms after one week. Probing the community DNA plasmid to detect *Alcaligenes* A5 pSS50 plasmid indicated that this plasmid was lost from the microcosms.

Selective plate counts based on inclusion of nalidixic acid indicated an initial decline followed by persistence of the *P. cepacia* DC102 over a four-week period. During the first week of incubation the total microbial count increased by several orders of magnitude, probably due to the glucose-mineral salts addition, and colony hybridization with the colonies on the general growth medium failed to detect the *P. cepacia* DC102 at this time. This failure was likely due to masking by the increased total bacterial numbers. *Pseudomonas cepacia* was still detected on the nalidixic acid plates at this time, and colony hybridization against these colonies confirmed that the target organism persisted. After two weeks the total microbial count declined to approximately original levels, and colony hybridization with the general growth medium again was able to detect the presence of *P. cepacia* DC102. The MPN hybridization results were comparable to the selective media counts and the colony hybridization results from the selective medium. Probing the total community DNA by the dot blot procedure, using a 1.3-kb repetitive restriction fragment to detect *P. cepacia* DC102, indicated that this organism, or at least the target genes, persisted over the month-long period of the experiment. Thus, the colony hybridization with a general medium had a problem in detecting *P. cepacia* DC102 when the general population increased, whereas the other methods were still able to detect the organism. Otherwise, the other methods gave similar results.

Detection of Gene Migration by Restriction Enzyme Digestion

Although the direct extraction of DNA and probe hybridization method does not directly discriminate between target genes that still reside within the introduced organism and those that have been transferred to other organisms, this method can be used with restriction endonuclease analysis (DNA fingerprinting) to detect and monitor GEMs and specific rDNA sequences. For use in monitoring, chromosomal or plasmid DNA is isolated from the cultivated GEM and cut with a battery of appropriate restriction enzymes.

Restriction enzyme digestion can be used to determine whether the genes are still in the same location or if they have moved (17). After restriction endonuclease analysis and separation by gel electrophoresis, the DNA fragments from the gels can be transferred and fixed onto nitrocellulose filters. The DNA is converted into single strands by denaturation, and the strands then are hybridized with appropriately labeled probe DNA. The labeled double-stranded DNA can be visualized by autoradiography, and the size of the labeled fragment can be compared with the restriction map produced from the original GEM. If the rDNA sequences and the probed sequences are in the same position on the gel, there is strong evidence that they are homologous.

Southern blot hybridization is very specific in picking out the fragment(s) carrying rDNA sequences, whether it is inserted into the plasmid or the chromosome. One would be able to answer the questions of whether the rDNA sequences have been altered or deleted, and whether the sequences are present in their original positions. An exact comparison of the two profiles or fingerprints would verify the identity of an introduced rDNA insert. Inserted and deleted DNA could be recognized through the changes in size of particular DNA fragments. However, this technology also has some drawbacks. For example, it is not very sensitive to minor sequence changes (point mutations), and it is relatively expensive.

SUMMARY

Proposed intentional releases of GEMs into the environment necessitate the development of appropriate methodologies for tracing organisms and their genes in various environmental samples. Studies have been conducted to determine the sensitivities and limitations of various methods for determining the fate of GEMs and their genes in the environment. Selective viable plate count procedures can be designed to detect the introduced organisms with high sensitivity; but they are restricted by potential mutations affecting the expression of the selective characteristic in the introduced organism, the occurrence of the particular selective characteristic in the indigenous organisms, and the need to culture the organism. The accuracy of this approach is greatly improved by colony hybridization procedures that use a specific gene probe to detect the introduced genes, but this approach is still only as sensitive as the plating procedure.

Direct extraction of DNA from environmental samples, coupled with dot blot hybridization with radiolabeled probe DNA or solution hybridization, gives a high degree of both sensitivity and precision. This approach does not require culturing of the organism; and even if an introduced gene moves into a new organism or if the introduced organism is viable but nonculturable, the gene probe methods will detect the persistence of the introduced genes in the environment. Efficient direct DNA extraction methods have been developed and tested following in vitro experimental additions of GEMs to sediment and water samples.

ACKNOWLEDGEMENT

Portions of this work were supported by cooperative agreement CR 81248801 from the United States Environmental Protection Agency, Environmental Research Laboratory, Gulf Breeze, Florida.

REFERENCES

1. Atlas, R.M. (1982) Enumeration and estimation of biomass of microbial components in the biosphere. In Experimental Microbial Ecology, R.G. Burns and J.H. Slater, eds. Blackwell Science Publishers, Oxford, England, pp. 84-102.
2. Barkay, T., and G. Sayler (1987) Gene probes as a tool for the detection of specific genomes in the environment. American Society for Testing and Materials special publication on Biotechnology Risk Assessment (in press).
3. Barkay, T., D.L. Fouts, and B.H. Olson (1985) The preparation of a DNA gene probe for the detection of mercury resistance genes in Gram-negative bacterial communities. Appl. Environ. Microbiol. 49:686-692.
4. Berent, S.L., M. Mahmoudi, R.M. Torczynski, P.W. Bragg, and A.P. Bollon (1985) Comparison of oligonucleotide and long DNA fragments as probes in DNA and RNA dot, Southern, Northern, colony and plaque hybridizations. Biotechniques 3:208.
5. Blackburn, J.W., R.K. Jain, and G.S. Sayler (1987) The molecular microbial ecology of a naphthalene-degrading genotype in activated sludge. Environ. Sci. Technol., Vol. 21 (in press).
6. Bohlool, B.B., and E.L. Schmidt (1980) The immunofluorescence approach in microbial ecology. In Advances in Microbial Ecology, Vol. 4, M. Alexander, ed. Plenum Press, New York. pp. 203-241.
7. Brown, T.H., R.K. Colwell, R.E. Lenski, B.R. Levin, M. Lloyd, P.J. Regal, and D. Simberloff (1984) Report on workshop on possible ecological and evolutionary impacts of bioengineered organisms released into the environment. Bull. Ecol. Soc. Am. 65:436.
8. Curtiss, III, R.(1976) Genetic manipulation of microorganisms: Potential benefits. Annu. Rev. Microbiol. 30:507-533.
9. Drahos, D.J., B.C. Hemming, and S. McPherson (1986) Tracking recombinant organisms in the environment: β-galactosidase as a selectable non-antibiotic marker for fluorescent pseudomonads. Bio/Technology 4:439-444.
10. Ford, S., S.C. Simpson, and B.H. Olson (1987) Evaluation of strategies for the detection of specific DNA sequences by colony hybridization in Gram-positive and Gram-negative microorganisms. Abstracts of the Annual Meeting, American Society for Microbiology, Atlanta, Georgia, Q129.
11. Forster, A.C., J.L. McInnes, D.C. Skingle, and R. H. Symons (1985) Non-radioactive hybridization probes prepared by the chemical labelling of DNA and RNA with a novel reagent, photobiotin. Nucl. Acids Res. 13:745-761.

12. Frederickson, J.K., D.F. Bezdicek, F.E. Brockman, and S.W. Li (1987) Enumeration of Tn5 mutant bacteria in soil by MPN-DNA hybridization and antibiotic resistance. Applied Environ. Microbiol. (in press).
13. Glaser, D., T. Keith, P. Riley, G. Chambers, J. Manning, S. Hatting, and R. Evans (1985) Monitoring techniques for genetically engineered microorganisms. In Research Needs in Biotechnology and the Environment, NTIS Order PB86-158979/REB, U.S. Environmental Protection Agency, Washington, D.C.
14. Grunstein, M., and D.S. Hogness (1975) Colony hybridization: A method for the isolation of cloned DNAs that contain a specific gene. Proc. Natl. Acad. Sci., USA 72:3961-3965.
15. Halvorson, H.O., D. Pramer, and M. Rogul (1985) Engineered Organisms in the Environment: Scientific Issues, American Society for Microbiology, Washington, D.C.
16. Hanahan, D., and M. Meselson (1980) Plasmid screening at high colony density. Gene 10:63-67.
17. Holben, W.E., J.R. Jannson, B.K. Chelm, and J.M. Tiedje (1987) Methods for assessing the fate of genetically engineered microorganisms in soil. Abstracts of the Annual Meeting, American Society for Microbiology, Atlanta, Georgia, Q131.
18. Jain, R.K., and G.S. Sayler (1987) Problems and potential for in situ treatment of environmental pollutants by engineered microorganisms. Microbiol. Sci. 4:59-63.
19. Jain, R.K., R.S. Burlage, and G.S. Sayler (1987) Methods for detecting recombinant DNA in the environment. CRC Crit. Rev. Biotechnol. (in press).
20. Jain, R.K., G.S. Sayler, J.T. Wilson, L. Houston, and D. Pacia (1987) Maintenance and stability of introduced genotypes in groundwater aquifer material. Appl. Environ. Microbiol. 53:996-1002.
21. Jannson, J.R., W. Holben, B.K. Chelm, and J.M. Tiedje (1987) Fate of genetically engineered Pseudomonads added to soil cores. Abstracts of the Annual Meeting, American Society for Microbiology, Atlanta, Georgia, Q132.
22. Leary, J.J., D.J. Brigati, and D.C. Ward (1983) Rapid and sensitive colorimetric method for visualizing biotin-labeled DNA probes hybridized to DNA or RNA immobilized on nitrocellulose: Bio-blots. Proc. Natl. Acad. Sci., USA 80:4045-4049.
23. Levin, M.A., R. Seidler, A.W. Bourquin, J.W. Fowle, III, and T. Barkay (1986) EPA developing methods to assess environmental release. Bio/Technology 5:38-45.
24. Levine, M.M., J.B. Kaper, H. Lockman, R.E. Black, M.L. Clements, and S. Falkow (1983) Recombinant DNA risk assessment studies in man: Efficacy of poorly mobilizable plasmids in biologic containment. Recomb. DNA Tech. Bull. 6:89-97.
25. Lewin, R. (1983) Genetic probes become even sharper. Science 221:1167.
26. Liang, L.N., J.L. Sinclair, L.M. Mallory, and M. Alexander (1982) Fate in model ecosystems of microbial species of potential use in genetic engineering. Appl. Environ. Microbiol. 44:708-714.
27. Lo, R.C., and L.A. Cameron (1976) A simple immunological detection method for the direct screening of genes from clone banks. Biochem. Cell Biol. 64:73.

28. McCormick, D. (1986) Detection technology: The key to environmental biotechnology. Bio/Technology 4:419-422.
29. Meinkoth, J., and G. Wahl (1984) Hybridization of nucleic acids immobilized on solid supports. Anal. Biochem. 138:267-271.
30. Miller, E.C., U. Juhl, and J.A. Miller (1966) Nucleic acid guanine: Reaction with the carcinogen N-acetoxy-2-acetylaminofluorene. Science 153:1125-1127.
31. Moseley, S.L., U. Huq, A.R.M.A. Alim, M. So, M. Samadpour-motalebi, and S. Falkow (1980) Detection of enterotoxigenic Escherichia coli by DNA colony hybridization. J. Infect. Dis. 142:892-898.
32. Ogram, A., G.S. Sayler, and T. Barkay (1988) DNA extraction and purification from sediments. J. Micro. Meth. (in press).
33. Omenn, G.S. (1986) Controlled testing and monitoring methods for microorganisms. In Biotechnology Risk Assessment: Issues and Methods for Environmental Introductions, J. Fiksel and V.T. Covello, eds. Pergamon Press, New York, pp. 144-162.
34. Pettigrew, C., and G.S. Sayler (1986) The use of DNA:DNA colony hybridization in the rapid isolation of 4-chlorobiphenyl degradative bacterial phenotypes. J. Microbiol. Meth. 5:205-213.
35. Renz, M. (1983) Polynucleotide-histone H1 complexes as probes for blot hybridization. EMBO J. 2:817-821.
36. Renz, M., and C. Kurz (1984) A colorimetric method for DNA hybridization. Nucl. Acids Res. 12:3435-3444.
37. Rigby, P.W.J., M. Dieckmann, C. Rhodes, and P. Berg (1977) Labeling deoxyribonucleic acid to high specific activity in vitro by nick translation with DNA polymerase I. J. Mol. Biol. 113:237-251.
38. Sayler, G.S., and G. Stacey (1986) Methods for evaluation of microorganism properties. In Biotechnology Risk Assessment: Issues and Methods for Environmental Introductions, J. Fiksel and V.T. Covello, eds. Pergamon Press, New York, pp. 35-58.
39. Sayler, G.S., C. Harris, C. Pettigrew, D. Pacia, A. Breen, and K.M. Sirotkin (1987b) Evaluating the maintenance and effects of genetically engineered microorganisms. Dev. Indust. Microbiol. 27:135-149.
40. Sayler, G.S., M.S. Shields, A. Breen, E.T. Tedford, S. Hooper, K.M. Sirotkin, and J.W. Davis (1985) Application of DNA:DNA colony hybridization to the detection of catabolic genotypes in environmental samples. Appl. Environ. Microbiol. 49:1295-1303.
41. Sayler, G.S., R.K. Jain, L. Houston, A. Ogram, C.A. Pettigrew, J.W. Blackburn, and W.S. Riggsby (1987a) Applications for DNA probes in biodegradation research. In Contemporary Microbial Ecology, Proceedings of the Fourth International Symposium on Microbial Ecology (in press).
42. Sharples, F.E. (1983) Spread of organisms with novel genotypes: Thoughts from an ecological perspective. Recomb. DNA Tech Bull. 6:43-56.
43. Steffan, R.J., and R.M. Atlas (1987) Use of specific gene probes for the detection of Pseudomonas cepacia DC102 in sediments. In Abstracts of the Annual Meeting of the American Society for Microbiology, Atlanta, Georgia, Abstr. Q130.
44. Steffan, R.J., J. Goksoyr, and R.M. Atlas (1986) Isolation of bacterial DNA from Ohio River sediments. Annual Meeting, American Society for Microbiology, Washington, D.C., Q-144.

45. Stotzky, G., and V.N. Krasovsky (1981) Ecological factors that affect the survival, establishment, growth, and genetic recombination of microbes in natural habitats. In Molecular Biology, Pathogenicity, and Ecology of Bacterial Plasmids, S.B. Levy, R.C. Clowes, and E.L. Koenig, eds. Plenum Press, New York, pp. 31-42.
46. Stotzky, G., and H. Babich (1986) Survival of, and genetic transfer by, genetically engineered bacteria in natural environments. In Advances in Applied Microbiology, Vol. 31, A.I. Laskin, ed. Academic Press, Inc., New York, pp. 93-138.
47. Tchen, P., R. Fuchs, E. Sage, and M. Leng (1984) Chemically modified nucleic acids as immunodetectable probes in hybridization experiments. Proc. Natl. Acad. Sci., USA 81:3466-3470.
48. Torsvik, V.L. (1980) Isolation of bacterial DNA from soil. Soil Biol. Biochem. 12:15-21.
49. Trevors, J.T., T. Barkay, and A.W. Bourquin (1987) Gene transfer among bacteria in soil and aquatic environments: A review. Can. J. Microbiol. 33:191-198.
50. Van Voris, P., J.K. Fredrickson, and M. Ligotke (1987) Terrestrial microcosms and containment systems to evaluate ecosystem effects of genetically engineered microorganisms. In Annual Meeting of the American Society for Microbiology, Atlanta, Georgia.
51. Yates, J.R., J.H. Lobos, and D.S. Holmes (1986) The use of genetic probes to detect microorganisms in biomining operations. J. Indust. Microbiol. 1:129-135.

DISCUSSION

McCARTY: In the experiments with anaerobic consortia of benzoate- and hydrogen-consuming organisms, it seems to me that a low hydrogen pressure would be unstable in the reaction you presented. The chlorobenzoate was actually acting as an electron receptor, not a donor.

TIEDJE: We have calculated the energetics of the dechlorination, particularly at low hydrogen concentration. The in situ hydrogen and chlorobenzoate concentrations give a $-\Delta G$ that is quite sufficient for energy for growth.

High hydrogen concentrations inhibit dechlorination. At first we thought more hydrogen is better; let's add more and speed up the reaction; but instead we got no dechlorination. What happens is that high hydrogen concentration shifts the metabolism of the dechlorinator to produce more reduced products (lactate) and stops the system from growing.

CHAPMAN: Is it appropriate to attribute the activity on the ortho-fluoro-benzoate by your 3-bromobenzoate culture to the dehalogenated reduction systems, or to the benzoate-oxidizing enzyme?

TIEDJE: I attribute it to the benzoate-oxidizing enzyme. I think it is analogous to work others have done with O-chlorobenzoate showing some dechlorination by other organisms.

UNTERMAN: Have you done free energy calculations for the reductive dechlorination of PCBs?

TIEDJE: We have for some of the likely redox couples that the General Electric (GE) results suggest might have occurred in sediments. Yes, there is significant energy available from dechlorination of PCBs.

ALEXANDER: I remember from some of your earlier work that you waited a long time for the system to begin to function. We and others have found that some highly chlorinated compounds in soil do not disappear for long periods and then do begin to disappear. Is it possible that, if one waited as long under aerobic conditions, one could obtain results similar to those under anaerobic conditions?

TIEDJE: Maybe, but aerobic microbiologists usually aren't as patient as anaerobic microbiologists! When we were studying this lag period to try to determine the mechanism of the anaerobic dechlorination reactions in sediment, we had to wait up to six months before dechlorination occurred for some compounds. When it happened, it was very active.

DEMAIN: In working with other anaerobes dealing with cellulose, we have *Clostridium thermocellum*, for example, which uses only a very small number of compounds, mainly cellobiose, unless some vitamins are added. Then the organism adapts to use glucose and fructose. Could you tell us how you do those experiments so as to conclude that only pyruvate is used? Did you add any additional nutritional material? If so, did that improve the substrate range for growth of the DCB-1?

TIEDJE: We didn't do an exhaustive nutritional study, but we did use rumen fluid, which is known as the elixir for most anaerobes. That complex source was present in all cases, as it did improve the growth of the organism. We believe that the reason it grows better in the consortium is that other members of the consortium supply some carbon compounds that are probably growth factors, such as supplied by rumen fluid. We just don't know what those compounds are at this stage.

ATLAS: In your anaerobic "zoo" you had many members, but you only need three to account for the reactions. Have you any idea what the other members are doing there or why the stable communities seem to be more complex than the minimum number of organisms?

TIEDJE: I believe that those other organisms are successful as scavengers. We tend in microbiology to think of the primary substrates that support growth of these organisms in the laboratory, not how they survive in nature. So the butyrate oxidizers and the sulfate reducers are probably living as scavengers on products from the primary organisms in the food chain. We're pretty sure that butyrate is not a substrate for the benzoate oxidizer in the pathway, because we've done isotope dilution studies and find no evidence of labeled butyrate. I think DCB-1 also lives as a scavenger in nature.

OMENN: The federal regulatory regime requires special studies for genetically engineered organisms. What kinds of tracking do you recommend for selected organisms that are not genetically engineered?

ATLAS: I would not recommend anything different if we want to try to track those organisms. There are several reasons for tracking, aside from the regulatory requirements or environmental concerns. If I were running a bioreactor with a particular organism, whether it was formulated through recombinant DNA technology or simply selected from a community, I'd want to know whether it persisted at the right levels in that bioreactor. If it were disappearing, I'd like to be able to maintain the appropriate levels. The simplest way to track this is by traditional selective plating. If you've put in an organism with a particular plasmid coding for a specific degradative capacity and your plating method selects for some other characteristic, you can be fooled. When the bioreactor has lost its capacity, you wonder why the reaction gets stuck at that point or decreases in efficiency. I'd prefer the gene probe approaches even for bioreactors with nonrecombinant organisms.

I think that methodology should be utilized in our basic tracking methods. In each situation, we need to determine the characteristics of the organism and the properties of the particular environment in order to decide which tests to use.

QUESTION: Could we add to an organism a selective marker that produces fluorescence, or some other visible means of tracking the organism, and then detect the organism through remote sensing?

ATLAS: There's no question we can engineer into an organism some visible tracking mechanisms. The luminescent (lux) genes which have been engineered into plants and microbes alike allow tracking of the organism by simple selective plating. We could put in other visible markers. However, I do not expect that we are going to introduce these organisms in concentrations high enough that remote sensing would be at all applicable to their detection. If we had, let's say, an ocean mass that contained a recombinant organism so dense that it was detectable from an aircraft, we'd be in trouble. We would have failed in our ability to regulate the organism in the first place. Visible means may seem to give the public a bit more comfort than other means of detection. But from a microbiological point of view, it doesn't much matter to me if I plate an organism on a medium for some selective markers that determine whether a colony is formed as opposed to a colony that lights up.

SUMMERS: Have you or Gary Sayler tried to amplify the genes to gain greater sensitivity?

ATLAS: We haven't done that yet. One of my graduate students has begun work with the *P. cepacia* AC1100 organism, which has a repetitive sequence offering multiple targets for gene amplification.

ROUNDTABLE DISCUSSION ON MICROBIAL ECOLOGY

Gordon Orians (Chairman), University of Washington
Martin Alexander, Cornell University
Ronald M. Atlas, University of Louisville
Al Bourquin, ECOVA Corporation
Rita Colwell, The University of Maryland
Dennis D. Focht, University of California
Morris Levin, U.S. Environmental Protection Agency
James Tiedje, Michigan State University

LEVIN: The Environmental Protection Agency (EPA) has a rapidly developing program on fate of chemicals and microorganisms and on potential ecosystem effects of introduction of biodegradative microorganisms. We are interested in risk assessment for any microbes used to enhance the environment, either microbial pest control agents or water treatment agents. Jim Tiedje mentioned the ecology within biodegradative consortia in the laboratory. Biotechnology methods, including DNA probes, could be useful to show which organisms were active in his consortia. With regard to tracking methods, we are producing a manual of standard methods in microbial ecology. This manual will describe and compare widely accepted methods. In general, I feel we need more discussion of survival, growth, persistence, and effects. We have to address these issues before the public will approve our use of biotechnology to enhance the environment.

ALEXANDER: I will start by paraphrasing from memory a poem by Hilaire Belloc, an English writer, about the microbe:

> "Microbes are so very small we can hardly see them at all. Scientists who ought to know say they must be so. Let no one ever ever doubt what nobody is sure about."

My point is that science is changing. Jim Tiedje elegantly summarized some work on anaerobic degradative processes. We used to think that if something did not occur aerobically, it wouldn't occur anaerobically. Now we know that there are certain processes that occur only anaerobically. Ten years from now, I predict, we will have new information indicating that aerobic processes work better sometimes, anaerobic processes work better other times, and often both work very well.

Dennis Focht questioned correctly my own error of many years back in claiming that there was a large group of completely recalcitrant or almost completely recalcitrant molecules. There are many reasons for the persistence of compounds in natural environments, including diffusion barriers and adsorption. The same problems will complicate efforts to track organisms.

BOURQUIN: From my early involvement in EPA's program on interactions of microbes and their ability to degrade toxic chemicals, in ecological risk assessment of genetically altered organisms, and most recently in trying to develop a research initiative for the use of microbes to enhance biodegradation, I recognize that advances in both microbial ecology and microbial physiology are absolutely necessary to achieve bioremediation in the environment.

Jim Tiedje showed that microbes interact in the environment and depend upon one another in order to achieve biodegradation. As Dennis Focht pointed out, we'll be able to exploit microbial functions in the environment as we learn more about the physiology of organisms and their pathways for degradation of toxic chemicals. Only after we understand this physiology and how the organisms interact in the environment can we devise plans for genetic engineering.

To track specific organisms through the environment, we can add unique catabolic functions or we can design unexpressed markers other than antibiotic resistance. I am confident that this work will be extremely useful.

COLWELL: Much can be discovered when one explores a new niche or a new habitat, as Jim Tiedje has demonstrated. We've had a couple of decades of very intensive molecular biology research and very good funding for molecular biology. But ecology and systematics have been left behind, sort of poor sisters. Ironically, we now find ourselves in a tizzy trying to figure out how to exploit the biotechnological advances that we've made.

The EPA has established excellent laboratories at Gulf Breeze and Corvallis, and the National Science Foundation (NSF) is providing funding through its Ecology Program, which is headed by Patrick Flanagan. Nevertheless, I will reemphasize a point that was raised by Gil Omenn in his review of the 1983 conference. We concluded that there was a need for national centers for studies of release, and I hope that some progress will be made for developing such centers.

For example, the Marine Ecosystems Research Laboratory at the University of Rhode Island is beautifully set up for doing mesocosms, the middle range of studies after microcosms and prior to release to an aquatic system. The University of Georgia's Ecology Institute and others have carried out very good ecology research for decades--mostly macroecology with some microecology. These would be systems that would provide us with the background information for the kind of release that we are talking about.

In systematics we're now into DNA hybridization and sequencing. However, as Jim Tiedje pointed out, we are dealing with new consortia, new organisms, niches that haven't been examined before, organisms that haven't really been cultured and that still may not be culturable.

Dennis Focht mentioned that we may have to add new organisms to an environment. We had concluded in 1983 that one should deal only (or first) with the organisms from the natural environment in which you wish to carry out some reaction--for example, treating a spill so that when the degradation has been completed, the organisms will resume their proper place in the system. But the naturally occurring organisms simply don't carry the function. We will need new knowledge in order to insert and express genes for biodegradation in natural organisms of the system.

For tracking organisms, the most powerful of the presently available tools are immunofluorescence and fluorescence microscopy, which can detect a single cell in a liter of solution. We have done this in tracking sources of capital bacteria, Legionella, and other organisms involved in cases of human outbreaks. In fact, one of my most recent findings, which will be presented at the United States-Japan Cholera Conference, is that the dormant stage of Vibrio cholerae is nonculturable on ordinary media, even by using heroic techniques, but it can be detected by immunofluorescent/epifluorescent microscopy. Under carefully controlled and National Institutes of Health-regulated procedures at the University of Maryland Center for Vaccine Development, we were able to feed nonculturable solutions of V. cholerae to human volunteers. Within 48 hours, one subject developed diarrhea, and the organism was culturable from stool specimens. In another case, symptoms appeared after five days, and the organism was then recovered from the stool specimen. These results have some significant implications both for infectious disease epidemiology and for risk assessment of the release of genetically engineered organisms.

The attachment capacity of organisms is part of their survival strategy. Medical microbiologists think that they discovered it; in fact, in the environment, microbial systems in biofilms on surfaces have been known for a long time. This powerful survival mechanism makes the organism difficult to detect and very hard to isolate and grow.

Finally, I think a major theme throughout this conference will be previously unexplored environments and previously uncharacterized organisms.

ORIANS: Gil Omenn has invited me to comment from the perspective of a macroecologist. I was struck by the use of the word "niche" for microorganisms with reference to a single molecule which is the metabolized substrate. When I talk about a niche, the metabolized substrates would come in the form of a rather lengthy list of Linnean binomials; an important part of the research on an organism's food niche, if we can call it that, consists of how different sources of energy are used, depending on the environment and changes in abundance. Hence, one is tremendously concerned about how the diets of an organism will switch with respect to ecological opportunities. To what extent is there information of this sort on potential niche-switching among the microorganisms whose energy sources may change as a result of the conditions in which they find themselves?

We were presented a simple diagram in which the appropriate substrate was depleted and then the population crashed. Is that necessarily so? To what extent are these organisms, although incredibly simple from my perspective, capable of rather subtle changes in the wild?

Also related to the concept of niche are the physical environmental requirements of the organisms, such as temperature, humidity, and various other parameters, in addition to diet. Quite clearly, the environments in which deliberately released organisms will have to function are dramatically different from those in the laboratory in which they are normally tested.

TIEDJE: I have a question for you, as a representative of macroecologists, about these anaerobic food chains. Generally, genetic engineers tend to think of putting all needed functions into one organism. Yet, in these anaerobic pathways, we have a whole series of specialists carrying out one step at a time. Specialists have evolved for each particular step. In biotechnology we say, "Why don't we shove all these genes into one organism and let one organism do the whole thing?" From an ecologist's and an evolutionary biologist's perspective, what are your comments?

ORIANS: Both perspectives raise very interesting problems. The simple organisms lacking brains that you work with would seem more difficult to package a complex array of things into. They may be unable to perform any one of the functions very effectively. It is an axiom in ecology that "a jack of all trades is master of none." You may find that the prime functions you wish to accomplish have been seriously compromised.

ALEXANDER: Microbiologists misuse not only the terms but also the concepts of "niche" and "habitat." Most microbiologists are trained in pure culture work; if an organism has a particular trait expressed in a laboratory study, that trait is assumed to be expressed when the organism is put back in nature. The equivalent is to study a parakeet in the zoo and expect it to be dominant in the field. The possession of a particular trait may be necessary, but it is not a sufficient characteristic to function in nature.

LANDIS: I would like to ask the prokaryotic chauvinists why fungi, algae, and protozoa have been neglected in these presentations thus far?

LEVIN: The EPA is concentrating on organisms that may come to us as part of an application for release into the environment. A search of patent literature in 1983 revealed very little research on eukaryotes, except for Saccharomyces cerevisiae (yeast).

ALEXANDER: Most of us by training are, in fact, prokaryotic chauvinists. We believe that most low-molecular-weight, readily biodegradable compounds are chiefly biodegraded by prokaryotic organisms. I think that view is wrong when it comes to conjugates. Fungi are known to metabolize conjugates as well as oligomeric materials. We think protozoa also are important, but as predators--not in biodegradation itself.

Our chauvinism is because many of us are pure culture bacteriologists working with soluble chemicals in the absence of reactive surfaces. Nature doesn't have pure cultures. Nature doesn't have many water-soluble compounds in environments without reactive surfaces.

LEWIS: The EPA/Cincinnati is supporting work on white-rot fungi and on yeast. Would these organisms be active on insoluble compounds (such as PCBs, dioxins, dibenzofurans) if emulsifiers are added or produced?

FOCHT: Yes, emulsifying agents or detergents might be useful for solubilization of such insoluble materials.

With some of these substrates, we really don't know how they are metabolized. The problem with adding emulsifiers, of course, is that you may add something that's toxic or, in fact, you may get the succession of an unwanted microbial community which grows on the emulsifying agent. It has always amazed me as to how microorganisms (we don't know if they're bacterial or fungal) metabolize synthetic polymers. How do they take polyethylene into the cell? Is it a spaghetti effect? We really don't know the answer, and I think these questions are important, particularly when you consider the tremendous spatial variability that one would encounter in treating these contaminants in soil.

I admit to being a bacterial chauvinist. Whereas plants have one form of metabolism, bacteria have at least four. The nitrogen cycle is exclusively the domain of bacteria.

TIEDJE: As time goes on I'm becoming convinced that availability of substrates to organisms is the key factor in degradation. I know that I can make an experiment on reductive dechlorination fail if I take an insoluble substrate and do not make it available for the cells. So, I think the availability problem is going to be a very tough one to overcome.

UNTERMAN: At GE, we have used surfactants to try to extract PCBs from soil. You can increase the solubility and make the rather insoluble PCBs available to the bacteria, but the biological activity drops off as we increase the concentration of surfactants. We have just scratched the surface, but we are trying to investigate this trade-off.

BULL: What happens to doubling times and biodegradation rates in mixed cultures? What kinds of model experiments can be done with regard to specialists and generalists?

TIEDJE: Kuenen and his colleagues in Holland have investigated mixed substrate utilization by mixed populations of a chemolithotroph, a heterotroph (Spirillium), and a facultative thiobacillus. Respective turnover rates of substrates determined the success of the organisms for the particular niche. In the anaerobic system, the carbon's got to come through fatty acids at some stage. So a specialist on fatty acids can be successful.

ORIANS: As a policy matter, if one's objective were to degrade some toxic substance and then hope that the organism would go away when it had done that job, then the more specialized it was, the less likely it would be to niche-switch and the better it would be. On the other hand, if one is expecting some repeated insults to the environment and you'd like the organism to persist so that you wouldn't have to reinoculate, you'd choose more of a generalist. In experimenting now with planned releases, and knowing of our inability to predict what's going to happen, it seems to me that one should put multiple ecological safeguards into the introduction. Therefore, the most specialized organism should be preferred, one that had been tested in a variety of ways.

TIEDJE: As a strategy, that plan makes both a better product--because you know it's going to be more successful in that particular niche--and a less risky product, because it's going to be a specialist. It helps both ways.

JONES: As a microbiologist, I would like to address the point that we don't know too much about effects when we release genetically engineered organisms. We have run some long-term microcosms in which we placed a recombinant DNA-containing common fresh-water organism, Pseudomonas putida, to see what the effect would be. In one experiment, the invertebrate stopped feeding. That preliminary observation is being retested.

In the United Kingdom, when biologists want to release marked viruses, they have to do checks, not just on potential target insects but on a whole range of other (nontarget) organisms. What is the situation in the United States?

LEVIN: The United States agencies have a series of so-called "points to consider," which involve the information that different agencies consider most important for the release of microbes. At EPA, we do ask questions about effects on nontarget organisms--like honey bees or lace wings or other macroorganisms--as well as effects on other microorganisms. There's a whole series of points that have to be answered, as any industrialist here will tell you. This is all detailed in the Federal Register. For microbial pesticides, "Subdivision M" contains the relevant information.

LAUDER: Many of these chemicals, particularly some of the insecticides, are delivered into the environment in a solvent which, in fact, represents another carbon source. This raises the point that Dr. Colwell was making regarding the need for systematics. We may need to realize that there is a succession of organisms, one series degrading the solvent and a second series degrading the substrate we find offensive.

COLWELL: I'm going to say something that a lot of people won't like, because it isn't the fashionable thing to do. I think a lot of work needs to be done on bacterial-microbial community structure. From our work on the effects of ocean dumping, it's clear that you get detectable and persistent changes in species composition. In the trench dump site in Puerto Rico that we studied for about ten years, even after the dumping ceased there was sufficient persistence of the chemicals to alter the species composition. You could do a transect through that area of the ocean and determine the location of the residual pharmaceutical manufacturing wastes. A good deal of systematic research on species interactions is necessary in order to be able to determine what the effects will be in our natural environment. There are many naturally occurring forms about which we have no knowledge, but from which very important ecological consequences may be derived.

FOCHT: Hazardous waste dump sites rarely contain one compound. A multitude of organics typically are present. We need to know whether there is a whole succession or wave of different pioneer communities that arise, or whether there is some type of preferential induction of a single organism that, for example, won't do anything to compound A until compound B is gone.

ATLAS: Can we devise realistic models to study fate and effects? In many situations we need to reflect realistic microbial community composition at a particular site where a chemical or chemical mixture is to be biodegraded. Similarly, one has to consider the physical form of the chemical as well as its own chemical environment. The fate of compounds within mixtures can be very different from the fate of a compound itself.

COLWELL: We took the organisms that had become dominant in a deep ocean dump site, and we also took organisms that were no longer dominant but were in nearby unimpacted sites. When we mixed them together and added the pharmaceutical waste, the same organisms became dominant again. We could reproduce in the laboratory what had occurred in the field.

LEVIN: The regulatory agencies are all interested in understanding the how and why and the basic mechanisms. At EPA we have a tiered list of regulatory questions that must be answered. First, does the maximum dose that you are going to put into the field at a given time have any effect on the organisms that are present? Only if there is an effect do you have to go on to the next tier. If you release the organism at a particular level, what are the most likely environmental concentrations? What organisms are most likely to be present? What are the conditions under which those organisms are going to be present? Is there an effect at that level? We're looking for effects, not mechanisms. So our questions do not require answers in the depth that Dr. Colwell and Dr. Atlas have discussed.

RELEVANT ADVANCES IN MOLECULAR BIOLOGY

PROSPECTS FOR LABORATORY ENGINEERING OF BACTERIA TO DEGRADE POLLUTANTS

K.N. Timmis, F. Rojo, and J.L. Ramos

Department of Medical Biochemistry
University of Geneva
Geneva, Switzerland

INTRODUCTION

Over the last few decades enormous quantities of industrial chemicals have been released into the environment. A large number of them, particularly those structurally related to natural compounds, are readily degraded by soil and water microorganisms. However, a significant proportion, mainly those having novel structural elements or substituents rarely found in nature (xenobiotics), are only catabolized slowly and thus tend to persist and accumulate in the environment. Certain compounds, particularly those that exhibit some degree of toxicity, contribute substantially to environmental pollution. Recent environmental catastrophes have underscored the acute danger that industrial chemicals constitute for our biosphere. However, the existence of many waste dump sites containing highly toxic substances and large scale chronic pollution certainly represent a more important long-term hazard. Clearly, in addition to terminating current production of the more toxic and persistent industrial chemicals, it is essential to exploit more effectively the biodegradative capacities of soil microorganisms in order to diminish the consequences of existing and continuing environmental pollution.

Fortunately, soil and water microorganisms collectively exhibit remarkable capacities to degrade a wide range of noxious organic chemicals (6) and to evolve degradative activities toward new compounds (16). On the other hand, the evolutionary process can be extremely slow, particularly where the acquisition of multiple catalytic activities is necessary. In such cases, the evolution of new metabolic activities in the laboratory may be helpful (3,4,9,15), because the frequency and type of genetic events needed (e.g., mutation, alteration of gene expression, and gene transfer) can be carefully controlled and selective conditions can be optimized (19).

There are basically three experimental approaches to the laboratory evolution of metabolic pathways, namely:

(a) Long-term chemostat selection, which often involves progressive replacement of a mineralizable substrate by a recalcitrant analog (5).

(b) In vivo genetic transfers, in which genes of critical enzymes of one organism are recruited into a pathway of another organism through experiments involving natural genetic transfer processes, such as transduction, transformation, and especially conjugation (22). This approach is sometimes facilitated by the fact that the genetic information for recently evolved pathways is frequently located on transmissible plasmids or transposons (7,25), genetically promiscuous elements that readily move from one replication unit to another within the same cell. Thus, once a critical enzyme has evolved in one organism, its gene can be easily transferred to others and become recruited into related or unrelated existing or evolving pathways. Once a new pathway has emerged in one organism, it can be readily transferred to others.

(c) In vitro evolution, in which cloned and well characterized genes are selectively transferred into a different organism in order to evolve a new pathway (13,19,20).

While all three approaches constitute powerful experimental tools to effect desired phenotypic changes in microbes, in vitro evolution has the added advantage of being highly controlled, as well as of producing predictable changes and enabling multiple genetic changes to be effected in a single step, often without the need to directly select the final phenotype desired (18). However, in vitro evolution does require detailed information on the genetics and biochemistry of the key elements of the pathways and the enzymatic steps to be manipulated. In contrast, chemostat selection and in vivo evolution can often be carried out without prior extensive characterization. In this brief review, we will discuss the use of genetic manipulation strategies for the rational evolution of new catabolic phenotypes in soil bacteria.

STRATEGIES FOR IN VITRO EVOLUTION OF CATABOLIC PATHWAYS

In general, the evolution of new metabolic potential involves the acquisition of new or modified enzymatic activities (2,3,20). However, since the synthesis of many enzymes is carefully regulated and occurs only in response to the appearance of specific induction signals (e.g., the appearance in the medium of the cognate substrate of the enzyme or pathway), the acquisition of new specificities of regulators of gene expression is also important (3,19,20). De novo evolution of a protein with a radically new activity often involves a number of genetic events (mutations, recombinations, fusions, etc.), most of which cannot be predicted a priori. Thus, laboratory (and natural) evolution of new phenotypes will generally involve the acquisition of new enzymatic or regulator specificities (for substrates and effectors, respectively) through mutational alteration of existing proteins or through recruitment of new proteins from different organisms. The ability of enzymes and regulators to undergo a relaxation of their specificities without loss of function (3,20) and the existence of enzymes exhibiting catalytic activities toward a broad range of structurally distinct substrate molecules (10,13,14,30) are of critical importance to evolution.

Two general strategies have been developed for the evolution of catabolic pathways for recalcitrant compounds by means of genetic manipulation (18). If the chemical exhibits substantial structural analogy to compounds that are readily degraded, the initial strategy of choice is to identify the steps of the known pathway that are nonpermissive for the chemical in question, and then to modify these such that they become permissive. This approach generally leads to an expansion of the substrate profile of the existing pathway. The expansion can be horizontal, in that more analogs of a single class of compounds are metabolized, as a result of the recruitment of isofunctional enzymes from other pathways (Ref. 13; see below) or of the mutational alteration of the substrate specificities of existing key enzymes (2,3,20). Pathway expansion can also be vertical, in that the existing pathway is used as a base onto which are grafted additional enzymes that extend the pathway upward (13,26). Alternatively, if an existing pathway for related compounds is not known, new pathways can be conceived and appropriate component enzymes can be sought in soil bacteria (23). Once the feasibility of a new route is established, the corresponding enzymes can be recruited into a single bacterium (or a consortium) by the cloning of their structural and regulatory genes and by the combination of such genes in an appropriate host organism (26).

EXPANSION OF THE SUBSTRATE RANGE OF A PATHWAY BY RECRUITMENT OF ENZYMES EXHIBITING RELAXED SPECIFICITIES

Horizontal Expansion of Catabolic Pathways

The basis of horizontal expansion of catabolic pathways lies in (a) the existence of isofunctional routes for degradation of structurally related compounds, (b) the existence of enzymes having relaxed substrate specificities in some routes, and (c) the ability to alter the specificity of proteins for their substrates/effectors by mutagenesis.

Pseudomonas putida bacteria containing TOL plasmid pWW0 are able to degrade and grow on benzoate and a variety of alkylbenzoates such as 3-(3MB) and 4-methylbenzoate (4MB), 3,4-dimethylbenzoate (3,4DMB), and 3-ethylbenzoate (3EB) (27). The TOL plasmid-specified catabolic pathway involves dioxygenation of the ring, followed by decarboxylation to yield alkylcatechols, which are then subjected to *meta* cleavage and further oxidation to Krebs cycle intermediates (Fig. 1). The genes of the pathway form an operon, the transcription of which originates from promoter Pm and is positively controlled by a regulatory protein encoded by the *xylS* gene (Ref. 10; see below).

Pseudomonas sp. B13 is able to degrade 3-chlorobenzoate (3CB) via 3-chlorocatechol followed by *ortho*-ring cleavage of the aromatic ring (Ref. 5; Fig. 1). It is not, however, able to degrade analogs of 3CB, such as 4-chlorobenzoate (4CB) or 3,5-dichlorobenzoate (3,5DCB), due to the narrow substrate specificity of the first enzyme of the pathway, benzoate 1,2-dioxygenase (22). The substrate specificity of an isofunc-

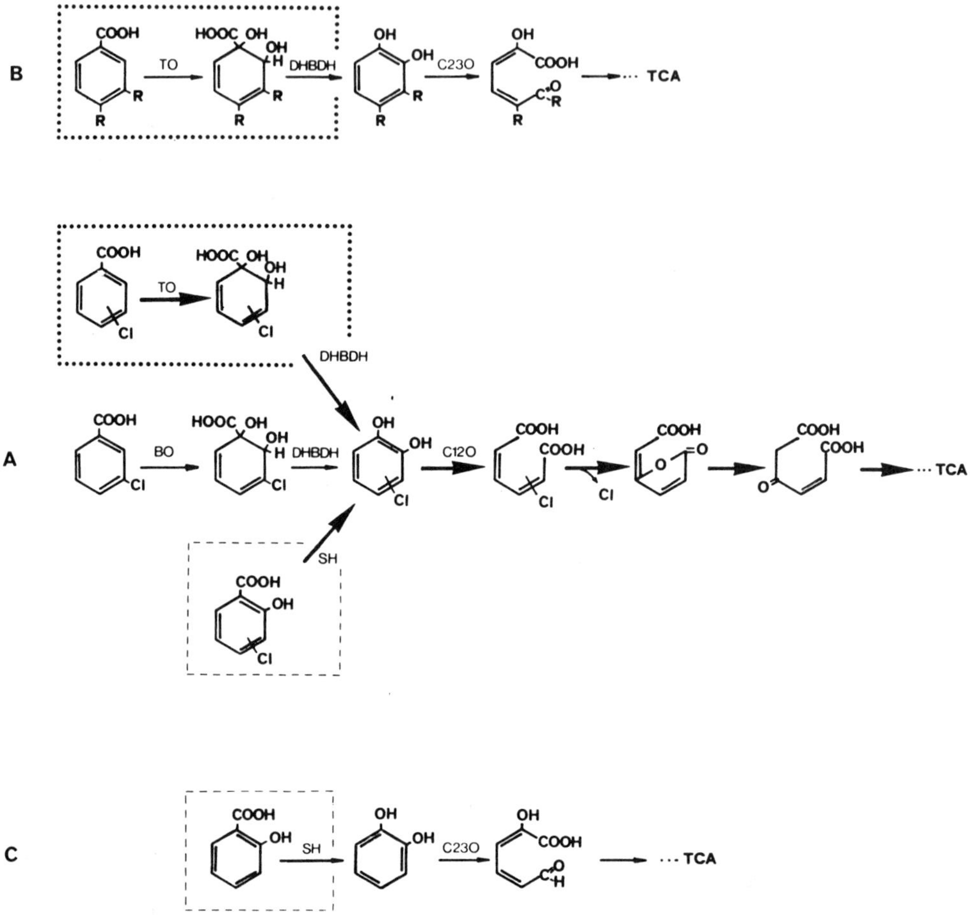

Fig. 1. Horizontal and vertical expansion of a catabolic pathway. The 3-chlorobenzoate catabolic pathway of Pseudomonas sp. B13 (A) involves transformation of the substrate to 3-chlorocatechol, which is ortho-cleaved, converted to 3-oxoadipate, and subsequently channeled into the tricarboxylic acid (TCA) cycle. The first enzyme of the B13 pathway, benzoate 1,2-dioxygenase (BO), has a narrow substrate specificity and is able to transform 3CB but not 4CB or chlorosalicylates (22). Recruitment of the genes coding for the relaxed substrate specificity enzyme toluate 1,2-dioxygenase (TO) and dihydroxycyclohexadiene carboxylate dehydrogenase (DHBDH) from the TOL plasmid meta-cleavage pathway (B), together with the promoter of these genes and the gene of the positive regulator of this promoter, allows B13 to degrade 4CB (13). Recruitment of the gene of the relaxed substrate specificity enzyme salicylate hydroxylase (SH) of the NAH7 plasmid-specified pathway for the degradation of salicylate (C) permits B13 to degrade and grow on 3-, 4-, and 5-chlorosalicylate (13).

tional enzyme, toluate 1,2-dioxygenase, which is specified by the TOL plasmid, is more relaxed and includes not only alkylbenzoates but also the chlorobenzoates 3CB, 4CB, and 3,5DCB (21,30). It is therefore anticipated that recruitment of this enzyme into strain B13 would allow the bacterium to degrade not only 3CB but also its analogs.

A DNA segment carrying the genes of toluate dioxygenase (xylD) and the following enzyme of the pathway, dihydroxycyclohexadiene carboxylate (benzoate cis-diol) dehydrogenase (xylL), plus the promoter Pm and the xylS gene (Fig. 3), were cloned into a broad host range plasmid vector and introduced into the B13 strain (13). As predicted, the B13 derivative thereby constructed could grow on 3CB and 4CB (Fig. 1), and synthesis of all catabolic enzymes involved in the metabolism of these substrates was fully regulated. On the other hand, the B13 derivative did not grow on 3,5DCB, even though the catabolic enzymes present in 4CB-grown bacteria are able to degrade this compound (22). The reason for this is discussed below.

Vertical Evolution of Catabolic Pathways

In contrast to the divergent routes of biosynthetic pathways, in which a large number of chemically distinct biomolecules are synthesized from a limited number of central intermediates, catabolic pathways are convergent and a wide variety of different substrates are channeled toward central intermediates and hence into central metabolic pathways. Pathways for the catabolism of aromatics converge at dihydroxy aromatic compounds, such as catechol, protocatechuate and gentisate, which are then subjected to dioxygenase-catalyzed ring fission prior to their entry into the Krebs cycle (Fig. 2).

Central pathways, such as the chlorocatechol ortho-cleavage pathway (Fig. 1), can be used as a base upon which to assemble additional enzymatic steps in order to permit the catabolism of more complex compounds. A simple example of vertical expansion of the chlorocatechol pathway of Pseudomonas sp. B13 was provided by construction of a derivative capable of degrading chlorosalicylates (13). B13 is not able to degrade either salicylate or chlorosalicylates, and bacteria that are able to catabolize chlorosalicylates are not readily isolated from soil. Plasmid NAH7 of P. putida specifies a pathway for the catabolism of naphthalene via salicylate and catechol (28). The salicylate hydroxylase, encoded by the NAH7 plasmid, exhibits a relaxed substrate specificity and oxidizes salicylate and methyl- and chlorosalicylates to the corresponding catechols (13). A DNA fragment of the NAH7 plasmid containing the gene encoding salicylate hydroxylase, as well as the promoter of this gene, and the gene of the positive regulator of this promoter (29), was cloned into a wide host range vector, and the hybrid plasmid was introduced into Pseudomonas sp. B13, which thereby acquired the ability to grow on 3-, 4-, and 5-chlorosalicylates (Ref. 13; Fig. 1). Thus, recruitment of the NAH7 salicylate hydroxylase into a bacterium capable of degrading chlorocatechols extended vertically the chlorocatechol degradation pathway and resulted in the construction of an organism that was able to catabolize compounds not readily degradable by soil isolates.

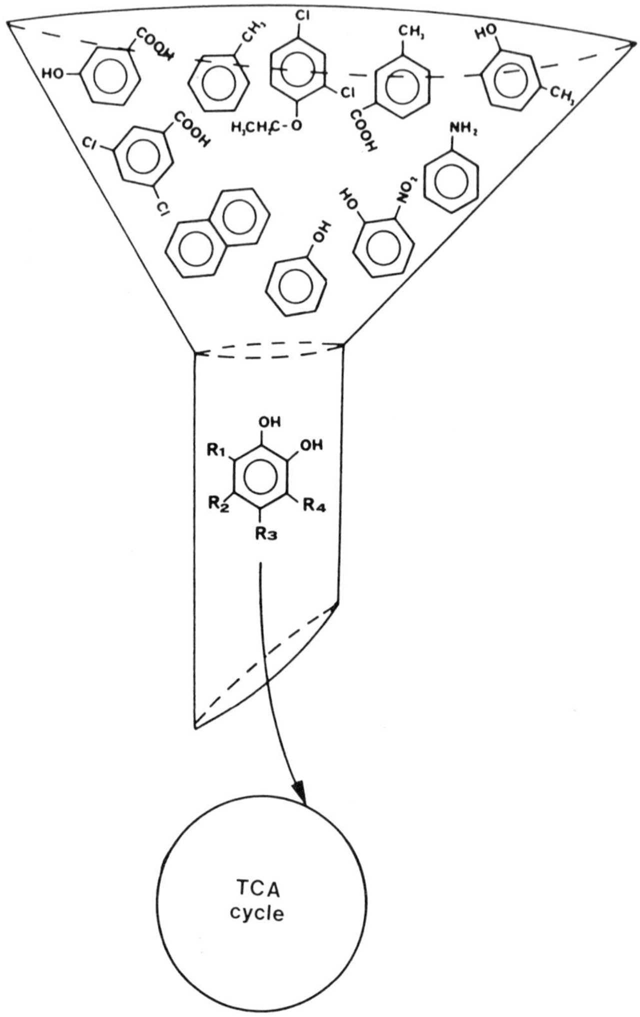

Fig. 2. Channeling of aromatic compounds to central pathways. Most aromatic compounds are transformed to dihydroxy intermediates, such as catechols, and thus channeled to central pathways.

EXPANSION OF THE SUBSTRATE RANGE OF A PATHWAY BY MUTATION

As indicated above, recruitment of toluate dioxygenase into B13 expanded its substrate range to 4CB but not to 3,5DCB, even though the catabolic enzymes of the pathway are able to process 3,5DCB. This suggested that 3,5DCB or one of its metabolites fails to induce synthesis of either toluate dioxygenase or the chlorocatechol catabolic enzymes of the 3CB pathway. In order to examine the inducer specificity of the positive regulator of expression of the xylD gene, the Pm promoter was fused in a cloning vector to lacZ, which encodes β-galactosidase, a readily-assayed test enzyme used extensively in gene expression studies (19). Escherichia coli K-12 carrying this hybrid plasmid, plus another containing xylS, the gene of the positive regulator of transcription from Pm, synthesizes β-galactosidase when benzoate is present in the growth medium but not when a noninducer such as salicylate is present. Using this system, a large number of benzoate analogs were tested for their ability to provoke synthesis of β-galactosidase through activation of the XylS regulator. The structural features of benzoate analogs critical for their interaction with the regulator were thereby analyzed (Tab. 1). From this study, it became clear that whereas 3CB and 4CB activate the XylS protein and induce synthesis of toluate dioxygenase, 3,5DCB does not (19). Thus, the failure of 3,5DCB to be metabolized is due to the fact that toluate dioxygenase is not produced by B13 derivatives containing xylDLS.

Tab. 1. Activation of XylS and XylS mutant proteins by benzoate analogs.

Benzoate analog	Induction ratio (relative increase in β-galactosidase)		
	xylS	xylS4	xylS352
None	1	1	1
2MB	18	37	10
3MB	17	14	14
4MB	4	10	3
2,3MB	10	12	-
3,4MB	5	8	-
2,4MB	1	5	1
2,5MB	1	1	2
3,5MB	1	1	4
4EB	1	8	1
3,5CB	1	1	3

Note: Escherichia coli K-12 bacteria containing two plasmids, one carrying a Pm::lacZ fusion and the other carrying either xylS or a mutant xylS allele (xylS4 or xylS352), were cultured in L-broth containing or lacking the benzoate analog indicated in the left-hand column. β-Galactosidase levels were subsequently measured, and the ratio plus analog/minus analog (induction ratio) was determined. A value of 1 indicates that the analog did not induce synthesis of β-galactosidase, and thus does not serve as an effector of the XylS or mutant XylS proteins, whereas values greater than 1 indicate that the analog served as an inducer. Abbreviations: B, benzoate; M, methyl; E, ethyl; C, chloro.

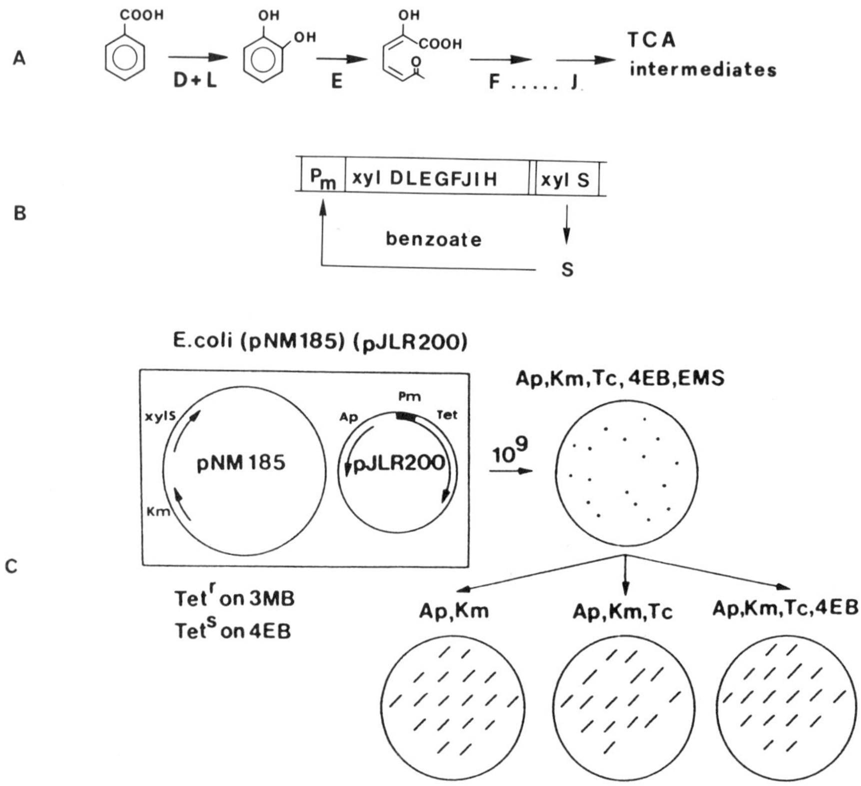

Fig. 3. Regulation of the TOL plasmid-specified meta-cleavage pathway, and selection of mutants exhibiting altered regulatory specificities. (A) Genes of the TOL plasmid-encoded meta-cleavage pathway (8). The gene xylD encodes toluate dioxygenase (TO; Fig. 1); xylL encodes dihydroxycyclohexadiene carboxylate dehydrogenase (DHBDH; Fig. 1); xylE is the gene of catechol 2,3-dioxygenase (C23O), etc. (B) Organization and regulation of the TOL genes. The genes of the meta-cleavage pathway are organized in an operon whose transcription originates at promoter Pm. Transcription from Pm is positively regulated by the product of the xylS gene, the XylS protein, which is activated by effector molecules, such as benzoate, that are usually substrates of the catabolic pathway. (C) Selection of mutant XylS proteins activated by new effector molecules such as 4-ethylbenzoate. Escherichia coli K12 (pNM185; pJLR200) bacteria were spread on nutrient agar plates containing ampicillin, kanamycin, tetracycline, 4-ethylbenzoate, and the mutagen ethylmethane sulfonate (EMS). After incubation of the plates, two classes of clones appeared: one that grew on tetracycline plates lacking 4EB (most of these contained a mutation in the Pm promoter, so that the tet gene was expressed constitutively), and another that grew only on tetracycline plates containing 4EB. Mutants of the latter type contained plasmids that carry mutant xylS genes whose products are activated by 4EB (Tab. 1).

In order to select xylS mutants that produce regulators having broader inducer specificities, a hybrid plasmid was constructed in which the Pm promoter was fused to a gene specifying resistance to tetracycline (Fig. 3). This gene is expressed, and thus host bacteria are resistant to tetracycline, only when the XylS protein is activated by benzoate or one of its analogs. Bacteria spread on a nutrient agar plate containing tetracycline and 3,5DCB, or some other ordinarily noninducing benzoate analog such as 4-ethylbenzoate, are killed because the XylS protein is not activated. The plating of a large number of such bacteria and the inclusion of a strong mutagen such as ethylmethane sulfonate in the plate result in the isolation of a few tetracycline-resistant colonies (Fig. 3). Some of these colonies produce mutant XylS proteins that are activated by new effector molecules, such as 3,5DCB or 4EB (19). Using this procedure, a number of different xylS mutants have been selected. Among them, xylS352 had acquired the ability to be activated by 3,5DCB and several other disubstituted benzoates (Tab. 1). When this mutant xylS gene, together with xylDL, was introduced into B13, the bacterium acquired the ability to degrade completely and grow at the expense of 3,5DCB (19).

RATIONAL RESTRUCTURING OF CATABOLIC PATHWAYS BY SEQUENTIAL MODIFICATION

Pseudomonas putida bacteria containing TOL plasmid pWW0 are able to grow on a variety of alkylbenzoates, including 3- and 4MB and 3EB but not 4EB (Fig. 4). Other bacteria able to degrade methylbenzoates are readily isolated from soil, but they also cannot generally degrade 4EB (20). 4-ethylbenzoate thus seems inherently more resistant to microbial attack than other alkylbenzoates. Biochemical and genetic studies were carried out to identify the steps in the TOL pathway that are nonpermissive for 4EB. First, an analysis of the inducer specificity of the XylS protein revealed that it is activated by 3- and 4MB, 3,4DMB, and 3EB but not 4EB (Ref. 20; Tab. 1). Thus, 4EB does not induce synthesis of the catabolic enzymes (Fig. 4). Use of the procedure outlined above enabled selection of XylS regulator mutants that are activated by 4EB. One such mutant xylS gene was then transferred to P. putida bacteria carrying the TOL plasmid (Fig. 5). This derivative also failed to grow on 4EB but did degrade the compound to 4-ethylcatechol (Fig. 4). Thus, the catabolic enzymes are synthesized in this derivative, and 4EB is transformed to 4-ethylcatechol, but the meta-cleavage enzyme does not permit further metabolism. Characterization of the meta-cleavage enzyme, catechol 2,3-dioxygenase, revealed that 4-ethylcatechol is in fact a suicide substrate which causes irreversible inactivation of the enzyme (Fig. 5).

It was reasoned that if a single amino acid change in the catechol dioxygenase could render it resistant to inactivation by 4-ethylcatechol, and that if this was the only further change needed to permit the complete degradation of 4EB via the TOL plasmid-encoded pathway, then the appropriate catechol dioxygenase mutants might be obtained by selecting directly for growth of bacteria on 4EB. A large number of P. putida bacteria carrying the TOL plasmid and the mutant xylS gene were plated on minimal medium containing ethylmethane sulfonate and 4EB as the sole source of

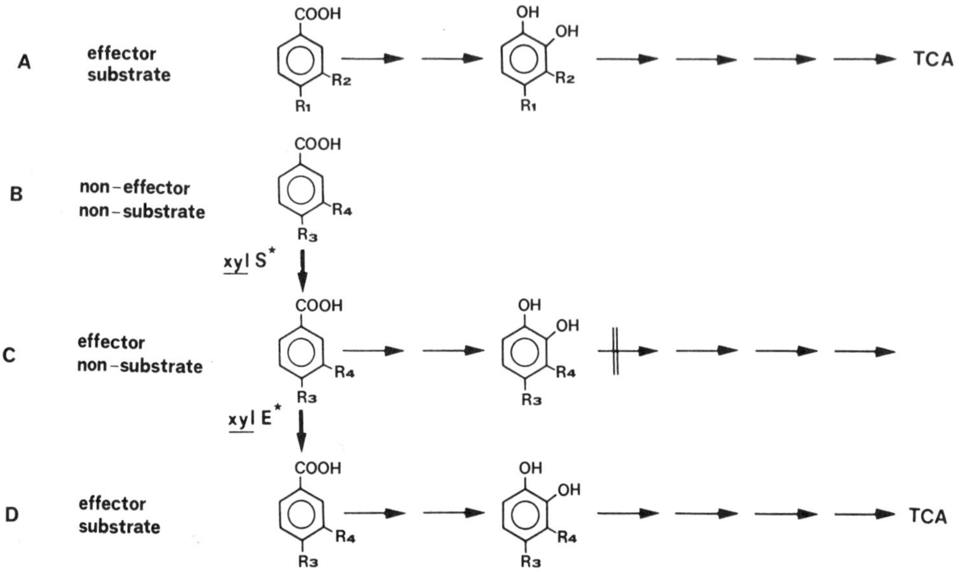

Fig. 4. Sequential restructuring of the TOL plasmid-specified meta-cleavage pathway to enable catabolism of 4EB. The pWW0 plasmid-encoded meta-cleavage pathway can catabolize benzoate, 3- and 4MB, 3,4DMB, and 3EB (R_1 = H or CH_3; R_2 = H, CH_3, or CH_2CH_3) (A) but not 4EB (R_3 = CH_2CH_3; R_4 = H). The latter does not serve as an activator of the xylS protein regulator of the catabolic operon promoter Pm (Tab. 1) and thus fails to induce synthesis of the catabolic enzymes (B). Isolation of a mutant that produces a XylS protein analog that is activated by 4EB (Fig. 3) results in the synthesis of all catabolic enzymes in response to the presence of 4EB in the bacterial culture medium (C). However, 4EB is only metabolized as far as 4-ethylcatechol, because this intermediate inactivates the ring cleavage enzyme C23O. The isolation of a mutant that produces a C23O analog resistant to inactivation by 4-ethylcatechol (Fig. 5) eliminates the final metabolic block and permits complete degradation of 4EB through the meta pathway (D). (TCA, tricarboxylic acid cycle.)

carbon and energy (Fig. 5). Several colonies grew under these conditions and all were subsequently shown to be able to metabolize 4EB, to use it as a source of carbon and energy, and to produce altered catechol dioxygenases that exhibited increased resistance to inactivation by 4-ethylcatechol (Ref. 20; Fig. 5). The TOL plasmid-encoded pathway for the degradation of alkylbenzoates was thereby rationally restructured by sequential modification to enable it to process 4EB by selective mutational alteration of two elements that relaxed the effector specificity of the XylS protein and increased the resistance of catechol 2,3-dioxygenase to inactivation by 4-ethylcatechol.

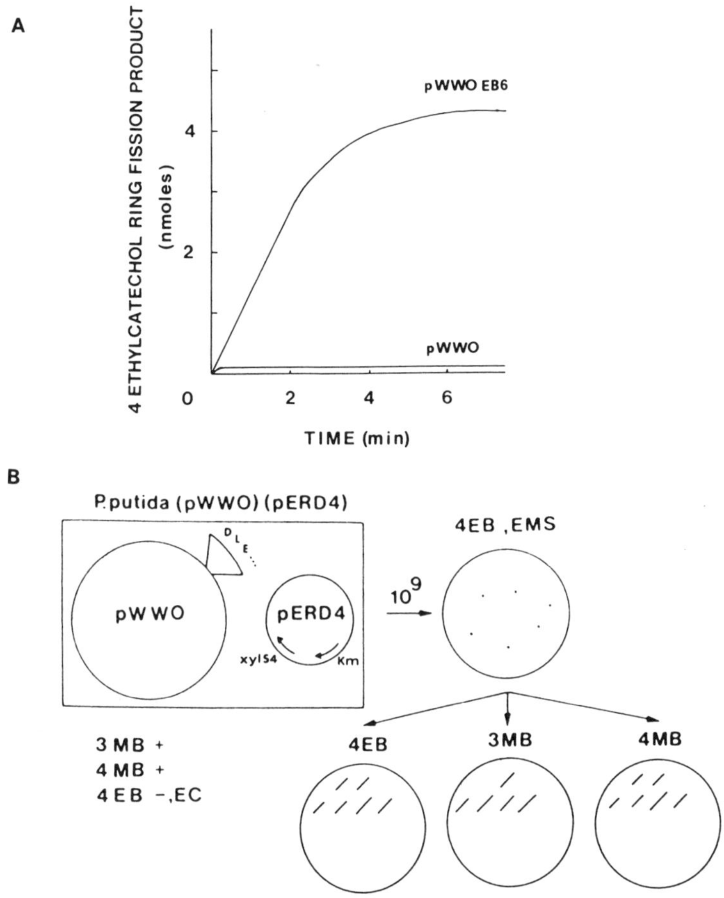

Fig. 5. Demonstration that 4-ethylcatechol is a suicide substrate for C23O and selection of C23O mutant enzymes resistant to substrate inactivation. (A) Influence of 4-ethylcatechol on the activities of C23O and mutant C23O. Cell-free extracts were prepared from Pseudomonas putida containing either TOL plasmid (pWW0) or mutant TOL plasmid (pWW0-EB6) that permits host bacteria to grow on 4EB. 4-Ethylcatechol was added at time zero. The amount of ring cleavage product formed was followed spectrophotometrically as a function of time. (B) Isolation of mutants producing C23O enzymes resistant to inactivation by 4-ethylcatechol. Pseudomonas putida (pWW0; pERD4) bacteria were spread on minimal medium containing 4EB as sole source of carbon and energy and the mutagen ethylmethane sulfonate (EMS). After incubation of the plates, six clones appeared, five of which grew on 4EB, 3MB, and 4MB, while the other grew only on 4EB and 4MB. All clones contained mutant xylE genes (xylE*) that encode C23O enzyme analogs more resistant to inactivation by 4-ethylcatechol than that encoded by the wild-type xylE gene.

DESIGNING NEW CATABOLIC PATHWAYS: PATCHWORK ASSEMBLY OF ENZYMES AND REGULATORY SYSTEMS

The rational redesigning of an existing pathway is a relatively straightforward experimental approach to evolve a pathway for the degradation of a recalcitrant compound. However, this approach will not always be sufficient, in which case it may be appropriate to construct a new pathway by rationally combining in patchwork fashion a series of catalytic activities derived from different pathways and different organisms.

Individual substituted phenols and benzoates have a significant level of toxicity, but nevertheless can generally be degraded via catechol by microorganisms in soil and in wastewater treatment plants. Industrial wastes, however, frequently contain mixtures of chloro- and methyl-substituted phenols and benzoates. Such mixtures are not only difficult to degrade but also tend to destabilize phenol- and benzoate-degrading microbial communities. The problem lies in the existence of both _ortho_- and _meta_-cleavage routes for degradation of catechols. Catechols and chlorocatechols are generally subjected to _ortho_ fission, whereas methylcatechols suffer _meta_ fission (Fig. 6). Although both pathways may exist in individual microorganisms, only one is usually functional at any given moment, according to the substrate that is available. However, when both chloro- and methylcatechols are formed from mixtures of chloro- and methylaromatics, both types of pathway are functional, often in the same organism, and the catechols will be subjected to both types of cleavage. Whereas _ortho_ cleavage of chlorocatechols leads to their productive metabolism, _ortho_ cleavage of methylcatechols leads to the formation of dead-end products. Similarly, whereas the _meta_ cleavage of methylcatechols leads to their productive metabolism, the _meta_ cleavage of chlorocatechols leads to the formation of either dead-end products or products that inactivate C23O, the ring cleavage enzyme (1,11,24). The nonproductive routing of catechol cleavage products during simultaneous metabolism of chloro- and methyl-substituted aromatics eventually perturbs the productive metabolism of aromatics by the cell or the community to such an extent that the community is destroyed (Ref. 11; see below).

One potential solution to this problem is the construction of catabolic routes for chloro- and methylaromatics that employ only one type of catechol ring fission mode. _Pseudomonas_ sp. B13 possesses only _ortho_-cleavage routes for catechols; it can grow on 3CB and acquires the ability to grow on 4CB when it receives the _xylDLS_ genes from the TOL plasmid (13,22). In order to create a stable B13 derivative that is able to degrade 4CB, the cloned TOL genes were inserted into transposon Tn_5_, and the hybrid element was subsequently transposed into the B13 chromosome. The derivative thereby obtained, FR1 (Fig. 7), grew on 3CB and 4CB but not on 3MB or 4MB. The latter two compounds were, however, co-metabolized via _ortho_ cleavage to the dead-end products 2- and 4-methyl-2-enelactone, respectively (Ref. 11 and 24; Fig. 7). In order to effect mineralization of the dead-end products of _ortho_ cleavage of 3MB and 4MB, it was necessary to identify and recruit additional enzymes. The methyl-2-enelactones that accumulated during metabolism of methylbenzoates were, therefore, isolated and used to select organisms that are able to use such compounds for growth. This resulted in the isolation of an _Alcaligenes_ sp. capable of growing on 2- and 4-methyl-2-enelactone (17). Growth on 4-methyl-2-enelactone involves isomerization of 4-methyl-2-enelactone to

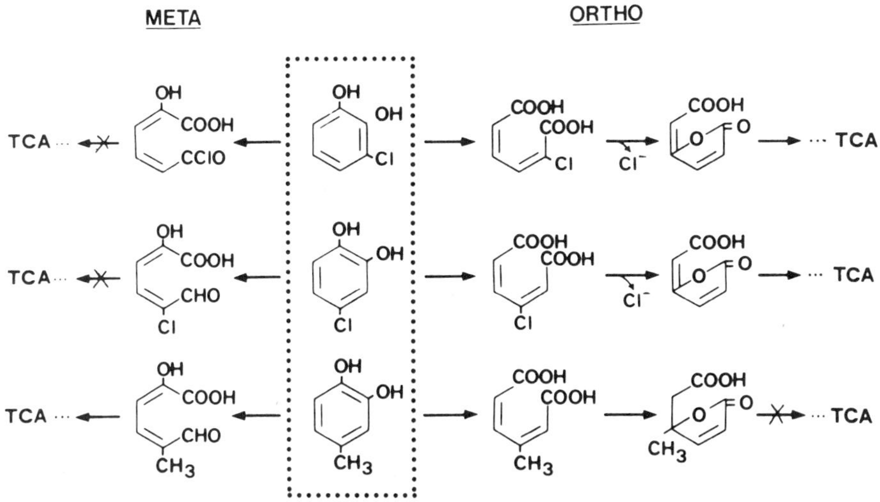

Fig. 6. Unproductive misrouting of substituted catechols by ortho- and meta-cleavage enzymes. Degradation of chloro- and methyl-substituted aromatics often involves, as a first step, their conversion to corresponding catechol derivatives. Chlorocatechols are degraded through ortho-cleavage pathways, but may be misrouted into a meta-cleavage pathway in bacteria containing a meta-cleavage enzyme. As a result, they are channeled to dead-end metabolites which, in those cases such as 3-chlorocatechol, are highly reactive and inactivate the meta-cleavage enzyme. Methylcatechols are normally degraded through meta-cleavage routes, although they can be co-metabolized by ortho-cleavage pathways, eventually forming methyllactone derivatives as dead-end metabolites. This is illustrated for 4-methylcatechol, from which 4-carboxymethyl-4-methylbut-2-en-1,4-olide (4-methyl-2-enelactone) is formed.

3-methyl-2-enelactone. Since Pseudomonas sp. B13 is able to grow on 3-methyl-2-enelactone, it was reasoned that recruitment of the isomerase function of the Alcaligenes sp. into FR1 would allow it to grow on 4MB. An Alcaligenes gene bank was, therefore, prepared in a wide host range, mobilizable cosmid vector and mass-transferred into FR1 by conjugation. Transconjugants that were able to grow on 4MB were readily isolated; the hybrid cosmid present in one of these was designated pFRC20P (Ref. 23; Fig. 7). Cell-free extracts of these derivatives exhibited enzymatic activities that converted purified 4-methyl-2-enelactone to 3-methyl-2-enelactone. High levels of activity were measured both in 4MB-grown cells and in acetate-grown cells, although highest levels were obtained in 4MB-grown cells. This indicates that expression of the isomerase is regulated in these constructions, but that the basal level of synthesis is high.

Pseudomonas sp. B13 is able to grow on phenol and, after adaptation, on 4-chlorophenol as sole sources of carbon and energy; catabolism is via an ortho-cleavage route, with catechol or 4-chlorocatechol as intermediates (12). The phenol hydroxylase involved also transforms 3-(3MP) and 4-

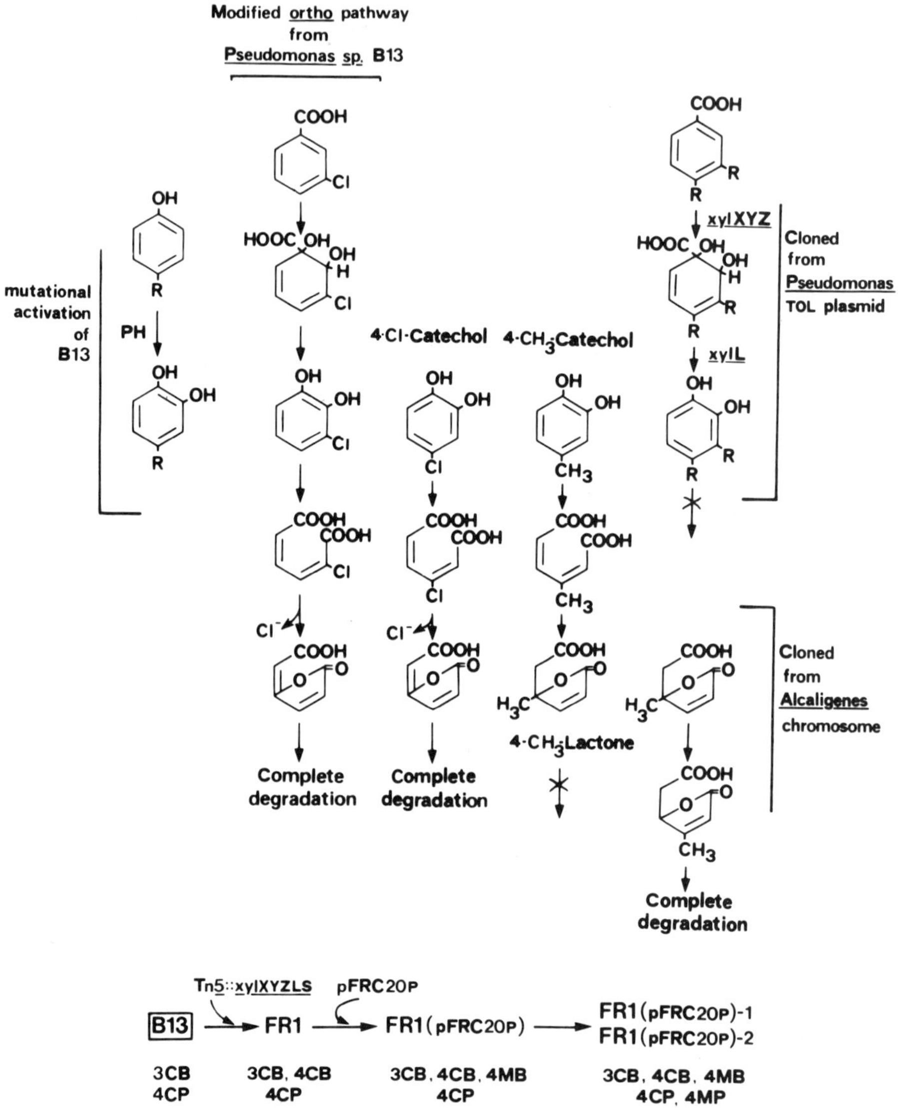

methylphenols (4MP) in cell-free extracts and produces from these substrates the corresponding methylcatechols (12). In principle, therefore, a derivative of B13, such as FR1(pFRC20P), that can mineralize 4-methylcatechol possesses all of the enzymes necessary to grow on 4MP as sole source of carbon and energy. However, although phenol is a growth substrate for this derivative, 4MP is not (23). Spontaneous mutants could, nevertheless, be selected that grow on 4MP (frequency 10^{-7} to 10^{-8}). Measurement of phenol hydroxylase levels in such mutant bacteria, grown in succinate, phenol, or 4MP, revealed little or no enzyme in succinate-grown cells and high activities in bacteria grown on phenol or 4MP. Synthesis of this enzyme is, therefore, specifically regulated.

Fig. 7. Constructed hybrid pathway for the simultaneous degradation of chloro- and methylaromatics. (A) The pathway. The route is based on the modified ortho pathway for 3CB of Pseudomonas sp. B13. Introduction into B13 of the TOL plasmid genes coding for toluate 1,2-dioxygenase (xylD) and dihydroxycyclohexadiene carboxylate dehydrogenase (xylL), together with that of the positive regulator of the xylDL operon (xylS), expands the degradation range to include 4CB and permits transformation of 4MB to 4-methyl-2-enelactone, which is accumulated as a dead-end metabolite (strain FR1). Recruitment of a 4-methyl-2-enelactone isomerase from Alcaligenes eutrophus allows transformation of 4-methyl-2-enelactone to 3-methyl-2-enelactone (bottom right), which is degraded by other enzymes of B13 [strain FR1-(pFRC20P)]. Mutational modification of expression of the phenol hydroxylase of B13 further extends the degradation capacities to chloro- and methylphenols [strains FR1(pFRC20P)-1 and -2]. The final route is thereby composed of five different pathway segments derived from three different organisms. (B) The genetic steps leading to the formation of FR1(pFRC20P)-1 and -2. Further abbreviations: 4CP, 4-chlorophenol; 4MP, 4-methylphenol.

Figure 8 shows that addition of a chloroaromatic such as 3CB to bacteria [e.g., P. putida KT2440(TOL) or Pseudomonas sp. B13(TOL)] that are actively degrading a methylbenzoate through a meta ring-fission pathway resulted in inhibition of cell growth as a result of misrouting of 3CB into the meta pathway and irreversible inactivation of the key enzyme of the pathway, catechol-2,3-dioxygenase (1), by the ring-fission product of 3-chlorocatechol, which is a highly reactive acylhalide. Addition of 4CB to such cells had a less severe impact on growth because the dead-end product, which forms as a result of misrouting of 4CB into the meta pathway, is less reactive (Fig. 7 and 8). Addition of 3CB or 4CB to FR1(pFRC20P) bacteria growing on 4MB had no inhibitory effect and, in fact, promoted further growth of the cultures (Fig. 8). Moreover, the B13 derivative degraded simultaneously the chloro- and methylbenzoates in the mixtures of 4MB plus 4CB and 4MB plus 3CB.

Thus, a novel ortho-cleavage pathway for the degradation of 4-methylbenzoate has been constructed by the patchwork assembly of four pathway segments, namely, (a) the TOL plasmid toluate dioxygenase and following cis-diol dehydrogenase, which transform methylbenzoates to methylcatechols; (b) the B13 chlorocatechol dioxygenase and chloromuconate cycloisomerase, which convert methylbenzoates to methyl-2-enelactones; (c) the Alcaligenes eutrophus 4-methyl-2-enelactone isomerase, which converts 4-methyl-2-enelactone to 3-methyl-2-enelactone; and (d) the B13 3-methyl-2-enelactone pathway, which completes the catabolic route. This pathway is further expanded through activation of the synthesis of a relaxed substrate specificity phenol hydroxylase (e). The newly evolved strain metabolizes chloro- and methyl-substituted phenols and benzoates exclusively via ortho-cleavage routes, tolerates shock loads of one type of substituted benzoate or phenol while utilizing the other as a carbon source, and degrades both types simultaneously when present together (23).

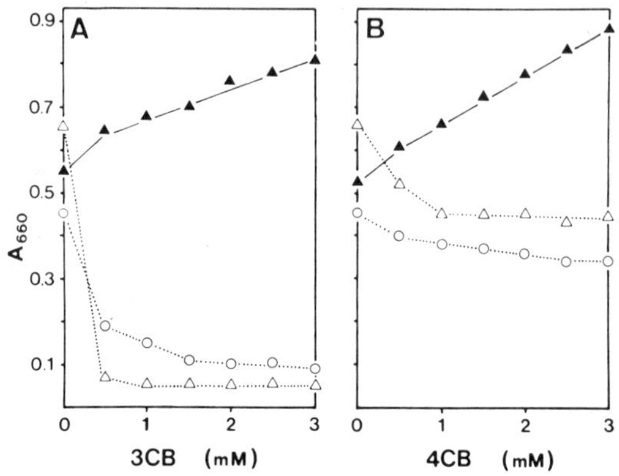

Fig. 8. Effect of chlorobenzoates on growth of bacteria-degrading methylbenzoates. Bacteria were grown in minimal medium containing 4-methylbenzoate (5 mM) as sole source of carbon and energy. When the cultures reached an absorbance of $A_{660} = 0.5$, they were diluted ten-fold with fresh medium containing both 4MB and either 3CB (A) or 4CB (B) at the indicated concentrations, and the incubation was continued for a further 20 hr. The final A_{660} was measured and plotted against the concentration of 3- and 4CB initially present in the fresh medium. The strains used were: *Pseudomonas putida* KT2440(TOL) (○) (possesses *meta* pathway for degradation of methylbenzoates); *Pseudomonas* sp. B13(TOL) (△) (possesses TOL plasmid *meta* pathway and an *ortho* pathway for 3CB); B13 derivative FR1(pFRC20P) (▲) (lacks a *meta* pathway, but possesses *ortho* pathways for degradation of chloro- and methylbenzoates).

CONCLUDING REMARKS

Current levels of environmental pollution urgently require, among other measures, much greater exploitation of microbial degradative activities. Unfortunately, some chemicals possess structural elements or substitutents that confer upon the molecule a high degree of resistance to enzymatic attack, or that are present in mixtures that are incompatible for the effective degradation of the toxic component (26). Soil microorganisms have the capacity to evolve enzymes capable of attacking most chemical groups, but the evolution of catabolic pathways proceeds very slowly where multiple genetic changes are required, and where the selection pressures may only be effective in selecting the last genetic change to occur.

A major advantage of experimental evolution of pathways is that laboratory selection conditions [e.g., antibiotic treatment when selecting acquisition of a hybrid plasmid containing the gene of an enzyme to be recruited or a mutant regulatory protein (see above)] may be totally unrelated to the ultimate phenotypic change desired (e.g., catabolism of an aromatic). Thus, effective selection procedures can be *custom designed* for each of the individual genetic changes required (18). Successive changes can either be effected in the organisms to be evolved, or be effected in different organisms and subsequently combined sequentially or simultaneously in the selected organisms. In this way, the evolutionary process can be enormously accelerated. The cloning of genes of proteins having useful properties, such as relaxed substrate specificities, for the purpose of genetic and functional analysis and for their transfer into organisms to be evolved, generates "evolutionary modules" for further experiments. Growth in the number of different modules available will increasingly facilitate the design and experimental evolution of new and more complex pathways.

In the experiments described above, existing pathways were restructured by enzyme recruitment through gene cloning or by mutational alteration of protein specificities, so that the newly acquired proteins recognize new substrates/inducers. In some cases, novel metabolic routes were created by the patchwork assembly of enzymes and regulators from different pathways and from different organisms. These experiments confirm that experimental evolution of metabolic pathways is feasible and holds considerable potential for accelerating the evolution of microbes that will be able to degrade particularly recalcitrant and toxic compounds.

Although we have focused largely on the manipulation of genes of enzymes and regulators, it is evident that other targets for experimental intervention will become relevant in the near future. The recruitment of appropriate membrane transport systems for unusual growth or transformation substrates may, in some cases, be important, as will the isolation of mutants resistant to the toxicity of certain compounds. In addition, however, there may be applications in which the efficacy of evolved activities can be considerably increased if the manipulated bacterium is able to undertake vectoral movement towards a nondiffusible or poorly diffusible target (e.g., plant root, hydrophobic chemical). Thus, experimental manipulation of bacterial taxis and motility should receive increasing attention during the next few years.

ACKNOWLEDGEMENTS

It is a pleasure to acknowledge the many fruitful and stimulating discussions with our colleagues, H.-J. Knackmuss and his group at the University of Stuttgart, and S. Harayama and N. Mermod at the University of Geneva, and the excellent secretarial assistance of F. Rey. F.R. and J.L.R. are recipients of postdoctoral fellowships from the European Molecular Biology Organization and the Consejo Superior de Investigaciones Cientificas de Espana, respectively.

REFERENCES

1. Bartels, I., H.-J. Knackmuss, and W. Reineke (1984) Suicide inactivation of catechol 2,3-dioxygenase from _Pseudomonas putida_ mt-2 by 3-halocatechols. _Appl. Environ. Microbiol._ 47:500-505
2. Campbell, J.H., J.A. Lengyel, and J. Langridge (1973) Evolution of a second gene for β-galactosidase in _Escherichia coli_. _Proc. Natl. Acad. Sci., USA_ 70:1841-1845.
3. Clarke, P.H. (1978) Experiments in microbial evolution. In _The Bacteria_, Vol. 4, L.N. Ornston and J.R. Sokatch, eds. Academic Press, New York, pp. 137-218.

4. Cocks, G.T., J. Aguilar, and E.C.C. Lin (1974) Evolution of L-1,2-propanediol catabolism in Escherichia coli by recruitment of enzymes for L-glucose and L-lactate metabolism. J. Bacteriol. 118:83-88.
5. Dorn, E., M. Hellwig, W. Reineke, and H.-J. Knackmuss (1974) Isolation and characterization of a 3-chlorobenzoate degrading Pseudomonad. Arch. Microbiol. 99:61-70.
6. Gibson, D.T. (1984) Microbial Degradation of Organic Compounds (Microbiology Series, Vol. 13), Marcel Dekker, Inc., New York.
7. Harayama, S., and R.H. Don (1985) Catabolic plasmids: Their analysis and utilization in the manipulation of bacterial metabolic activities. In Genetics Engineering: Principles and Methods, Vol. 7, J.K. Setlow and A. Hollaender, eds. Plenum Publishing Corporation, New York, pp. 283-307.
8. Harayama, S., P.R. Lehrbach, and K.N. Timmis (1984) Transposon mutagenesis analysis of meta-cleavage pathway operon genes of the TOL plasmid of Pseudonomas putida mt-2. J. Bacteriol. 160:251-255.
9. Harayama, S., J.L. Ramos, and K.N. Timmis (1986) Experimental evolution of plasmid specified functions. In Antibiotic Resistance Genes: Ecology, Transfer and Expression, Banbury Report 24, S.B. Levy and R.P. Novick, eds. Cold Spring Harbor Laboratory, Cold Spring Harbor, New York, pp. 389-402.
10. Harayama, S., R.A. Leppik, M. Rekik, N. Mermod, P.R. Lehrbach, W. Reineke, and K.N. Timmis (1986) Gene order of the TOL catabolic plasmid upper pathway operon and oxidation of both toluene and benzylalcohol by the xylA product. J. Bacteriol. 167:455-461.
11. Knackmuss, H.-J. (1983) Xenobiotic degradation in industrial sewage: Haloaromatics as target substrates. In Biotechnology, Biochemical Society Symposium No. 48, C.F. Phelps and P.H. Clarke, eds. Biochemical Society, London, pp. 173-190.
12. Knackmuss, H.-J., and M. Hellwig (1978) Utilization and cooxidation of chlorinated phenols by Pseudomonas sp. B13, Arch. Microbiol. 117:1-7.
13. Lehrbach, P.R., J. Zeyer, W. Reineke, H.-J. Knackmuss, and K.N. Timmis (1984) Enzyme recruitment in vitro: Use of cloned genes to extend the range of haloaromatics degraded by Pseudomonas sp. strain B13. J. Bacteriol. 158:1025-1032.
14. Mermod, N., S. Harayama, and K.N. Timmis (1986) New route to bacterial production of indigo. Bio/Technology 4:321-324.
15. Mortlock, R.P. (1982) Metabolic acquisitions through laboratory selection. Ann. Rev. Microbiol. 36:259-284.
16. Pemberton, J.M., B. Corney, and R.H. Don (1979) Evolution and spread of pesticide degrading ability among soil microorganisms. In Plasmids of Medical, Environmental and Commercial Importance, K.N. Timmis and A. Puhler, eds. Elsevier/North-Holland Biomedical Press, Amsterdam, pp. 287-299.
17. Pieper, D.H., K.H. Engesser, R.H. Don, K.N. Timmis, and H.-J. Knackmuss (1985) Modified ortho-cleavage pathway in Alcaligenes eutrophus JMP134 for the degradation of 4-methylcatechol. FEMS Microbiol. Lett. 29:63-67.
18. Ramos, J.L., and K.N. Timmis (1987) Experimental evolution of catabolic pathways of bacteria. Microbiol. Sci. 4:228-237.
19. Ramos, J.L., A. Stolz, W. Reineke, and K.N. Timmis (1986) Altered effector specificities in regulators of gene expression: TOL plasmid xylS mutants and their use to engineer expansion of the range of aromatics degraded by bacteria. Proc. Natl. Acad. Sci., USA 83: 8467-8471.

20. Ramos, J.L., A. Wasserfallen, K. Rose, and K.N. Timmis (1987) Redesigning metabolic routes: Manipulation of TOL plasmid pathway for catabolism of alkylbenzoates. Science 235:593-596.
21. Reineke, W., and H.-J. Knackmuss (1978) Chemical structure and biodegradability of halogenated aromatic compounds. Substituent effects on 1,2 dioxygenation of benzoic acid. Biochim. Biophys. Acta 542:312-423.
22. Reineke, W., and H.-J. Knackmuss (1979) Construction of haloaromatics utilizing bacteria. Nature 277:385-386.
23. Rojo, F., D.H. Pieper, K.H. Engesser, H.-J. Knackmuss, and K.N. Timmis (1987) Assemblage of ortho cleavage route for simultaneous degradation of chloro- and methylaromatics. Science 235:1395-1398.
24. Schmidt, E., I. Bartels, and H.-J. Knackmuss (1985) Degradation of 3-chlorobenzoate by benzoate or 3-methylbenzoate-utilizing cultures. FEMS Microbiol. Ecol. 31:381-389.
25. Timmis, K.N., M.I. Gonzalez-Carrero, T. Sekizaki, and F. Rojo (1986) Biological activities specified by antibiotic resistance plasmids. J. Antimicrob. Chemothe. Vol. 18, Suppl. C, pp. 1-12.
26. Timmis, K.N., P.R. Lehrbach, S. Harayama, R.H. Don, N. Mermod, S. Bas, R. Leppik, A.J. Weightman, W. Reineke, and H.-J. Knackmuss (1985) Analysis and manipulation of plasmid encoded pathways for the catabolism of aromatic compounds by soil bacteria. In Plasmids in Bacteria, D.R. Helinski, S.N. Cohen, D.B. Clewell, D.A. Jackson, and A. Hollaender, eds., Plenum Publishing Corporation, New York, pp. 719-739.
27. Worsey, M.J., and P.A. Williams (1975) Metabolism of toluene and the xylenes by Pseudomonas putida (arvilla) mt-2: Evidence for a new function of the TOL plasmid. J. Bacteriol. 124:7-13.
28. Yen, K.M., and I.C. Gunsalus (1982) Plasmid gene organization: Naphthalene/salicylate oxidation. Proc. Natl. Acad. Sci., USA 79: 874-878.
29. Yen, K.M., and I.C. Gunsalus (1985) Regulation of naphthalene catabolic genes of plasmid NAH7. J. Bacteriol. 162:1008-1013.
30. Zeyer, J., P.R. Lehrbach, and K.N. Timmis (1985) Use of cloned genes of Pseudomonas TOL plasmid to effect biotransformation of benzoates to cis-dihydrodiols and catechols by Escherichia coli cells. Appl. Environ. Microbiol. 50:1409-1413.

CONSTRUCTING MICROBIAL STRAINS FOR DEGRADATION OF HALOGENATED AROMATIC HYDROCARBONS

Peter J. Chapman*

Department of Biochemistry
University of Minnesota
St. Paul, Minnesota 55108

INTRODUCTION

Studies of the biodegradation of organic compounds have gained increased interest in the last few years as their relevance to environmental pollution and the treatment of chemical wastes has acquired a wider appreciation. This area of research, once described in some microbial physiology texts as "peripheral metabolism," has become central to our concerns about the persistence of noxious organic chemicals in the environment and to the development of rational biological approaches for their removal.

Among those chemicals that prompt concern are a variety of halogenated hydrocarbons, of which the halogenated aromatic hydrocarbons represent a subcategory. In recent years it has been possible to isolate or construct in the laboratory bacteria capable of growth with some of the simpler members of this series of chemicals which have a reputation for both toxicity and persistence.

This chapter will examine some of the methods that have been used to isolate and to construct bacteria capable of growing aerobically with chlorinated aromatic compounds, including hydrocarbons. We will also describe some recent work in this area of research.

NOVEL AND UNUSUAL SUBSTRATE DEGRADATION

Laboratory approaches that have successfully yielded aerobic bacterial cultures capable of growth with unusual or novel substrates have usually involved long-term batch incubations, soil perfusions, or chemostat

---

*The author is presently on leave as a visiting scientist at the United States Environmental Protection Agency, Environmental Research Laboratory, Gulf Breeze, Florida 32561.

operation. During such procedures, selection pressures can be changed gradually so that concentrations of growth substrates can be altered; one substrate can be changed at the expense of another; and, in the case of chemostat use, dilution rates can be altered to permit growth of organisms with changed division rates (Tab. 1). Cultures developed in this way may possess catabolic reaction sequences that are unusual, if not unique, and can lend themselves to further selection and alterations that broaden or expand already novel degradative traits to include a wider range of organic compounds. This is possible because one general principle of biodegradation recognizes that an economical use of catabolic enzymes is ensured by their organization into pathways that converge on the tricarboxylic acid cycle. Given a novel pathway in a newly isolated strain, it may then be possible to expand its degradative potential with the interaction of additional genes by selection for the desired strains from suitable combinations of organisms in chemostats, by straightforward conjugation experiments using degradative plasmids, or by introduction of the requisite genes cloned into suitable plasmid vectors.

To illustrate some of the above approaches, one can cite the impressive body of work pioneered by H.-J. Knackmuss and his colleagues. Initially, by use of a chemostat selection, a strain of *Pseudomonas*, B13, was obtained from a diverse population of benzoate-utilizing bacteria that could eventually utilize 3-chlorobenzoate as the sole source of carbon for growth (5). This was accomplished by a gradual replacement of benzoate by 3-chlorobenzoate in the inflowing medium over a period of several weeks. Other investigators have isolated 3-chlorobenzoate utilizing bacteria with similar properties (2,11). From an extensive study of the biochemistry of the 3-chlorobenzoate pathway, it is now evident that *Pseudomonas* B13 relies upon the enzymes of benzoate catabolism to convert 3-chlorobenzoate to a mixture of 3- and 4-chlorocatechol (Fig. 1). This conversion also occurs in various bacteria by a co-metabolic process, since the accumulating chlorocatechols cannot be further degraded to provide carbon or energy for cell division (13,34). Strain B13 differs from these other strains in its capacity to effect complete degradation of both isomers of chlorocatechol, with concomitant release of inorganic halide.

Novel enzymes comprising a chlorocatechol pathway present in *Pseudomonas* B13 have been shown (Fig. 2) to be responsible for attack on both the chlorocatechol isomers (pyrocatechase II) (4) and for the lactonization

Tab. 1. Approaches to novel strain construction and selection.

1. Chemostat isolation/Long-term batch operation

   Gradual variation of selection pressures so that growth substrates, concentrations, and dilution rates can be altered.

2. Novel hybrids of different pathways

   A. Chemostat selection from combinations of utilizing/co-oxidizing organisms (plasmids) for utilization of required substrate.

   B. Laboratory construction by plasmid transfer of requisite genes (cloned) to appropriate backgrounds.

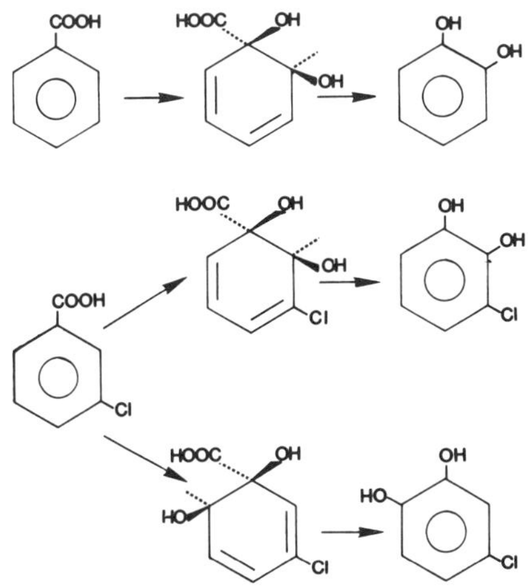

Fig. 1. Co-metabolism of 3-chlorobenzoate by enzymes of benzoate degradation can lead to formation of 3-chloro- and 4-chlorocatechol (13,34).

of the resulting 2- and 3-chloromuconic acids (cycloisomerase II) (27,28). Subsequent enzymes required are a specific hydrolase for the isomeric enol lactones (diene lactone hydrolase) and a maleylacetate reductase to catalyze conversion to β-ketoadipate. These reactions comprise a central degradative sequence, with steps essentially the same as those previously described by others studying the degradation of chlorinated phenoxyacetic acid herbicides (Fig. 3) by bacterial isolates (6,33).

Because catechols represent convergence points for the biodegradation of a variety of aromatic structures, the presence in Pseudomonas B13 of a sequence for the complete degradation of chlorocatechols makes it a candidate for use in expanding degradative capabilities to include chloroaromatics other than 3-chlorobenzoate. This has been accomplished in a number of cases by combining in hybrid strains the ability of one organism to co-oxidize chloroaromatics to chlorocatechols with the ability of Pseudomonas B13 to complete the degradation of chlorocatechols (see below).

Several organisms have been described recently with the ability to degrade dichloro-substituted benzenes (3,30,31). These isolates, like Pseudomonas B13, appear to have developed analogous chlorocatechol-degrading enzyme sequences in response to a selection pressure set up by a co-metabolic attack. Consider the known mechanisms by which aromatic hydrocarbons undergo initial attack by aerobes (Fig. 4). Three mechanisms are evident. One involves dioxygenation of the aromatic ring, forming cis-dihydrodiol products to be further oxidized to catechols (9,10). Thus, toluene can be converted to 3-methylcatechol. Where an alkyl substituent, such as methyl or a larger alkyl substituent, is present, its oxidation via initial hydroxylation may lead to the formation of a carboxylic acid intermediate. This is the case, for example, with reactions encoded by genes of the TOL plasmid (16,36). Finally, ring mono hydroxylation, as in the conversion of toluene to p-cresol by Pseudomonas mendocina (24), represents an alternative route which leads to 4-hydroxy benzoic acid by subsequent methyl group attack.

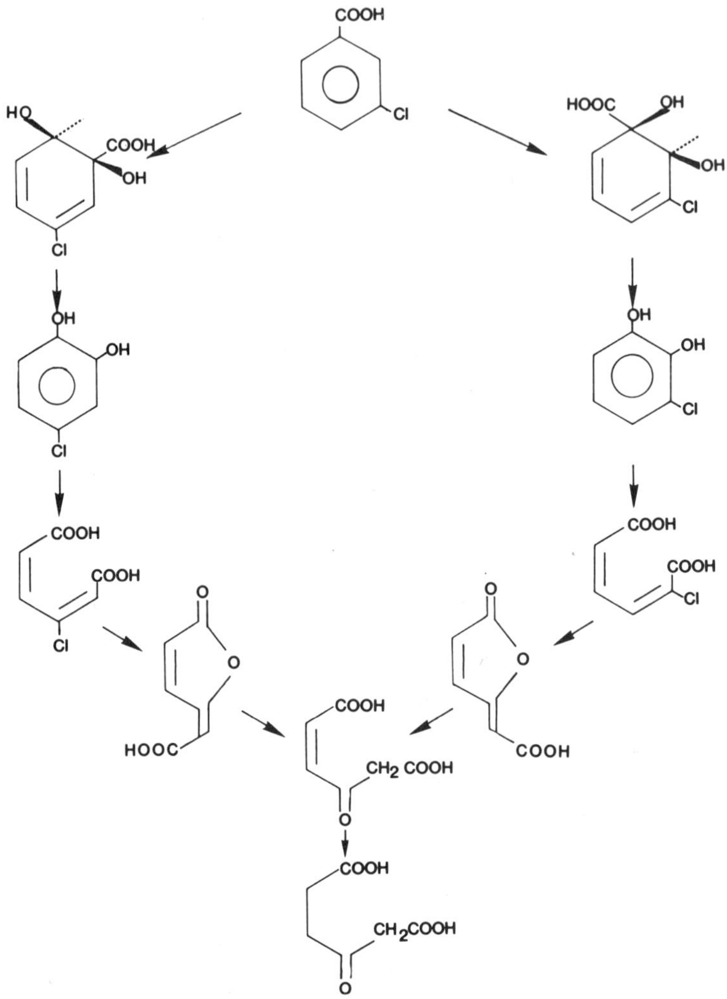

Fig. 2. Pathways by which Pseudomonas B13 degrades 3-chlorobenzoate (top). After the diverging initial reactions that form 3-chloro- and 4-chlorocatechol, dioxygenase cleavage by pyrocatechase II gives 2-chloro- and 3-chloromuconic acid, respectively. Cyclo-isomerase II catalyzes formation of the trans and cis isomers of γ-carboxymethylene butenolide (diene lactones), respectively, with the sequence completed by diene lactone hydrolase and maleyl acetate reductase (4,27,28).

Gibson and co-workers have shown that pseudomonads possessing the dioxygenase responsible for ring attack on toluene also act on a number of disubstituted benzenes (7). Thus, m-xylene, p-xylene, and a number of other 4-substituted toluenes are converted via cis-dihydrodiols to catechols by a toluene-utilizing strain of Pseudomonas putida (8). All dichlorobenzene isomers are similarly converted to diols by a mutant of this strain, but mono- and dichlorocatechols are not degraded (15). Interestingly, these same types of reactions are those responsible for the initial attack on 1,3- and 1,4-dichlorobenzene in recently described isolates (Tab. 2).

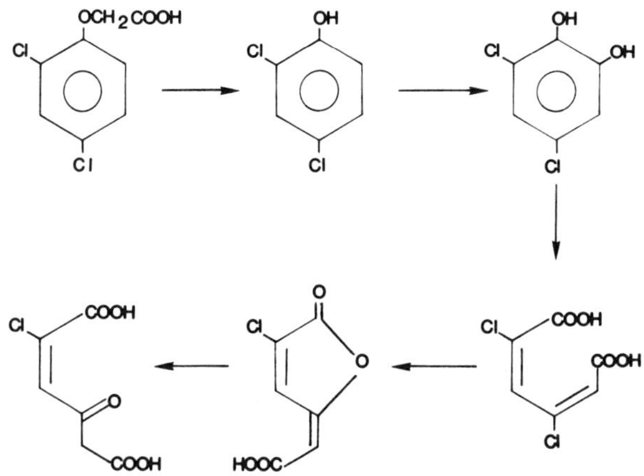

Fig. 3. Biodegradative pathway of the herbicide, 2,4-dichlorophenoxyacetic acid (2,4-D), showing cleavage (ortho) of 3,5-dichlorocatechol, lactonization to a chlorine-substituted diene lactone, and hydrolysis (6,33).

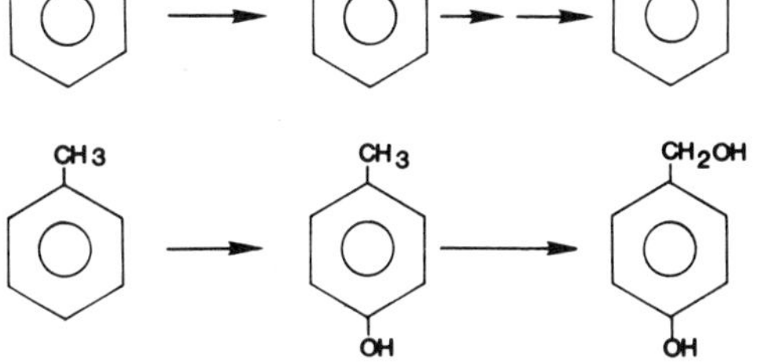

Fig. 4. Different mechanisms by which bacteria attack the aromatic hydrocarbon, toluene. Direct dioxygenase attack on the ring leads to cis-dihydrodiol formation; methyl group oxidation leads to benzoic acid formation; and monohydroxylation of the ring gives p-cresol (9,10,16,24,35).

The organisms described in Tab. 2 are able to grow with the listed substrates and were isolated in long-term procedures using batch, semi-continuous, or continuous approaches over a number of months. In the degradative pathways employed by these organisms, ring dioxygenation leads to the formation of 3,5-dichlorocatechol from 1,3-dichlorobenzene and of 3,6-dichlorocatechol from 1,4-dichlorobenzene, followed by *ortho* cleavage, lactonization, and diene lactone hydrolysis (3,30,31). These organisms apparently employ essentially the same types of reactions for catabolism of the dichlorocatechols as *Pseudomonas* B13 uses for the catabolism of the monochlorocatechols. In other words, *ortho* cleavage of halocatechols and lactonization with halide elimination can be employed for catechols with different patterns of dichloro-substitution. While this conclusion could have been drawn from studies utilizing 4-chloro- and 2,4-dichlorophenoxyacetic acid (2,4-D), these more recent studies show that the chlorocatechol route can be employed for different dichloro-substituted catechols. One may speculate that these organisms developed in response to selection pressures exerted through the formation of dichlorocatechols by the co-metabolic action of aromatic hydrocarbon dioxygenases on dichlorobenzenes. Comparison of the enzymes and the genes of these dichlorocatechol degradative sequences with those from *Pseudomonas* B13 will be of interest.

EXPANDING PRE-EXISTING CATABOLIC TRAITS

Because *Pseudomonas* B13 relies on the chromosomally encoded benzoate dioxygenase and benzoate diol dehydrogenase to attack 3-chlorobenzoate, its growth is limited by the specificity of those enzymes toward compounds that serve as substrates. 4-Chlorobenzoate is not usually a substrate for benzoate dioxygenase in pseudomonads. If it were, subsequent enzymes in strain B13 would have the ability to completely degrade the 4-chlorocatechol that would be formed by its action, combined with that of benzoate diol dehydrogenase. On the other hand, enzymes encoded by the genes of the TOL (pWWO) plasmid and others include a sequence for the conversion of variously alkyl-substituted benzoic acids to the corresponding catechols (Fig. 5). These acids are metabolites of the different substituted toluenes that are growth substrates for TOL-carrying strains (16,35).

Tab. 2. Recent examples of isolation of microorganisms utilizing dichlorobenzene substrates.

Growth substrate	Organism	Conditions	Reference
1,3-Dichlorobenzene	*Alcaligenes* sp. OBB65	Batch culture	3
1,4-Dichlorobenzene	*Alcaligenes* sp. A175	Batch culture, two to ten months	30
1,4-Dichlorobenzene	*Pseudomonas* sp. J36	Semicontinuous (ten months); continuous (six weeks); increasing hydrocarbon concentration	31

Accordingly, Reineke and Knackmuss (20,21) introduced the TOL plasmid into Pseudomonas B13, selecting for exconjugants of B13 that were able to grow with 3-methylbenzoic (m-toluic) acid. Such constructs were unable to grow with hydrocarbon substrates such as toluene, evidently because of a structural modification of the TOL plasmid in Pseudomonas B13. They were also unable to grow with 4-chlorobenzoate, but did, however, form an unproductive metabolite or co-metabolite, a yellow, meta-cleavage ring fission product of 4-chlorocatechol. After several days, spontaneously arising mutants growing with 4-chlorobenzoate made their appearance on plates. These mutants, when purified, grew with both 3- and 4-chlorobenzoate, but not with m-toluic acid or other toluic acids or their hydrocarbon precursors. This is apparently because productive metabolism of 4-chlorocatechol can occur via the chlorocatechol sequence only if preceded by a mutation that prevents unproductive metabolism, namely, a mutation rendering the TOL-encoded catechol 2,3-dioxygenase inactive (23). In this way, 4-chlorocatechol is shunted entirely via the chlorocatechol route at the expense of losing the ability to degrade toluic acids. Further spontaneous forward mutations, as described by Reineke and Knackmuss (21), permit growth with 3,5-dichlorobenzoate. These mutations further expand the degradative capabilities of strain B13. Chemostat selection for 4-chlorobenzoate and 3,5-dichlorobenzoate utilizers from combinations of Pseudomonas B13 and donors of the TOL plasmid has also been used, with similar results (12).

Recognition by Timmis and co-workers (18) of the role of toluate dioxygenase and toluate diol dehydrogenase in this expansion of the catabolic potential of strain B13 has prompted cloning of the precise genes encoding these functions [xylD (toluate dioxygenase; now xylX, Y, and Z) and xylL (toluate diol dehydrogenase)] into strain B13 to show their specific role (Tab. 3). Similarly, the gene for salicylate hydroxylase from the NAH plasmid encoding for naphthalene degradation, when cloned into strain B13, expands the nutritional range of B13 to include chlorosalicylates as growth substrates (18).

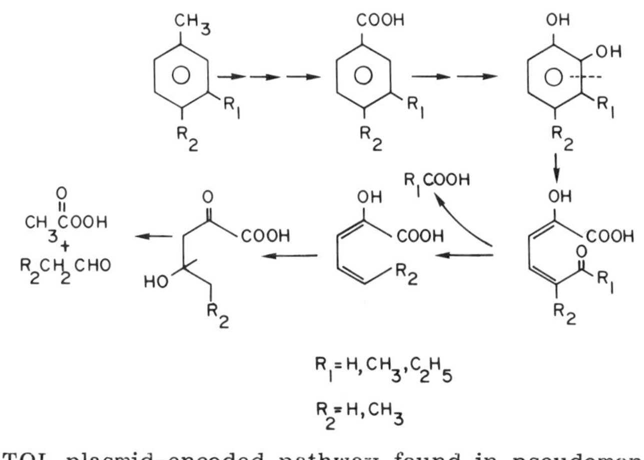

Fig. 5. TOL plasmid-encoded pathway found in pseudomonads for substituted toluenes and toluic acids. Where R_1 = H and R_2 = CH_3, the hydrocarbon designated is p-xylene, which is degraded via p-toluic acid, 4-methylcatechol, and meta ring cleavage (16,35).

Tab. 3. Expanding degradative range of the chlorobenzoate pathway.

1. Introduction of TOL plasmid (pWWO) into Pseudomonas B13: Broad specificity enzymes convert 4-chlorobenzoate and 3,5-dichlorobenzoate to halocatechols (20).

2. Cloning of requisite TOL genes, xylD and xyl, in broad host range vector introduced into Pseudomonas B13: 4-Chlorobenzoate and 3,5-dichlorobenzoate utilized (18).

3. Cloning of salicylate hydroxylase via broad host range vector into Pseudomonas B13: Chlorosalicylates utilized (18).

Our own interest in this area developed from recent work on transformations effected by the TOL reaction sequence. To obtain catechol 2,3-dioxygenase-defective mutants of the TOL plasmid, we employed the approach just described to select spontaneous 4-chlorobenzoate-utilizing mutants from cells that are able to use 3-chlorobenzoate and that carry the TOL (pWWO) plasmid.

Bacteria were isolated in our laboratory by modifications of the chemostat approach (Tab. 4). 3-Chlorobenzoate was gradually introduced at increasing concentrations into three chemostats, two of which contained a second growth substrate that was gradually eliminated (experiments 1 and 2, Tab. 4). Inocula used were mixtures of soil, water, and activated sludge. Establishment of 3-chlorobenzoate-utilizing populations required about ten to 12 weeks in all cases. Cultures of P. putida were always obtained, even when no second substrate was present. This indicates that at low concentrations of 3-chlorobenzoate degradation, growth may be possible, and that this method may be selecting for organisms that are able to grow at higher substrate concentrations. Experimental results with one of our isolates, P. putida 633, are described here; essentially the same results were obtained with strains 635 and 637.

Tab. 4. Isolation of 3-chlorobenzoate utilizers by chemostat selection.

Experiment	Conditions: Start	Conditions: Finish	Isolates
1	5 mM BA	-	Pseudomonas putida 633
	0.25 mM 3CB	5 mM 3CB	Pseudomonas putida 634
2	5 mM glucose	-	Pseudomonas putida 635
	0.25 mM 3CB	5 mM 3CB	Pseudomonas putida 636
3	0.25 mM 3CB	5 mM 3CB	Pseudomonas putida 637

Note: Room temperature, 18° to 20°C; dilution rate, 0.1 vol hr^{-1}.

Abbreviations: BA, benzoate; 3CB, 3-chlorobenzoate.

By conjugal transfer of the TOL (pWWO) plasmid from an auxotrophic *Pseudomonas aerginosa* donor (Fig. 6) and selection for prototrophic utilization of *m*-toluic acid, transconjugants were readily obtained in simple plate matings. After purification, such transconjugants were found to grow with 3-chlorobenzoate, *m*-xylene, and other hydrocarbon substrates characteristic of the TOL pathway (Tab. 5), indicating that in these isolates the TOL plasmid was maintained in stable and entirely functional fashion. In addition, 3- and 4-chlorotoluene served as growth substrates; growth with the latter was accompanied by an accumulation of the yellow ring fission product formed from 4-chlorocatechol. Interestingly, no growth was found with 4-chlorobenzoate, although spontaneous mutants utilizing this substrate arose on plates at frequencies of 10^{-6} to 10^{-7} and then grew on 4-chlorotoluene without accumulation of yellow ring fission product. This selection favored catechol 2,3-dioxygenase-defective mutants of the TOL plasmid, as indicated by the mutants' failure to grow with TOL substrates (e.g., *m*-toluic acid and *m*-xylene) and by the failure of induced cells to turn yellow when sprayed with catechol. Additionally, when the TOL plasmid was mated into other pseudomonads lacking both a *meta*-fission pathway for catechols and a chloro-catechol sequence (using as selection, complementation of a defective benzoate dioxygenase in the recipient), then substituted catechols were accumulated from *m*-toluic acid, *m*-xylene, and other TOL substrates.

The 4-chlorobenzoate-positive mutants of the TOL-carrying *P. putida* 633 show other interesting properties. They grow readily with 3-bromotoluene supplied as a vapor, whereas their parent does not (Fig. 7). Spontaneous mutants of the latter arise on 3-bromotoluene. (These mutants may just be seen in Fig. 7 in areas corresponding to the third line of the 3-bromotoluene-grown plates.) Slow growth on 3-iodotoluene by all TOL-carrying strains indicates that the TOL-encoded toluate-degrading enzymes supply a less specific system. The results with 4-bromotoluene are similar to those seen with 4-chlorotoluene. No growth or spontaneous mutation was observed by any strain on 4-iodotoluene. Evidently, in these *P. putida* exconjugants the TOL plasmid is maintained in a fully functional form and allows for an extensive broadening of their

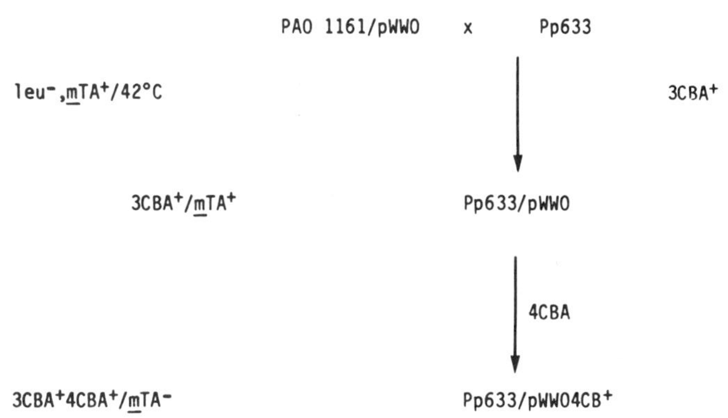

Fig. 6. Procedure by which TOL (pWWO) plasmid was introduced into *Pseudomonas putida* 633.

Tab. 5. Growth of Pseudomonas putida derivatives with haloaromatics.

Organism	Growth					
	m-Toluate	m-Xylene	3CB	3CT	4CB	4CT
P. putida 633 (WT)	-	-	+	-	-	-
P. putida 633/pWWO	+	+	+	+	-(m)	+(y)
P. putida 633/pWWO4CB1	-	-	+	+	+	+
P. putida 633/pWWO4CB2	-	-	+	+	+	+

Abbreviations: 3CB, 3-chlorobenzoate; 4CB, 4-chlorobenzoate; 3CT, 3-chlorotoluene; 4CT, 4-chlorotoluene.

biodegradative potential against halogenated toluenes. Why immediate growth occurs with 4-chloro- and 4-bromotoluenes and not with 4-chloro- or 4-bromobenzoate is not yet clear, but may be related to the two different TOL regulatory systems exerted through the xylR and xylS gene products (19,36). This finding and the capacity of these organisms to grow with and to co-oxidize other substituted toluenes are presently under investigation.

OTHER EXAMPLES OF EXPANSION OF DEGRADATIVE FUNCTIONS

Some examples have been assembled (Tab. 6) to show how chlorobenzene degraders, chlorophenol degraders, chloroaniline degraders, and chlorosalicylate degraders can be developed in chemostats by combining appropriate bacterial strains supplying the requisite genes. In all these examples, Pseudomonas B13 supplies the chlorocatechol reaction sequence, while other organisms selected for growth with benzene, phenol, aniline, or salicylate have the capacity to co-oxidize chlorinated derivatives at least as far as the corresponding catechols. By selecting for growth with the appropriate chloroaromatic compound, hybrid organisms have been isolated in which the chlorocatechol sequence now extends the co-oxidation abilities of these organisms beyond product accumulation into complete degradative pathways. The same results have also been achieved by direct conjugal transfer of the plasmid-encoded chlorocatechol genes into those strains capable of co-oxidation.

CONCLUSIONS

Strain construction of versatile degradative organisms is now a realistic option for the degradation of a number of halogenated aromatic hydrocarbons. Organisms can be selected for desirable traits in long-term experiments with gradual changes of substrates and their concentrations. The limitations of such approaches appear to be those of enzyme availability and the requisite breadth of enzyme specificity.

For example, it should be evident from the work considered here that the general mechanisms for monochlorocatechol degradation can be extended to include certain, if not all, dichlorocatechol isomers. At this time, however, it is not clear whether trichlorocatechols can serve as substrates for

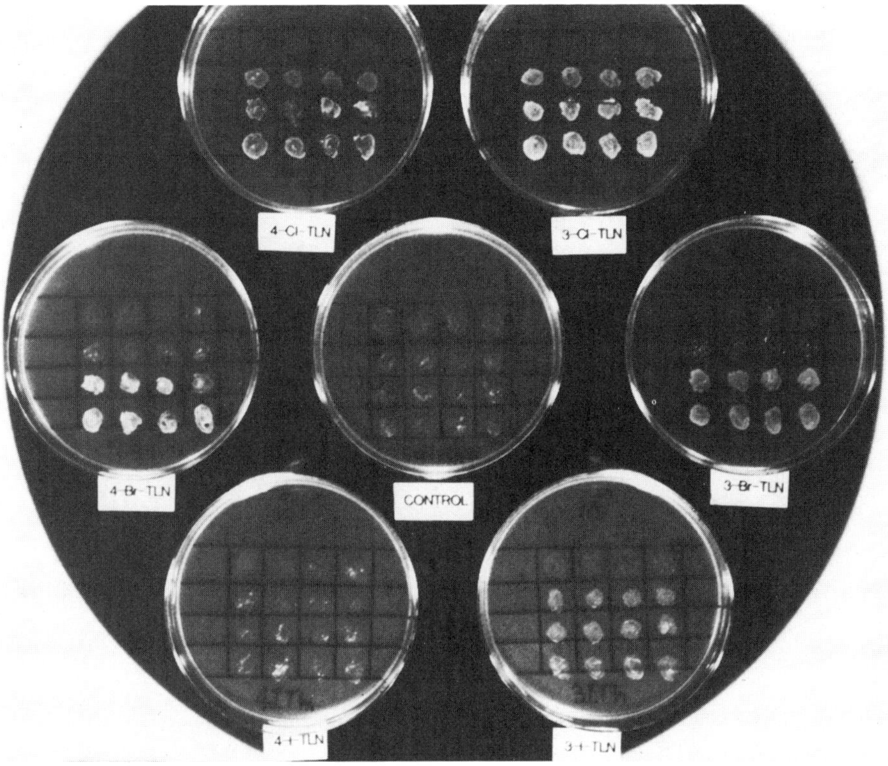

Fig. 7. Growth of Pseudomonas putida 633 and its derivatives on halogenated toluenes. On each plate the top line of patches is wild-type strain 633, the second line is 633/TOL, and lines 3 and 4 are independent mutants of 633/TOL selected for ability to grow with 4-chlorobenzoate. All plates contained vapors of the hydrocarbons shown; control plates contained no hydrocarbons.

this chlorocatechol sequence. Other enzymic mechanisms appear to be required for effective degradation of more highly chlorinated compounds, such as pentachlorophenol and 2,4,5-trichlorophenoxyacetic acid (1,14,32).

As these pathways are defined by additional research, they too may provide enzyme sequences that can be recruited for construction of strains for biodegradation of more highly chlorinated aromatic hydrocarbons.

ACKNOWLEDGEMENT

This work has been financed, in part, with Federal funds from the United States Environmental Protection Agency (EPA) under cooperative agreement number CR-812660 with the Gulf Breeze Environmental Research Laboratory. The contents do not necessarily reflect the views and policies of the EPA nor does mention of trade names or commercial products constitute endorsement or recommendation for use.

Tab. 6. Approaches used in constructing microorganisms with hybrid pathways.

Hybrid	Organisms used	Procedure	Reference(s)
Chlorobenzene-utilizing bacterium WR1306	Benzene-utilizing bacterium plus *Pseudomonas* B13	Chemostat selection	22
Mixed chlorophenol-utilizing *Alcaligenes* A7-2	Phenol-utilizing *Alcaligenes* A7 plus *Pseudomonas* B13	Continuous culture (also conjugation)	29
Chloroaniline-degrading *Pseudomonas* JL3	Aniline/2-methylaniline-utilizing *Pseudomonas* JL1 plus *Pseudomonas* B13	Chemostat selection (also conjugation)	17
Chlorosalicylate-utilizing *Pseudomonas* WR401B	Methylsalicylate-utilizing *Pseudomonas* WR401 plus *Pseudomonas* B13	Continuous culture (also conjugation)	25, 26

REFERENCES

1. Apajalahti, J.H.A., and M.S. Salkinaja-Salmen (1987) Dechlorination and para-hydroxylation of polychlorinated phenols by Rhodococcus chlorophenolicus. J. Bacteriol. 169:675-681.
2. Chatterjee, D.K., S.T. Kellog, S. Hamada, and A.M. Chakrabarty (1981) Plasmid specifying total degradation of 3-chlorobenzoate by a modified ortho pathway. J. Bacteriol. 146:639-646.
3. DeBont, J.A.M., M.J.A.W. Vorage, S. Hartmans, and W.J.J. van den Tweel (1986) Microbial degradation of 1,3-dichlorobenzene. Appl. Environ. Microbiol. 52:677-680.
4. Dorn, E., and H.-J. Knackmuss (1978) Chemical structure and biodegradability of halogenated aromatic compounds. Two catechol 1,2-dioxygenases from a 3-chlorobenzoate-grown pseudomonad. Biochem. J. 174:73-84.
5. Dorn, E., M. Hellwig, W. Reineke, and H.-J. Knackmuss (1974) Isolation and characterization of a 3-chlorobenzoate degrading pseudomonad. Arch. Microbiol. 99:61-70.
6. Evans, W.C., B.S.W. Smith, H.N. Fernley, and J.I. Davies (1971) Bacterial metabolism of 2,4-dichlorophenoxyacetate. Biochem. J. 122:543-551.
7. Gibson, D.T., V. Mahadevan, and J.F. Davey (1974) Bacterial metabolism of para- and meta-xylene: Oxidation of the aromatic ring. J. Bacteriol. 119:930-936.
8. Gibson, D.T., J.R. Kock, C.L. Schuld, and R.E. Kallio (1968) Oxidative degradation of aromatic hydrocarbons by microorganisms. II. Metabolism of halogenated aromatic hydrocarbons. Biochemistry 7:3795-3802.
9. Gibson, D.T., M. Hensley, H. Jashioka, and T.J. Mabry (1970) Formation of (+)-cis-2,3-dihydroxy-1-methylcyclohexa-4,6-diene from toluene by Pseudomonas putida. Biochemistry 9:1626-1630.
10. Gibson, D.T., B. Gschwendt, W.K. Yeh, and V.M. Kobal (1973) Initial reactions in the oxidation of ethylbenzene by Pseudomonas putida. Biochemistry 12:1520-1528.
11. Haller, H.D., and R.K. Finn (1979) Biodegradation of 3-chlorobenzoate and formation of black color in the presence and absence of benzoate. Eur. J. Appl. Microbiol. Biotechnol. 8:191-205.
12. Hartman, J., W. Reineke, and H.-J. Knackmuss (1979) Metabolism of 3-chloro-, 4-chloro- and 3,5-dichlorobenzoate by a pseudomonad. Appl. Environ. Microbiol. 37:421-428.
13. Horvath, R.S., and M. Alexander (1970) Cometabolism of m-chlorobenzoate by an Arthrobacter. Appl. Microbiol. 20:254-258.
14. Kilbane, J.J., D.K. Chatterjee, J.S. Karns, S.T. Kellog, and A.M. Chakrabarty (1982) Biodegradation of 2,4,5-trichlorophenoxyacetic acid by a pure culture of Pseudomonas cepacia. Appl. Environ. Microbiol. 44:72-78.
15. Kilbourn, R.G., and D.T. Gibson (1979) Metabolism of chlorinated benzenes by Pseudomonas putida. In Annual Meeting of the American Society for Microbiology, Los Angeles, California, Abstr. N70, p. 191.
16. Kunz, D.A., and P.J. Chapman (1981) Catabolism of pseudocumene and 3-ethyltoluene by Pseudomonas putida (arvilla) mt-2: Evidence for new functions of the TOL (pWWO) plasmid. J. Bacteriol. 146:179-191.
17. Latorre, J., W. Reineke, and H.-J. Knackmuss (1984) Microbial metabolism of chloroanilines: Enhanced evolution by natural genetic exchange. Arch. Microbiol. 140:159-165.

18. Lehrbach, P.R., J. Zeyer, W. Reineke, H.-J. Knackmuss, and K.N. Timmis (1984) Enzyme recruitment in vitro: Use of cloned genes to extend the range of haloaromatics degraded by Pseudomonas sp. strain B13. J. Bacteriol. 158:1025-1032.
19. Nakazawa, T., S. Inouye, and A. Nakazawa (1984) Positive regulation and transcription initiation of XYL operons on TOL plasmid. In Plasmids in Bacteria, D.R. Helinski, S.N. Cohen, D.B. Clewell, D.A. Jackson, and A. Hollaender, eds. Plenum Press, New York, pp. 415-429.
20. Reineke, W., and H.-J. Knackmuss (1979) Construction of haloaromatics utilising bacteria. Nature (London) 277:385-386.
21. Reineke, W., and H.-J. Knackmuss (1980) Hybrid pathway for chlorobenzoate metabolism in Pseudomonas sp. B13 derivatives. J. Bacteriol. 142:467-473.
22. Reineke, W., and H.-J. Knackmuss (1984) Microbial metabolism of haloaromatics: Isolation and properties of a chlorobenzene-degrading bacterium. Appl. Environ. Microbiol. 47:395-402.
23. Reineke, W., D.J. Jeenes, P.A. Williams, and H.-J. Knackmuss (1982) TOL plasmid pWWO in constructed halobenzoate-degrading Pseudomonas strains: Prevention of meta pathway. J. Bacteriol. 150:195-201.
24. Richardson, K.L., and D.T. Gibson (1984) A novel pathway for toluene oxidation in Pseudomonas mendocina. In Annual Meeting of the American Society for Microbiology, St. Louis, Missouri, Abstr. K54, p. 156.
25. Rubio, M.A., K.-H. Engesser, and H.-J. Knackmuss (1986) Microbial metabolism of chlorosalicylates: Accelerated evolution by natural genetic exchange. Arch. Microbiol. 145:116-122.
26. Rubio, M.A., K.-H. Engesser, and H.-J. Knackmuss (1986) Microbial metabolism of chlorosalicylates: Effect of prolonged subcultivation on constructed strains. Arch. Microbiol. 145:123-125.
27. Schmidt, E., and H.-J. Knackmuss (1980) Chemical structure and biodegradability of halogenated aromatic compounds. Conversion of chlorinated muconic acids into maleoylacetic acid. Biochem. J. 192:339-347.
28. Schmidt, E., G. Remberg, and H.-J. Knackmuss (1980) Chemical structure and biodegradability of halogenated aromatic compounds. Halogenated muconic acids as intermediates. Biochem. J. 192:331-337.
29. Schmidt, E., M. Hellwig, and H.-J. Knackmuss (1983) Degradation of chlorophenols by a defined mixed microbial community. Appl. Environ. Microbiol. 46:1038-1044.
30. Schraa, G., M.L. Boone, M.S.M. Jetten, A.R.W. van Neerven, P.J. Colberg, and A.J.B. Zehnder (1986) Degradation of 1,4-dichlorobenzene by Alcaligenes sp. strain A175. Appl. Environ. Microbiol. 52:1374-1381.
31. Spain, J.C., and S.F. Nishino (1987) Degradation of 1,4-dichlorobenzene by a Pseudomonas sp. Appl. Environ. Microbiol. 53:1010-1019.
32. Steiert, J.G., and R.L. Crawford (1986) Catabolism of pentachlorophenol by a Flavobacterium. Biochem. Biophys. Res. Commun. 141:825-830.
33. Tiedje, J.M., J.M. Duxbury, M. Alexander, and J.E. Dawson (1969) Metabolism pathway of degradation of chlorocatechols by an Arthrobacter species. J. Agric. Food Chem. 17:1021-1026.
34. Walker, N., and D. Harris (1970) Metabolism of 3-chlorobenzoic acid by Azotobacter species. Soil Biol. Biochem. 2:27-32.

35. Williams, P.A., and K. Murray (1974) Metabolism of benzoate and methylbenzoates by Pseudomonas putida (arvilla) mt-2: Evidence for the existence of a TOL plasmid. J. Bacteriol. 120:416-423.

36. Worsey, M.J., F.C.H. Franklin, and P.A. Williams (1978) Regulation of the degradative pathway enzymes coded for the TOL plasmid (pWWO) from Pseudomonas putida mt-2. J. Bacteriol. 134:757-767.

STABLE INHERITANCE OF BACTERIAL PLASMIDS: PRACTICAL CONSIDERATIONS IN THE RELEASE OF ORGANISMS INTO THE ENVIRONMENT

Stanley N. Cohen,[1,2] Christine A. Miller,[1]
Serge Beaucage,[1] and Donald P. Biek[1]

[1]Department of Genetics
[2]Department of Medicine
Stanford University School of Medicine
Stanford, California 94305

INTRODUCTION

Most current applications of biotechnology involve the use of genetic engineering for the production of biologically active proteins in cells grown in fermenters. When such products are made in bacterial cells, gene constructs that yield large quantities of the desired product are employed using multicopy bacterial plasmid replicons. The continuing presence in the cell of a plasmid carrying the desired genes commonly can be assured by the addition of a selectable gene to the plasmid replicon and by growth under selective conditions. For fermentations of short duration, continued selection for the presence of the plasmid during the fermentation usually is not required.

When microorganisms are introduced into the environment, there generally is not an opportunity after introduction to apply active selection for extrachromosomal replicons and the genes that they carry. Thus, when persistence of a particular gene within a population of environmentally dispersed microorganisms is desired, stability of the carrier molecule in the microorganism is a practical consideration that must be addressed. Conversely, sometimes it may be desirable to avoid stable inheritance of the introduced gene; in such cases, organisms containing unstable constructs could quickly carry out the biodegradation task desired, and subsequently the genetic element encoding the degradative capacity could be eliminated from the population. Such scenarios require consideration of plasmid stability in terms of the desired biotechnological objective of a particular intervention.

During the past seven years, regions that accomplish the partitioning of extrachromosomal replicons to daughter cells and consequently ensure the stable inheritance of these plasmids in populations of dividing cells have been identified on plasmids pSC101 (8), NR1 (9), R1 (11), F (12), and P1 (1). Certain other plasmids, such as CloDF13 (6), P1 (2), and ColE1 (14), contain stability-related loci that are involved in the resolution of multimers into monomeric units; site-specific recombinations within these loci reverse the plasmid instability that occurs as a consequence of the end-to-end joining of plasmid replicons to yield multimers.

Mathematical models predict that random distribution of a plasmid present at cell division at a copy number of 20 or more per cell will ensure that each daughter cell receives at least one copy of the plasmid. As long as even one plasmid DNA molecule exists in a cell, the plasmid's copy-number control mechanism is expected to bring the copy number to a level characteristic for that plasmid. Thus, each daughter cell will, prior to cell division, contain the same number of plasmids as the parent cell had prior to its cell division. Plasmids having a copy number too low to ensure that each daughter cell receives at least one plasmid by a random distribution process presumably must employ an active mechanism for this purpose. Loci that encode such a mechanism have been termed *par* (partitioning) (8,11) or *stb* (stability) (9). While the precise molecular nature of partitioning mechanisms is incompletely understood for most plasmids, it appears to vary for different replicons--in some cases causing the death of cells that have lost plasmids (5,7), while in others employing specific *cis*-acting DNA genetic functions to accomplish plasmid distribution and stable inheritance.

It has also been known that transcription of DNA sequences in the region of a plasmid's replication origin can affect plasmid stability; introduction of a strong promoter onto certain high-copy-number plasmids can interfere with plasmid propagation, decrease copy number, and consequently reduce stability (3). The addition of a terminator downstream from the promoter reverses these effects. In other instances, transcription of DNA sequences in and around the origin of replication region may be required for normal plasmid replication to occur.

My laboratory's investigations of plasmid stability have focused on elucidating the mechanisms of partitioning of the pSC101 plasmid. In 1980, we (8) discovered that deletion of a region of the pSC101 plasmid derivative resulted in loss of the plasmid from some members of the population of actively growing cells. In our initial studies of the pSC101 *par* locus, we showed that a 375-bp *EcoRI/AvaI*-generated DNA fragment adjacent to the essential replication region of the plasmid (Fig. 1) carries genetic information sufficient to accomplish the stable inheritance of pSC101-derived replicons. In the absence of antibiotic selection, plasmid loss occurred from about 75% of cells growing for 100 generations when the DNA segment between the *EcoRI* site at the left of the map in Fig. 1 and the proximal *HaeII* site was deleted. Instability of pSC101 plasmids occurred even with mutant plasmids that had a copy number of 50 per cell (suggesting that random distribution was not operative), although the instability observed with high-copy-number mutants was less than was seen with wild-type pSC101 replicons (15).

The originally observed properties of the *par* region suggested that the determinant necessary for stable inheritance of the plasmid is physically distinct from the functions that control the replication and incom-

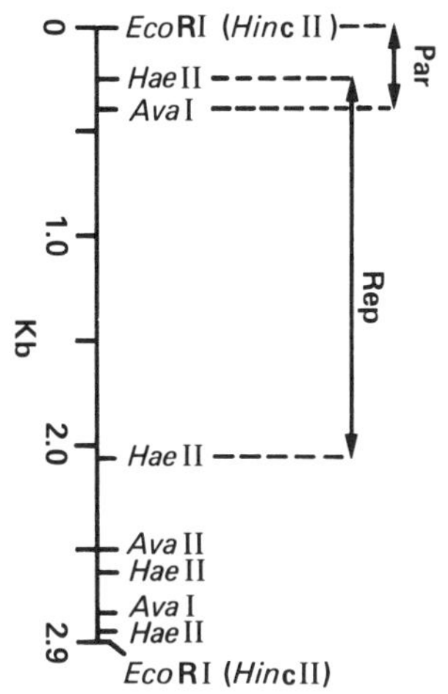

Fig. 1. Map of the HincII-A fragment of pSC101. The locations of the cleavage sites of various restriction endonucleases are shown; the HincII sites in pSC101 have been made into EcoRI sites in the pTU1 plasmid (8), which served as the source of the DNA fragment studied. The positions of the partition (par) and replication (rep) regions are shown. (The diagram is from Ref. 8.)

patibility functions of the plasmid. However, our subsequent discovery that certain mutants within the par region can affect the ability of co-existing plasmids carrying the same replication region to compete with each other suggested that the par region has at least some functional link to plasmid replication.

Sequencing of the 375 base pair EcoRI/AvaI DNA fragment that contains the pSC101 par locus revealed three segments that showed direct or inverted homology with each other (Fig. 2) (10). Deletion of segments a or a' resulted in the appearance of the Cmp⁻ phenotype. This phenotype affects the plasmid's ability to compete with a related replicon on equal terms. It was discovered during studies that produced deletion of one of the three PR segments included within the DNA par region. Cmp⁻ plasmids are entirely stable when in a cell lacking an unrelated replicon but are always at a disadvantage when a Cmp-related plasmid is present (the Cmp⁻ phenotype is summarized in Tab. 1).

Deletion of both of the peripheral segments, or of one of the segments plus the intervening b segment, yielded an unstable par⁻ phenotype, while deletion of all three segments resulted in a plasmid that was lost at a rate greater than would be expected from a random distribution model, given the copy number of the plasmid. Segments a, b, and a' were thus termed PR (for "partition-related") segments. Plasmids deleted for all three PR segments, which are lost from the majority of cells after forty generations of such growth in nonselective media, and from more than 99% of the cells after sixty generations of such growth, have been assigned the Super-par⁻ phenotype. These phenotypes are summarized in Tab. 2.

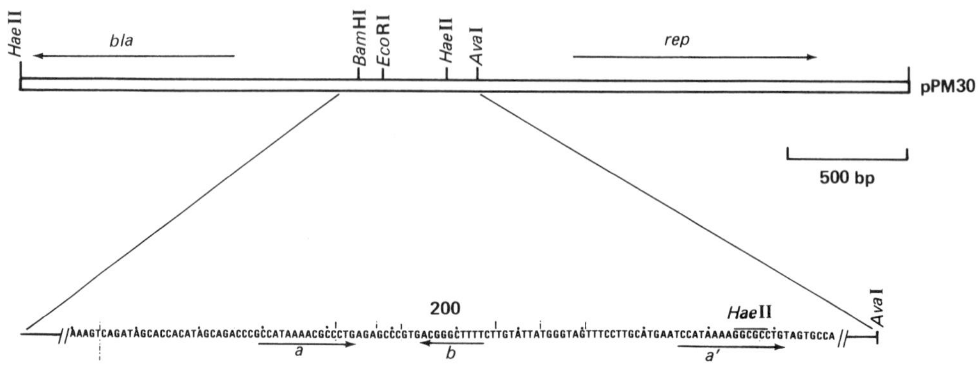

Fig. 2. Map of pPM30 and the sequence of the par region. The upper diagram depicts linearization of the par$^+$ pSC101 derivative pPM30 at one of its HaeII sites. The bla (β-lactamase) gene and rep (replication) region are indicated. The EcoRI/AvaI fragment containing the par locus was sequenced (10), and by deletion analysis (15) the PR segments were defined. They are indicated in this figure by a, b, and a'. (Figure modified from Ref. 15.)

Table 3 summarizes the properties of the par locus of pSC101. Stabilization occurs in cis but not in trans, and can occur also when the pSC101 par locus is placed on certain unrelated plasmids, such as pACYC184, R1, and F (Ref. 8, 11, and 13; C.A. Miller and S.N. Cohen, unpubl. data). The latter effect indicates that the pSC101 partitioning system is not replicon-specific. This cis-acting effect of the par region suggested that the pSC101 par locus encodes a site that accomplishes partitioning, rather than a protein. Consistent with this view is our observation that cloning of the par locus on a high-copy-number replicon such as pACYC177 has no effect on the stability of a co-existing pSC101 plasmid replicon; thus, individual copies of the par locus do not appear to compete with each other in the same way that differently tagged copies of the same replication region are incompatible with each other. In an analogous experiment carried out earlier (4), placement of the pSC101 replication/incompatibility region on a multicopy ColE1 plasmid led to extremely rapid loss of a co-existing wild-type pSC101 replicon.

Tab. 1. Characteristics of the Cmp$^-$ phenotype.

1. Cmp$^-$ plasmid is totally stable in the absence of co-existing Cmp$^+$ plasmid of the same incompatibility type.
2. Reduced ability to compete with Cmp$^+$ co-existing plasmid of the same incompatibility type.
3. Normal incompatibility is expressed against other Cmp$^-$-incompatible plasmids.
4. Copy number is normal.

Tab. 2. Summary of structural and functional interrelationships of phenotypes associated with par region deletions.

Deletion*	Phenotype
a, b, and a'	Super-par$^-$
a' and b	Par$^-$
a' and b	Par$^-$
a and a'	Par$^-$
a	Cmp$^-$
a'	Cmp$^-$

*Complete deletion of a PR segment yields greater effect than partial deletion (see Ref. 15 for details).

Recent experiments (S. Beaucage and S.N. Cohen, unpubl. data) have further localized the minimal sequence required for par function. These findings, which will be published elsewhere, have shown that sequences comprising only part of the PR segments are required for partitioning, and that in addition certain nucleotides in one of the inter-PR regions are essential.

Mechanism of Action of the par Region

Evidence obtained early in our studies of the par region of pSC101 suggested that this region might function by allowing individual plasmids in the intracellular pool to become recognized as separate replicons. Super-par$^-$ derivatives of pSC101 are lost at a rate consistent with random distribution of a replicon having a copy number of 4 per cell at division, whereas the actual copy number was measured at several times this value (15). Thus, the number of copies in the intracellular pool of plasmids

Tab. 3. Summary of major properties of pSC101 par locus (8).

1. Independent of location in cis; nonfunctional in trans.
2. Functional in either orientation.
3. Par$^-$ phenotype is reduced but not eliminated by elevation in copy number.
4. Stabilizes unrelated plasmids.
5. No instability of pSC101 results from multiple copies of the pSC101 par locus on compatible co-existing plasmid.

does not seem to be taken into account in the distribution of Super-par plasmids. To test the notion that the par region of pSC101 is required for plasmids to be recognized and counted as individual molecules within the intracellular pool, the rate of loss of a Super-par$^-$ derivative of pSC101 that also carries a temperature-sensitive (ts) mutation in the replication region was compared with the rate of loss of a ts par$^+$ derivative. As Fig. 3 shows, half of the cell population lost the ts Super-par$^-$ plasmid in the first generation following a shift to a temperature nonpermissive for replication, whereas a ts par$^+$ derivative was lost from similarly treated cells at a rate consistent with the plasmid copy number averaging 10 per cell. Thus, the pool of ts Super-par$^-$ plasmids appears to act as a single segregation unit (15).

In summary, this review of certain of our investigations of plasmid partitioning is intended to point out that plasmid stability is an important practical consideration in the genetic manipulation of microorganisms that are to be introduced into the environment. While the mechanisms that af-

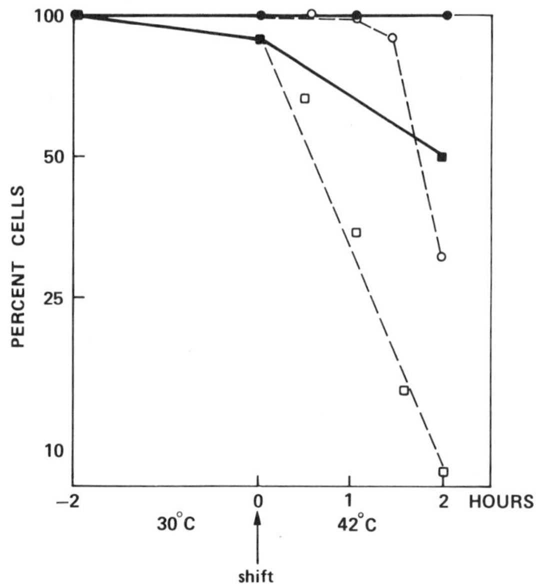

Fig. 3. Rate of loss of par$^+$ vs Super-par$^-$ ts pSC101-derivative plasmids upon shift to the nonpermissive temperature. PM191 (a recA$^-$ derivative of C600), carrying plasmids pCM301 (par$^+$) or pCM302 (Super-par$^-$), which are ts for replication, were grown at 30°C without antibiotics. After two hours, half of each culture was shifted to 42°C. Samples were taken at 30-min time points, plated on nonselective media at 30°C, and the resulting colonies were tested for resistance to Ap to determine the presence of the plasmid. Circles are cells containing pCM301; squares are cells containing pCM302; solid lines and closed symbols represent growth at 30°C; dashed lines are growth at 42°C. The generation time at 42°C was approximately 30 min. (Figure from Ref. 15.)

fect plasmid stability are still incompletely understood at the molecular level, the current state of knowledge allows manipulations that can achieve either the goal of stable inheritance or the goal of unstable inheritance, if desired. It is hoped that the next few years will lead to more information at the mechanistic level.

ACKNOWLEDGEMENTS

These studies were supported by grant GM26355 to SNC. DB is the recipient of fellowship awards from the Damon Runyon/Walter Winchell Cancer Fund and the National Institutes of Health. SB is the recipient of a Senior Postdoctoral Fellowship Award from the California Division of the American Cancer Society.

REFERENCES

1. Austin, S., and A. Abeles (1983) Partition of unit-copy miniplasmids to minidaughter cells. I. P1 and F miniplasmids contain discrete, interchangeable sequences sufficient to promote equipartition. J. Molec. Biol. 169:353.
2. Austin, S., M. Ziese, and N. Sternberg (1981) A novel role for site-specific recombination in maintenance of bacterial replicons. Cell 25:729.
3. Bujard, H., D. Stueber, R. Gentz, U. Deuschle, and U. Peschke (1985) Insertion of transcriptional elements outside the replication region can interfere with replication, maintenance, and stability of ColE1-derived plasmids. In Plasmids in Bacteria, Donald R. Helinski et al., eds. Plenum Press, New York, pp. 45-52.
4. Cabello, F., K. Timmis, and S.N. Cohen (1976) Replication control in composite plasmid constructed by in vitro linkage of two distinct replicons. Nature 259:285.
5. Gerdes, K., P.B. Rasmussen, and S. Molin (1986) Unique type of plasmid maintenance function: Postsegregational killing of plasmid-free cells. Proc. Natl. Acad. Sci., USA 83:3116.
6. Hakkaart, M.J.J., E. Veltkamp, and H.J.J. Nijkamp (1982) Maintenance of the bacteriocinogenic plasmid Clo DF13 in Escherichia coli cells. II. Specific recombination functions involved in plasmid maintenance. Molec. Gen. Genet. 188:338.
7. Hiraga, S., A. Jaffé, T. Ogura, H. Mori, and H. Takahashi (1986) F plasmid ccd mechanism in Escherichia coli. J. Bacteriol. 166:100.
8. Meacock, P.A., and S.N. Cohen (1980) Partitioning of bacterial plasmids during cell division. A cis-acting locus that accomplishes stable plasmid inheritance in cell populations. Cell 20:529.
9. Miki, T., A.M. Easton, and R.H. Rownd (1980) Cloning of replication incompatibility and stability functions of R plasmid NR1. J. Bacteriol. 141:87.
10. Miller, C.A., W.T. Tucker, P.A. Meacock, P. Gustafsson, and S.N. Cohen (1983) Nucleotide sequence of the partition locus of Escherichia coli plasmid pSC101. Gene 24:309.
11. Nordström, K., S. Molin, and H. Aagard-Hansen (1980) Partitioning of plasmid R1 in Escherichia coli. I. Kinetics of loss of plasmic derivatives deleted of the par region. Plasmid 4:215.
12. Ogura, T., and S. Hiraga (1983) Partition mechanism of F plasmid: Two plasmid gene-encoded products and a cis-acting region are involved in partition. Cell 32:351.

13. Skogman, G., J. Nilsson, and P. Gustafsson (1983) The use of a partition locus to increase stability of tryptophan-operon-bearing plasmids in _Escherichia coli_. _Gene_ 23:105.
14. Summers, D.K., and D.J. Sherratt (1984) Multimerization of high copy number plasmids causes instability: ColEI encodes a determinant for plasmid monomerization and stability. _Cell_ 36:1097.
15. Tucker, W.T., C.A. Miller, and S.N. Cohen (1984) Structural and functional analysis of the _par_ region of the pSC101 plasmid. _Cell_ 38:191.

BIOTRANSFORMATIONS OF MERCURY COMPOUNDS

Anne O. Summers

Department of Microbiology
The University of Georgia
Athens, Georgia 30602

INTRODUCTION

The ability of bacteria to convert inorganic and organic mercury compounds to less toxic, volatile elemental mercury is among the most widely found plasmid-determined bacterial phenotypes. It is the best understood microbial metal transformation at levels of analysis from molecular genetics and biochemistry to population biology (4,14). Mercury compounds are important contaminants in some environmental sites.

Narrow- and Wide-Spectrum Resistances

Early studies demonstrated that the majority of gram-negative bacteria which were resistant to any mercury compound were resistant only to inorganic mercury, merbromin, and fluorescein mercuric acetate (narrow-spectrum resistance). These strains converted inorganic mercury to volatile metallic mercury; they did not biotransform the few organomercurials to which they were resistant (4,14). The basis for their resistance to these few organomercurials has never been precisely determined. From 10-30% of these isolates were also resistant to many other organomercurials including phenylmercuric acetate and merthiolate (broad-spectrum resistance). The broad-spectrum resistance strains could biotransform these and other organomercurials to volatile metallic mercury and a reduced organic moiety. Gram-positive bacteria such as the staphylococci typically have the broad-spectrum resistance phenotype (7). Narrow-spectrum resistance has also been found in Streptomyces and Streptococci (11). Bacterial strains carrying either narrow- or broad-spectrum mercury resistance have been found in most eubacterial genera and in a great variety of pristine and impacted natural settings (4,14).

The locus conferring mercury resistance (mer) is typically found on large self-transferrable plasmids. However, since the locus is usually part of a transposable element, mercury resistance can be found on any prokaryotic replicon, including the chromosome and bacteriophages in addition to plasmids (4,14).

The Mechanism of Bacterial Mercury Resistance

The biotransformation of mercury compounds is effected by a *mer*-encoded Hg(II) uptake system which brings mercury through the cell wall and delivers it to an intracellular enzyme which reduces mercuric ion to metallic mercury. Volatile elemental mercury is relatively inert and leaves the cell simply by passing through the cell wall. Organomercurials are split enzymatically, yielding Hg(II) which is then reduced, as above. The expression of these functions is regulated; only when the cell is exposed to mercury compounds are the genes for this detoxification system expressed (4,6,9,14).

The components of the mercury detoxification system are:

(a) the mercuric ion reductase, encoded by the *mer*A gene;
(b) the mercuric ion transport system, encoded by the *mer*TPC genes;
(c) the organomercurial lyase, encoded by the *mer*B gene;
(d) a polypeptide with no defined function, encoded by the *mer*D gene; and
(e) the regulatory protein, encoded by the *mer*R gene.

The functions of these proteins are described below. In gram-negative bacteria, the genes are arranged within the *mer* locus as an operon (linear array of co-ordinately expressed coding regions) with the structure (4,14):

<--->
*mer*R (OP) *mer*T *mer*P [*mer*C] *mer*A [*mer*B] *mer*D

The (OP) region includes the operator(s) at which MerR acts to repress its own transcription and that of the structural genes. It also contains divergent promoter sites at which MerR and RNA polymerase act (in the presence of mercury) to transcribe the mRNA for either *mer*R (leftward arrow) or for the structural genes *mer*TPCAD (rightward arrow). *mer*C and *mer*B do not occur in every example of the operon; when found, they occur in the indicated positions. The *mer* operon of a staphylococcal (gram-positive) plasmid has been sequenced; it contains reading frames which resemble *mer*A and *mer*B, as well as several reading frames whose identification awaits relevant physiological and biochemical studies in this system (7).

The Mechanism of Mercury Biotransformations

Induction. In naturally occurring systems, the autoregulated protein MerR is synthesized at a constant low level and, in the absence of mercury, it represses the synthesis of the structural genes. When mercury enters the cell, it interacts with the MerR protein, which then activates transcription of the structural genes, *mer*TPCAD (6,9).

Purified MerR protein binds Hg(II) (12). It also binds to a specific DNA sequence with dyadic symmetry in the *mer* OP region both in vitro (6,12) and in vivo, though only in the latter does addition of Hg(II) alter its binding position (Heltzel, unpubl. observ.). Genetic studies implicate cysteine residues in MerR's carboxy-terminus in induction (Park and Ross, unpubl. observ.).

Hg(II) transport. The first portion of the mRNA in the mer operon to be synthesized encodes the Hg(II) uptake system, merTP[C]. MerT and MerC are very hydrophobic inner membrane proteins of approximately the same size (4,14). Since some naturally occurring loci do not have a merC gene (14, Gilbert, unpubl. observ.), this protein is not absolutely required for either broad- or narrow-spectrum mercury resistance. Hg(II) initially binds to the periplasmic MerP protein, which delivers it to MerT[C] in the inner membrane (14). MerP is present in considerable excess over the other mer membrane proteins; it may sequester Hg(II) in the periplasm so that the metal ion passes through the membrane principally via the mer detoxification system rather than through some other uncontrolled and deleterious route (14).

Both MerT and MerC proteins have potential membrane-spanning regions (4,8,14) and are believed to move the Hg(II) to the inner face of the cytoplasmic membrane where it can be reduced by the soluble, cytosolic enzyme Hg(II) reductase, the product of the merA gene. Mutants lacking either merT, merP, or merC or any combination are sensitive to Hg(II) unless the Hg(II) reductase is overexpressed by some artificial means (8, Hamlett, unpubl. observ.). Surprisingly, it is necessary to bring Hg(II) through the inner membrane in order for this system to work (as it occurs naturally).

Reduction of mercury. The Hg(II) reductase enzyme is composed of two identical subunits of about 60,000 daltons. Each subunit binds a single FAD molecule. Electrons from NADPH travel through the FAD to a pair of cysteine residues in the active site which effect the reduction of the substrate Hg(II) to Hg(0) (10,13). The primary structure and the reaction mechanism of the Hg(II) reductase are very similar to other disulfide oxidoreductases, including glutathione reductase. However, there are subtle differences, likely involving peripheral cysteine residues (Moore, pers. comm.), between Hg(II) reductase and these related enzymes for which Hg(II) is not a substrate but a potent inhibitor. Once reduced, volatile and lipid soluble Hg(0) probably passes directly through the cell membranes into the growth medium and, then, into the atmosphere.

The Hg(II) reductase is very heat tolerant and, in vivo, is capable of reducing Au(III) and Ag(I) salts to their respective metallic colloids (14). Its amino-terminal 80 residues are proteolytically removed in vivo without apparent effect upon its enzymatic activity (13). The physiological relevance and cellular agent(s) of this processing are unknown. This amino-terminal portion of the Hg(II) reductase bears a striking resemblance to the MerP protein. Thus, the Hg(II) reductase appears to have resulted from a primordial fusion between a merP gene and a gene encoding a disulfide oxidoreductase (8,14). The role of this naturally occurring evolutionary step in developing an enzyme uniquely able to handle Hg(II) is currently being examined to discover the generalizability of such events to other metals.

Lysis of organomercurial compounds. When present, the organomercurial lyase (OL) gene, merB, occurs near the end of the mer operon (5, Totis, unpubl. observ.). Such operons also differ from those conferring narrow-spectrum resistance in being inducible by organomercurials as well as by inorganic mercury (5).

The OL enzyme consists of a single polypeptide chain of 22,000 daltons. It does not require NADPH for activity in vitro and does not have bound co-factors. Its substrate range is very broad and it functions by an ES2 reaction mechanism (2,3). The OL is a soluble enzyme, although whether it is compartmentalized in the cytosol or the periplasm has not been unequivocally established.

Population Biology

Two-thirds of the naturally occurring mercury-resistant gram-negative bacteria which have been examined contain DNA which hybridizes to that of one or both of the two related and well-characterized transposable Hg-resistance loci, Tn21 and Tn501. Mercury resistance in the remaining strains is conferred by some less related locus(i) (P. Gilbert, unpubl. observ.). These isolates were obtained either from human or animal flora or from other terrestrial or freshwater sources. Interestingly, very recent work on mercury-resistant saltwater isolates suggests that few if any contain sequences which hybridize to previously described loci (N. Hamlett and T. Barkay, pers. comm.). Thus, there is considerable natural variety in this ubiquitous detoxification system.

REFERENCES

1. Barkay, T., L. Fouts, and B.H. Olson (1985) A method for the detection of mercury resistance genes in natural bacterial isolates. Appl. Env. Microbiol. 49:686-692.
2. Begley, T.P., A.E. Walts, and C.T. Walsh (1986) Bacterial organomercurial lyase: Overproduction, isolation, and characterization. Biochemistry 25:7186-7192.
3. Begley, T.P., A.E. Walts, and C.T. Walsh (1986) Mechanistic studies of a protonolytic organomercurial cleaving enzyme: Bacterial organomercurial lyase. Biochemistry 25:7193.
4. Foster, T.J. (1987) Genetics and biochemistry of mercury resistance. CRC Crit. Rev. in Microbiol.
5. Griffin, H.G., T.J. Foster, S. Silver, and T.K. Misra (1987) Cloning and DNA sequence of the mercuric and organomercurial resistance determinants of plasmid pDU1358. Proc. Natl. Acad. Sci., USA 84:3112-3116.
6. Heltzel, A., D.G. Gambill, W.J. Jackson, P.A. Totis, and A.O. Summers (1987) Overexpression and DNA-binding properties of the mer-encoding regulatory protein from plasmid NR1 (Tn21). J. Bact. 169: 3379-3384.
7. Laddaga, R.A., L. Chu, T.K. Misra, and S. Silver (1987) Nucleotide sequence and expression of the mercurial-resistance operon from Staphylococcus aureus plasmid pI258. Proc. Natl. Acad. Sci., USA 84:5106-5110.
8. Lund, P.A., and N.L. Brown (1987) Role of the merT and merP gene products of transposon Tn501 in the induction and expression of resistance to mercuric ions. Gene 52:207-214.
9. Lund, P.A., S.J. Ford, and N.L. Brown (1986) Transcriptional regulation of the mercury-resistance genes of transposon Tn501. J. Gen. Microbiol. 132:465-480.
10. Miller, S.M., D.P. Ballou, V. Massey, C.H. Williams, Jr., and C.T. Walsh (1986) Two-electron reduced mercuric reductase binds Hg(II) to the active site dithiol but does not catalyze Hg(II) reduction. J. Biol. Chem.
11. Nakahara, H., J.L. Schottel, T. Yamada, Y. Miyakawa, M. Asakawa, J. Harville, and S. Silver (1985) Mercuric reductase enzymes from

Streptomyces species and group B Streptococcus. J. Gen. Microbiol. 131:1053-1059.
12. O'Halloran, T., and C. Walsh (1987) Metalloregulatory DNA-binding protein encoded by the merR gene: Isolation and characterization. Science 235:211-214.
13. Schultz, P.G., K.G. Au, and C.T. Walsh (1985) Directed mutagenesis of the redox-active disulfide in the flavoenzyme mercuric ion reductase. Biochem. 26:6840-6848.
14. Summers, A.O. (1986) Organization, expression, and evolution of genes for mercury resistance. Ann. Rev. Microbiol. 40:607-634.

DISCUSSION

Ananda M. Chakrabarty (Chairman), The University of Illinois

FOCHT: Could we introduce the cycloisomerase into a meta-cleaving organism and create another route?

TIMMIS: My intuition is to say, "No." I think Peter Chapman could probably answer. However, in terms of 3-chlorobenzoate, acylhalide kills the meta-cleavage enzyme, so you'll never get adequate amounts of the semialdehyde produced.

CHAPMAN: I concur with Dr. Timmis. The structures that are produced by metafissionary actions are not cis-cis dicarboxylic acids and, therefore, do not lend themselves to cycloisomerization. So there would not be a rational basis for introducing the cycloisomerase enzyme to extend a metafission pathway.

Would you speculate on the suicide properties of the ring fission product of 4-methylcatechol?

TIMMIS: Such a mechanism has been proposed for the inactivation of the ring fission enzyme. The ring fission enzyme has been one of the formations of the acylhalide in the case of the chloro ring. It may be true. I didn't go into the other mutants, but one of our mutants lost the ability to process one of the methylcatechols, the 3-methylcatechol, I believe. In fact, the wild-type enzyme is inactivated by the catechol itself, if you look at it carefully. Our feeling is that here you have a deoxygenase working by producing active oxygen species. It's already a rather uncomfortable situation for the enzyme producing those active oxygen species. If you now provide a substrate that doesn't quite fit, you've got a problem of geometry. Where do you put the active oxygen species? Do you put it on the catechol or do you put it onto a residue of the protein? Our gut feeling at the moment is that inactivations occur by oxidation of the enzyme.

OMENN: In using plasmids, we worry less that the engineered organism will lose its function and thereby be ineffective, but more that the plasmid will be transferred to another organism. That's a different kind of instability.

COHEN: In general, nonconjugative plasmids are used in order to limit interbacterial transfer. Although some nonconjugative plasmids can be mobilized by other self-transmissible plasmids, there are schemes to ensure that the mobilized plasmid will not be able to perpetuate itself in an organism other than its original host. One can also envision ways to engineer an organism so that when the plasmid is lost, the organism will not survive. By using appropriate genetic approaches, there are rather straightforward ways to accomplish such goals. In addition, potentially there are ways to ensure that when an environmentally introduced organism has carried out its task, it loses the plasmid that encodes whatever degradative genes are being employed, and that the organism without the plasmid will no longer be viable.

CHAKRABARTY: You said that par locus confers stability on other plasmids, yet it must act in cis.

COHEN: The stability is conferred only when the par locus has been inserted into the other plasmid. We have isolated a par-minus mini F, which the pSC101 par region stabilizes. Kurt Nordstrom has used the pSC101 par region to stabilize a derivative of the large antibotic resistance plasmid R1. I've shown a slide that the pSC101 par region can stabilize pACYC184, which is not stable by itself. It doesn't stabilize all plasmas that have been looked at, but it does stabilize some.

CHAKRABARTY: Have you actually tried to stabilize any plasmid that's not from *Escherichia coli*?

COHEN: Yes. We have no evidence that the pSC101 par region can stabilize plasmids from other species, but we haven't tried extensively. At the time that we first isolated the pSC101 par region, Mervyn Bibb and Janet Schottel in my laboratory were doing experiments with *Streptomyces* plasmids, and they had isolated a par-minus derivative of the SCP2 plasmid. They tested whether the pSC101 par region stabilized SCP2, but it did not. The organisms are very different, and I wouldn't have expected it to. However, it was an experiment that we felt we should do anyway.

POGELL: When you have the par locus deleted in *E. coli*, and you're selecting for beta-lactamase resistance, do you ever get integration or do you just lose the plasmids?

COHEN: Integration ordinarily doesn't occur with pSC101. If you are selecting for resistance, the par-minus plasmid isn't lost from the population. You select for propagation of those cells in the population that retain the plasmid. The plasmid continues to replicate, and it is perpetuated in the population.

TIEDJE: What is the relationship between long generation times and plasmid copy number?

COHEN: That's a good question. We've wondered about it ourselves. However, we haven't looked at partitioning under variable growth rate conditions.

TIMMIS: There has been some work done on this relationship. The longer the generation time, the slower the growth rate and the higher the plasmid copy number. To isolate some of these large degradative plasmids, the golden rule is that you put the organism under the worst possible growth conditions in order to get the best plasmid yields. Stability has not been assessed independently; but the general conclusion would be that the higher the copy number, the better the chance of survival for the plasmid.

COHEN: There have been a number of such studies of copy number, but we know that stability is not entirely or necessarily dependent on copy number.

FALKOW: Since plasmids exist in naturally occurring organisms, and since the generation time of naturally occurring organisms is actually much lower than we usually deal with in the laboratory, the stability is very good. The generation time in the gut for E. coli is on the order of six to eight hours, rather than 20 minutes. Plasmids have no difficulty existing under those conditions.

LEVIN: I have a problem when you say the plasmid is lost. Chakrabarty's data indicated that, as the substrate is depleted, the plasmid activity is lost. But when you put the substrate back, there is a rapid increase in that ability which indicates that the capability is still there. How does that fit your concept of stability?

COHEN: They are lost from a certain fraction of the cell population. If one considers "stability" as the propagation of a plasmid in 100% of the individuals in the population, plasmids that are "lost" are not stably propagated in every member of the population. Eventually, they will, for all practical purposes, disappear from the population because cells lacking the plasmid will outgrow them in the absence of selection. There will be an increasingly small percentage of the population that carries the plasmid. However, if then one reinstitutes selection, these cells will be selected for so long as some cells in the population are there that retain the plasmid.

FURLONG: Do you know if periplasmic proteins have any mercury-binding affinity?

SUMMERS: We don't know that yet. We've got the purified protein but we have some worries about the fact that we got it by overexpression. In our hands, it does not appear to have a high affinity for mercury. But, we also have seen some processing artifacts in our overexpressed system.

CHAPMAN: I get the impression, Anne, that there are no gram-positive organisms in your system.

SUMMERS: What I have worked with is not a gram-positive system; but, of course, the mercury-resistance genes occur in both gram-positive and gram-negative systems. In the cases where they have been examined in detail or the sequence is known, at least the reductase and the organomercurial lyase are quite similar. For the other genes, I think that's certainly not the case.

CHAPMAN: The reason I ask is that a number of gram-positive organisms have no glutathione and so presumably have no glutathione reductase as the mer system might have predicted.

SUMMERS: That's a very interesting point.

FOCHT: Why is there a sudden flurry of literature reports of organisms capable of degrading compounds such as chloro- and dichlorobenzenes?

CHAPMAN: I think that with regard to the evolution of organisms with specific biodegradative activities, the time one may have to wait is related to how different xenobiotic substances are from those that occur naturally in the environment of the organism. Where there are fairly close similarities, we can expect to wait relatively short periods of time for evolution, either in the lab or in the wild. On the other hand, when we're confronted with chemicals that are heavily chlorinated and contain carbon skeletons that are completely different from those that occur naturally, we can expect very long waits. I'm thinking particularly of the chlorodiene insecticides, for which there is very little evidence of complete disappearance (perhaps partial reduction under anaerobic conditions).

SPAIN: I think the reason is that we've just now begun to be patient enough to wait for up to a year to isolate them from our chemostats. Also, all of us have found ways to supply chlorobenzene at low concentrations. The chloro- and dichlorobenzenes are toxic; if you just put them in the culture medium, they will kill any potential degraders. So we've all devised ways to introduce it in the vapor phase and be very patient for a long time. I wish I knew whether the organisms have been there all along or whether they've evolved in our chemostats. I'm looking for ways to try to answer that question. They've recently been isolated readily from the Rhine River, suggesting that they've been there all along.

QUESTION: When you isolated mutant organisms that produced catechol dioxygenase resistant to ethylcatechol, what happened to their kinetic properties?

TIMMIS: In both instances, there was a slight increase in the Km and a slight decrease in Vmax, so that the enzymes were less efficient at doing what they should do, but not dramatically so.

TIEDJE: We have a dilemma. Most of our examples of biodegradation utilize model compounds, rather than the problem chemicals in our environment. How promising are these methods for high-priority contaminants?

CHAPMAN: Practical applications will not be achieved overnight, I'm afraid, but I don't think dramatic progress is beyond the bounds of possibility. The time element may depend on whether you're going to be an impatient aerobic microbiologist or an anaerobic microbiologist who exercises patience. Pentachlorophenol, for example, is an almost completely chlorinated compound. Yet there are now numerous reports of organisms capable of growth with this compound at various concentrations. We're beginning to elucidate the catabolic pathways for degradation of these compounds (see Frick, et al., this Volume). Presumably some of those enzyme systems will be useful in strain construction, in the same way that the enzymes of chlorocatechol degradation have been useful to expand the potential to mono-, di-, and possibly trichlorinated aromatic compounds. I think it is possible to look at chemicals that may have a simple mono- or dichlorinated dibenzofuran structure and obtain organisms capable of their degradation. On the other hand, as a cautionary note, there is no guarantee that those organisms will degrade dibenzofurans or TCDDs that have large numbers of chlorine substituents. But it's a step in the right direction.

TIEDJE: Isn't it a big leap from enzymatic conversions to the specific genes responsible for those enzymes?

CHAKRABARTY: We need to develop an understanding about the evolution of genes. If we are introducing a new compound for which there are microorganisms that have the genetic capability to degrade an analogous compound, then it's presumably easier. However, if the compound is quite different, then you may be asking the microorganisms to evolve a whole set of completely new genes. Many priority compounds have the unusual property of being very hydrophobic. Our 2,4,5-T-degrader seems to produce surfactants or emulsifiers for 2,4,5-T. Basically, we are asking the microorganisms to evolve not only the structural genes, but also the pathway and the genes for biosynthesis of a particular emulsifying compound. The major constraint has been a lack of understanding as to how new genes evolve for the degradation of these compounds.

CHAPMAN: As Stan Cohen expressed eariler, many people think that genetic engineering is simply slapping a gene into a vector and that's the end of the story. It very often is if you're working with E. coli. But, in Pseudomonas, you're working with pathways, not simple genes, a whole batch of enzymes, and a whole batch of subunits. It's an interaction of genetic engineering and soft physiology. When they are taken out of the laboratory and put into some model ecosystem or, ultimately, into the real world, completely different sets of environmental conditions operate on the genetics and the physiology of the organism.

There's a whole world of new experimentation. Researchers need to establish a dialogue with the organisms they are going to manipulate. Establishing the dialogue--"Would you accept this piece of genetic information and express it for us?"--in our hands, at least, never takes less than a year; very often it takes much longer. And that's in the laboratory, not in the environment. So, there's an enormous amount of spade work to be done. That's why most of us are working with model compounds and laboratory strains just to work out the ground rules. Then we can attack real problems.

FOCHT: Would you comment on the limited availability of a number of compounds that are crucial to the investigation of the chlorocatechol pathway?

CHAPMAN: You've identified an important and difficult area. The distribution of these compounds is limited to those individuals who have been making them, usually by microbial transformations and by the use of purified enzyme systems. As a result, relatively few groups can pursue these investigations.

TIMMIS: In our lab, the bottleneck does not lie at the level of having the intermediate compounds, many of which are provided by Hans Knackmuss. We are cloning genes of enzymes which we believe are going to be useful in assembling new pathways. We freeze away the clones so that we have a whole set of modules that we can assemble at will. Our rate of progress depends on having the widest possible number of useful modules, and this number is increasing.

COHEN: If one knew what the reactive site of a particular enzyme looked like, and if one knew how gene structure determined this reactive site, then one could in fact design the gene sequence to one's needs. That's not possible at the present time, but molecular biology and protein chemistry are clearly moving in that direction. I expect that such experiments will be possible in the not too distant future.

SUMMERS: When you get a change in function, either a regulatory function or an enzymatic function, is the change occurring in a large block of the sequence or are you looking at point mutations?

TIMMIS: The changes that we have generated and examined by sequence analysis have always involved point mutations.

SUMMERS: There's a lot of evolutionary biology that says evolution doesn't occur only via point mutations, but by exchanging and recombining large segments of DNA.

TIMMIS: I tried to give an overview of the strategies involving genetic engineering techniques that we're using to accelerate the evolution of new pathways. I did emphasize that other approaches are valid. We and others will generate something that works in the laboratory and then use chemostats to modulate or to optimize the activity. With long-term selection in the chemostat, more complex genetic changes surely may occur.

KINGSBURY: In the regulatory agencies and public interest organizations, there's an aversion to genetic constructions involving plasmids, especially if they're going to be introduced for environmental applications. What is our present knowledge of regulation and the ability to put boundary conditions around plasmids that are designed for these kinds of applications?

COHEN: I recall some of the discussions that took place during the early days of genetic engineering, when a number of people believed that plasmids, particularly antibiotic-resistance plasmids, should not be used for DNA cloning experiments, but that phages were in some way "safe." Presumably the concern about using plasmids was transmissibility. There are nonconjugative plasmids which are, for all intents and purposes, nonmobilizable. (Conceivably they could be transduced by a bacteriophage.) I think it is artificial to distinguish plasmid genes from chromosomal genes. We're talking about genetic material, and genetic material is fluid and mobile in nature.

Transposons exist on plasmids and go to chromosomes. Phages can pick up segments of DNA within cells and transduce those segments to other cells no matter where those segments are. Now it is practical to introduce segments of DNA into chromosomes. In fact, workers in my laboratory are carrying out experiments to introduce large pieces of DNA into the *Streptomyces* chromosome using site-specific recombination systems indigenous to the chromosomes of *Streptomyces*. If regulatory agencies are concerned about using plasmids as carriers, there are ways around that issue. But I personally believe that their special concern about plasmids is not necessarily well-founded.

OMENN: Dr. Summers, do you feel that the kinds of resistance for which you have described mechanisms have practical value in ecological settings? Would they have practical value in detoxifying metal content or protecting introduced microbes against toxicity from heavy metal concentrations that are in the same medium?

SUMMERS: Yes, to both questions. I believe these resistance mechanisms are there because they are sensing metals in the environment. There's a reasonable correlation to the presence of the resistance phenotype in situations where selection for the metal has been exacted. The only transformations that we might be interested in, at the present time, would be the demethylation of methylmercury compounds and the conversion of methylmercury or organomercurial neurotoxic compounds to less toxic compounds, i.e., metallic mercury. You're not going to get rid of mercury. At least in the mercury biodegrading systems, it's not a matter of converting mercury to something else. We're not talking about alchemy here, though I have joked that these are alchemical enzymes. You're going to have mercury, and it's going to be accessible to systems within the host that will convert it back to toxic forms. It is conceivable that organisms capable of the mercury detoxifications could be employed right now.

MINIMIZING RISKS OF RELEASE OF MICROORGANISMS INTO THE ENVIRONMENT

USING KNOWLEDGE OF VIRULENCE FACTORS TO SELECT OR DESIGN ORGANISMS WITH LOW RISK OF PATHOGENICITY[1]

Stanley Falkow

Department of Microbiology
Stanford University
Stanford, California 94305

As Omenn and others have emphasized, applying our knowledge of ecology, physiology, and genetics of microbes to the pollution cleanup challenges of our society may require that microorganisms be deliberately released into the environment. Naturally, there are scientists, legislators, and laymen who view such an action with concern. Among their concerns are what the potential danger to human and animal health might be. Fortunately, a fair amount is known about the factors that are involved in microbial pathogenicity. Can we use this knowledge to design genetically engineered microorganisms (GEMs) destined for release into the environment to have a minimal impact on animal or plant health? A medical microbiologist will immediately respond, "Of course. It has already been done for decades in human and veterinary medicine."

A prominent example of release of genetically engineered microorganisms into the environment is live polio vaccine, which was developed in the 1950s, genetically engineered to be safe, and given in a living form to millions of people. A genetically-engineered live typhoid bacterial vaccine, licensed widely around the world, is another example. Thus, medical and public health knowledge about the transmissibility and the pathogenesis of human infectious diseases provide relevant lessons for discussing the risks associated with the release of GEMs.

An Introduction to Bacterial Pathogenicity

Bacteria are commonly transmitted from one person to another via respiratory aerosols or by fecal oral spread (best appreciated by watching toddlers at play in a sandbox). Similarly, people may become infected by interactions with animals (Zoonosis). In recent years, the sexual transmission of infectious agents has become more common; this has been characterized as the mixing and matching of mucous membranes.

---

[1]Presented as a lecture for the interested general public.

The goal of any microorganism is simply defined: to multiply. Several terms must be defined. There is an important distinction between "infection" and "disease," although we often use these terms synonymously. Infection is the capacity of a microorganism (bacterium, virus, fungus) to multiply regularly on or within a host. In contrast, disease is overt damage resulting from multiplication of the microorganism on or within a host. We are often infected, but, fortunately, only rarely do we suffer disease.

One of the terms used in my title, "pathogenicity," is the capacity of a microorganism to cause disease, or overt damage, in a significant proportion of normal, susceptible individuals who are exposed to that organism. This definition excludes opportunistic organisms, which ordinarily do not cause disease in normal individuals, although I grant that "normal" is difficult to determine. "Virulence" is the degree of pathogenicity. Thus, the organism called *Streptococcus pneumoniae* causes a common pneumonia in humans. But type 3 pneumococcus is more virulent than type 1 pneumococcus, because type 3 kills a larger proportion of those who are infected. Both are pathogenic.

A pathogen is a specially adapted microorganism that has a certain set of characteristics we might call the "pathogenic personality." Similarly, one could specify the microbial personality important for a microorganism to live in soil and fix nitrogen.

Characteristics of Pathogens

For a pathogen to be successful, it first must gain access to the host--by sneezing, fecal-oral transfer, or other routes. Once within the host, the pathogen must find a unique place in which it can establish itself and begin to grow and multiply. It must avoid the normal host protection mechanism. It must then multiply and, if it can, cause overt damage to the host.

There are corollary features of pathogens, having to do with special adaptations. Pathogens adapt to a host's biological and social behavior. Therefore, the evolution of organisms that are capable of causing harm is a continuous, dynamic process. Within the last decade or so, some of the most important medical problems that have been called "new" have been those caused by infectious agents. Toxic shock syndrome and Legionnaire's disease are notable examples. Herpes simplex, *Chlamydia trachomatis*, and the AIDS virus all reflect changes in sexual mores. These agents existed before. Changes in human behavior have led to their increased dissemination. They are not more virulent; they are not new creations. The diseases result as microorganisms adapt to changes in our behavior.

The toxic shock syndrome may be viewed as an impingement by humans on the ecology of microorganisms. Women preferred tampons and then less frequent changes of tampons, which were made to be more absorbent. To meet these preferences, manufacturers modified tampons chemically. The chemical changes resulted in increased Mg^{++} chelation. Removal of magnesium from the vaginal fluid led to the proliferation of a particular kind of staphylococcus and permitted that staphylococcus to elaborate a particular toxin. Thus, toxic shock syndrome. The organism always had that capability, but humans made it possible for a particular organism to exploit our foibles.

Pathogens also use elaborate genetic means to modulate their free-living life from their parasitic life. Most organisms that are capable of causing disease do not stay strictly within the host. For example, *Escherichia coli* is highly adapted to live in the mammalian large bowel. At least some strains are pathogenic. For *E. coli*, the bowel must seem like a microbial Florida--warm and comfortable. Suddenly, there's a light at the end of the tunnel, and where does *E. coli* find itself, but out into the cold, cruel world! Under those circumstances an organism must be able to use very different sets of genes, coding for very different attributes, to survive in the cold, cruel world, compared with being on holiday with very different competitors in microbial Florida. This phenomenon applies to any organism that has a special adaptation.

The Evolution of Pathogenicity

It is important to note that pathogens are clonal. With the growth of microbial genetics, there has been a view that microorganisms enjoy a kind of primitive sex life. You would think from some textbooks that the microorganisms inhabiting your bowel are in a state of high sexual excitement and constantly exchanging genes. This is not true. Most of the evidence indicates that bacteria rarely exchange chromosomal genes. Indeed, they shouldn't, because they are haploid--they only have one set of genes. Why should an organism like *E. coli* give up millions of years of evolution for a few seconds of conjugative bliss? Organisms have set up a way to protect individuality at the genetic level. They just can't have genes coming in and recombining randomly. At the same time, total isolation genetically would eventually be biological suicide. So organisms must have some kind of flexibility.

As an example, *E. coli* can cause meningitis in newborns. Only a small subset of particular *E. coli* strains is found, whether the infant is in Africa, the United States, or Southeast Asia. The causative agent of *E. coli* meningitis is basically a single bacterial clone. An organism had a good idea, developed, and became widespread.

Clonal organisms maintain their flexibility to adapt and become pathogenic because essential determinants of pathogenicity and microbial specialization are often found on dispensable mobile genetic elements--phages, plasmids, and transposons. These genetic elements are maintained separate from the bacterial chromosome. They represent a kind of genetic currency that microorganisms spend freely back and forth with each other; they share information, to some extent, on survival. In doing so, they preserve their individuality and specialization on chromosomal genes and maximize their flexibility on the added extrachromosomal genes. If they don't need the dispensable elements, they get rid of them; phages, plasmids, and transposons are not essential to microbial survival under most circumstances.

We often indict the microorganism for the sins of its plasmids or its viruses. The examples are many. Diphtheria toxin is made by a bacterial virus that infects the host bacterium. The same is true of tetanus toxin and botulism. The capacity of most enteric bacteria to cause diarrheal disease and fevers is determined by extrachromosomal elements, not by chromosomal genes.

For a long time we felt that microorganisms became adapted to their hosts by slowly growing in them and by slowly becoming adapted to them.

It appears, from the experimental data, that that is not true. Rather, it looks as if organisms that become pathogens are more highly adapted because they inherited large blocks of genes. There was a macroevolutionary step. Microbial specialization, including, I think, pathogenicity, seems to evolve by quantum jumps rather than by slow adaptive evolution. We still do not know how these genes come to be and how they became organized in blocks. However, the rather broad answer--why pathogens are different from organisms that are not pathogenic in the same group of bacteria--is that the pathogens have blocks of extra, different genes. It is these extra genes that contribute the most to the "pathogenic personality."

Using Genetic Techniques to Eliminate Pathogenicity

With this view about pathogenicity, one can try to dissect away different layers of adaptation of microorganisms to a host. The challenge is to focus upon those few genes that are essential for pathogenicity.

Bubonic plague, the "black death," killed about half the known human population in the 14th Century. A molecular fingerprint of pathogenic strains of *Yersinia pestis*, the organism that causes plague, reveals three plasmids. Nonpathogenic *Y. pestis* strains have lost one of the plasmids spontaneously. Thus, the capacity of *Y. pestis* to cause fatal disease is dependent upon, at least in part, possession of this dispensable element. When I was here at the University of Washington several years ago, we demonstrated (1) that the presence of the plasmid was associated with remarkable effects on human cells cultured in flasks in the laboratory.

The wild-type strain totally destroyed human cells, while a strain without the plasmid did not. The phenomenon is complex, we learned (2). The plasmid serves as a kind of beacon or sensor for the organism. *Y. pestis* likes to live in rodents, in fleas, and in people--wherever it can multiply. But it has to know where it is at any given time.

Genes on the plasmid code for toxins that will kill cells and for enzymes that will enable the bacterium to resist being engulfed by pus cells and killed by the normal serum antibodies. Remarkably, this plasmid seems to sense temperature and the concentration of calcium in the host. If *Y. pestis* senses that it is at ambient temperature (roughly 20°C) and is surrounded by reasonably high levels of free calcium, the organism is as innocuous as a pussycat. If, however, the organism, through this plasmid, senses that the temperature is 37°C (the temperature inside a mammal) and that the concentration of free calcium is low (as it is inside a human cell), the plasmid then begins to produce all of the virulence determinants.

This is a handy device for a microorganism. For the plague bacillus, it is clearly useful to be able to recognize if it is free in the environment or id within a flea's foregut (presumably at a lower temperature and high concentration of free Ca^{++}), versus being within a blood cell of a rat or (really by accident) a human. The comparative behavior of *Yersinia* strains with or without a plasmid can be strikingly different. The loss of the plasmid removes its capacity to cause disease. It does not remove, however, its capacity to cause infection. A plasmid-free *Yersinia* strain can still multiply freely in the environment; it can still enter a flea or a rodent host and multiply to some extent. But the capacity to cause disease is gone. The only way that the organism can regain its virulence is to regain the plasmid, which is an unlikely event in the "real" world.

Preventing Infection

The next step in understanding the pathogenicity of *Y. pestis* is to learn how to stop it from causing infection. What is involved here is the capacity of the organism to attach to and enter a mammalian cell.

Ralph Isberg and I were able to isolate a gene from *Y. pestis* which, when put into any other microorganism, gave that organism the capacity to attach to and enter human cells--or at least cultured human cells (3). Thus, we conferred on a common nonpathogenic laboratory strain, *E. coli* K-12, one of the properties of pathogenicity: the capacity to gain entry into the host by invasion. It does so by initially interacting at the surface of an animal cell. It's not really an invasion, which is a kind of forced entry; rather, the eukaryotic cell embraces the bacterium and willingly brings it inside.

We isolated the gene--which we called *Inv* (for invasion)--and modified it by recombinant DNA techniques. The modified gene was returned to the chromosome of *Y. pestis*; thus, we replaced a good gene for invasion with a bad gene for invasion. The result was that we totally removed the capacity of the *Yersinia* to invade animal cells. We have now learned not only how to remove the capacity for *Yersinia* to cause disease, to remove the capacity to cause infection.

It might be useful, in some circumstances, to have an organism that can infect an animal but not cause disease, as a means to deliver certain kinds of substances into animal hosts or for release into the environment. Indeed, if you take the *Yersinia* organism that is devoid of a plasmid and feed it to an animal, the animal will not get sick, but it will become immune to subsequent infection. This kind of genetic manipulation of pathogenicity traits forms the basis for a number of vaccine strategies.

Designing and Producing Safer Vaccines

How minimal a change can one make in the capacity of an organism to act as a pathogen? One no longer sees many cases of whooping cough, once a scourge of young children. However, as you may be aware, there is considerable controversy about the current vaccine for whooping cough, which is thought to cause untoward neurological effects in a very small percentage of those who receive the vaccine. As a result, the organism that causes whooping cough, *Bordetella pertussis*, is being investigated anew. How does this organism cause disease? On the basis of new knowledge, can we come up with a safer vaccine?

When *B. pertussis* enters a susceptible host, it localizes to the trachea, which is covered with ciliated cells. The microorganism adheres to the cilia and to the surface of the ciliated cells. Indeed, during the entire infection cycle, *B. pertussis* stays on the surface of cells; it never enters them. *Bordetella pertussis* elaborates a number of toxic substances as it multiplies on the surface of respiratory cells. These toxins can be absorbed by cells and distributed systematically. Thus, a major component of whooping cough is intoxication by bacterial products; in fact, when the symptoms of the disease are most florid, it is quite difficult to isolate living *Bordetella* bacteria from patients. The bacteria leave a toxin souvenir, however. The major *Pertussis* toxin has the capacity to affect a variety of normal cellular functions, including the normal response to hormonal stimulation.

From even this elementary view of pertusis, one can see that there should be two fundamental strategies to prevent whooping cough. One could prevent the microorganism from sticking to respiratory cells in the first place and, thus, abort the infection. Alternatively, one could inhibit the function of the toxin and prevent disease. We have cloned the gene for pertussis toxin (4). By analyzing the sequence of these genes, we deduced that a minimal change in the structure of the toxin might have a profound effect on toxin activity. By manipulating the nucleotide sequence of the toxin gene we added four new amino acids (12 nucleotide pairs) to the toxin molecule, a minimal change. Then, using further recombinant methods, we created a B. pertussis strain identical to the wild type except that it contained an additional four amino acids in this particular molecule.

The addition of these four amino acids had an extraordinary effect on the behavior of the microorganism. Ordinary B. pertussis evokes a marked white cell response; the genetically engineered strain had no adverse effect whatsoever on white cells. The neurological damage associated with pertussis immunization is thought to be a reflection of the toxin's capacity to induce anaphylaxes. An animal model has been developed in mice to simulate this event. The wild-type organism formulated as a vaccine sensitized up to 90% of animals; a vaccine prepared from the organism, identical except for the four additional amino acids, did not sensitize a single mouse. By all the parameters that were measured, the genetically engineered organism retained the capacity to cause infection, but not disease. In fact, it brought about perfectly good immunity and probably could be the foundation for a safer vaccine.

Concluding Remarks

How can we apply the lessons learned from the study of microbial pathogenicity to the concerns that have been voiced about the release of GEMs into the environment? One scenario that is often voiced is that some minor genetic change in a nonpathogenic microbe would result in the creation of a "new" pathogen. I submit that the risk of this happening is very slight, indeed. Pathogenicity is multigenic and not accidental. In addition, it seems to me that the study of the ways in which microorganisms bring about disease has led us to design microorganisms at the genetic level to behave in expected, rather than unexpected ways. It is a simple lesson. Understanding biological mechanisms, and manipulating them in the most minimal way possible, is the key to reducing any risks associated with the release into the environment of genetically modified organisms. The release of GEMs is not to be taken lightly, but we need not unnecessarily fear doing so, either.

REFERENCES

1. Portnoy, D.A., S.L. Moseley, and S. Falkow (1981) Characterization of plasmids and plasmid associated determinants of Yersinia enterocolitica pathogenesis. Infect. Immun. 32:775-782.
2. Portnoy, D.A., and R.J. Martinez (1985) Role of a plasmid in the pathogenicity of Yersinia species. Curr. Top. Microbiol. Immun. 118:29-51.
3. Isberg, R.R., D.L.K. Voorhis, and S. Falkow (1987) Identification of invasin: A protein that allows enteric bacteria to penetrate cultured mammalian cells. Cell 50:769-778.
4. Black, W.J., and S. Falkow (1987) Construction and characterization of Bordetella pertussis toxin mutants. Infec. Immun. 55:2465-2470.

ROUNDTABLE DISCUSSION ON MINIMIZING RISKS OF RELEASE OF MICROORGANISMS INTO THE ENVIRONMENT

Gilbert S. Omenn (Chairman), University of Washington
Ernesta Ballard, Pacific Celebration '89
Ananda M. Chakrabarty, The University of Illinois
Stanley N. Cohen, Stanford University School of Medicine
Stanley Falkow, Stanford University
Gwyn Jones, Fresh Water Biological Association
P.H. Pritchard, U.S. Environmental Protection Agency
Kenneth N. Timmis, University of Geneva

TIMMIS: Stanley Falkow made the comment that, although plasmids are a means of providing genetic flexibility or a means of transferring nonessential information, the frequencies of transfer between organisms in the gut under normal circumstances are very low. Have these been measured in a definitive way under a variety of conditions, in particular, under a variety of selection conditions?

FALKOW: During the first years of recombinant DNA, I was one of the members of the Risk Assessment Committee of the National Institutes of Health (NIH). We were asked to design experiments to actually measure transmission of plasmids within eight human volunteers. Even if one took a plasmid that had a very high level of transfer in the laboratory, its transmission in the normal gut was not measurable. That is, it was less than 10^{-10}. If, on the other hand, you fed the individual an antibiotic for which the plasmid encoded resistance and did the same experiments, transmission was measurable, but still only about 10^{-7}.

To answer your question: yes, you can do it. E.S. Anderson looked at another ten subjects in England. Around the world these experiments were done with different human volunteers and different strains, with essentially identical results. Mutations and transfer of genetic information do occur in nature. Indeed, those are very important aspects of evolution, not to be underestimated. The question is simply, "Is transmission at a low level?" In my opinion, it is much lower than most people had thought.

TIMMIS: When I said "selection," what I meant was conditions, such as diarrhea, that might be selective for the plasmid functions.

FALKOW: Some of you may have encountered what has been called "Montezuma's Revenge" or some variety of traveler's diarrhea. These are caused by E. coli strains with a particular kind of plasmid--in fact, several different plasmids. These plasmids are transmissible in the laboratory from one organism to another, to a variety of different species. The fact remains that these plasmids are virtually clones of E. coli. They are not widely disseminated. It appears that pathogenicity is not dependent on one, two, or even three determinants, but on many. Any kind of microbial specialization is largely dependent on the interaction and cooperation of many different genes, and pathogenicity is no different. You can't get a plasmid to give the capacity to produce toxin to any organism and have it suddenly become a pathogen. It just doesn't happen.

COHEN: It is possible to transfer a chromosomal gene and get invasiveness of a previously noninvasive organism. If you then take the gene for pathogenicity and transfer it to the organism that has received the gene for invasiveness, is the recipient organism likely to become invasive?

FALKOW: If you take E. coli K12 and put the genes for invasion in the plasmid into it, nothing occurs. The organism is not pathogenic. On the other hand, it is possible to take the plasmid from plague and interchange it with other groups of bacteria. There are three species of Yersinia: one is Y. pestis, which causes plague; another is the more common Y. enterolytica, which caused an outbreak in Seattle some years ago involving contaminated tofu. The plasmids are essentially identical in those organisms. Again, Y. enterolytica does not give you plague. The two species have other properties that are different under those circumstances.

There has been an attempt by Sam Formal at Walter Reed to convert a harmless E. coli into an organism that causes dysentery by using genetic manipulation. He was successful. In doing so, he had to transfer about a 140-kD plasmid from Shigella. This plasmid had the capacity to encode for probably close to 100 different proteins. He had to change large segments in, I believe, six parts of the chromosome in order to get something that was almost as good as the organism that causes watery dysentery.

TIMMIS: That's absolutely right, but that organism was not pathogenic for man.

FALKOW: I mean "pathogenic" in the sense that it could cause infection, it could infect humans. It did not necessarily cause disease, but it could certainly infect.

JONES: I feel a little out of place here. I feel a little uncomfortable at 37°C; I prefer to work at 20°C, which is what my environment is. Listening to Dr. Falkow, I wondered why those bacteria were not transferring their information. There is evidence that large plasmids in cold waters are transferring frequently at an optimum of 20°C. In one stream in our part of the United Kingdom, a cryptic plasmid--cryptic, meaning we're ignorant of what it does--is found in a variety of species at different times and at different sites in the same stream. So it's moving around. There is evidence that certain large plasmids in other rivers are transferring at temperatures of up to 20°C and frequently at 10°C.

FALKOW: The most common plasmid found in typhoid (Salmonella) is a plasmid that has an optimum of transfer at 20°C. In fact, if one looks at the dissemination of an R plasmid in a hospital environment, as was done here at the University of Washington, one finds that there is large dissemination. As I remember the data, it went into twelve different species of Enterobacteriaceae, but there was always an underlying selection. I don't want to belittle transfer that can occur in nature. I can simply say that microbial populations, whether or not they're pathogens, are largely clonal, which means that horizontal gene transfer must be relatively rare.

COHEN: However, evolution requires variability, which occurs through the process of mutation. Mutations don't occur at a high enough frequency in any one individual to produce the kind of variability that is necessary to accomplish evolution. Accumulations of mutations are interchanged by recombination, whereby gene segments from one source are able to recombine with gene segments from another source. This isn't a man-made procedure; this is a natural process. Recombination requires gene exchange; genes must move from one organism into another organism. This gene exchange occurs in higher organisms, in humans and in others, by sexual activity. Sexual reproduction provides an opportunity for recombination in microbes that are not diploid and that have only one set of genes on their chromosome. The very fact that evolution has occurred implies per se that gene transfer and recombination have occurred and are occurring on an ongoing basis in microbes.

PRITCHARD: At EPA, we deal with a lot of environmental issues in good, old-fashioned microbial ecology. We have a concern, analogous to the study of disease, to determine whether an organism will actually infect the environment. Will it reproduce and become a part of the natural communities? The difference is that we don't have good criteria upon which to decide whether that's good or bad. Is it all right if a genetically engineered organism, when added to the environment, becomes a normal component of that community?

FALKOW: Is your concern related to return of an organism back to its environment? Some people have the idea that if you genetically modify a microorganism, somehow a priori that is going to be evil. Risks have nothing whatsoever to do with the genetic manipulation, only the final product(s). In many cases, there has been undue emphasis on the process of doing genetic engineering, which, after all, has been done in agriculture for a long time. If you take an organism that already exists in an environment, make subtle changes to it, and put it back into the same environment, it should be safe. I do understand that if one takes an organism and puts it into an environment in which that organism ordinarily does not exist, there is a high degree of suspicion. Each case should be reviewed.

I think the bowel is no less complex than the soil, and I think some of the same rules apply.

PRITCHARD: That may be. If you apply a selective pressure to the bowel and add an organism, it may very well be that you will get some undesirable organism taking over, becoming a common member. When we are dealing with the presence of various kinds of chemicals that man has added, we have the idea of working with niche. Can organisms, in fact, adapt to using available resources and, therefore, become established? Is that something we are concerned about? Are there similar kinds of analogies in the human gut?

COHEN: For the purpose of being concerned about alteration of the natural environment, how would the EPA, for example, define the natural environment? Is it defined as the current environment? It's not clear what it is that we specifically want to avoid changing.

PRITCHARD: There are numerous possibilities, many of which are worth considering. There may be relatively small, insignificant changes. The question is, "Can we relate them to anything of significance?" Are we concerned about preserving a relatively unimportant species in the environment? This issue, it seems to me, might help define what we consider to be a natural environment. It may be that we have some precise handles on what we consider natural and unnatural in talking about relationships in disease. I doubt that any consensus currently exists as far as the environment is concerned.

COHEN: I wonder whether one can, in fact, term any present environment "natural," because man and animals have been altering the environment for years. Every time a field is plowed with cow manure, it's an alteration of the environment. Whenever an antibiotic is used, it's altering the occurrence of bacteria in the environment. Whenever one gives a vaccine, one is altering the environment by creating a disadvantage for that organism against which one is giving the vaccine.

I can think of some things that we don't want to do, but not on that list is "alteration" of the "current" environment, which I don't believe to be any more "natural" than the environment that would result from some of the environmental introductions being discussed.

BALLARD: The EPA is the only regulator represented on the panel, but I hope that we speak for other regulatory agencies as we respond. The EPA, unlike the public's perception of it, is not an agency empowered by Congress to protect the environment. It is a permitting agency; it permits discharges. That's a very important distinction, because EPA is not burdened with determining what a natural environment is. It is burdened with permitting discharges into the air, the water, or the earth. The difficulty for the regulators--and this panel has been a perfect example--is that regulation has to occur now. It can't wait for the next iteration. It can't wait for the next batch to hatch. It has to happen today and it has to be effective for five years or 18 months. The permit has to be measurable, and the protection increasingly has to be absolute.

The kind of expert witness testimony provided today has become increasingly unintelligible, even to the well-educated public. The many years of detailed work necessary to achieve the kinds of genetic manipulation that the speakers at this conference have been discussing are met by a general lack of trust.

Since the Second World War, since the inevitable marriage of high technology and science with weapons production and defense, and as a result of several unfortunate presidencies, the level of public trust in this country has declined markedly. Science has been dragged right down with politics. The regulators are poised with one foot each on the horns of science and politics, needing to listen and understand.

One other factor complicates the use of this technical information in current "real-time" regulation: the role of lawyers. Our entire public process has taken on the trappings of a courtroom. Basically, the whole demeanor and the whole tactic of courtroom litigation have been adopted by all public bodies. Science is at great pains to be useful in public policy formation, because litigators will use scientific uncertainty as a tool to polarize and, therefore, to press their point of view.

Scientists adopt exactly the opposite approach to uncertainty. They seek together to resolve the uncertainty and to find some protocol that they can use to prove that either one or the other is right. There's far less loss of face.

The regulator sits in an atmosphere in this country that is increasingly charged by the expectation of clash and conflict and the expectation of succeeding at the sake of someone else with information that is not currently useful. The purpose for which your research is being conducted, which is clearly to enhance the quality of life, is not useful today to EPA, which has to decide on a permit that might deal with whether or not the by-products of your lab can be flushed down the toilet or whether they have to be treated as a hazardous waste.

FALKOW: Those are political decisions.

BALLARD: But they're real decisions, whether they're political or not.

FALKOW: One of the things I learned from testifying for the Cambridge Committee on Recombinant DNA and a variety of bodies was a principle which I hold. That is, if a university or an industry operates in the community and has some effect on that community, the public has the right essentially of informed consent to say whether or not they want that activity to go on. I accept that, but that's not a scientific judgment. I spent a great deal of my life on an issue which I consider much more dangerous than recombinant DNA--the use of antibiotics to supplement animal feeds. When all of the evidence was in, the head of the Food and Drug Administration (FDA), who is now president of Stanford University, made a judgment; then the Congress overrode his decision and said we will continue using antibiotics in feed, due to intense lobbying. Those are political decisions.

How do you look at the environment? You do exactly what we do. You use DNA probes. We follow plague and fleas with a DNA probe. We squish the fleas and we take a DNA probe that's specific for plague and see whether or not it's in there. It's easy enough to tell you scientifically how to define what's going to happen in an environment. Clearly, the people who work in microbial ecology are going to do the same thing as the people who work in pathogenic bacteriology: they're going to use recombinant techniques and new genetics to understand more about their discipline. That's going to make these decisions a lot easier. But, then, if you want the answer on release, ask your Congressman, don't ask me.

PRITCHARD: From my standpoint in research at EPA, we're much more concerned with the scientific aspects than the political aspects. We follow the same kinds of experimental approaches as colleagues outside the government. We use DNA probes to follow organisms in the environment. We're obviously going to use our understanding about technology to learn more about microbial ecology, and that's part of our objective. But by the same token, we have to be mindful of this question as to what do we consider significant. It keeps coming up all the time.

COHEN: I do think that there's one kind of relevant experience. If you take an organism out of its environment and put it back into the same environment, even with a genetic change (usually a deletion), I think there is sufficient evidence to indicate that nothing unpredictable will happen. In contrast, I don't believe in a predictable ecology at this point in time.

OMENN: Before we turn to the general discussion, I want to ask Dr. Jones to describe experiments and questions pertaining to the release of genetically engineered organisms in the United Kingdom.

JONES: There have been relatively few releases. The release that has attracted the most publicity has been of viruses to control insect pests. A genetically manipulated virus has a relatively limited host range. The people concerned at the Institute of Virology have had to go through a series of tests to ensure that the target species the virus might hit is specific and that a whole range of nontarget species are not affected. We don't think that this should be a very controversial situation.

There has been some discussion over the release planting of transgenic potatoes, but I don't think that comes into the scope of this symposium. More recently, there is the release of Rhizobium that contains Tn5. A lot of discussion has gone on among geneticists and macroecologists, the modelers, on the committee. People like me are in the middle. I'm a microbial ecologist, not a geneticist and not a modeler. On the whole, I think the decisions have been unanimous.

There are certain matters we have discussed at the political end of the sphere, namely prediction and perception. The first involves predicting what would happen when we've looked at a single Pseudomonas species into which a common recombinant plasmid has been put. When we have taken auxotrophic mutants of that species and put a variety of these into the same environment, at the end of the day the Pseudomonas species that we obtained from that environment into which we had put the plasmid was the least successful survivor. The auxotrophic was the best survivor. So predictability was certainly difficult.

Perception--the public perception of what goes on--is more difficult, especially in the United States, where your legal system seems to be able to stop what's going on if one has the right activist in the right quarter. What has had the biggest effect on our committee, surprisingly, has been Chernobyl, the fact that science and technology can have such a major effect across national boundaries. We still are not eating lamb in our part of the country because of radiation levels from Chernobyl. This has had an enormous effect on communities, particularly rural areas. Thus, anybody planning a release of organisms in the United Kingdom is advised to take it up with the parish councils, or local authorities. The environmental health officer has to be informed. It's not often realized that he's a man of considerable power and can stop it just like that, if he just feels that way. He doesn't have to go to court. So I suppose that gives you a feel for the British scene. On the European scale, there is now a German moratorium on release which may affect us in terms of European controls.

OMENN: There has been a series of meetings in the United States, organized by the NSF and also by EPA, about risk assessment with regard to deliberate or inadvertent releases of organisms into the environment. The key questions come back to those that Stan Cohen was asking. What questions can be framed with enough specificity to generate measurable and meaningful responses? Can we tell whether a change is or will be a problem? Without focusing the questions, it's impossible to focus the measurement. We've had valuable information presented by Ron Atlas about tracking organisms, genomes, or specific genes into the environment and by Morris Levin about following the organisms that are introduced to see their fate--whether they survive, whether they grow, whether they proliferate and transfer their gene, how far they're dispersed.

Many of these parameters are susceptible to direct measurement. It is fair to ask whether the likelihood of these events is sufficient to warrant the expenditure in general. But it is certain that some well-selected experiments could be worth their weight in gold in demonstrating what is and is not possible to measure. As Dr. Pritchard asked, there must be analogies to the experiments described by Dr. Falkow on extent of transmissibility of plasmids in human volunteers. There are many opportunities for "environmental volunteers," given that there are so many contaminated sites undergoing site assessments and awaiting some kind of redress.

ALEXANDER: Fortunately, most of the biosphere has not yet been catastrophically affected by humans. (Give us time!) Biosphere function requires a large number of species, some of which are essential and some of which are probably not particularly essential. Granting that it is very unlikely that an organism returned to its original habitat with or without small genetic changes would cause a problem, I think the issue is apart from genetic engineering. Suppose we introduce an organism into a different environment, which is what we are concerned about in environmental improvement. We are moving the organisms from place to place. The issue of human health, I think, is a minor one; playing just a numbers game, the probability that an organism that is transported to a new place will affect the one species that is studied in medical schools may be, purely on a statistical basis, one divided by the total number of species that exist. So let us go back to ecosystem function. Given the tremendous value of biotechnology and genetic engineering, how do we minimize the risk, which may be very small, to natural populations that are important to the biosphere? Let us assume that the organisms are moved to another environment and that we have to look at significant species that function in nature. How do we minimize these risks?

FALKOW: The only answer I could give is to understand what you are modifying as precisely as you can and make the smallest change.

COHEN: I would go one step further, Dr. Alexander, and challenge the presumption of special risks. I question again how to define what are natural populations in the context of prior interventions in the biosphere. It was natural populations of bacteria and viruses that gave us bubonic plague and smallpox. We wiped out the organism that gave us smallpox, although that was a product of natural evolution. There are many human, animal, and plant diseases that we have attempted to treat or prevent by modifying the populations that nature has handed down by the process of evolution. Mankind has been modifying natural populations of plants and animals by breeding and crop modification for years.

As a geneticist, I have been concerned about the consequence of losing gene pools in nature by normal plant breeding activities. Plants are being bred that are more successful, but some of the earlier species are not being propagated. Various United Nations agencies have attempted to set up seed pools in locations around the world. The point is that mankind has been so long altering natural populations of various organisms that it's hard to determine, in many instances, what is natural or unnatural at this juncture.

BALLARD: I'd like to give a slightly different perspective, not speaking from any scientific point of view but simply observing what the people of this country have determined to be the relative role of humans in what you call natural populations. I think Dr. Alexander prefaced his remarks by saying that impacts on humans were less of concern than impacts on the environment. In the 17 years since 1970 that environmental protection laws have been written, revised, and administered, there's been what I consider an alarming shift away from some balance of protection for all inhabitants of this planet toward an enormously expensive public health program with what are clearly demonstrably diminished benefits to human beings' livers, lungs, and spleens. Meanwhile, there is an enormous opportunity lost in neglect of environmental or ecological programs. But the public is not letting up. The fear of the unknown chemical contaminant or toxin is so much greater than the certainty of being maimed by a chainsaw, for example, that the public persists in demanding health protection. The industrial and capital costs of reducing these pollutant loads are phenomenal. Until the public is convinced that what these gentlemen are doing is benign and possibly has a beneficial public health gain, we will continue moving toward living in some sort of a cubicle for ourselves, letting the other species cope however they can.

VERHEY: I have a philosophical question. Doesn't thinking about the changes that we're talking about make you feel like slowing down? Doesn't it make you wonder whether we know enough about the way things work to be making the sort of progress that we seem to be making? I'm not advocating anti-progress. I just want to point out that biological mistakes have been made through ignorance and that we don't know everything yet.

TIMMIS: I think we must think about the time bomb that we're sitting on in terms of pollutants in the dumpsites and the catastrophes that have occurred over recent years. Think of the accident which just sterilized the upper part of the Rhine River for a considerable period of time. Research in environmental pollution related to cancer is not nearly as glamorous as oncogene research, for example, but perhaps it is more important. We are polluting the environment at an accelerating rate--accidentally, deliberately, chronically. If we don't find solutions, we have a danger of creating an environment that is irredeemably hostile to life forms.

COHEN: I think your question is whether the status quo is better than what might be as a result of knowledge that we obtain. I would argue, as Ken Timmis has, that the status quo is fraught with hazard. Imagine, for example, on the subject of acquired immunodeficiency syndrome (AIDS), if recombinant DNA techniques were not available for AIDS diagnosis and for creating a vaccine. We can't simply accept what fate has handed to us. It is the nature of science to acquire knowledge, and to apply it when appropriate.

FOGEL: I'd like the panel to make a guess as to whether, since the dawn of life itself, the biomass has increased, decreased, or remained constant?

JONES: The human biomass has increased. That's the one thing that's creating the biggest problem.

FALKOW: In natural environments, regardless of what we do, the numbers in a species remain relatively constant over time.

JONES: There's plenty of evidence from microbial ecology that as a population becomes overcrowded, then self-regulation comes in.

FALKOW: AIDS is not an accident. I'm not sure we'll ever know precisely, but AIDS is likely the result of the incursion of humans into an area in which humans and certain animals had not really interacted before. Indeed, there's a very interesting book called Plagues and Peoples, written by McNeil about ten years ago, describing how the evolution of humans and the evolution of disease have gone hand in hand. We're going to lose many people in Central Africa for the first time in history from AIDS. A million people a year have been dying from malaria for a very long time, but now a million more people are going to die from AIDS just in that region. We've always known how to prevent sexually transmitted disease; it doesn't take any great genius to understand. However, implementation has been much more difficult.

COMMENT: I have a comment on this debate about what a natural ecosystem is, what natural populations are, and whether we are really worried about it. When the ecologist or ecotoxicologist tests daphnia, algae, and other organisms, we don't worry about just one organism. We are interested in preserving the functionality of that ecosystem. What a natural population is, God only knows, and He's not telling any more, not at least since the inception of agriculture.

Just look at the Chesapeake Bay. An entire rock bass fishery has been completely destroyed by too many nutrients and too many that are bad. The anthropogenic lowering of the pH in the Adirondacks on the Canadian shield is a more subtle effect. It has taken 15 years for people to really see it. Again, the functionality of those ecosystems has been destroyed; we see the alteration and loss of populations. If too many of these functional ecosystems are impaired, then we'll have something serious to worry about. It's one biosphere.

TIMMIS: I agree, except that we are trying to eliminate an environmental pollutant from a particular system or a particular site; elimination of that pollutant will alter the nature of that site and will inevitably, I would think, alter the ecosystem. To return to Dr. Alexander's point, there are two ways that one could significantly alter the ecology of a particular system. First, the organisms that one releases could multiply or take over in a certain fashion and, thereby, disadvantage one or more of the existing organisms. As we have seen, all the evidence would suggest that released organisms would be at a disadvantage, and would be less able to compete in the system once the particular selection pressure is gone--namely, the substrate that one wants to eliminate. The second type of alteration is elimination of the target pollutant, which is our objective.

COMMENT: The whole point is that the reason we want to get rid of the pollutant is because it has already altered the function of the ecosystem. Most people think in what seems to me a micro sense, and I don't mean on a scale of microorganism. I'm talking about most Superfund sites, whether you're talking about tundra or the Chesapeake Bay. The Chesapeake Bay is an enormous, very complex ecosystem. Certain places, such as the Baltimore Harbor, are particularly polluted. Cleaning up the harbor will certainly improve the functionality of that ecosystem. I don't think anyone would say that that's the wrong thing to do. But when they tried to move some of the polluted sludge, they resuspended a lot of the heavy metals and other contaminants in the Bay. Preserving the functionality of ecosystems involves thinking on an ecosystem level instead of, for example, a 100-acre Superfund site.

SATISH WALIA: Do organisms with a cryptic plasmid really have pathogenic genes? Can we develop criteria to determine whether these organisms containing the cryptic plasmid are human pathogens? Is there any difference between an environmental organism and a human pathogen?

FALKOW: *Pseudomonas aeruginosa* is not a pathogen. It does not meet the definition that exposure of normal people to *P. aeruginosa* will make them sick. Nor do normal people get sick from *Pseudomonas cepacia*. Of course, medical advances have created populations of individuals who would not ordinarily have lived, who do not have normal host defense mechanisms. Even so, as far as I know, *P. aeruginosa* only causes swimmer's ear and skin irritation, in cystic fibrosis patients, for example. But the organism sometimes thinks like a pathogen. After it sets up the primary infection, it uses a genetic modulation system. Instead of causing a fairly severe disease, it reaches a happy medium by modulating its production of a number of toxins and other products. The last time I heard of a case with *Pseudomonas maltophilia*, infection was due to intravenous infusion from IV units.

WALIA: I'm working with *P. maltophilia* and PCB degradation.

FALKOW: There might be a problem in immunocompromised, hospitalized patients, but not in the community at large.

OMENN: Any other comments about Pseudomonas? It's likely to be a major class of organisms for many of these environmental applications. If there are other comments either about risk to humans or about any environmental effects, we would welcome them now.

CHAKRABARTY: I agree with Stan Falkow that *Pseudomonas aeruginosa* or other Pseudomonas species presumably will never cause any problem in healthy human beings. Nevertheless, people who are compromised may get infected with these sorts of bugs. We should also ask whether we will increase the number of *Pseudomonas aeruginosa* sufficiently through deliberate or incidental exposures to alter the likelihood of infection. Such speculations lead to endless arguments.

FALKOW: I don't think that Pseudomonas will be a problem for humans, but there are pathogens for plants. Discussion about human health is misguided; the emphasis should be on plants, not people.

OMENN: This comment comes back to Ernesta Ballard's point about our excessive focus on humans. I would add, however, that under existing statutes and regulatory practices, there is likely to be attention in the community, in hospital settings, and, perhaps, in workplaces, to any significant subgroup of identifiable persons who might have much higher risks. The Clean Air Act requires that EPA, at least for the criteria air pollutants, identify subgroups in the population that may be at high risk for adverse effects and set the standards so as to protect those subgroups with an adequate margin of safety. In fact, the agency has attempted to do that in the case of the National Ambient Air Quality Standards for ozone (subgroup with asthma, bronchitis, or emphysema) and for lead (young children). Similarly, the Occupational Safety and Health (OSH) Act requires OSHA to set workplace health standards such that no worker, even if exposed at the level of the standard for a full working lifetime, shall suffer any adverse effect. Then a few lines later the statute states, "... within technical feasibility." The OSH Act hasn't been implemented as explicitly as has the Clean Air Act.

SKLADANY: I have a hypothetical question. I think people are going to improve organisms. So the question concerns, how we make decisions to release them. Say I brought the individual members of the panel a modified organism to degrade PCBs in sludges, and I gave each $1,000,000, access to a technical staff, and one year to make a decision. How would you approach the problem? What experiments would you do, and how would you decide at the end of that year whether the organism in question would be safe for release into the environment?

PRITCHARD: I can sketch how this process currently takes place within EPA. For short time-fuse kinds of registration submissions, various outside specialists are brought together with experienced people within EPA. They go through the available data and make their evaluations without additional experimental studies. The EPA can require the manufacturer to perform (or may ask its own research laboratories to perform) specific scientific experiments. It depends on what we know, what we are capable of doing, and whether we feel we can obtain a significant result within the given time frame.

SKLADANY: It would be nice if we could wait for all the theoretical understanding to be elucidated so that things can be engineered and the problems properly solved. But we don't have the time. We are going to have to make decisions and take action based on limited information. What are the key factors that will allow us to make those decisions so that we can move forward even with limited information?

BALLARD: Your hypothetical year becomes a matter of historical interest to those involved in the regulatory action, because usually three or four more years will be consumed in the dual process of a citizens' suit brought up by an environmental group challenging the agency for taking too much prosecutorial discretion, and simultaneous litigation from the permit recipient claiming you have been too restrictive.

OMENN: The experimental data required are far less extensive, at present, if the organisms have not undergone genetically engineered manipulations. If they have, EPA review under TSCA or FIFRA Subdivision M provisions is still case-by-case, with points to consider which generate a general scheme for laboratory investigations and, later, field observations.

TAUB: I would like to explore the hypothesis that it is probably going to be safer taking an organism from the local environment, modifying it slightly, and then putting it back in the local environment. The point has been made several times at this conference that it is difficult to evolve a pathogen. A pathogen has to have a lot of abilities to interact with the potential host. Imagine that, say, earthworms have some opportunistic diseases. Sometimes something quite subtle, such as magnesium ion concentration or the ability of an organism to retain iron, may allow the organism to increase its dominance or its pathogenicity. If you take an organism from the environment, isn't it more likely that one might take an opportunistic pathogen and increase its pathogenicity in that particular environment? The earthworm has evolved a cuticle that helps keep bacteria out. If you modify an organism that has many properties of an earthworm pathogen and give it extra biodegradative capabilities, might you accidentally create a pathogen? Isn't this more likely to happen with an organism, modified slightly, from the earthworm's environment than from, say, a deep marine environment where the organism hasn't been in contact with the earthworm for a long time?

FALKOW: To my knowledge, the only way that you can make an organism an instant pathogen is if you give it the ability of poisonous intoxication. Even in that circumstance, it's not clear that a pathogen would be "created." Let me be more specific. If you were to clone the gene for the toxin of botulism and put it into *E. coli*, I do not think you would see a major increase in botulism disease, because *Clostridium botulinum* is found already in the intestines of a number of people. With the exception of infants, no symptoms occur. In certain parts of the world, most people carry *Clostridium tetani*. So even if you put the tetanus toxin gene into *E. coli*, toxicity would be unlikely. To come up with an effective organism that would have the capacity to interact with a complex species, find a particular niche, and be able to compete with an organism already there, I think, is very unlikely. Adherence is initially one of the most important aspects. You're better off working with an organism that already occupies that niche than you are taking one from an unknown niche.

COHEN: Dr. Taub, as I understand it, you're arguing that a genetic change in an organism already in a particular niche would accomplish two things. First, it would make it pathogenic. Second, it would give it a selective advantage over the organisms that are already there. I would argue from an evolutionary point of view that, if there were a high likelihood of such an occurrence, it probably would have occurred in the course of natural evolution. The organisms that currently exist in a particular niche are usually the most highly evolved organisms. It is unlikely that one could accomplish both things through the process of a single genetic change that could make an organism compete better in a particular niche. That doesn't mean, however, that all possible genetic modifications are likely to have occurred at some point in time. Nevertheless, one can argue that the reason that *E. coli* doesn't make insulin unless you genetically alter it is not because it has never been exposed to the insulin gene. Rather it is because that insulin gene doesn't provide a selective advantage. In the course of evolution there have been widespread opportunities for genetic exchange.

PRITCHARD: We seem to be talking about all-or-nothing kinds of phenomena. We may have to consider the rate at which some of these things occur. Say we add to the environment an organism that has the ability to break down a particular chemical that is a contaminant, and it turns out that the organism also breaks down the mucilaginous ooze that comes out of an earthworm. This might not be a unique event, because there may be other organisms breaking down this material or we would be up to our ears in it. On the other hand, if you add an organism that has a particular controlling function, and if you add enough of those organisms, you may speed up the degradation of that material such that this situation becomes a problem.

Ecologists who perceive certain kinds of threats have the ability to test these questions. If you can come up with a scenario as to why a particular organism might affect, say, an earthworm, you could develop a bioassay to see what happens. The question is, "How many bioassays are necessary before we feel fairly certain that there will be no ecological effects?"

FALKOW: It is very hard for me to conceive of making a primary pathogen. Nevertheless, polymicrobic and multifactorial infections do exist. If you take in a toxic substance like alcohol, lose consciousness, and aspirate some saliva into your lungs, then you can get a mixed infection, because you introduce a lot of organisms into an adjacent site that is ordinarily sterile. You possibly could have an untoward cooperative effect that might not have been totally predicted.

TAUB: I explicitly assumed that you took an organism out of its environment that was essentially an opportunistic pathogen, and that you gave it a new property, not unrelated to its pathogenicity.

FALKOW: I have trouble with the term "opportunistic," because it usually implies that there is something wrong with the host. If you inhale saliva into the normally sterile lung, you can get an infection by organisms that ordinarily would not cause a disease. That is opportunistic. But you have to breach a normal host mechanism.

ALEXANDER: I have been involved with several score of risk analyses. One of the undiscussed risks of being in risk analysis is that people who represent a technology think the risk analyst is against the technology. I am particularly uncomfortable in biotechnology because I am dealing, for the first time, with people in my own discipline. I have never done that before for the EPA. But let me try to focus on the problem as I see it without saying I'm against biotechnology, because I have been indirectly releasing improved organisms for some time. My question arises from the fact that diseases appear episodically and catastrophically in populations in nonagricultural, nonurban areas. There have been many cataclysmic changes in freshwater, marine, and higher plant populations. If the organisms did not appear de novo, then they were transported from one lake to another lake, from one part of the ocean to another part of the ocean. If we are moving organisms, whether they are genetically engineered or not, how do we reduce the incidence of those cataclysms that do occur? I think this is the objective question from an ecological viewpoint. Most of us are not trying to be obstructionists, though I admit there are some who are. We are trying to evaluate and minimize the problems in order to promote the technology. That is very different from being obstructionists.

APPLICATIONS TO CHEMICAL AND ENVIRONMENTAL TARGETS, COMBINING MICROBIAL AND ENGINEERING APPROACHES

BIOENGINEERING ISSUES RELATED TO IN SITU REMEDIATION OF CONTAMINATED SOILS AND GROUNDWATER

Perry L. McCarty

Department of Civil Engineering
Stanford University
Stanford, California 94305

INTRODUCTION

Biological processes have been used well over 100 years for the treatment of organic-bearing municipal and industrial wastewaters and municipal landfills. Aerobic biological processes, such as activated sludge and trickling filters, are used in virtually every city in the United States for treating dilute municipal wastewaters. Anaerobic processes are nearly as widely used to convert the concentrated sludges resulting from dilute waste treatment into methane gas, a useful form of fuel. About 30 years ago, it was first realized that by using completely mixed treatment processes, industrial wastewaters can be readily treated even though they contain organic compounds, such as phenol, that are toxic to microorganisms at the concentrations present. This is possible because the degrading microorganisms maintain the concentration of the chemical within the reactor below the toxic threshold. Biological treatment systems for such wastewaters are now common. Thus, a wide variety of biological processes for converting many forms of hazardous chemicals to innocuous end products are well established and extensively used throughout the world.

While prevalent, conventional treatment systems have not been widely applied to the difficult problems of treating xenobiotic compounds that are (a) relatively refractory to biological transformation, (b) highly toxic, (c) either so concentrated that bacteria cannot achieve the intimate contact required or so dilute that reaction rates are low, or (d) so strongly sorbed or mixed within a soil matrix that biological access to them is limited. However, in recent years, many compounds that were once believed to be refractory to biological action are now recognized as being transformed naturally by native microorganisms in the environment. This has led to major efforts to capture this ability in engineered systems where the type of transformation can be controlled and transformation rates can be increased.

Many contaminants of concern are present in surface soils (Fig. 1); others have penetrated into the vadose zones below the surface where the soil pores are filled partly with water and partly with air. Frequently,

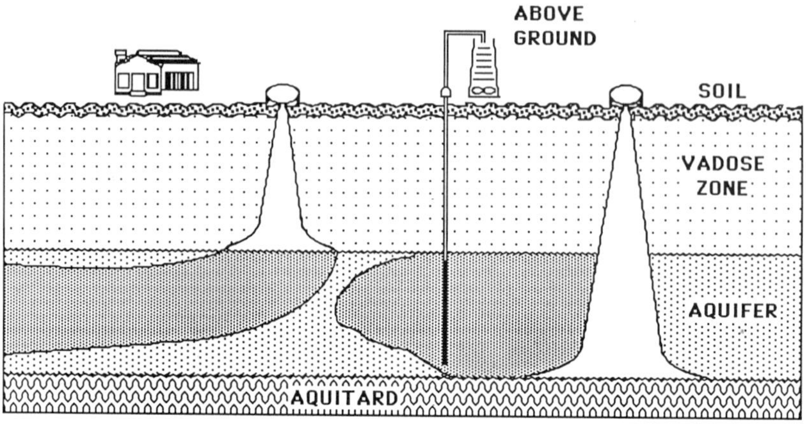

Fig. 1. Treatment domains. Different locations of soil contamination and possible locations for above-ground or in situ treatment.

contaminants have penetrated farther into the saturated zones where the pores are filled with water and where the major aquifers containing the groundwater supplies occur. Fluids that are lighter than water, such as gasoline, tend to mound and spread on the surface of groundwater and from there dissolve; fluids that are denser than water, such as trichloroethylene (TCE), tend to sink to the bottom of the aquifer. Conventional above-ground treatment systems are difficult to implement because the contaminants are usually sorbed strongly to the soils. Cleanup procedures often employ leaching with water so that chemicals can dissolve and be taken to the surface for such treatment. Because of unfavorable partitioning between the soil and water phases, such procedures are generally inefficient and slow. This has caused many to consider in-situ biological treatment as an alternative approach. Here, indigenous or introduced microorganisms would be encouraged to grow in the subterranean environment where the contaminants reside, and thus would have better access to the sorbed as well as dissolved contaminants.

In-situ biological transformation in the soil is in itself not new. In fact, soil application was the first engineered method for treatment of human wastes centuries ago, and occurs naturally throughout the animal kingdom. Soil application is a well-recognized method of industrial wastewater treatment. Treatment on the surface is relatively easy and lends itself fairly well to engineering control. One can plow, disc, and introduce microorganisms, nutrients, oxygen, and other chemicals into the surface soil relatively easily. Treatment in the unsaturated and saturated groundwater zones below the surface is more challenging because these zones are hard to characterize and difficult to access for introduction of chemicals and microorganisms. This, coupled with a lack of adequate understanding of rates, mechanisms, and pathways for transformation of many of the hazardous chemicals of concern, makes in situ biotransformation one of the most challenging bioengineering problems of our time.

This chapter addresses a few issues of importance to in-situ biotransformation of hazardous chemicals. These issues include a discussion of relevant biotransformation processes, reaction kinetics, and reaction stoichiometry.

BIOTRANSFORMATION PROCESSES

In general, biological wastewater and solid waste treatment processes take advantage of the microorganisms' normal function of metabolizing organic materials to obtain energy for synthesis and maintenance. Here, a portion of the organic material, serving as a primary energy source, is converted to end products through oxidation/reduction processes, and a portion of the organic carbon is synthesized into cellular material. This natural process is largely responsible for the turnover of dead organic material in the environment. Such conversions take place in aerobic environments in which oxygen serves as the terminal electron acceptor, or in anaerobic processes where nitrates, sulfates, carbon dioxide, or the organic compounds themselves may serve as electron acceptors. In this case where bacteria use the compounds of concern for energy, the effects of different environmental conditions on pathways, end products, and rates of reaction are sufficiently well known so that engineered systems can be designed and operated to bring about the particular conversions desired.

Table 1 indicates that some organic compounds can serve as primary energy substrates under either aerobic or anaerobic conditions. For such compounds, there is an assimilative capacity in the subsurface within which they are likely to be degraded naturally by indigenous bacteria, provided that concentrations are low enough not to be inhibitory and that sufficient inorganic nutrients for biological growth are present. This represents the simplest case of subsurface contamination to be solved biologically. Other compounds (i.e., petroleum hydrocarbons) are readily degraded as primary substrates under aerobic conditions, but anaerobic pathways are not available, degradation rates are very slow, or specialized bacteria that are not ubiquitous are required. Here, oxygen must be present or supplied in order for degradation to proceed. This presents more of an engineering challenge than the first category, but applications where this approach has been used are common (23).

Tab. 1. Biological processes and environmental conditions under which different compounds may be transformed by bacteria.

<u>Primary substrates</u>	
Aerobic and anaerobic:	Glucose, acetone, isopropanol, acetate, benzoate, phenol
Aerobic primarily:	Alkanes, benzene, toluene, xylene, vinyl chloride, 1,2-dichloroethane, chlorobenzene
<u>Co-metabolism (secondary substrates)</u>	
Oxidations:	Trichloroethylene, dichloroethylene, dichloroethane, vinyl chloride, chloroform
Reductions:	1,1,1-Trichloroethane, trichloroethylene, tetrachloroethylene, dichloroethylene, dichloroethane, carbon tetrachloride, chloroform, DDT, lindane, polychlorinated biphenyls

Also listed in Tab. 1 is another well-documented biological process (15) that results in biotransformation of organic materials. However, little advantage has yet been taken of this process in engineered systems for organic degradation, and the factors affecting rates of transformation are, by comparison, not well known. This process is commonly called "co-metabolism." Co-metabolism has recently been defined (8) as "... the degradation of a compound only in the presence of other organic material which serves as the primary energy source." Here, enzymes produced by bacteria in order to metabolize given substrates for energy can interact with other organic molecules that may be present, bringing about transformations that appear to be fortuitous. This process was first termed "co-oxidation" by Leadbetter and Foster (22), but Jensen (17) later suggested the more general term, co-metabolism, to include dehalogenation reactions which are often reductions.

Oxidative co-metabolic processes include transformations of alkanes to form organic acids and the oxidation of TCE by methanotrophs. Reductions might include the conversion of TCE to dichloroethylene by methanogenic bacterial consortia. Whether these transformations all fall within the limited sphere defined as co-metabolism is not clear. However, in general, the halogenated compounds are often so low in concentration that organisms bringing about the transformation must be supplied with a primary substrate for energy and growth. The trace contaminants then become secondary substrates (27), which may or may not provide some energy for growth. In this sense, co-metabolism may be considered as a subcategory of secondary substrate utilization.

The organic products of secondary substrate utilization may, in themselves, be resistant to further biological transformation, may be transformed further by co-metabolism, or may be used by the same or other organisms for energy. In mixed bacterial consortia that collectively can oxidize a mixture of petroleum hydrocarbons, initial stages of some oxidations are likely brought about through co-metabolism by some of the organisms present. The oxidations are then completed by other organisms that use the co-metabolites for energy.

Some microorganisms that initially transform a chemical by co-metabolism may adapt themselves to using the transformation product for energy. The chemical then becomes an energy source for the organism. This adaptive process may be of some significance, in an evolutionary sense, for making formerly refractory compounds into ones that are readily processed in the environment as primary substrates. There is ample evidence of compounds that may be transformed by co-metabolism by one organism, but are readily used as an energy source by another. Thus, the designation of a given transformation as being either fortuitous co-metabolism or true energy metabolism is often difficult to make. The distinction between co-metabolism and other forms of secondary substrate utilization is perhaps not too important in engineering applications. What is important to know is that if a group of organisms carrying out a desired transformation cannot obtain sufficient energy from it, then they must be supplied with an appropriate energy substrate to grow and function. Indeed, a given organism may need to be supplied with the correct energy substrate in order to induce the transforming enzyme of interest.

Co-metabolism, as such, has been identified as a major mechanism in the transformations of aliphatic hydrocarbons and pesticides (15). The literature here is extensive. Many analogies exist between metabolism of halogenated compounds in mammals and in microorganisms (37). A variety of enzymes and co-factors exist both in mammals and in microorganisms, and transformation products of given organic chemicals are similar.

TRANSFORMATIONS OF HALOGENATED ALIPHATIC COMPOUNDS

Transformations of aromatic hydrocarbons and pesticides have been described in some detail in the literature over the past 30 years (18). Studies on these compounds began quite early because of the demonstrated hazard they presented to humans and the environment. Recently, halogenated aliphatic compounds have also become of great concern. They represent an important group of industrial chemicals in terms of production volume, usage for a broad range of benefits, broad distribution in the environment, and wide exposure of the population. Also, they cause a variety of toxicological effects in species ranging from bacteria to humans. The halogenated aliphatic compounds are among the most frequent contaminants in groundwater, especially 1,1,1-trichloroethane (TCA), tetrachloroethylene (PCE), and TCE (38). While just a few years ago these compounds were believed to be refractory to biological degradation, it is now known that many, if not all, of the halogenated aliphatic compounds can be either oxidized or reduced by bacteria.

Reductive Dehalogenation

In 1981, biological transformation of halogenated methanes under anaerobic methanogenic conditions was demonstrated (7). Subsequently, chlorinated ethenes and ethanes were found to be transformed by reductive dechlorination, and methanogenic conditions were shown to favor such transformations (4). However, some transformations, such as that of carbon tetrachloride, also took place under denitrification conditions (5). Several investigators have identified intermediate products in these transformations (2,3,4,10,31,35,36) and have confirmed that reductive dehalogenation of alkanes and alkenes is favored under methanogenic conditions (3,10,35,36). Recently, Belay and Daniels (3) observed the reductive dehalogenation of chlorinated ethylenes and ethanes with pure cultures of methanogens. With mixed cultures, Vogel and McCarty (35) reported reductive dehalogenation of PCE to TCE, dichloroethylene, and vinyl chloride. But in the process, 12% to 24% of the PCE carbon was oxidized to carbon dioxide. Thus, some oxidation also appears to have occurred. Other studies have indicated that chemical transformations can also take place slowly (36), resulting in the overall picture of the transformations of TCA, PCE, and their intermediates as illustrated in Fig. 2. In general, rates tend to be higher with increased degree of compound halogenation (37).

Since groundwater environments are often anaerobic--or are made so through the co-contamination with other solvents, such as acetone and isopropanol, that can be used as primary substrates for methanogenic microbial consortia--reductive dehalogenation often occurs in groundwater, re-

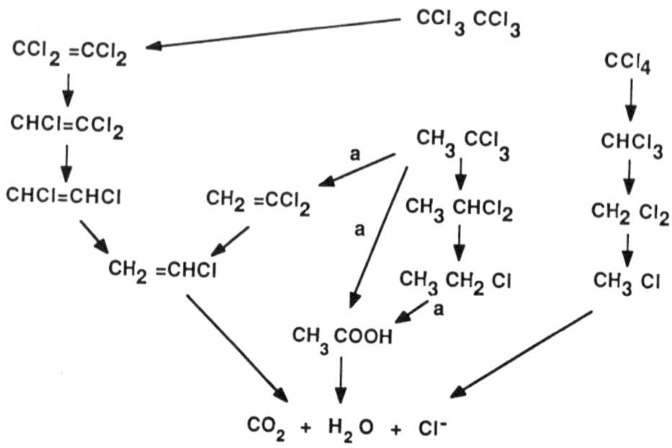

Fig. 2. Anaerobic transformation pathways for selected chlorinated aliphatic compounds [after Vogel et al. (37)]. Arrows with "a" indicate abiotic transformations; other arrows represent biotic transformations.

sulting in a variety of the intermediates depicted in Fig. 2. This suggests that the microorganisms involved are frequently present in subsurface environments, including quite deep aquifers, and that some opportunity exists for in situ management to enhance the rate of biotransformation. The concern is over how completely the process can be made to proceed, as some of the intermediates, such as vinyl chloride, are a greater hazard to humans than are some of the parent compounds. This is an example of a particular difficulty that must be resolved in the engineering application of in-situ biological degradation of organic compounds.

Oxidative Dehalogenation

The ability of methane-oxidizing bacteria (methanotrophs) to cometabolize alkanes oxidatively is well known; in fact, it was observation of this characteristic that led to the formulation of the concept of co-oxidation by Leadbetter and Foster (22). Wilson and Wilson (40) recently demonstrated that aerobic enrichment cultures grown on methane can transform TCE. Since then, there has been a flurry of activity to obtain better information on the ability of methanotrophs to transform the range of important halogenated aliphatic compounds in order to make use of this process. Janssen et al. (16) have now demonstrated that pure cultures of methanotrophs can indeed oxidize TCE and 1,2(t)-dichloroethylene. With the latter, 1,2(t)-dichloroethylene oxide is formed as an intermediate, confirming that epoxidation is a first step in the oxidation of chlorinated ethenes, as has been speculated (9,13,24) and as found with other alkenes (32,33). Little et al. (24) also demonstrated TCE oxidation by pure cultures of methanotrophs, and indicated that the co-metabolism occurred when the organisms were oxidizing methanol as well as methane. They could not detect the epoxide of TCE, but they did observe what tentatively appears to be glyoxylic acid and dichloroacetic acid. These compounds may not be found as end products in mixed cultures if they are used by other microorganisms for energy and growth. Based upon these observations, a mechanism was proposed for oxidation of TCE that includes epoxidation as the first step followed by either one of two chemical transformations leading to the above products.

Studies by Fogel et al. (9) and Hensen et al. (14) have demonstrated the range of chlorinated aliphatic compounds co-metabolized by methanotrophic enrichment cultures and their relative rates of oxidation. Several chlorinated methanes, ethanes, and ethylenes could be oxidized and, in general, the rates of oxidation increased with a decrease in the degree of chlorination. The highly chlorinated compounds, such as tetrachloroethane and PCE, were not utilized. Vinyl chloride is utilized faster than cis- and trans-1,2-dichloroethylene, which are oxidized faster than 1,1-dichloroethylene and TCE. Dichloromethane is utilized faster than chloroform. Rates of utilization for chlorinated ethanes decreased in the following order: 1,2-dichloroethane, 1,1-dichloroethane, 1,1,2-trichloroethane, and TCA.

Thus, the potential for degradation differs for the various halogenated aliphatic compounds. As with the alkane analogs, oxidation probably results in the production of the respective aldehydes and acids prior to complete mineralization, perhaps by other microorganisms.

Other bacterial groups appear capable of oxidizing halogenated aliphatic compounds. A bacterial strain metabolizing either phenol or toluene was recently reported to degrade TCE, 1,1-dichloroethylene, and cis-1,2-dichloroethylene, but not PCE (28). Another strain growing on citrate was reported to degrade PCE as well as TCE (11).

Hartmans et al. (12) isolated a Mycobacterium strain L1 from soil and found it capable of using vinyl chloride as a sole substrate for energy and growth. Thus, its conversion was not by a co-metabolic process. Stucki et al. (34) reported on the isolation of a strain that could oxidize 1,2-dichloroethane as a sole carbon source for energy. There have also been several isolates with the ability to utilize methylene chloride as a sole source of carbon for energy (19,20).

In summary, the methanotrophic microorganisms appear to have the capability to oxidize at least partially many, but not all, of the short-chain chlorinated aliphatic compounds. Some bacteria that use compounds other than methane for energy have similar abilities for oxidizing halogenated aliphatics. In addition, some bacteria can even oxidize simple chlorinated alkanes and use them as the sole carbon source for energy. Thus, recent studies have indicated that broad possibilities exist for biodegradation of chlorinated aliphatic compounds.

REACTION RATES

The engineering design of a biological treatment system for degradation of hazardous chemicals requires some knowledge of reaction rates. These affect the size of the treatment reactors and the length of time required for a given degree of degradation. Factors affecting the rates of degradation of organic compounds that can be used as primary energy sources for bacteria are well known and form the basis for design of biological treatment systems now in common usage (21). However, there is relatively little known about factors affecting the rates of degradation by co-metabolism or by secondary utilization in general. This is a particular lack that needs to be rectified. In the following, a brief description of the kinetics of transformation of primary substrates is presented, followed by a discussion of some features of secondary utilization. Then, the effect of sorption on transformation rates is reviewed.

Primary Substrate Utilization

The rate of utilization of substrates that can serve as primary energy sources for bacterial growth is generally given by the Monod relationship (21):

$$-\frac{dS}{dt} = kX\frac{S}{S + K_s} \tag{1}$$

where:

S = concentration of primary substrate, mg/l
X = concentration of active organisms, mg/l
t = time, days
k = coefficient of maximum rate of substrate utilization, day^{-1}
K_s = affinity or half-velocity coefficient, mg/l

At high primary substrate concentration, the rate of utilization is maximum and first order with respect to organism concentration, but zero order with respect to substrate concentration:

$$-\frac{dS}{dt} = kX, \quad (S>>K_s) \tag{2}$$

At low primary substrate concentration, the rate of utilization is first order with respect to both organism concentration and substrate concentration:

$$-\frac{dS}{dt} = \frac{k}{K_s}XS, \quad (S<<K_s) \tag{3}$$

Here, the ratio of k/K_s is a constant which is commonly called the second-order rate constant for biological degradation (29,30). This value is frequently useful when considering degradation of contaminants at low mg/l concentrations.

It is obvious from Equations 1 to 3 that in order to increase the rate of substrate utilization, the organism concentration, X, should be increased. Organisms do grow in response to consumption of primary substrate, according to:

$$\frac{dX}{dt} = Y(-\frac{dS}{dt}) - bX \tag{4a}$$

or

$$\frac{dX}{dt} = YkX\frac{S}{S + K_s} - bX \tag{4b}$$

where:

Y = yield coefficient, mg bacteria/mg substrate
b = decay coefficient, day^{-1}

Using Equation 4b, the substrate concentration (S_{min}) at which no net growth can result (dX/dt = 0) (27) is given by:

$$S_{min} = \frac{bK_s}{Yk - b} \tag{5}$$

If a single substrate enters the subsurface environment, but its concentration is below S_{min}, then the substrate acts like a secondary substrate and there can be no net growth of microorganisms. In this case, the rate of substrate utilization will not increase over that governed by the concentration of active organisms present. Since microbial populations in the soil are low, half-lives for such compounds may be measured in months to years, if the compounds decompose at all. However, if the concentration is greater than S_{min}, then the microbial population can increase in size, and the rate of substrate utilization will then increase with time.

For a typical case under aerobic conditions at 20°C where k = 10 day^{-1}, K_s = 10 mg/l, Y = 0.5, and b = 0.1/day, then S_{min} would equal 0.2 mg/l. With decrease in temperature, k and b decrease and K_s usually increases. The net effect on S_{min} is not obvious; it may either increase or decease. S_{min} is likely to vary by an order of magnitude or more above or below 0.2 mg/l, depending upon the organism, the substrate, and the environmental conditions under which the organism grows. In addition, oligotrophic bacteria that adapt to low substrate concentrations may have a lower S_{min} value than bacteria that grow in more nutrient-rich environments. In any event, there appears to be a minimum value under steady-state conditions to which a substrate concentration can be reduced when this substrate is the sole source of energy.

Frequently, groundwater contaminants are present in concentrations in the low μg/l range, perhaps well below the S_{min} value for the microorganisms. However, it has been demonstrated that bacteria can grow if they are provided with a primary substrate at a concentration greater than S_{min}. They may also be capable of degrading a trace contaminant simultaneously, while using it as a secondary substrate (20). Thus, indigenous microorganisms might be stimulated to grow in numbers by addition of a nonharmful primary substrate such as acetate, so that their rate of utilization of a hazardous substance such as dichlorobenzene can be increased, even though the latter is present in the low μg/l or ng/l range (4).

Rate of Secondary Substrate Utilization

While knowledge of factors affecting primary substrate utilization is quite good, there are few data upon which to generalize about factors affecting the rate of secondary substrate utilization by microorganisms. By some definitions, co-metabolism, as a subcategory of secondary substrate utilization, occurs only while the primary energy substrate is being consumed. Primary substrate utilization is required to maintain in an active state the requisite enzyme for co-metabolism. With some organics, in the absence of a primary substrate, energy generated through endogenous metabolism may be adequate to maintain requisite enzymes in an active state so that transformations may still occur. There may be competition between the primary substrate and the secondary substrate for an enzyme required for both processes. Here, there would tend to be a significant interaction between the rates of metabolism, so that reaction kinetics would be more complex, as discussed below.

A useful way to consider the rate of secondary substrate utilization is to compare this rate with that for primary substrate utilization alone. Table 2 lists rates of utilization of low concentrations (<1 mg/l) of several organics in comparison with the rate of utilization of the primary energy substrate, acetate, under both aerobic and methanogenic conditions as determined with acetate enrichment cultures (6). Here, reaction rates are indicated by the ratio of k/K_s, or the second-order rate constant. The results indicate that the utilization rate so expressed was higher in comparison with acetate for some, about the same for others, and much less than for acetate with still others.

In general, based upon the data available for transformations of chlorinated aliphatic compounds by methanotrophs, the rate is higher for compounds containing fewer chlorine atoms per carbon; while for anaerobic systems, the reverse is true (37). Thus, vinyl chloride is transformed very rapidly in the aerobic system, but very slowly in the anaerobic system. On the other hand, in the anaerobic methanogenic system, carbon tetrachloride and PCE are reduced quite rapidly; while in the aerobic methanotrophic system, these compounds do not appear to be readily oxidized. It thus appears that for mixtures containing highly chlorinated aliphatic compounds, a combination of anaerobic treatment followed by aerobic treatment may be a way to obtain more complete degradation of a mixture of halogenated aliphatic compounds.

Tab. 2. Measured values of k/K_s for simultaneous utilization of primary substrate (acetate) and several secondary substrates under aerobic and anaerobic conditions (6).

	Aerobic culture		Methanogenic culture	
	Concentration (μg/l)	k/K_s (l/mg·d)	Concentration (μg/l)	k/K_s (l/mg·d)
Primary substrate	1,000	3.8	100,000	0.63
Secondary substrates				
Chlorobenzene	11	2.5		
1,2-Dichlorobenzene	10	10		
1,4-Dichlorobenzene	10	11		
1,2,4-Trichlorobenzene	11	5		
Ethylbenzene	9	35		
Styrene	8	50		
Naphthalene	14	40		
Chloroform			28	0.85
Carbon tetrachloride			17	2.1
1,1,1-Trichloroethane			18	0.46
Tetrachloroethylene			15	0.08

Substrate Interactions

In bacterial kinetics, the interactions between the primary substrate used for energy and the halogenated organics that serve as secondary substrates need more clarification in order to adequately engineer biological treatment systems for decomposition of low-concentration contaminants. For example, LaPat-Polasko et al. (20) compared rates of utilization of methylene chloride and acetate by a pseudomonad that could use both for energy. Methylene chloride was used at a faster rate as a primary substrate than acetate. In mixtures of the two, whether acetate or methylene chloride served as the primary substrate, the two compounds were used simultaneously. However, with acetate as the primary substrate, acetate utilization decreased in the presence of methylene chloride. In contrast, when methylene chloride was the primary substrate, trace concentrations of acetate increased the methylene chloride utilization rate. Such interactions make formulation of reaction rates difficult. Laboratory studies are needed to help better understand these interactions.

In some cases, there is competition between substrates. This occurs with methanotrophs for some substrates. The methane monooxygenase that is used as the first step of methane oxidation is also used as the first step of co-metabolism of other compounds. Reaction kinetics might be modeled based upon enzyme kinetics for competitive inhibition (1):

$$S + E \rightleftharpoons SE \longrightarrow E + P_s \tag{6}$$

$$I + E \rightleftharpoons IE \longrightarrow E + P_i \tag{7}$$

where S represents the primary substrate and I, the inhibitor, each of which competes for enzyme E, forming substrate- and inhibitor-enzyme complexes, SE and IE. These complexes then decompose to form products P_s and P_i, from the primary substrate and the inhibitor, respectively. Here, the co-metabolized compound would be considered the inhibitor. From this model, the primary substrate utilization rate is similar to Monod kinetics (Equation 1), but with a term that includes the effect of inhibitor concentration:

$$-\frac{dS}{dt} = kX \frac{S}{S + K_s\left(1 + \frac{I}{K_i}\right)} \tag{8}$$

where K_i represents an affinity constant for the inhibitor. Figure 3 shows how the ratio of actual reaction rate to maximum reaction rate (R/R_{max}) varies with relative substrate concentration (S/K_s) and relative inhibitor concentration (I/K_i). The inhibitor does not affect the maximum rate of primary substrate utilization, but does decrease the rate for a given primary substrate concentration.

When the rate of utilization of the secondary substrate is of interest, then the roles might be considered reversed. The primary substrate would become the inhibitor, with higher concentrations reducing the rate of utilization of the secondary substrate. This, however, complicates the

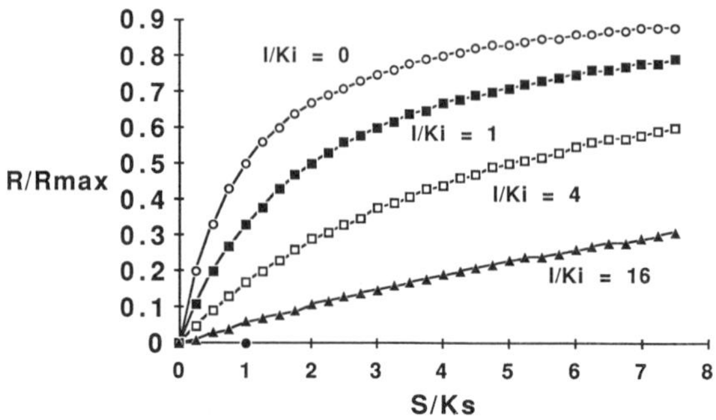

Fig. 3. Relationship between relative substrate concentration (S/K_S) and relative reaction (R/R_{max}) as a function of relative inhibitor concentration (I/K_i).

picture, since the primary substrate must still be present to provide energy for growth and, perhaps, for generation of the active form of the enzyme. Yet, its presence would reduce the rate of consumption of the co-metabolized compound of interest. This suggests that the reaction rate for utilization of the co-metabolized compound would be optimized by maintaining some intermediary concentration of the primary substrate. Obviously, this is a fairly complex situation, one that needs to be well understood for application. More fundamental studies in the laboratory are needed involving the most common chemical contaminants and appropriate mixtures of transforming microorganisms, as well as pure cultures.

Sequential Utilization

Another aspect of reaction rates occurs when co-metabolized compounds are transformed in a sequential fashion. This occurs, for example, in the reductive dehalogenation of chlorinated aliphatics under methanogenic conditions, such as the conversion of PCE to TCE to dichloroethylene to vinyl chloride and then, perhaps, to ethylene or carbon dioxide. The relative rates of each of these steps differ, even though they may all be induced by the same enzyme system. Thus, there could be competitive interactions between the compounds as well. This is another area of reaction kinetics that needs better resolution, perhaps through basic laboratory research.

Sorption

Most of the hazardous organic chemicals of concern tend to sorb onto soils and aquifer materials in such a way that only a small fraction of the material in the contaminated soil, aquifer, or vadose zone may actually be in solution (Fig. 4). An important question is how the sorption or partitioning into the air space affects the rate of decomposition, an area on which the literature is somewhat controversial.

There is evidence that sorbed materials may be either easier or more difficult to consume. However, the simplest model would assume that the rate is determined by the solution concentration alone. Then, since sorption or volatilization reduces the solution concentration, these processes

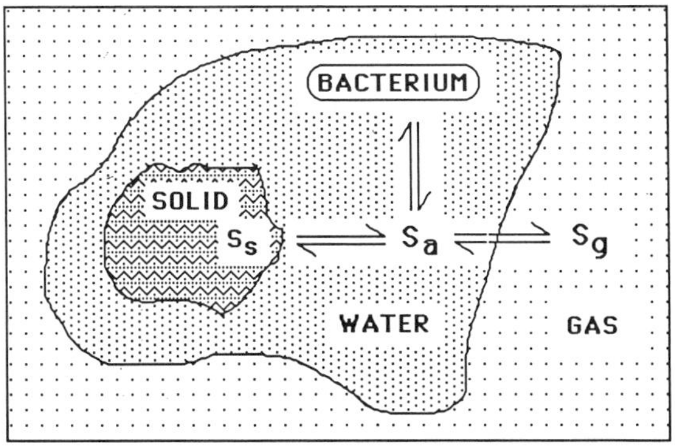

Fig. 4. Mass transfer considerations. Phases in which contaminant (S) may be located that affect mass transfer and overall rate of bacterial utilization.

would reduce the overall rate of the reaction. In addition, as the solution concentration is reduced by consumption, then a gradient would exist such that sorbed and volatile material would move into the solution phase. The overall rate of compound utilization would then be a function of rates of transfer from the sorbed and gaseous phases, as well as of the bacterial utilization rate. If facilitated bacterial transport from the sorbed phase were possible, then the rate of utilization would be faster than indicated by this simple model.

Few studies are available to help in the understanding of rates of bacterial reactions in the presence of solid and gaseous phases. What information is available suggests that sorption and volatilization tend to adversely affect reaction rates. The exact mechanism involved is probably different depending upon the chemical of concern, the subsurface matrix to which it is sorbed, and the microorganisms involved in its decomposition. Also, sorption makes difficult the extraction of hazardous compounds from subsurface environments for treatment by any above-ground system, whether physical, chemical, or biological. Thus, these issues need to be much better understood in order to adequately evaluate the relative feasibility of different approaches for remediation of a contaminated site.

In summary, relatively little is known about the multitude of factors affecting the rates of transformation of hazardous materials. There are simple models available for bacterial transformation of primary substrates, but there are many other complexities that make such models difficult to apply. Some factors whose effects are little known are interactions between the primary substrates used for energy and the hazardous compounds generally used as secondary substrates, sequential decomposition, and reaction rates as affected by sorption and volatilization. In addition, an important factor not yet discussed is toxicity, which can be a factor even with low mg/l concentrations of many of the halogenated compounds. These factors are probably different for different compounds, subsurface environments, and microorganisms. Widespread application of in-situ biological treatment for hazardous compounds will depend upon obtaining a better basic understanding of the many factors affecting transformation rates. At this stage, more basic research is in order, using fairly well-defined systems. Also, applied laboratory studies are needed with natural microbial populations and environmental samples of subsurface materials. Field demonstrations would then be in order to confirm laboratory findings and to address engineering issues associated with natural-system complexities and scale-up.

STOICHIOMETRY

The last issue to be addressed is the stoichiometry of the reactions involved in transformation of hazardous organic contaminants. Bacteria obtain energy for growth through oxidation/reduction reactions. These generally require an electron donor, which often is an organic compound, and an electron acceptor. The electron acceptor under aerobic conditions is oxygen, but under anoxic conditions it may be nitrates, sulfates, carbon dioxide, or, in some cases, organic compounds. There must be a balance between the transformation of the electron donors and the electron acceptors. In addition, there must be carbon, nitrogen, phosphorus, and other trace nutrients required for cellular growth. These all need to be in proper balance for the reaction to proceed. This balance is given by the reaction stoichiometry. In the engineering of a biological treatment system for hazardous chemicals, the nature of the biological system to be used needs to be defined; the necessary chemicals that are not part of the chemical matrix being treated must be supplied for the organisms to maintain proper balance. The physical transportation of the various chemicals required to the location where needed is another critical issue in in situ transformation of hazardous chemicals.

If an organic contaminant is used as the sole source of energy for bacterial growth, then a stoichiometric equation for the overall reaction can be readily established (26). The necessary components of such a reaction are provided in Tab. 3, which lists a generic half-reaction for organic oxidation (R_d) together with half-reactions for possible electron acceptors (R_a). In addition, two possible reactions for cell synthesis (R_c) are indicated: one when ammonia serves as the nitrogen source for cell synthesis, and the other when nitrate is used. A balanced equation for an overall reaction can be obtained by combining the three components so that an electron balance is maintained. In general terms, the overall reaction (R) is given by the following:

$$R = R_d + f_e R_a + f_s R_c, \qquad f_e + f_s = 1 \tag{9}$$

where f_e represents the fraction of the electron donor oxidized for energy and f_s represents the fraction associated with conversion to bacterial cells. Table 4 contains an illustration for the oxidation of a hypothetical hydrocarbon with a molecular formula represented by C_7H_{12}. In this case, the value of f_s was assumed to be 0.5.

For aerobic oxidations, f_s typically varies between 0.12 and 0.6 depending upon the particular organic compound, the bacteria conducting the oxidation, and the relative rate of the reaction (25,26). The slower the reaction, the lower the value for f_s, because here more of the organic is

Tab. 3. Generic half-reactions for construction of a balanced equation for an overall bacterial reaction.

Overall reaction (R)

$$R = R_d + f_s R_c + f_e R_a$$

where: $f_s + f_e = 1$

Electron donor half-reaction (R_d)

$$\frac{1}{e} C_aH_bO_cN_d + \frac{2a - c}{e} H_2O = \frac{a}{e} CO_2 + \frac{d}{e} NH_3 + H^+ + e^-$$

where: $e = 4a + b - 2c - 3d$

Electron acceptor half-reactions (R_a)

Aerobic: $\frac{1}{4} O_2 + H^+ + e^- = \frac{1}{2} H_2O$

Anaerobic: $\frac{1}{6} NO_3^- + H^+ + \frac{5}{6} e^- = \frac{1}{12} N_2 + \frac{1}{2} H_2O$ Denitrification

$\frac{1}{8} SO_4^{2-} + H^+ + e^- = S^{2-} + \frac{1}{2} H_2O$ Sulfate reduction

$\frac{1}{8} CO_2 + H^+ + e^- = \frac{1}{8} CH_4 + \frac{1}{4} H_2O$ Methane fermentation

Cell synthesis equations (R_c)

$$\frac{1}{4} CO_2 + \frac{1}{20} NH_3 + H^+ + e^- = \frac{1}{20} C_5H_7O_2N + \frac{2}{5} H_2O$$

$$\frac{5}{28} CO_2 + \frac{1}{28} NO_3^- + \frac{29}{28} H^+ + e^- = \frac{1}{28} C_5H_7O_2N + \frac{11}{28} H_2O$$

oxidized to provide energy for maintenance rather than growth. With denitrification, values for f_s are about 80% of the above; while with sulfate reduction or methanogenesis, f_s is much lower, generally in the range of 0.04 to 0.2.

Based upon the balanced equation for the hypothetical hydrocarbon listed in Tab. 4, 96 kg of the hydrocarbon would require 160 kg of oxygen and 14 kg of ammonia nitrogen for the reaction to proceed properly. In this process, 113 kg of bacterial cells [represented by the empirical formula $C_5H_7O_2N$ (26)] is produced. Not included in the balanced equation are the other nutrients required for bacterial growth. The phosphorus requirement is about one-sixth of that for nitrogen. Thus, in the above example, 2.3 kg of phosphorus would be required, usually in the form of orthophosphate.

On the basis of the above estimates from stoichiometry, one can estimate the chemical needs for rectifying a groundwater contaminated with, say, 1,000 gallons (6,000 pounds) of the hydrocarbon. These calculations are summarized in Tab. 5. An engineering issue is how to convey the chemicals needed to the location of contamination for an in situ clean-up. With contaminated groundwater, this may be done by dissolving oxygen

Tab. 4. Example of stoichiometry for oxidation of a hypothetical hydrocarbon (C_7H_{12}) (based upon equations in Tab. 3).

Assumptions

$$f_s = 0.5, \quad f_e = 1 - 0.5 = 0.5$$

$$a = 7, \quad b = 12, \quad c = 0, \quad d = 0, \quad e = 4(7) + 12 = 40$$

Equation construction

$$R_d: \ \frac{1}{40} C_7H_{12} + \frac{14}{40} H_2O = \frac{7}{40} CO_2 + H^+ + e^-$$

$$f_s R_a: \ \frac{1}{8} O_2 + \frac{1}{2} H^+ + \frac{1}{2} e^- = \frac{1}{4} H_2O$$

$$f_e R_c: \ \frac{1}{8} CO_2 + \frac{1}{40} NH_3 + \frac{1}{2} H^+ + \frac{1}{2} e^- = \frac{1}{40} C_5H_7O_2N + \frac{2}{10} H_2O$$

$$R_d: \ \frac{1}{40} C_7H_{12} + \frac{1}{8} O_2 + \frac{1}{40} NH_3 = \frac{1}{40} C_5H_7O_2N + \frac{2}{40} CO_2 + \frac{4}{40} H_2O$$

or

$$C_7H_{12} + 5\ O_2 + NH_3 = C_5H_7O_2N + 2\ CO_2 + 4\ H_2O$$

MW: 96   160   14(N)   113 (cells)

and ammonia in water and injecting the water so that it comes into contact with the contaminants. In order to supply the 10,000 pounds of oxygen required, 30 million gallons of water saturated with pure oxygen to a concentration of 40 mg/l would be required. Because of this large quantity of water, hydrogen peroxide is sometimes used as a source of oxygen, since it is soluble and can provide a higher equivalent concentration of oxygen. Stabilizers are then used to prevent too rapid conversion of the hydrogen peroxide to oxygen:

$$2H_2O_2 = O_2 + 2H_2O \tag{10}$$

For the above case, 21,000 pounds of hydrogen peroxide would be required to supply the needed oxygen.

As a result of hydrocarbon oxidation, 7,062 pounds of bacteria would be produced, a mass larger than that of the original hydrocarbon contamination. This large mass has a potential for clogging the aquifer. This is another engineering issue that must be considered. With time, and if additional oxygen is supplied, the bacterial mass will decompose perhaps to about 20% of that indicated in the table.

Next, consider the case in which a primary substrate must be supplied for degradation of a contaminant acting as a secondary substrate. Here, the knowledge of chemicals required is not as good, and experience is very limited. From their various studies on use of methanotrophic bacteria for the co-metabolic oxidation of TCE, Wilson and White (39)

Tab. 5. Example of chemical requirements for hypothetical clean-up of hydrocarbon spill (based upon equations in Tab. 4).

Quantity of hydrocarbon spill
1,000 gallons (6,000 lb)
Oxygen required
10,000 lb (or 21,000 lb hydrogen peroxide) Equivalent to 30 million gallons of water containing 40 mg/l dissolved oxygen
Ammonia-nitrogen required
875 lb
Bacteria produced
7,062 lb

indicated that the ratio of the mass of the primary substrate used, methane, to the mass of TCE oxidized varied between about 100/l and 1,000/l. Similarly, in the studies on reduction of TCE by methanogenic bacteria, in which acetate was used as the primary energy source, Bouwer and McCarty (4) used a mass of acetate to a mass of TCE transformed in a similar range. These high ratios indicate that large quantities of chemicals would have to be injected into an aquifer system for treatment of even a relatively small contamination. Not only does the primary substrate for energy have to be added, but also oxygen for aerobic reactions as well as appropriate bacterial nutrients. In any event, the quantity of chemicals to be added is large, and moving the chemicals to the location of need is difficult. Obviously, there is a major need to learn how to minimize the chemical requirements for contaminant control. Degradation of chemicals by co-metabolism certainly presents challenging problems for the engineer.

While biological approaches represent significant challenges, they also offer the potential for complete destruction of hazardous chemicals when present at dilute concentrations. Here, other approaches may be even more costly. In cases where laboratory studies have shown feasibility, field demonstrations are needed to help work out the engineering issues involved in system design and operation. These should first be carried out in well-characterized environments that are sufficiently simple so that definitive conclusions can be drawn.

SUMMARY

This chapter presents some of the engineering challenges in biological degradation of organic contaminants in surface soils and the subsurface environment. Extraction of contaminants from the subsurface is generally costly, slow, and difficult. This has led to an interest in using in situ techniques for biodegradation of contaminants. For some contaminants that can be readily used as primary energy sources for bacteria with or without the presence of oxygen, bacterial removal is relatively simple and, in some situations, occurs naturally. In other cases, such as hydrocarbon spills, in situ treatment is still attractive, but has the added cost and difficulty of supplying required oxygen and nutrients for the fairly efficient aerobic oxidation. There are some indications that hydrocarbons may be biodegraded in the absence of oxygen; but, in these cases, the rates of degradation appear to be slow. More research in this area is desirable.

The major challenge to both engineers and scientists lies in the decomposition of hazardous chemicals that appear to be transformed by the process of co-metabolism, or else are so low in concentration that they can only serve as secondary substrates. In either case, primary substrates are required to supply the major energy requirements for bacterial growth and/or for activation of enzymes necessary for the transformation. Engineering experience in the utilization of such processes for degradation of contaminants in the environment is very limited. Little is currently known about substrate interactions and how to optimize the bacterial transformations. Chemical requirements for achieving co-metabolism are high. In addition, knowledge is lacking about methods for getting the correct amounts of chemicals to the locations where needed and in the form needed. The solution of these important issues presents a significant challenge to the engineering and science communities, requiring both basic laboratory studies and field demonstrations in well-characterized environments.

ACKNOWLEDGEMENTS

This material is based on work supported, in part, by the National Science Foundation (Grant ECE 8519243) and, in part, by the United States Environmental Protection Agency (Grant EPA CR-81220). It has not been subject to EPA review and, therefore, should not be assumed to reflect the views or bear the endorsement of that agency.

REFERENCES

1. Bailey, J.E., and D.F. Ollis (1986) Biochemical Engineering Fundamentals. McGraw-Hill, New York.
2. Barrio-Lage, G., F.Z. Parsons, R.S. Nassar, and P.A. Lorenzo (1986) Sequential dehalogenation of chlorinated ethenes. Environ. Sci. Technol. 20:96-99.
3. Belay, N., and L. Daniels (1987) Production of ethane, ethylene, and acetylene from halogenated hydrocarbons by methanogenic bacteria. Appl. Environ. Microbiol. 53:1604-1610.
4. Bouwer, E.J., and P.L. McCarty (1983) Transformation of 1- and 2-carbon halogenated aliphatic organic compounds under methanogenic conditions. Appl. Environ. Microbiol. 45:1286-1294.
5. Bouwer, E.J., and P.L. McCarty (1983) Transformations of halogenated organic compounds under denitrification conditions. Appl. Environ. Microbiol. 45:1295-1299.
6. Bouwer, E.J., and P.L. McCarty (1985) Utilization rates of trace halogenated organic compounds in acetate-supported biofilms. Biotech. Bioengineer. 27:1564-1571.
7. Bouwer, E.J., B.E. Rittmann, and P.L. McCarty (1981) Anaerobic degradation of halogenated 1- and 2-carbon organic compounds. Environ. Sci. Technol. 15:596-599.
8. Brock, T.D., D.W. Smith, and M.T. Madigan (1984) Biology of Microorganisms, 4th ed. Prentice-Hall, Englewood Cliffs, New Jersey.

9. Fogel, M.M., A.R. Taddeo, and S. Fogel (1986) Biodegradation of chlorinated ethenes by a methane-utilizing mixed culture. Appl. Environ. Microbiol. 51:720-724.
10. Gossett, J.M. (1985) Anaerobic Degradation of C1 and C2 Chlorinated Hydrocarbons, U.S. Air Force Report ESL-TR-85-38, National Technical Information Service, Springfield, Virginia.
11. Hadden, C.T., S.B. Benson, and N.W. Revis (1987) Degradation of perchloroethylene and trichloroethylene by an environmentally isolated Bacillus. Annual Meeting of the American Society for Microbiology, Atlanta, Georgia, Abstr. Q-103, p. 299.
12. Hartmans, S., J.A.M. de Bont, J. Tramper, and K.C.A.M. Luyben (1985) Bacterial degradation of vinyl chloride. Biotech. Lett. 7:383-388.
13. Henry, S.M., and D. Grbic-Galic (1986) Aerobic degradation of trichloroethylene (TCE) by methylotrophs isolated from a contaminated aquifer. Annual Meeting of the American Society for Microbiology, Washington, D.C., Abstr. Q-64, p. 294.
14. Henson, J.M., M.V. Yates, and J.W. Cochran (1987) Metabolism of chlorinated aliphatic hydrocarbons by a mixed bacterial culture growing on methane. Annual Meeting of the American Society for Microbiology, Atlanta, Georgia, Abstr. Q-97, p. 298.
15. Horvath, R.S. (1972) Microbial co-metabolism and the degradation of organic compounds in nature. Bacteriol. Rev. 36:146-155.
16. Janssen, D.B., G. Grobben, and B. Witholt (1987) Toxicity of chlorinated aliphatic hydrocarbons and degradation by methanotrophic consortia. In Proceedings of the Fourth European Congress on Biotechnology, Vol. 3., O.M. Neijssel, R.R. van der Meer, and K.C.A.M. Luyben, eds. Elsevier Science Publishers, Amsterdam, pp. 515-518.
17. Jensen, H.L. (1963) Carbon nutrition of some microorganisms decomposing halogen-substituted aliphatic acids. Acta Agr. Scand. 13:404-412.
18. Kobayashi, H., and B.E. Rittmann (1982) Microbial removal of hazardous organic compounds. Environ. Sci. Technol. 16:170A-183A.
19. Kohler-Staub, D., S. Hartmans, R. Galli, F. Suter, and T. Leisinger (1986) Evidence for identical dichloromethane dehalogenases in different methylotrophic bacteria. J. Gen. Microbiol. 132:2837-2843.
20. LaPat-Polasko, L.T., P.L. McCarty, and A.J.B. Zehnder (1984) Secondary substrate utilization of methylene chloride by an isolated strain of Pseudomonas Sp. Appl. Environ. Microbiol. 47:825-830.
21. Lawrence, A.W., and P.L. McCarty (1970) Unified basis for biological treatment design and operation. J. Sanitary Engineer. Div., Am. Soc. Civil Engineers 96(SA3):757-778.
22. Leadbetter, E.R., and J.W. Foster (1959) Oxidation products formed from gaseous alkanes by the bacterium Pseudomonas methanica. Arch. Biochem. Biophys. 82:491-492.
23. Lee, M.D., and C.H. Ward (1984) Reclamation of contaminated aquifers: Biological techniques. In Proceedings of the 1984 Hazardous Material Spills Conference, pp. 98-103.
24. Little, D., A.V. Palumbo, S.E. Herbes, M.E. Lidstrom, R.L. Tyndall, and P.J. Gilmer (1987) Trichloroethylene biodegradation by pure cultures of a methane-oxidizing bacterium. App. Env. Microbiol. (submitted for publication).
25. McCarty, P.L. (1971) Energetics and bacterial growth. In Organic Compounds in Aquatic Environments, S.D. Faust and J.V. Hunter, eds. Marcel Dekker, New York, pp. 495-531.

26. McCarty, P.L. (1975) Stoichiometry of biological reactions. Progress in Water Technology 7:157-172.
27. McCarty, P.L., M. Reinhard, and B.E. Rittmann (1981) Trace organics in groundwater. Environ. Sci. Technol. 15:40-51.
28. Nelson, M.J.K., S.O. Montgomery, W.R. Mahaggey, and P.H. Pritchard (1987) Biodegradation of trichloroethylene and involvement of an aromatic biodegradative pathway. Appl. Environ. Microbiol. 53:949-954.
29. Paris, D.F., D.L. Lewis, and N.L. Wolfe (1975) Rates of degradation of malathion by bacteria isolated from aquatic systems. Environ. Sci. Technol. 9:135-138.
30. Paris, D.F., W.C. Steen, G.L. Baughman, and J.T. Barnett, Jr. (1981) Second-order model to predict microbial degradation of organic compounds in natural waters. Appl. Environ. Microbiol. 41:603-609.
31. Parsons, F., and G.B. Lage (1985) Chlorinated organics in simulated groundwater environments. J. Am. Water Works Assoc. 77(5):52-59.
32. Patel, R.N., T. Hou, A.I. Laskin, and A. Felix (1982) Microbial oxidation of hydrocarbons: Properties of a soluble methane monooxygenase from a facultative methane-utilizing organism, Methylobacterium sp. strain CRL-26. Appl. Environ. Microbiol. 44:1130-1137.
33. Stirling, D.I., and H. Dalton (1979) The fortuitous oxidation and cometabolism of various carbon compounds by whole-cell suspensions of Methylococcus capsulatus (Bath). FEMS Microbiol. Lett. 5:315-318.
34. Stucki, G., U. Krebser, and T. Leisinger (1983) Bacterial growth on 1,2-dichloroethane. Experimentia 39:1271-1273.
35. Vogel, T.M., and P.L. McCarty (1985) Biotransformation of tetrachloroethylene to trichloroethylene, dichloroethylene, vinyl chloride, and carbon dioxide under methanogenic conditions. Appl. Environ. Microbiol. 49:1080-1083.
36. Vogel, T.M., and P.L. McCarty (1987) Abiotic and biotic transformations of 1,1,1-trichloroethane under methanogenic conditions. Environ. Sci. Technol. (in press).
37. Vogel, T.M., C.S. Criddle, and P.L. McCarty (1987) Transformations of halogenated aliphatic compounds. Environ. Sci. Technol. 21:722-736.
38. Westerick, J.J., J.W. Mello, and R.F. Thomas (1984) The groundwater supply survey. J. Am. Water Works Assoc. 76(5):52-59.
39. Wilson, B.H., and M.V. White (1986) A fixed-film bioreactor to treat trichloroethylene-laden waters from interdiction wells. In Proceedings of the Sixth National Symposium and Exposition on Aquifer Restoration and Groundwater Monitoring, National Water Well Association, Columbus, Ohio.
40. Wilson, J.T., and B.H. Wilson (1985) Biotransformation of trichloroethylene in soil. Appl. Environ. Microbiol. 49:242-243.

BIOREMEDIATION OF CHEMICAL SPILLS*

Gaylen R. Brubaker and Jurgen H. Exner

International Technology Corporation
Edison, New Jersey 08818

INTRODUCTION

Although microbial processes have been used for decades to destroy industrial and municipal wastes, it has only been within the last ten years that scientists and engineers have seriously examined the use of microorganisms to treat organic contaminants that become entrapped within soils and aquifers (4). The techniques that are now emerging in this area involve the integration of chemical, microbial, geological, and engineering know-how. These methods are now being used to enhance nature's ability to biodegrade a wide variety of organic materials. In some instances, this involves modification of the microbial environment to accelerate ongoing microbial processes; while in others, preselected microbes are added that have the tolerance or enzymes to degrade the contaminants of concern. Both strategies will be necessary to overcome the many complex matrices of materials that are found currently in our environment.

This chapter will discuss two case histories that involve microbial degradation of chemical contaminants. The first involves a combination of chemical and microbial methods used to clean up a sudden release of formaldehyde in a railroad yard. The second involves the use of in situ bioreclamation to clean up a more gradual leak of gasoline at a service station. In each instance, the in situ microbial approach avoided extensive excavation of soils, roadways, and/or railroads. Both instances also involved other remediation tools in addition to microbial remediation, emphasizing the need to examine complementary and synergistic remediation techniques wherever possible.

---

*Portions of this chapter were previously reported by Sikes, McCallock, and Blackburn at the 1984 Hazardous Materials Spills Conference in Nashville, Tennessee.

SURFICIAL FORMALDEHYDE CLEANUP

An emergency response situation was created by the release of 20,000 gallons of a concentrated (35%) formaldehyde solution into drainage and irrigation ditches in a railyard in northern California. An earthen dam was created down-slope from the spill area, and a fleet of trucks was mobilized to remove runoff from the area. Unusually high rainfall over the next five days required the removal of almost three million gallons of rainwater before it was decided that dilution effects were sufficient to protect down-gradient receptors. After the rainfall subsided, a variety of options were examined to treat the residual contamination.

The area of concern included roadbed, ballast, subgrade, and drainage canals along the railroad right-of-way for a distance of about 100 feet in either direction from the spill. Although about 250 cubic yards of soils had been excavated prior to the rain, further excavation was considered unacceptable, as it would have disrupted rail traffic along the single track line. As an alternative, a novel combination of in situ chemical and microbial remediation was selected.

Laboratory studies showed that formaldehyde could be degraded rapidly with hydrogen peroxide in an alkaline solution (pH 9 to 10). Small-scale tests using soil from the site verified that the concentration of formaldehyde could be reduced from the 20,000-50,000 ppm present to 1,000 ppm within an hour.

Based on these data, the chemical oxidation of the formaldehyde-contaminated railroad roadbed and adjacent soils was initiated over an area of approximately 15,000 square feet (250 by 60 feet). The chemical oxidation was performed in two steps. First, the area was saturated with a buffered solution of sodium hydroxide and soda ash (pH 9.5). This material was applied from a 5,460-gallon vacuum truck, which had transported the solution from an off-site staging area. The solution was delivered by a portable gasoline-powered pump, at a rate of approximately 50 gpm, through a 1.5-inch fire hose fitted with an adjustable fog nozzle. Two technicians were required to perform the task: one controlled the nozzle and one assisted in moving the hose. Personal protective gear was required and included full-face respirators with organic vapor cartridges, full vinyl rain suits, boots, and gloves.

Following the saturation of the area with the alkaline solution, a 5% hydrogen peroxide solution was sprayed onto the same area in the same manner. The peroxide was delivered in a 6,000-gallon stainless steel trailer. The sequential application of reagents was required because hydrogen peroxide is unstable at high pH. If the alkaline and hydrogen peroxide solutions had been mixed offsite and then transported to the site, much of the hydrogen peroxide would have decomposed prior to arrival.

The application of alkaline solution and 5% hydrogen peroxide was alternated over two days to assure that the proper pH was maintained and that sufficient peroxide was available for oxidation of the organics. The total quantity of materials used was approximately 6,000 gallons of alkaline solution and 12,000 gallons of 5% hydrogen peroxide. The reaction was "visible," with significant effervescence occurring because of the evolution of oxygen and carbon dioxide. Samples of contaminated soil and leachate were collected periodically for formaldehyde analyses. The concentration dropped quickly from a 30,000-50,000 ppm level to approximately 5,000 ppm (four to six hours), and reached 500-1,000 ppm by the end of the second day. At this point, continued chemical treatment did not appear cost-effective, and an alternative microbial "polishing" method was initiated.

Based on the literature available on the biodegradation of formaldehyde (1,2,3), as little as 150 ppm could be toxic to unacclimated bacteria, but acclimated cultures were known to degrade up to 1,750 ppm. Based on these findings, it was decided to try in situ biological oxidation to complete the remediation of the spill site. An inoculum of specially cultured microorganisms that were acclimated for formaldehyde degradation and guaranteed to be free of human pathogens was purchased (Polybac, Inc.).

Figure 1 illustrates the spill site. A sump was excavated adjacent to the track such that runoff could be contained and pumped to an on-site biological treatment tank. A 1,600-gallon treatment tank was equipped with a compressed air generator and a coarse bubble diffuser to provide continuous agitation and aeration. Bacterial inoculum and nutrients were added to this tank and then sprayed over the contaminated area using a diaphragm pump and a series of spray nozzles. Water was pumped from the sump into the aerated tank and then sprayed onto the soil to achieve continuous treatment of the soil and water. Inoculum and nutrients were added periodically and the system was monitored regularly for pH, temperature, dissolved oxygen, ammonia, and formaldehyde.

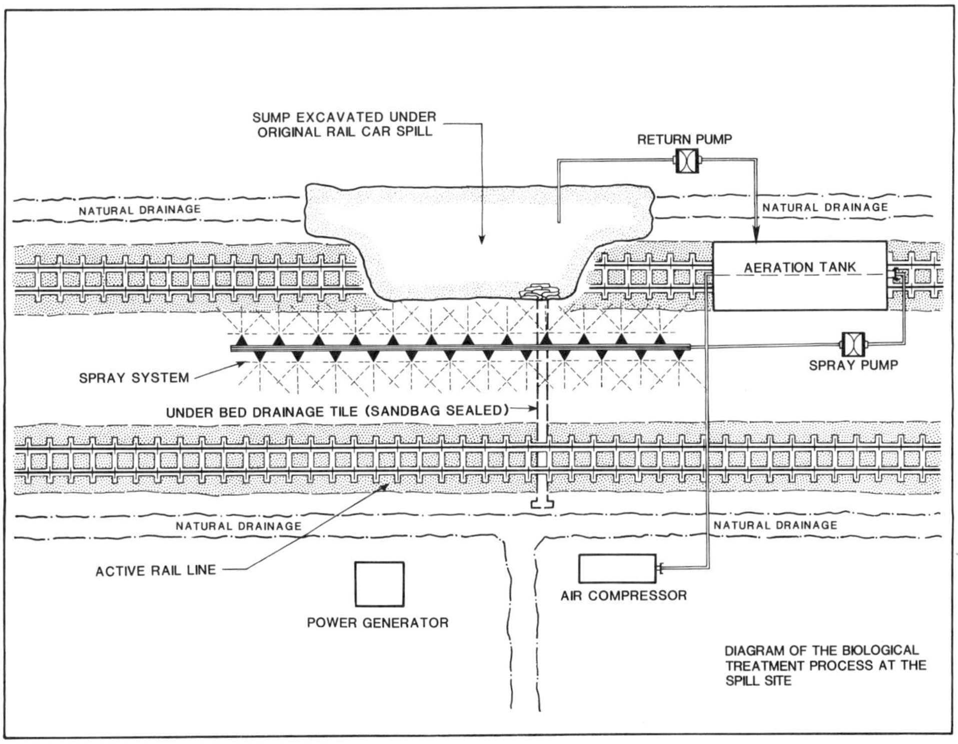

Fig. 1. Diagram of the biological treatment process at the spill site.

It rained continuously during the initial five days of treatment. This may have been partially responsible for keeping the temperature of both the sump water and the tank between 10° and 15°C. During this time, the concentration of formaldehyde in the tank remained about 600 ppm. At day 6, the rain stopped and a 1.5-million BTU steam cleaner was put into service as a source of heat for the aeration tank. The temperature in the tank was maintained at 20° to 30°C for the next 17 days, during which the concentration decreased to less than 1 ppm.

After several days at this level, a culvert below the tracks was cleaned out; it had been sealed off since early in the spill response. During this process, the concentration of formaldehyde and the biological activity in the aeration tank increased sharply and then returned to their previous low levels (Fig. 2).

Although the type of rapid response required in this kind of environmental cleanup never allows for the type of preplanning or analytical analysis desired as one assesses the project years later, this case history represents an early example of a creative on-site remediation strategy using microbial degradation.

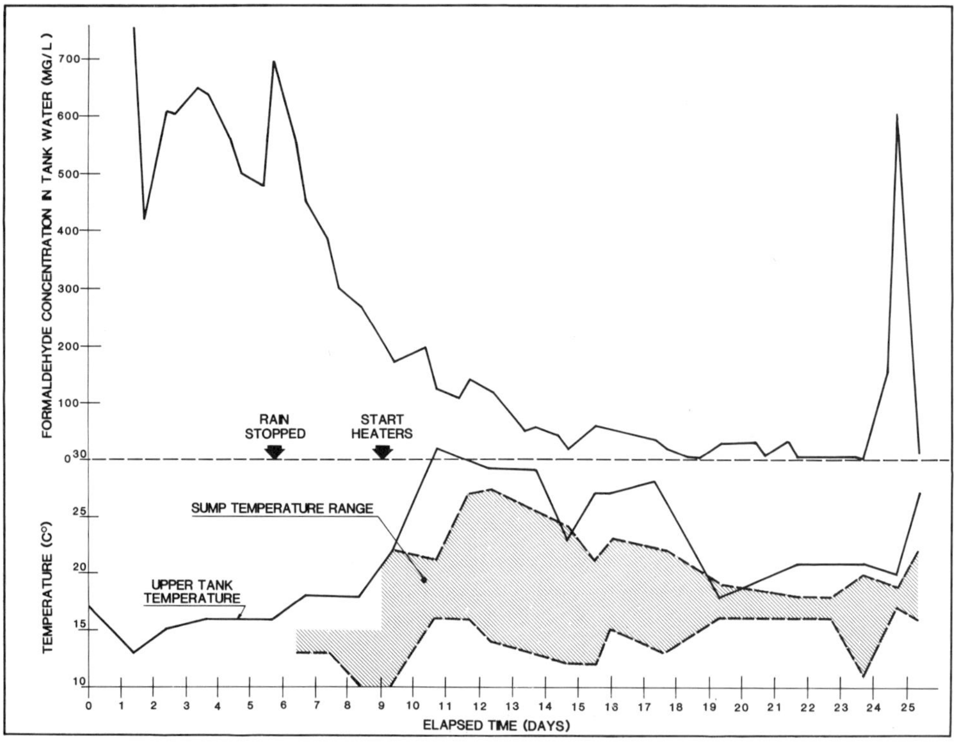

Fig. 2. Results of microbial treatment of formaldehydes at the spill site.

SUBSURFACE GASOLINE REMEDIATION AT A SERVICE STATION

The contamination of interest, located at a service station in southern California, resulted from a gasoline leak from an underground storage tank. A series of corings and monitoring well installations delineated the extent of contamination in the soil and water under and around the station. These initial observations indicated that the soil contamination was confined to a zone of 30,000 square feet covering the southeastern portion of the station and part of the adjacent intersection (Fig. 3). The vertical extent of the contamination varied, but did not extend below about 35 feet in depth.

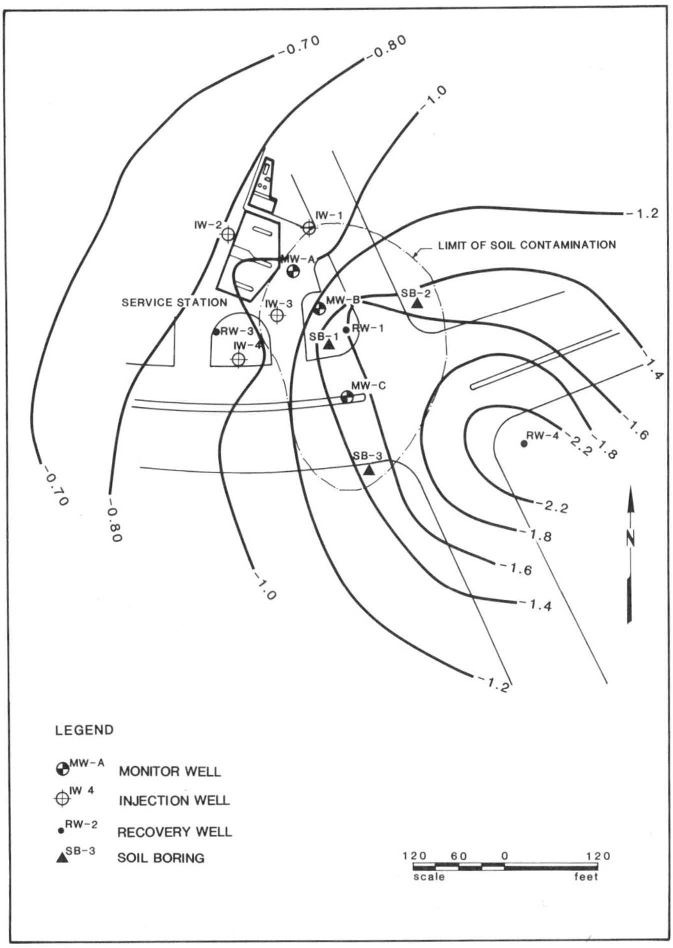

Fig. 3. Hydrogeological model of injection/recovery design for leakage from an underground storage tank at a service station. Contours show draw-down (in feet) relative to the natural gradient.

This evaluation also determined that both the water gradient and the water table vary considerably as a result of rainfall and fluctuations in the water flow of a nearby river. The piezometric gradient across the site was 1.5 to 2.0 feet per 100 feet, with the water table at approximately 15 to 20 feet below surface. However, gradients of two to three feet per 100 feet were observed when the water table was at 25 feet below the surface. The formation consists of sand, gravel, and cobble with some clay. The upper 40 feet consists of coarse, poorly sorted sand. Clays, although not abundant, were found in discontinuous layers 1 to 12 inches thick, interbedded with thin layers of silt or as rounded modules. Below 40 feet, the formation consists of gravel and cobble gravel interlaid with coarse sand. The steep gradient and highly permeable sand result in a very rapid groundwater flow, at times estimated to be more than 25 feet per day.

In order to contain the leached hydrocarbons, recovery wells were installed with an activated carbon system for discharge to a nearby irrigation canal according to regional regulatory criteria. Total reliance on this "pump-and-treat" method would have required many years of treatment, and, although the contamination zone was not extensive, excavation and transfer of the soil to a landfill would have been very expensive. Therefore, bioreclamation was considered as an option for remediating the site.

The site assessment determined that the soil and groundwater contained substantial populations of bacteria capable of degrading gasoline and that the addition of various nutrients and an oxygen source resulted in a dramatic increase in the microbial populations. Soil compatibility studies with microbial nutrient solutions indicated that the nutrients could be readily transported through the formation.* The evaluation concluded that enhanced bioreclamation was an appropriate strategy for this site.

The initial addition rates of these microbial nutrients were based on the results of laboratory studies which determined that only minimal concentrations of nitrogen and phosphorus were necessary to maximize biodegradation. The studies also demonstrated that transport of the required nutrients could be readily accomplished in the formation. The process design concluded that hydrogen peroxide should be added continuously to the injection water to maintain a 500 ppm concentration. A blend of ammonium chloride and phosphates was provided as a solid, dissolved on-site, and added to the injection water as a concentrate. The use of batch additions of this nutrient formula assured better transport of phosphate and avoided calcium and phosphate precipitation. This material was added in daily batches of 230 lb for the first four days, after which the frequency was reduced to two or three times per week.

---

*All work on this site used Restore® microbial nutrients. Restore® 375 microbial nutrient is a proprietary formulation which supplies nitrogen and phosphorus for microbial utilization while minimizing the precipitation of calcium and magnesium phosphate in the groundwater. Restore® 105 is a stabilized 35% solution of hydrogen peroxide. The natural conversion of hydrogen peroxide to oxygen and water provides an ideal method of supplying oxygen to subsurface microbes. Restore® is a registered trademark of the International Technology Corporation.

The remediation plan involved pumping water from two recovery wells located within the station property and reinjecting the water through four wells located along the up-gradient side of the contaminated zone. Microbial nutrients were to be added to the recirculated water prior to distribution into the injection wells. A third recovery well located down-gradient of the contaminated zone recovered water for treatment with activated carbon prior to discharge into the nearby irrigation canal. This groundwater flow system was based upon a computer model and was designed to sweep the entire contaminated area with the nutrient solutions. Figure 3 shows the locations of these injection and recovery wells and the cone of depression that these wells create in the water table.

Immediately prior to system start-up and throughout project operation, water at various observation wells was analyzed for nutrient and dissolved oxygen concentrations. Results from these analyses were used to regulate the rates of addition of microbial nutrients to the system, so that optimal subsurface conditions for microbial biodegradation could be maintained. Also, the results of the analysis were used to adjust pump flow rates to ensure that the microbial nutrients were being adequately distributed throughout the contaminated zone. In addition, microbial analyses were conducted on a regular basis to monitor and assess the viability of the subsurface microbial populations throughout the period of project operation. The results of these analyses confirmed an increase in the populations of hydrocarbon-degrading bacteria within the contaminated areas of the site.

Progress of the remediation was documented by periodic analysis of both water (from monitoring wells) and soil (from split-spoon coring samples) for total and specific contaminant concentrations. A summary of the results of these analyses is shown in Tab. 1 and 2. These tables show the changes in the range and average concentrations of volatile fuel hydrocarbons (determined from EPA methods 8015, 8020, and 5030) in the soil and water from the contaminated zone with respect to time. Changes in the concentrations of volatile fuel hydrocarbons are indicative of the remediation progress at the site, because they represent the major component of the contaminant and include many specific compounds of regulatory interest. The tables show a substantial decline in the average concentration of the contamination in both soil and water samples taken from the site.

Tab. 1. Soil contamination during bioreclamation.

Coring location	Days following process initiation					
	0		100		200	
	Range	Average	Range	Average	Range	Average
1	0.6- 51.3	13.0	0-14.0	2.8	0-5.0	1.0
2	0.6- 141.0	25.0	0	0	0	0
3	46.0-1,287.0	245.0	0-16.0	4.0	0-5.0	0.8

Note: All concentrations as ppm volatile fuel hydrocarbons.

Tab. 2. Groundwater contamination during bioreclamation.

Well location	Days following process initiation			
	0	30	100	200
A	3.0	35.0	5.2	0.1
B	24.0	44.0	28.0	7.1
C	2.3	40.0	11.0	4.2

Note: All concentrations as ppm volatile fuel hydrocarbons.

Table 1 shows the change with respect to time in the concentration of volatile fuel hydrocarbons present in the soil at three different locations (Fig. 3) within the contaminant plume. At each of these locations, three separate cores were made and samples from each core were taken at five-foot intervals between the depths of 15 and 30 feet. The data for each location and time interval are therefore derived from nine individual soil samples. The precipitous decline in the soil contamination resulted from the stimulation of the indigenous bacteria that utilize the hydrocarbons as a food source.

Changes in the concentration of volatile fuel hydrocarbons in groundwater samples taken from three monitoring wells at the site are shown in Tab. 2. Data from these wells (of the many monitoring wells located on and around the station property) are shown because the wells are located in the most contaminated region and are positioned directly over or hydraulically down-gradient of the area where the gasoline leak occurred. The initial increases in contamination levels of the groundwater were the result of the stimulation of the naturally occurring bacteria. These bacteria frequently produce biosurfactants that increase the release of petroleum hydrocarbons from the soil. Because the groundwater contamination is the result of the dissolution of hydrocarbons sorbed onto the soil, overcoming its persistence is dependent upon the remediation of the contaminated soil. Hence, the ultimate decline in groundwater contamination is further evidence of the remediation of the contaminated soil.

Unfortunately, in December of 1986, a gasoline spill from a neighboring service station resulted in temporary shutdown of the bioreclamation system. It was feared that the pumping systems that were used to transport nutrients into the original contamination zone might draw the new contamination onto the site.

SUMMARY

It is evident from the data collected to date that substantial progress was made in the remediation of the site prior to the shutdown of the bioreclamation system. Extrapolation of the data suggests that completion of the project was imminent. Further remediation at the site, including the possibility of expanding the original area treated with in situ bioreclamation, is pending further definition of the new sources of contamination. The success of the two projects described here demonstrates the efficacy and potential of enhanced bioreclamation in remediating contamination problems both in soils and in groundwater.

REFERENCES

1. Gellman, Isaih, and H. Heukelekian (1950) Biological oxidation of formaldehyde. Sewage and Industrial Wastes 22:1321-1325.
2. Marion, C.V., and G.W. Malaney (1963) The oxidation of aliphatic compounds by *Alcaligenes faecalis*. J. Water Poll. Control Fed. 35:1269-1284.
3. McKinney, Ross E., and John S. Jeris (1955) Metabolism of low molecular weight alcohols by activated sludge. Sewage and Industrial Wastes 27:728-735.
4. Wilson, J.T., L.E. Leach, M. Henson, and J.N. Jones (1986) In situ biorestoration as a ground water remediation technique. Ground Water Monitoring Rev. Fall, pp. 56-64.

MICROBIOLOGICAL CLEANUP OF GROUNDWATER CONTAMINATED BY PENTACHLOROPHENOL

Thomas D. Frick,[1] Ronald L. Crawford,[1]*
Michael Martinson,[2] Tom Chresand,[2]
and George Bateson[2]

[1]University of Minnesota
Gray Freshwater Biological Institute
Navarre, Minnesota 55392

[2]BioTrol, Inc.
Chaska, Minnesota 55318

INTRODUCTION

Background

Pentachlorophenol (PCP) was first produced during the 1930s as a biocide for use by the wood-preserving industry. Since then, it also has been used as a fungicide, bactericide, herbicide, molluscicide, algicide, and insecticide. Its primary use today, however, is still as a wood-preserving chemical. About 80% of the approximately 2,000 tons of PCP that are produced annually in the United States is used for commercial wood treatment (3). The widespread use of PCP and PCP-treated products has led to its broad distribution in the environment. Groundwater and drinking water contamination by PCP is becoming increasingly common (1). Environmental contamination with PCP is of concern both as an ecological threat and as a public health menace.

Pentachlorophenol contamination may be particularly severe in the vicinity of wood-treating plants, where spills and contaminated wastes that were disposed of on-site may leach into groundwater. There are sites throughout the United States where groundwater contamination is severe enough to require remediation. The most common remedial treatment has been to pump the water to the surface and pass it through a bed of activated charcoal. Pentachlorophenol is not destroyed by this process, but simply transferred to the charcoal. The PCP-laden charcoal must then be

---

*Present address: Department of Bacteriology and Biochemistry, University of Idaho, Moscow, Idaho 83843.

handled as a hazardous waste. Alternative technologies that actually destroy PCP would be highly advantageous.

Pentachlorophenol Biodegradation

Although PCP is highly toxic to most organisms, it is possible to isolate microorganisms that can utilize PCP as a sole source of carbon and energy (2,5,11,14,15,18,19). Such cultures have potential for use as chlorinated phenol detoxification agents. Work by Edgehill and Finn (5) as well as work in our laboratory (9) suggests that inoculation of PCP-contaminated soils or waters with PCP-degrading bacteria may be a feasible method for detoxifying such environments.

There has been considerable interest in developing microbiological methods to treat wastewater contaminated with PCP and other chlorinated phenols. Most of the effort has been aimed at adapting conventional wastewater treatment systems using aerated, activated sludge (4,10,12,17, 20) or anaerobic digesters (7,8). Such systems are capable of good results under optimal conditions; however, they are quite sensitive to moderate changes in hydraulic load or the level of PCP in the influent. The wastewater treated in these types of systems typically is high in biological oxygen demand (1,000 mg/l) and relatively low in chlorinated aromatic compounds (1 to 10 mg/l). For example, this is the type of wastewater produced by kraft pulp mills that use chlorine bleach to decolorize wood pulps and by sawmill operations that spray cut timber with dilute PCP solutions to prevent surface blemishing by molds. Surface waters and groundwaters that have been contaminated with PCP as a result of the wood-preserving industry typically are an order of magnitude lower in biological oxygen demand and an order of magnitude higher in PCP. This type of water is difficult to treat using conventional wastewater-treatment technologies. We are aware of only one report where an activated sludge system was investigated to test treatability of PCP-contaminated groundwater (16). The system was not very effective, and only 36% of the PCP was removed from the water.

In previous work, we demonstrated that inoculation of PCP-contaminated natural waters with cells of a PCP-degrading *Flavobacterium* was an effective method for decontaminating such waters (9). Numerous types of water were treated, including river water, lake water, and groundwater. Decontamination was most effective between 15° and 30°C, and between pH 7.5 and 9.0. Inoculation of water with as few as 10^5 cells/ml resulted in effective PCP removal. Pentachlorophenol concentrations were reduced to undetectable levels from initial levels ranging from 10 ppb to 100 ppm, usually within 48 hr after inoculation.

In other studies, we observed that microbial degradation of PCP in outdoor experimental streams was accomplished largely by microorganisms attached to surfaces, particularly rock surfaces (11). This led us to establish a small bench-top bioreactor containing stream rocks that had been colonized by a natural microflora of PCP-degrading organisms. After the reactor had been equilibrated for several weeks, PCP solutions containing 100 to 300 ppm of PCP could be recirculated through the column, with complete degradation of PCP to carbon dioxide and inorganic chloride. This bioreactor was very tolerant to abrupt changes in the influent PCP concentration or the hydraulic load; it was maintained for many months without significant decline in PCP removal capacity.

Purpose and Scope of Study

This study was undertaken to determine the feasibility of designing a large bioreactor, modeled after our small bench-top bioreactor, to decontaminate groundwater at a site where an aquifer had been contaminated with PCP. The PCP-degrading microorganisms in the original bioreactor had been immobilized in a biofilm covering the rock surfaces. Use of such an immobilized biofilm allowed the flow rate through the column to be greater than the growth rate of the microorganisms in the column, without causing washout of the microorganisms.

The study was carried out in four phases. In the first phase we examined a number of materials for their suitability as support matrices for immobilizing the biofilm. In the second phase, the active biofilm (a microbial consortium capable of degrading PCP) was transferred from our bench-top bioreactor to a large volume of support material. The bioreactor vessel was designed and constructed during this phase. During the third phase of the study the bioreactor was assembled in our pilot plant, and the biofilm was developed in situ. In the final phase of the study, the bioreactor was installed in the field and tested with "real world" PCP-contaminated groundwater.

MATERIALS AND METHODS

Selection of Support Material

Four materials [1-inch stones, 1/4-inch glass beads (Aldrich, Milwaukee, Wisconsin), 3/4-inch ceramic saddles (Norton, Akron, Ohio), and 1-inch polypropylene, tri-pack spheres (Jaeger, Houston, Texas)] were investigated as support matrices for the bioreactor biofilm. A bed volume of 0.5 to 1.5 liters of each material was packed into separate glass columns equipped with airstones at the bottom. The columns were connected so that each maintained its liquid level above the packing material, and the effluent of the first column became the influent of the second column, etc. The first column was our bench-top bioreactor containing the 1-inch stone packing. A thick biofilm had already developed on these stones. Material sloughed by this column became the inoculum for the other columns.

Fifty liters of medium containing 5 mM potassium phosphate, 5.9 mM sodium nitrate, 0.45 mM magnesium sulfate, 20 mM ferrous ammonium sulfate, and 100 ppm PCP at pH 7.4 were recirculated through the columns at a flow rate of 3 liters per hour. Aeration was started so that each column received a steady stream of bubbles. The pH was adjusted with NaOH or HCl, and the PCP concentration restored to 100 ppm on a daily basis. The biofilms were allowed to develop for six weeks. Colonization of the different packing materials during this period was judged by determining the difference in PCP concentration at the influent and effluent of each column. The packings also were judged as to how well they allowed air bubbles to be distributed through the bioreactor volume.

Bioreactor Vessel Design

The major design considerations for the bioreactor vessel were: (a) the vessel should be large enough to minimize wall effects and to treat reasonably large volumes of groundwater so that the data from this study would be useful for scale-up calculations; (b) the vessel should utilize modular construction to ease assembly and disassembly; and (c) the vessel should have many sampling ports so that distributions of PCP, dissolved oxygen levels, and pH variations throughout the column could be determined.

We chose to make a 200-liter reactor vessel. The design details are shown in Fig. 1. The vessel was fabricated from clear acrylic so that biofilm development and air distribution could be monitored visually. Sampling ports were made from 1/4-inch stainless-steel tubing which extended 3 to 6 inches into the interior of the packing. The tubes penetrated the wall of the bioreactor through nylon Swagelok fittings and were capped with small shut-off valves.

Figure 2 shows the bioreactor installed in our pilot plant. Four 55-gallon polyethylene drums were used as influent and effluent reservoirs. Later in the study, when the bioreactor was more active, these drums were used only for feed, and two 400-gallon tanks were used to collect the effluent. A variable-speed Moyno pump provided metered flow through the bioreactor.

A second bioreactor, with a volume of approximately 15 liters, also was fabricated. This second bioreactor was to be used in the laboratory to study the effects of nutrients and temperature on PCP biodegradation while the larger bioreactor was being tested in the field. The smaller

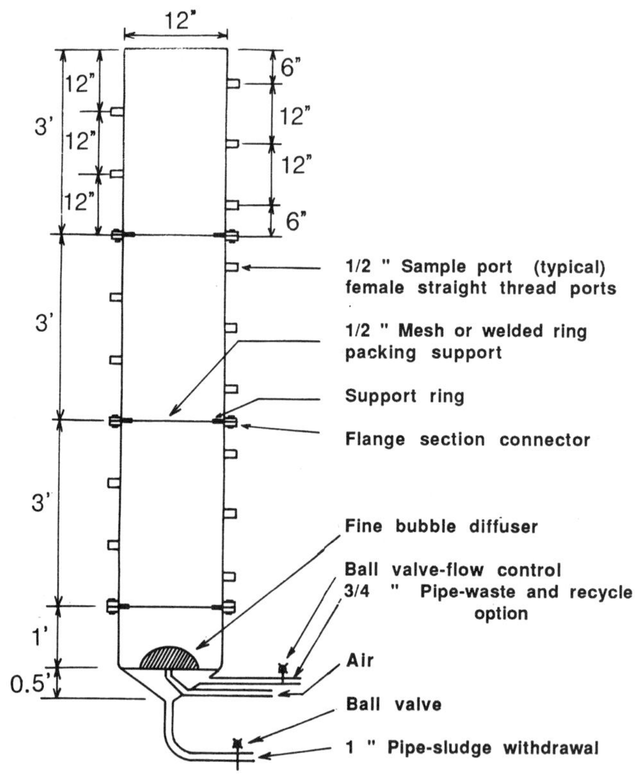

Fig. 1. Schematic diagram of the 200-liter bioreactor.

bioreactor vessel, shown in Fig. 3, included many of the design features of the larger column.

Production of Biofilm

The active biofilm was transferred from the stones in our small bench-top bioreactor to approximately 150 liters of ceramic saddles. The ceramic saddles were placed in a 55-gallon polyethylene drum and flooded with a medium containing 5 mM potassium phosphate, 5.9 mM sodium nitrate, 0.45 mM magnesium sulfate, 20 mM ferrous sulfate, 100 ppm monosodium glutamate, and 20 ppm PCP at pH 7.4. The PCP used throughout this study was technical grade, similar to that used in the wood-preserving industry. It contained 93.7% PCP, 5.8% tetrachlorophenols, and traces of lesser chlorinated phenols and nonpolar chlorinated aromatic compounds. The medium was circulated through the bed via a submersible pump which pumped water from the bottom of the drum to about four inches above the liquid surface. Air entrained at the pump discharge provided aeration.

Medium pH and PCP concentrations were monitored daily. The pH was adjusted as necessary. When the PCP concentration had decreased to nearly undetectable levels, the system was resupplied with original levels of PCP and glutamate. In succeeding feedings, sodium acetate was used instead of monosodium glutamate. The biofilm was cultured this way for three weeks.

Fig. 2. Large bioreactor as installed in the pilot plant.

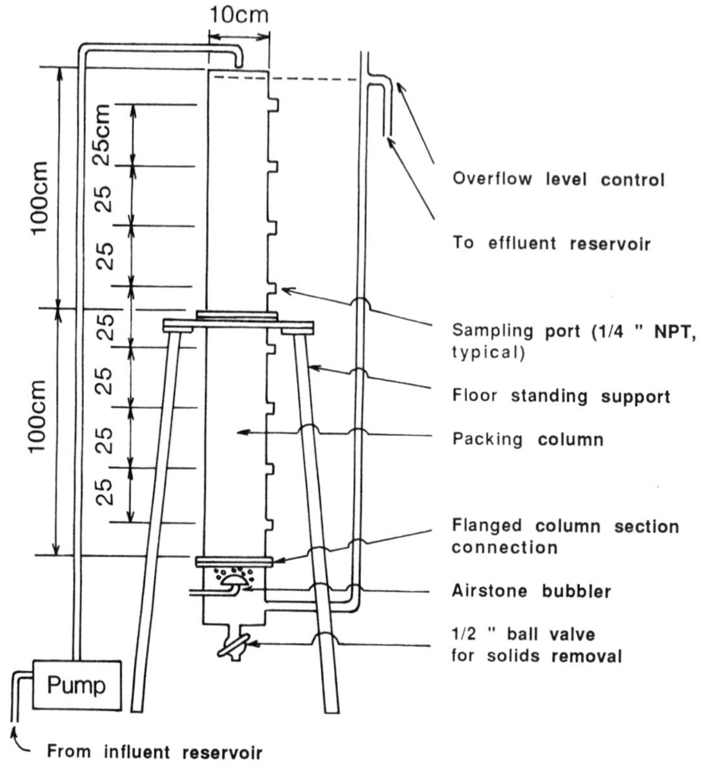

Fig. 3. Schematic diagram of the small laboratory bioreactor.

Both bioreactor vessels were packed with these precolonized saddles. The procedures used and results obtained with the two bioreactors have been quite similar, so the following discussion applies only to the 200-liter bioreactor. The bioreactor vessel was packed by filling each section three-quarters full with a mixture of one-half tap water and one-half liquid used for the batch precolonization. Saddles were added and allowed to settle through this medium into the column. Fresh saddles were mixed with precolonized saddles in the ratio of 1:3 in order to fill the reactor volume. A reservoir equal to the void volume of the column was prepared and contained 20 ppm PCP, 100 ppm sodium acetate, and ammonium nitrate and potassium phosphate in the ratio C/N/P equal to 100/5/0.5. Air was supplied at 5 liters per minute, which was just sufficient to give good bubble distribution throughout the packing.

The biofilm was matured in situ by recirculating medium through the column. The original pumping rate exchanged the medium every four hours. Pentachlorophenol biodegradation was monitored daily; when the PCP was consumed, fresh PCP and acetate were added. The bioreactors initially required three days to consume all the PCP in the reservoir, but after ten days of operation only one day was required. As the PCP biodegradation rate increased, the amount of PCP in the reservoir was increased to 50 ppm and then 100 ppm. The acetate concentration was not increased; so, at this stage of development, the bioreactors were being supplied with equal amounts of carbon as PCP and acetate. As the PCP concentration was increased, the pumping rate also was increased to the point where the bioreactor volume was exchanged approximately every two hours. During 20 days of in situ development, the PCP biodegradation rate increased to the point where virtually all of the PCP (100 ppm) was consumed in six hours. At this point the bioreactor was changed to single pass (continuous feed) operation.

Single pass operation was started using a medium containing 60 ppm PCP and 60 ppm sodium acetate plus ammonium nitrate and potassium phosphate sufficient to make the C/N/P ratio 100/5/0.5. The initial flow rate was 1.5 gallons per hour over a 30-day period. This progression was not linear because the bioreactor suffered two serious setbacks. The first problem was caused by competition from a microorganism that started growing on the acetate in the feed reservoirs. This problem was overcome by discontinuing acetate in the feed. The second problem resulted from a leak which developed in one of the welded plastic joints. Repairing this leak required the bioreactor to be drained and reflooded several times. This process caused a significant amount of biomass to be dislodged from the column packing. After each of these episodes, the flow through the column was reduced by one-half to allow the biofilm to recover.

The effluent generated during this phase of the study was collected in two 400-gallon tanks. The water in each tank was analyzed for PCP concentration, pH, chemical oxygen demand, and total suspended solids before being discharged into the sewer. Most of the time all of the PCP was consumed in the bioreactor; however, during several episodes, 10 to 20 ppm of PCP carried into the effluent for brief periods. On these occasions the effluent reservoirs were inoculated with pure cultures of our PCP-degrading Flavobacterium (9) and aerated until the PCP was removed. All of the water discharged into the sewer contained less than 0.1 ppm PCP (13).

Field Installation

The field-trial phase of this study was carried out with the cooperation of a wood-treating company at their pole-treating facility in Minnesota. This site has been used for wood-preserving activities since the 1930s, and has extensive soil and groundwater contamination. The site has been under remediation for several years. The site owner has excavated the contaminated soil and mapped the groundwater contamination. The contaminated groundwater is a large pool of slowly moving water. The highest PCP concentrations run along a line roughly southwest from the former disposal area. Pentachlorophenol concentrations in groundwater samples from this area range between 5 ppm and 100 ppm. A purge well was installed near the center of the property and used as the source of contaminated groundwater during this study. Details of the well are shown in Fig. 4.

Preliminary field tests of the bioreactor used contaminated process water which the site owner had accumulated in a large storage tank. This water contained the same types of contamination as the groundwater and was employed to acclimate the bioreactor to the host of compounds present in the "real world" situation. Table 1 shows a comparison of this process water with groundwater from a 1985 purge well test.

Before the bioreactor was moved to its field installation, it was acclimated to the contaminated water for ten days. The normal feed water was

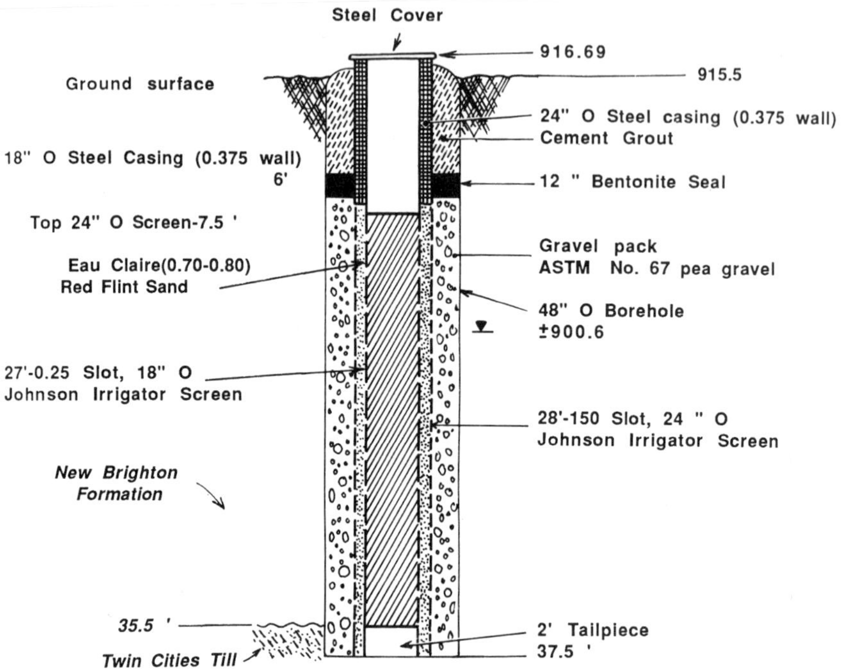

Fig. 4. Details of purge well design at the field site.

augmented with 12.5% contaminated water for five days and then raised to 25% contaminated water for five days. The bioreactor was drained and disassembled during the move to its field installation. When it was reassembled, the middle section was left out, decreasing the volume by one-third. Construction of a shelter at the field installation was greatly simplified by shortening the length of the bioreactor. Our data to this point indicated that most of the PCP-degrading microorganisms were in the top section of the bioreactor, so we did not sacrifice much activity by shortening the bioreactor.

The equipment used during this phase of the study is diagrammed in Fig. 5A. Water was recirculated from the 5,000-gallon reservoir for four weeks, during which time we demonstrated the effectiveness of the bioreactor at removing PCP from water. With these data, we obtained permission from the Twin Cities Metropolitan Waste Control Commission to switch the influent to groundwater and to discharge the effluent directly into the sewer. Figure 5B shows the configuration of the equipment during this final phase of the field study.

Analytical Methods

Pentachlorophenol was assayed either spectrophotometrically or by high performance liquid chromatography (HPLC). In the initial phases of the study, PCP was measured spectrophotometrically. Ultraviolet-visible spectra were recorded on a Hewlett Packard Model 8451 Diode Array Spectrophotometer. The machine was programmed to measure PCP concentration from the absorbance at 329 nm using a baseline constructed between

Tab. 1. Characteristics of contaminated surface and well water at the field demonstration site.

	Surface water	Well water*
pH	6.8	7.3
Chemical oxygen demand (COD)	410	200 ppm
Total organic carbon (TOC)	326	nd† ppm
Dissolved solids	nd	740 ppm
Oil and grease	66	9 ppm
Alkalinity as $CaCO_3$	nd	450 ppm
Pentachlorophenol	77	36 ppm

*Fall, 1985 test pumping.
†Not determined.

276 nm and 360 nm, as shown in Fig. 6. Accurate PCP concentrations could be determined in slightly turbid solutions and in the presence of small amounts of other UV-absorbing material. Later in the study, when samples contained larger amounts of UV-absorbing material, PCP concentrations were determined by chromatography on a Hewlett Packard Model 1084b HPLC. Samples were filtered through a 0.8-micron membrane filter and separated on a C-18 reverse phase column (Brownlee, Santa Clara, California) with an isocratic eluant of 60% acetonitrile and 40% 7 mM phosphoric acid. These conditions resulted in baseline separation of PCP

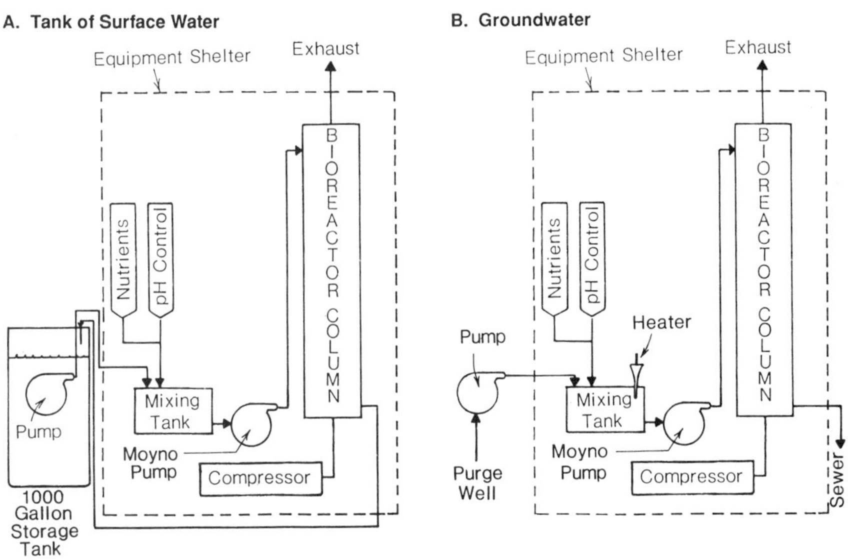

Fig. 5. Schematic diagram of the field installation of the large bioreactor.

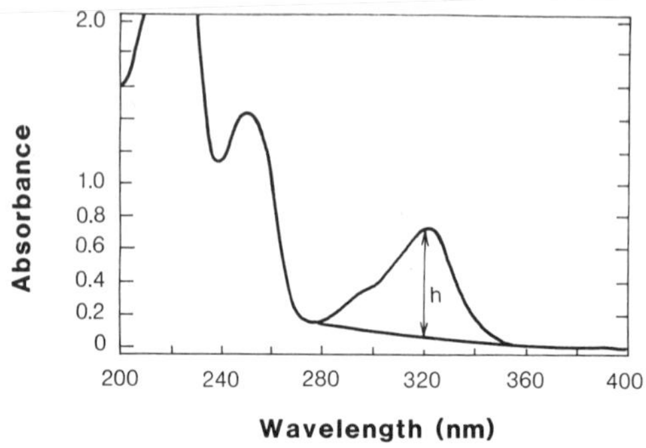

Fig. 6. Ultraviolet-visible spectrum of pentachlorophenol. Spectrophotometric determinations of PCP concentrations were made from the calculated peak height (h) rather than from the total absorbance at 320 nm.

from other components in the samples. During the early phases of the study, the spectrophotometric assay was checked with the HPLC assay. The two methods agreed within 5% of each other.

Polynuclear aromatic hydocarbons were assayed by HPLC using EPA method 610, with analysis done on a C-8 column using 40% acetonitrile in water. Individual compounds were detected using a fluorescence or UV detector.

Dissolved oxygen as ppm was measured with a Cole Parmer Model 5513 dissolved oxygen meter. A special measuring cell was constructed so that a sample could be drawn from one of the sampling ports directly into the measuring chamber without being exposed to the atmosphere. The first portion of the sample was allowed to fill and overflow the measuring chamber before the dissolved oxygen was measured. Ammonium nitrogen and orthophosphate were determined colorimetrically by the phenate method and ascorbic acid method, respectively (6). Chemical oxygen demand (COD) was determined by titration after refluxing with dichromate (6). Suspended solids (dry weight) were measured as that material which was held on a 0.45-micron porosity, polycarbonate membrane filter. A portable meter and glass double-junction combination electrode were used to measure pH.

RESULTS AND DISCUSSION

Production and Activity of the Biofilm

Of the four materials investigated as support matrices for developing a PCP-degrading biofilm, the ceramic saddles were most suitable. They were colonized faster than any of the other materials. After only two weeks of colonization, they had developed significant PCP biodegradation activity and were approaching the activity of the well-developed biofilm in our stone-filled bioreactor. The saddles also had excellent flow and air distribution characteristics. Our bioreactors were, therefore, designed to use ceramic saddles as packing.

The techniques of batch precolonization and in situ growth were successful in transferring the PCP-degrading microbial consortium from our small bench-top bioreactor to our 15- and 200-liter bioreactors. A second carbon source was included in the media during this phase of the study, because PCP provides very little carbon and biofilm growth on PCP alone would be very slow. Acetate was used because it is similar to carbon sources that might be found in the field. Water contaminated by wood-preserving operations contains a wide variety of organic compounds of which oils are a major constituent. Oils typically are biodegraded via fatty acid metabolic pathways that produce acetate as an intermediate breakdown product. It is likely, therefore, that acetate will be present in water contaminated by wood-preserving operations.

Adding acetate to the medium benefited the precolonization and initial in situ growth phases; but during the later phase of in situ growth, it caused a serious problem. The bioreactors were in the single-pass, continuous-feed stage of development, and were consuming all of the PCP in the medium, when their performances suddenly deteriorated and 20 to 40% of the PCP was carried through into the effluent. This occurred simultaneously in both bioreactors. A possible cause could have been residual chlorine in the tapwater, or some similar phenomenon. Fresh media were being prepared daily, and this might account for the simultaneous occurrence of the problem in both bioreactors. The flow through each system was reduced by one-half, and the systems were observed for three days to see if they would recover, but they did not.

We noted that a bacterial film had formed on the walls of the feed reservoirs. The PCP concentrations in the reservoirs did not decrease, which indicated that the acetate utilizers were depleting the medium of a nutrient necessary for the PCP-degrading microorganisms. In order to test this hypothesis, the reservoir for the 200-liter bioreactor was cleaned and fresh medium was prepared without sodium acetate. The 15-liter bioreactor was used as a control and operated as before with sodium acetate. Over a period of six days, the 200-liter bioreactor recovered PCP-degrading activity, but the 15-liter bioreactor did not. The nitrogen and phosphorus nutrients in the feed to the 15-liter bioreactor were doubled and then tripled, with no improvement in activity. However, when the acetate was removed from the feed to the 15-liter bioreactor, the reactor recovered activity, and in three days it was consuming all of the PCP in the feed.

Apparently the microorganisms growing on the acetate were depleting an essential nutrient from the medium, or they were secreting a compound antagonistic to the PCP-degrading microorganisms. We had anticipated that including an easily usable carbon source like acetate might cause problems with maintaining the PCP-degrading consortium. However, we expected the problem earlier, when the PCP concentrations were lower, rather than later. It was probably a matter of chance when an organism was able to become established and cause problems in biofilm development. Had this happened earlier, in the precolonization phase, adding acetate to the medium could have slowed rather than accelerated growth of the PCP-degrading biofilm.

Another phenomenon we observed also may be linked to the inclusion of sodium acetate in the medium. We anticipated that maintaining alkalinity in the bioreactors might be a problem, since PCP biodegradation releases hydrochloric acid:

$$C_6Cl_5O^-Na^+ + 4.5O_2 + 2H_2O \rightarrow 6CO_2 + NaCl + 4HCl$$

Groundwater in east-central Minnesota is quite hard; so, in the field study phase, the water had adequate alkalinity to neutralize this acid. In the laboratory, the softened water had hardness that was only marginally sufficient. Instead of acidifying the medium, however, the pH was raised one pH unit by passage through the bioreactor. This phenomenon occurred in all phases of development of the bioreactors. During the precolonization and early in situ colonization, we attributed this excess alkalinity to biodegradation of the sodium acetate, which releases sodium hydroxide:

$$CH_3CO_2^-Na^+ + 2O_2 \rightarrow 2CO_2 + H_2O + NaOH$$

While acetate was present at a substantial molar excess to PCP, the pH should and did rise. When the PCP concentration was increased to the point at which equal carbon was supplied by PCP and acetate, biodegradation of both compounds should result in a 50% molar excess of acid. The pH of the effluent from the bioreactors, however, did not decrease. Experiments with fresh saddles showed that they did not leach alkalinity into distilled water. It is possible that biological growth caused alkalinity to be released from the saddles, or that excess alkali from early growth on acetate may have been immobilized on the column by ion exchange. If this were the case, the alkalinity would start to drop as this excess was consumed.

Field Operations

After one week of operation in the field without sodium acetate, the alkalinity in the effluent of the 200-liter bioreactor started to decline. After the second week of operation, the pH of the effluent was reduced slightly by passage through the bioreactor. This is the behavior that we expected. The 15-liter bioreactor, however, had handled a proportionally similar flow and was still producing an alkaline effluent without sodium acetate supplementation.

The pH rise within the bioreactor did not appear to harm the development of PCP-degrading microorganisms. The rate of PCP consumption increased approximately 1,000-fold during the precolonization and early in situ growth phases. Only a portion of this activity, however, was immobilized in the biofilm. The remainder was free in the recirculating medium. A series of experiments comparing the activity in aliquots of reactor liquid with equal volume samples of saddles indicated that 30 to 50% of the activity was immobilized in the biofilm.

Switching the bioreactors from recirculating to single-pass, continuous-feed operation quickly flushed any nonimmobilized PCP-degrading microorganisms out of the system and allowed unambiguous observation of PCP biodegradation by microorganisms immobilized in the biofilm. The effectiveness of the biofilm at removing PCP can be seen clearly in the UV-visible spectra shown in Fig. 7. Note the very rapid decrease in the ab-

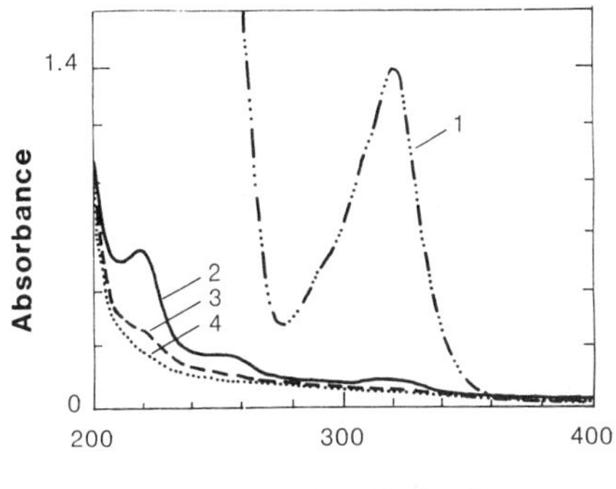

Fig. 7. Ultraviolet-visible spectra showing PCP biodegradation by the large bioreactor. 1, Influent PCP = 55 ppm; 2, first sampling port (6 inches depth), PCP = 1.5 ppm; 3, second sampling port (12 inches depth), PCP = 0.2 ppm; 4, third sampling port (18 inches depth), PCP = <0.1 ppm.

sorbance band at 320 nm. This absorbance band is characteristic of PCP, and its decrease indicates the disappearance of PCP. These spectra also show the loss of absorbance in the 240 to 280 nm region. Absorbance in this region is characteristic of all aromatic compounds. The lack of absorbance in this region indicates that the PCP was completely degraded and not transformed into other chlorinated aromatic compounds. It also indicates that aromatic impurities in the technical-grade PCP (primarily tetrachlorophenols) also were degraded in the bioreactor. The spectra in Fig. 7 also demonstrate how quickly the PCP was degraded in the bioreactor. These data were recorded near the end of the in situ development phase, but are characteristic of data throughout this phase.

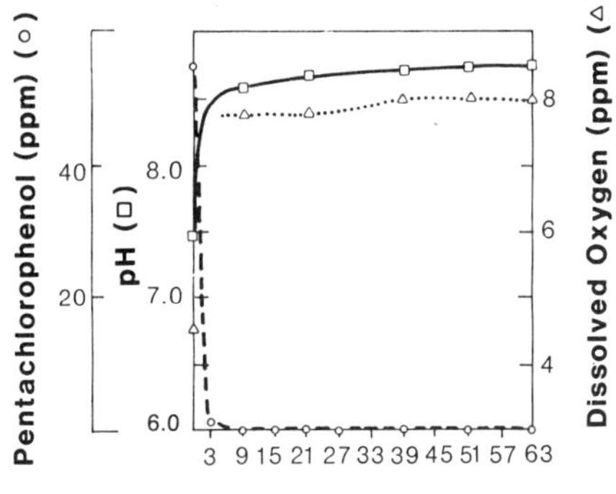

Fig. 8. Water chemistry profile within the large bioreactor (pH and dissolved oxygen).

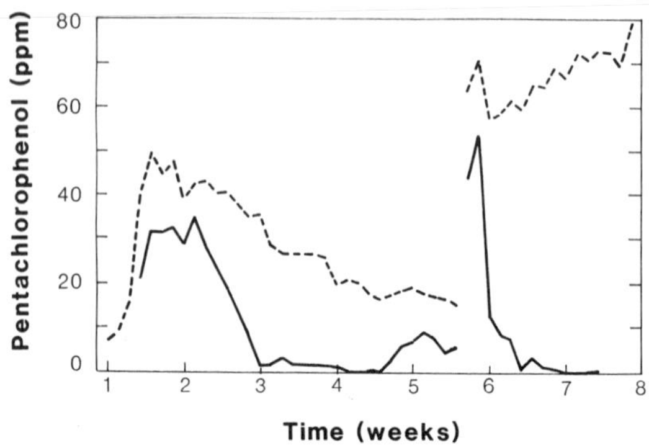

Fig. 9. Pentachlorophenol concentrations during the field study. Influent, -----; effluent, ——.

Virtually all of the PCP was consumed in the upper 10% of the 200-liter bioreactor. This is shown in the water chemistry profiles depicted in Fig. 8. These data show the rapid loss of PCP and the rapid increase in pH as the media passed through the bioreactor. As discussed previously, this increase continued even after the most likely source of alkalinity (sodium acetate) was discontinued in the feed. The air flow of 5 liters per minute, which was originally chosen to give good bubble distribution, was sufficient to maintain nearly saturating dissolved oxygen.

The large bioreactor activity suffered a setback during this phase of development because of competition from acetate-degrading microorganisms. The water chemistry profiles during this period were similar to those in Fig. 8, except that the PCP concentration dropped from 55 ppm to 10-20 ppm, and then leveled off through the rest of the column. This indicated that the lower part of the column was not being heavily colonized by PCP-degrading microorganisms. However, a background level of PCP-degrading colonization remained in the lower sections of the bioreactor. This was demonstrated when the bioreactor was disassembled and moved to the field. Only the top and bottom sections of the column were used in the field installation; the middle section was left in the laboratory. When this middle section was supplied with PCP, it degraded the PCP at the rate of 0.3 mg PCP consumed/hr/liter of reactor volume.

Further development of the biofilm was hampered by our inability to provide significantly higher flow rates to the large bioreactor in the pilot plant. This bioreactor, at the end of the in situ stage of development, consumed 6 mg PCP/hr/liter of reactor volume. Most of the PCP was consumed in the upper 10% of the bioreactor. The activity in this upper portion of the bioreactor was 60 mg PCP consumed/hr/liter. The upper portion of the 15-liter bioreactor achieved an activity of 90 mg PCP consumed/hr/liter; however, it too needed significantly higher flow to reach maturity.

Higher flow was provided by moving the 200-liter bioreactor to its field installation. Initially, a tank containing 5,000 gallons of contaminated water was used as a model for groundwater. This water was recirculated for four and one-half weeks before the system was changed over to treat actual contaminated groundwater. Data showing the influent and effluent

PCP concentrations during this phase of the study are shown in Fig. 9. The bioreactor was acclimated over a period of three days by increasing the concentration of the new water supply. Under these conditions, the bioreactor required about ten days to achieve optimal activity. Both moving the bioreactor and substitution of the new water supply were reversible shocks to the system. We discovered that, in addition to these problems, the nutrient-addition pump had malfunctioned, so that no nitrogen or phosphorus had been added to the water during the first five days of operation.

Once these reinstallation difficulties were corrected, the activity in the bioreactor increased until it degraded more than 95% of the influent PCP. When the bioreactor had achieved this level of activity, the feed rate was increased gradually from the initial rate of 19 gallons per hour to 30 gallons per hour. Visual inspection of the biofilm documented that a dark-brown biomass quickly spread throughout the column.

In addition to consuming PCP, a number of unidentified peaks in the HPLC chromatograms also were removed by the bioreactor. A general measure of this effect was the decrease in COD from 445 mg/liter in the influent to 378 mg/liter in the effluent. The light yellow color of the water (UV cutoff, 310 nm) was not decreased during the treatment. The process water, which was initially very clear, became more turbid after several weeks of recirculation. This undoubtedly was due to material being dislodged from the sides of the storage tank. During the third week of operation, the suspended solids (dry weight) in both the influent and effluent were 34 mg/liter.

The dramatic improvement in activity that occurred after the problems with the nutrient-addition pump were corrected made us wonder whether further increases in phosphorus and nitrogen levels might result in even better PCP removal activity. We estimated, based on the COD, that there were approximately 400 ppm of carbon in the water. This meant that the ammonium nitrogen and phosphorus levels should have been 10 ppm and 2 ppm, respectively. When these were assayed, the nitrogen and phosphorus levels were 9.5 ppm and 2.2 ppm in the influent, and 8.5 ppm and 2.0 ppm in the effluent. Although it did not seem likely that nutrients were limiting, we doubled the levels that were being added. Assays then showed 19 ppm ammonium nitrogen and 4 ppm phosphorus in both influent and effluent. The activity did not show any discernible improvement; so after four days the original levels of nutrients were resumed. The activity at this point was very good, but the PCP in the effluent was not undetectable as it had been in the pilot-plant installation.

After several weeks of good activity, the performance began to falter. This decrease in activity corresponded with an early cold spell. The water temperature, which had been 20° to 22°C, fell to 15°C. A heater was ordered and preparations were made for changing the system over to treat groundwater. During four weeks of treatment, the level of PCP in the 5,000 gallons of contaminated surface water was decreased by 70%.

The large discontinuity in the data in Fig. 9 indicates the point at which the system was transferred to groundwater. Pentachlorophenol concentration in the effluent increased sharply for a brief period due to cold temperature and high flow. The transition was made without lowering the flow from 30 gallons per hour, and the water temperature fell to 13°C. The flow was cut in half after the first day of operation, and a 1,000-watt

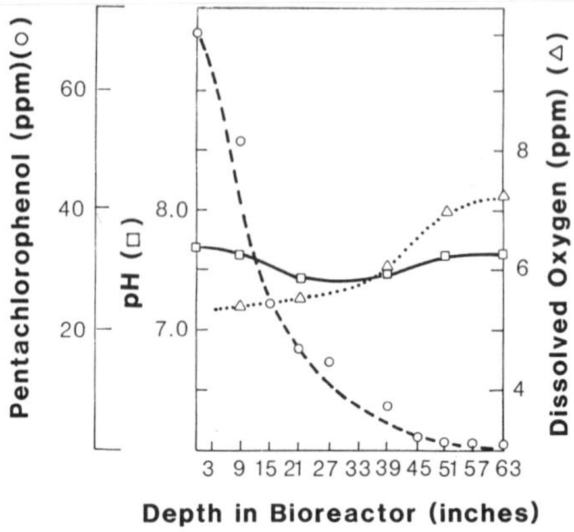

Fig. 10. Water chemistry profile in the large bioreactor (PCP, dissolved oxygen, and pH) after acclimation.

heater was installed on the second day. After the temperature was raised to 23° to 25°C, the PCP-degrading activity quickly recovered to the point where the effluent contained less than 1 ppm PCP. Over the next ten days, the flow was gradually increased to 30 gallons per hour while maintaining excellent effluent water quality.

A profile of the water chemistry in the large bioreactor during the second week of continuous feed on groundwater is shown in Fig. 10. The increased maturity of the bioreactor is evident from the movement of the PCP biodegradation profile deeper into the bioreactor. The overall activity of the bioreactor had reached 59 mg PCP consumed/hr/liter of reactor volume. The activity in the upper third of the bioreactor was 126 mg PCP consumed/hr/liter. The lowered level of dissolved oxygen in the upper part of the bioreactor was evidence of higher biological oxygen consumption, but was also partially the result of the higher flow of poorly oxygenated water. The pH profile in Fig. 11 was observed several times. The initial drop in pH is probably due to PCP degradation and partial biodegradation of nontarget compounds to fatty acids. Deeper within the bioreactor, the fatty acids would be consumed, resulting in the partial recovery of pH.

The bioreactor was developed with the primary goal of removing PCP from groundwater; however, it is also able to remove a number of nontarget compounds. This is shown by the decrease in COD from 270 mg/liter in the influent to 190 mg/liter in the effluent. It can also be seen from the decrease in the number of peaks in the HPLC chromatograms shown in Fig. 11. Except for the chlorinated phenols, most of these compounds have not been identified. One class of compounds, the polynuclear aromatic hydrocarbons, was measured in the influent and effluent from the bioreactor. The bioreactor was able to remove more than 95% of most of these compounds (proprietary data, not shown).

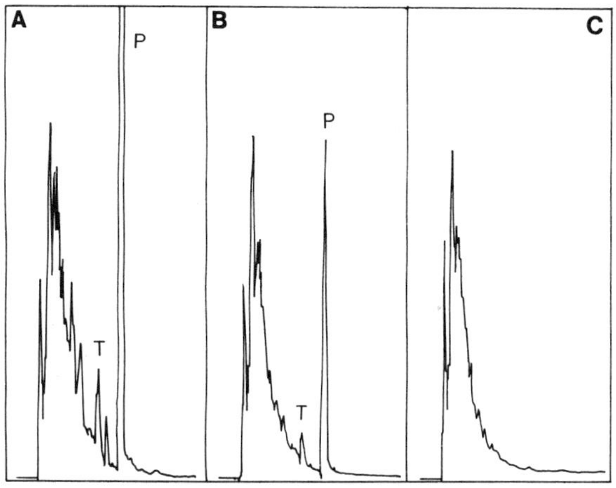

A: Influent (76 ppm penta)
B: Middle of Bioreactor (26 ppm penta)
C: Effluent (0.1 ppm penta)
Peak marked (P) is pentacholorophenol, and peak marked (T) is unresolved tetrachlorophenols.

Fig. 11. High performance liquid chromatography of groundwater during treatment in the large bioreactor.

CONCLUSIONS

This study demonstrated that an immobilized biofilm bioreactor can effectively decontaminate groundwater contaminated with PCP. Our bioreactor removed PCP from a synthetic PCP-containing waste, a tank of industrial process water that was highly contaminated with wood-preserving chemicals, and contaminated groundwater pumped from the aquifer beneath a wood-preserving operation. The many contaminants other than PCP in the "real world" contaminated water were not antagonistic to the PCP-degrading microorganisms. In fact, the activity of the bioreactor increased by an order of magnitude while treating these wastes. Many of the co-contaminants, particularly the polynuclear aromatic hydrocarbons and polychlorinated phenols, also were removed in the bioreactor.

There was considerable variability in PCP contamination in water pumped from the aquifer. The overall level was twice what we had expected from a 1985 test pumping, and during this study the level varied by plus or minus 10% on a daily basis. This type of variability is difficult for conventional bioreactors to handle (4,20). Our immobilized biofilm reactor, however, was able to maintain low PCP concentration (<1 ppm) in the effluent regardless of the variability of the PCP concentration in the influent. The bioreactor also accommodated abrupt increases of 10% to 15% in the flow rate without having significant PCP breakthrough in the effluent.

A commercially useful system at this site would be capable of handling a flow of 900 gallons per hour. Assuming the average packing activity of 60 mg/hr/liter that was achieved in this study, a bioreactor 24 inches in diameter and 20 feet tall would be required. A preliminary capital cost estimate for installing this reactor is $80,000, with the operating costs estimated at $3.00 per 1,000 gallons. These costs should be competitive with carbon treatment.

ACKNOWLEDGEMENTS

This work was supported by the United States Department of Interior, Geological Survey (USGS) Grant No. 14-08-0001-G1139, and by Bio-Trol, Inc., Chaska, Minnesota. These data have been incorporated previously into a final report to the USGS. Discussions here do not necessarily represent the policy of the USGS, and no endorsement of these data by the federal government should be assumed. Publication No. 8753 of the Idaho Agricultural Experiment Station.

REFERENCES

1. Buhler, D.R., M.E. Rasmusson, and H.S. Nakave (1973) Occurrence of hexachlorophene and pentachlorophenol in sewage and water. Environ. Sci. Technol. 7:929-934.
2. Chu, J.P., and E.J. Kirsch (1972) Metabolism of pentachlorophenol by an axenic bacterial culture. Appl. Microbiol. 23:1033-1035.
3. Crosby, D.G. (1981) Environmental chemistry of pentachlorophenol. Pure Appl. Chem. 53:1051-1080.
4. Edgehill, R.U., and R.K. Finn (1983) Activated sludge treatment of synthetic waste water containing pentachlorophenol. Biotechnol. Bioengineer. XXV:2165-2176.
5. Edgehill, R.U., and R.K. Finn (1983) Microbial treatment of soil to remove pentachlorophenol. Appl. Environ. Microbiol. 45:1122-1125.
6. Franson, M.A.H., ed. (1981) Standard Methods for the Examination of Water and Wastewater, 15th ed., American Public Health Association, Washington, D.C.
7. Guthrie, M.A., E.J. Kirsch, R.F. Wukasch, and C.P.L. Grady (1984) Pentachlorophenol biodegradation. II. Anaerobic. Water Res. 18:451-461.
8. Hakulinen, R. (1982) The Enso-Fenox Process for the treatment of kraft pulp bleaching effluent and other wastewaters of the forest industry. Paperi ja Puu-Papper och Tra 5:341-344.
9. Martinson, M.M., J.G. Steiert, D.L. Saber, and R.L. Crawford (1985) Microbiological decontamination of pentachlorophenol in natural waters. In Proceedings of the Sixth International Biodeterioration Symposium, Washington, D.C., August 5-10, 1984. Commonwealth Agricultural Bureaux, London.
10. Moos, L.P., E.J. Kirsch, R.F. Wukasch, and C.P.L. Grady (1983) Pentachlorophenol biodegradation. I. Aerobic. Water Res. 17:1575-1584.
11. Pignatello, J.J., M.M. Martinson, J.G. Steiert, R.E. Carlson, and R.L. Crawford (1983) Biodegradation and photolysis of pentachlorophenol in artificial freshwater streams. Appl. Environ. Microbiol. 46:1024-1031.
12. Reiner, E.A., J. Chu, and E.J. Kirsch (1978) In Pentachlorophenol: Chemistry, Pharmacology, and Environmental Toxicology, K.R. Rao, ed. Plenum Press, New York, pp. 67-81.
13. Renberg, L. (1974) Ion-exchange technique for the determination of chlorinated phenols and phenoxy acids in organic tissue, soil, and water. Anal. Chem. 46:459-461.
14. Stanlake, G.J., and R.K. Finn (1982) Isolation and characterization of a pentachlorophenol-degrading bacterium. Appl. Environ. Microbiol. 44:1421-1427.
15. Suzuki, T. (1977) Metabolism of pentachlorophenol by a soil microbe. J. Environ. Sci. Health B12(2):113-127.

16. Thompson, G.E., H. Husain, J. Parry, and P.J. Gilbridge (1978) Hydrogeological control and cleanup of soil and groundwater contaminants at Northern Wood Preservers Ltd. In Proceedings of the 25th Ontario Industrial Waste Conference, June 18-21.
17. Thuma, N.K., P.E. O'Neill, S.G. Brownlee, and R.S. Valentine (1983) Microbial degradation of selected hazardous materials: Pentachlorophenol, hexachlorocyclopentadiene and methyl parathion. EPA Project Summary, EPA-600/S2-83-117.
18. Trevors, H. (1982) Effect of temperature on the degradation of pentachlorophenol by Pseudomonas species. Chemosphere 11:471-475.
19. Watanabe, I. (1973) Isolation of pentachlorophenol decomposing bacteria from soil. Soil Sci. Plant Nutr. (Tokyo) 19:109-116.
20. White, J.T., T.A. Bursztynsky, J.D. Crane, and R.H. Jones (1976) Treating wood preserving plant waste waters by chemical and biological methods. EPA Report, EPA-600/2-81-043.

REMEDIATION OF MULTIMEDIA CONTAMINATION FROM THE WOOD-PRESERVING INDUSTRY

A. Dale Pflug and M. Boyd Burton

BioTrol, Inc.
Chaska, Minnesota 55318

INTRODUCTION

Microbiology is useful in remediating sites contaminated with toxic and hazardous wastes. This chapter will focus on the wood-preserving industry, an important and proud industry that has a number of challenging environmental problems which it is currently addressing, not unlike the problems faced by many other industries in this country.

We will examine briefly the historical practices of the wood-preserving industry, describe the problems that these practices have created, and address the role that microbiology can play in contaminant removal. Finally, we will describe the results that we are achieving or expect to achieve with the treatment systems that BioTrol is developing and using to permanently eliminate site contamination problems.

THE WOOD-PRESERVING INDUSTRY

Currently, there are approximately 700 plants or sites of former plants in the United States where wood preserving is, or has been, conducted. Wood preservatives extend the useful life of wood by using bactericides to treat poles, posts, railroad ties, and dimensional lumber. Wood preserving has been practiced on some sites in this country continuously since the late 1800s, and the preservation of wood goes back even further in history (4).

The most frequently used preservatives include pentachlorophenol (PCP) for poles and posts, creosote for poles, piers, and railroad ties, and metal-based preservatives for dimensional lumber, the most common of which are copper, chromium, and arsenate (CCA) (5). Wood preservatives are generally applied in liquid form, using treatment tanks or cylinders, and then the wood is stored in outside areas to dry. These preservatives are either toxic or carcinogenic in nature and have not traditionally been treated very successfully using microbiology (1).

Because of the age of this industry, many of the plants are quite old. It is not unusual to see plants that have been operating continuously since the 1930s or 1940s, long before environmental regulation or environmental awareness existed. As a result, unlined lagoons and treatment pits dug in the ground were used for many years. Treatment cylinders, which were later used to replace the treatment pits, did not provide protection against spills. Preservative chemical storage tanks were not diked or tested for leakage. In general, housekeeping practices associated with handling and storage of freshly treated wood were poor, resulting in significant deposits escaping into the environment.

Many of the plants have made major improvements to their operations, but these historical practices created potentially significant problems that site owners are still faced with solving today (3). Figure 1 shows a greatly simplified cutaway of the subsurface, from which one can follow the development of the problems. Sludge deposits are found in the treatment operation areas and in the storage areas where the wood is moved after treatment. Particularly high build-ups of sludge and heavily contaminated soils are visible near the surface. Migration of the preservatives from a variety of leaks and spills has occurred into the subsurface soil and water. Concentrations of heavily contaminated soils exist under the unlined lagoons. Leakage from treatment pits, treatment cylinders, storage tanks, and lagoons has migrated to significant depths in the subsurface and has resulted in contamination in both the saturated and unsaturated zones of the aquifer.

A moving contamination plume in the groundwater frequently necessitates an interim-response, groundwater pumping program to prevent off-site migration. The original contamination source continues to release

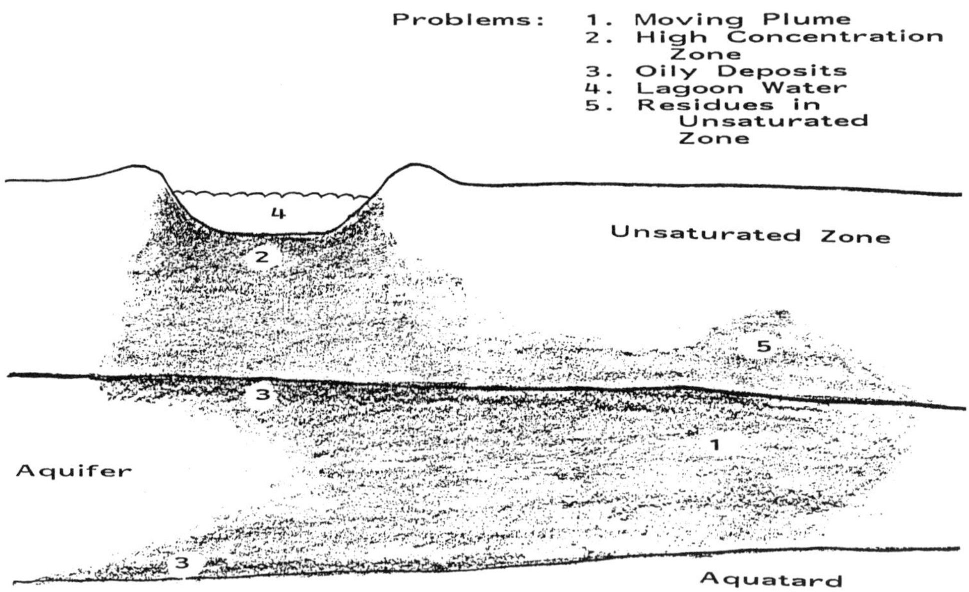

Fig. 1. Typical contaminated site.

materials to the subsurface. Groundwater contamination may be complicated by high concentrations of oil--both light and heavy--either floating on the surface or at the bottom of the aquifer, which can become new point sources. Seasonal fluctuations of the aquifer can create new sources away from the original source.

The problems to be solved include: (a) contaminated soils both at or near the surface and at significant depths; (b) contaminated soils and water in or around lagoons or old lagoon areas that have been filled; and (c) contamination in the groundwater. The traditional approach to these problems has been containment or removal and redisposal, practices that have proven to be less than satisfactory. However, these problems can be addressed using microbiologically based treatment systems.

WHY MICROBIOLOGY?

In reauthorizing Superfund in 1986, Congress recognized both the inadequacy and the risks associated with traditional waste management practices. It mandated the use of on-site, innovative, cost-effective treatment technologies to achieve permanent remedies in site clean-up. Microbiology satisfies the criteria established for an innovative technology because it can be used to accomplish both effective detoxification and volume reduction and can be particularly cost-effective in addressing contaminants of relatively low concentrations.

The traditional use of microbiology in land farming and in sewage treatment plants is one of the oldest and most widely used technologies to treat municipal wastewaters and sludges. Eagerness to demonstrate the field applications of microbiology in hazardous waste management without adequate engineering considerations, however, has led to premature and poorly planned field trials, many of which have failed to meet expectations for decontamination effectiveness or treatment rates (4).

Successful treatment of hazardous and toxic wastes requires both a proven microbial population <u>and</u> well-engineered support systems that optimize the microbes' inherently favorable characteristics while providing needed protection against the toxic effects of the contaminants. The microbes function as catalysts and can effectively be utilized and controlled in systems that frequently can be operated in an unattended manner. When the use of microbiology is more broadly accepted in environmental applications, even better results can be anticipated using engineered microbes. Today, however, the regulatory debate over genetic engineering precludes any practical use of modified organisms in the environment. Only naturally occurring bacteria are utilized in the systems being developed by BioTrol.

THE BIOTROL TREATMENT SYSTEMS

In 1985, BioTrol acquired an exclusive license from the University of Minnesota for a system that is uniquely effective in detoxifying PCP, one of the most difficult of the wood preservatives. Using microbiology originally developed at the University of Minnesota's Gray Freshwater Biological Institute by Dr. Ron Crawford (Ref. 2; Frick et al., this Volume), BioTrol has developed the engineered systems that optimize the use of Flavobacterium to treat the wood-preserving waste.

At its corporate headquarters in Chaska, Minnesota, BioTrol has laboratories and pilot-plant facilities to conduct sample analysis and treatability studies. These facilities are used to establish the design and performance data required for installing commercial treatment systems. In addition, an on-going commitment to research and development is maintained to support the development of the company's treatment systems and services. These services range from sample collection and analysis, through treatability studies, to the design and engineering of the systems that will be required to accomplish the on-site treatment services.

Field installation and startup of the biological systems are an integral part of performing on-site services. Although some of the systems will operate in an unattended manner, BioTrol also provides operations and maintenance personnel as required for its systems when desired by the site owner.

BioTrol uses a family of treatment systems to treat contamination in groundwater, lagoon water, and soils. At different wood-preserving sites, we have examples of the types of success that can be achieved. We have treated over 500,000 gallons of PCP- and creosote-contaminated groundwater in a small 2.5-gpm treatment system on a continuous basis for over 11 months, and anticipate upgrading this to a 20- to 50-gpm system as part of an interim-response groundwater pumping program in the near future. A 50-gpm lagoon water treatment system is currently being started up to treat over 3,000,000 gallons of lagoon waters. These water systems can also be used in effluent pretreatment applications. In the fall of 1987, we will demonstrate a soil treatment system at a third wood-preserving site as the next step in a program that has been successful in pilot-plant tests, and we are moving toward an in situ demonstration in the near future.

The BioTrol Aqueous Treatment (BAT) System has been under development in the laboratory for over three years. It is an immobilized bacterial system, scaled up in the spring of 1986 to field demonstration size. Following activation and trials in the Chaska pilot facilities, the system was installed in August, 1986, at the Bell Lumber and Pole Company wood-preserving site in New Brighton, Minnesota, where contaminated groundwater exists from extended use of PCP and creosote. It has been operating very successfully there ever since.

A simplified configuration of the continuous process is shown in Fig. 2. It is simple, reliable, and performs well. With heat conservation, the system requires very little energy, since typical groundwater is only about 10° below optimal operating temperature.

On the Bell Pole site, groundwater from the well is averaging 80 to 120 ppm of PCP, with additional contamination from a variety of creosote components. After a single pass through the BAT System, the contamination is biologically treated to levels that permit discharge to a POTW without the use of a final polishing step such as the use of carbon.

The BAT System effectively removes the contaminants, as the HPLC analysis in Fig. 3 depicts. Note that the scale on the upper graph is more than three times greater than that on the lower graph. Broad decontamination is achieved. The effectiveness of the treatment can be seen also by comparing the peaks on the two graphs. The treatment does not

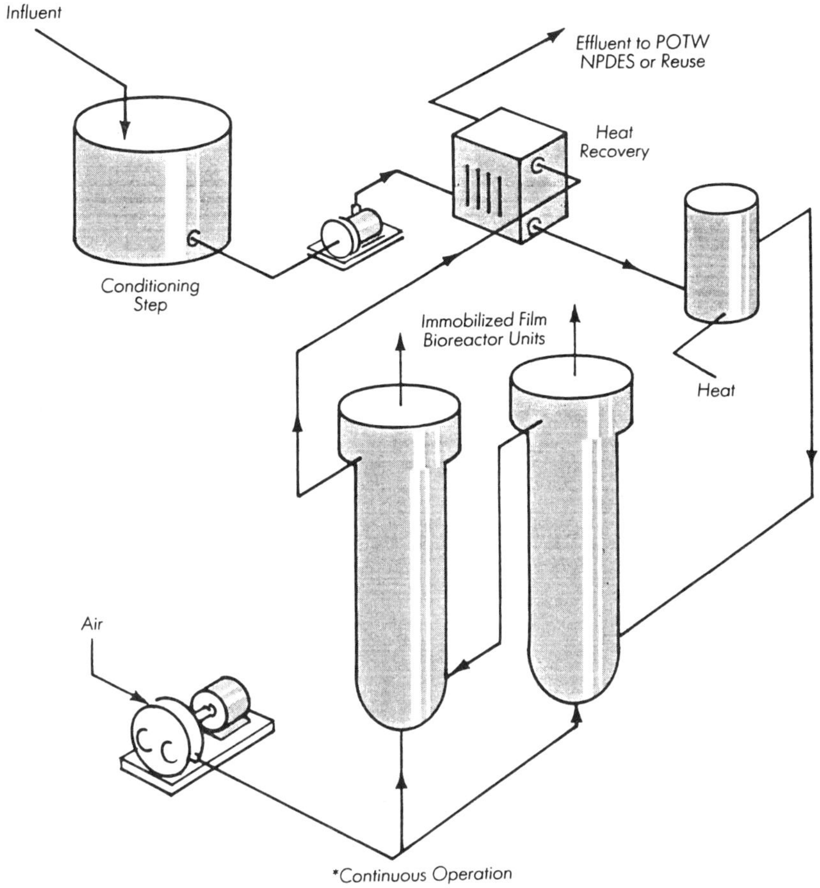

Fig. 2. Typical configuration of the BioTrol Aqueous Treatment System (BATS).

result in formation of intermediates, an important answer to a question that is frequently raised about the use of microbiology.

Table 1 presents comprehensive data for the BAT System performance, using HPLC analysis of the influent well water and the BAT System effluent. The limit of detection on the PCP is 10 ppb. This is exceptional performance for this type of contaminated water. Also note the removal of creosote components, and particularly the effective treatment of carcinogens.

With this type of performance, the system can be used to meet different discharge standards, depending upon the eventual disposition of the effluent [i.e., for plant reuse on-site, for release to surface waters (NPDES), or for discharge to a sewer system (POTW)]. Because this system does not build a large biomass, and because it has a low rate of slough, it is currently being considered for use in an aquifer up-gradient recharge system. The system is designed with both redundancy and feed-

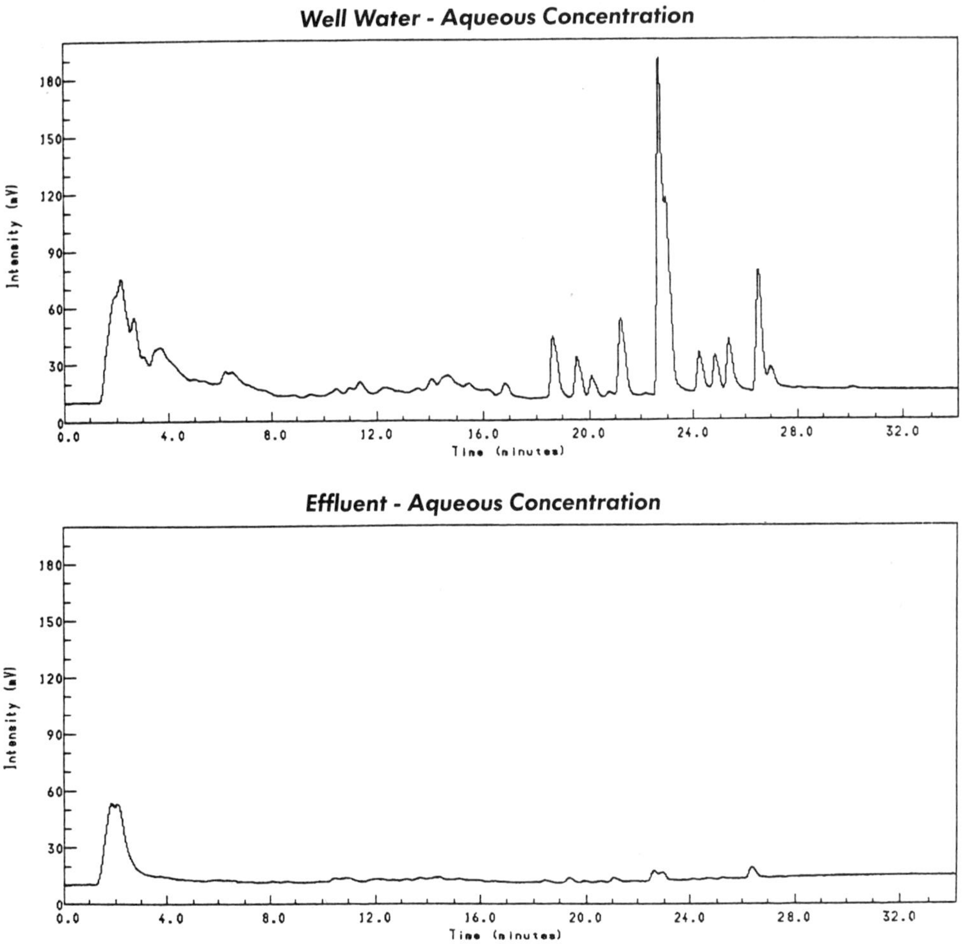

Fig. 3. High-pressure liquid chromatography analyses of polynuclear aromatic (PNA) concentrations by fluorescence detection.

back control to permit unattended operation. Cost is very competitive with either activated carbon or chemical oxidation technologies (ozone, hydrogen peroxide, etc.). This particular application has a total cost of less than one cent per gallon.

The available capacities of the BAT Systems will vary according to the type and concentrations of contaminants. Active proposals currently exist for systems that range from 10 gpm to over 2,000 gpm in capacity. This same engineered support system will also be used with other microbes to treat other contaminants.

A more complex system is being used for lagoon water treatment at another wood-preserving site. Because this water contains elevated levels of BOD and COD, in addition to PCP, creosote, and other wood-treating byproducts, the system being installed will use multiple steps to accomplish the biological treatment goals. BioTrol has a lagoon water treatment sys-

Tab. 1. High-pressure liquid chromatography analyses by ultraviolet and fluorescence detection.

	Influent well water (ppb)	BAT System effluent (ppb)
Pentachlorophenol	93,000	51
Acenaphthylene	4,400	<19
Naphthalene	1,930	81
Acenaphthene	2,040	140
Phenanthrene	260	38
Anthracene	250	20
Fluoranthene	470	150
Pyrene	230	15
Benzo(α)anthracene	280	9
Chrysene	170	8
Benzo(β)fluoranthene	450	8
Benzo(κ)fluoranthene	180	7
Benzo(α)pyrene	210	5
Dibenzo(α,h)anthracene	300	33
Benzo(ghi)perylene	320	4
Fluorene	540	19
Indeno(1,2,3-cd)pyrene	200	<3

tem (enhanced BAT System) currently being installed on a site in the southeastern United States that will operate at 50 gpm on a continuous basis. Following preconditioning of the water, the major portions of the COD and BOD are first removed in an enhanced sludge system and then treated to remove the PCP and creosote components in the BAT System reactors. As in the groundwater system, this system uses only biological treatment and does not require other polishing steps to achieve the desired treatment objectives.

Operating data are not yet available at this time, but preliminary data gathered during the treatability study (shown in Tab. 2) reflect the treatment results of the water, which will subsequently go to an NPDES-permitted outfall.

A third member of the treatment systems family will be used for soils. The BioTrol Soil Treatment System (BST System) is based upon excavation of the soil, decontamination of the soil in an on-site mobile treatment system, and return of the soil to the original point of excavation following decontamination confirmation. The process utilizes a new soil contaminant extraction concept developed in BioTrol's laboratories and includes, as an essential part, decontamination of the system's process water using the microbiological techniques incorporated in the BAT System.

Figure 4 illustrates the flexibility inherent in the system. Following the extraction step, which uses water as a process carrier, a major portion of the contamination in the soil has been removed, enabling an optional biological polishing step to be added if further contaminant removal is desired. The contaminants in the process water are separated into a concentrate stream and process water for subsequent treatment in the BAT

Tab. 2. Treatability of lagoons.

Parameter	Lagoon water (ppm)	
	Untreated	Treated
COD (total)	4,400-6,700	500
BOD (total)	1,300-1,600	240
Pentachlorophenol	10-36	0- 0.50
Total phenols	4	0- 0.60
Polyaromatic hydrocarbons	37	0- 0.075
Oil and grease	160	0-30
Suspended solids	2,450	0-20
pH	7.4	7.0- 8.0

System. The concentrate can be recycled back to the site wood-preserving process, further decontaminated microbially for use as a fuel, or sent to an incinerator, depending on the options that are available.

The BST System mobile unit will operate at a capacity of five to ten tons per hour. Following excavation and screening, the contaminated soil will move through the system on a continuous basis. In pilot studies on actual PCP- and creosote-contaminated soils containing 3% total organic contaminants, the contaminant levels were reduced to 20-25 ppm following the extraction step. These contaminant levels can be reduced further, using a final biological polishing step, if necessary.

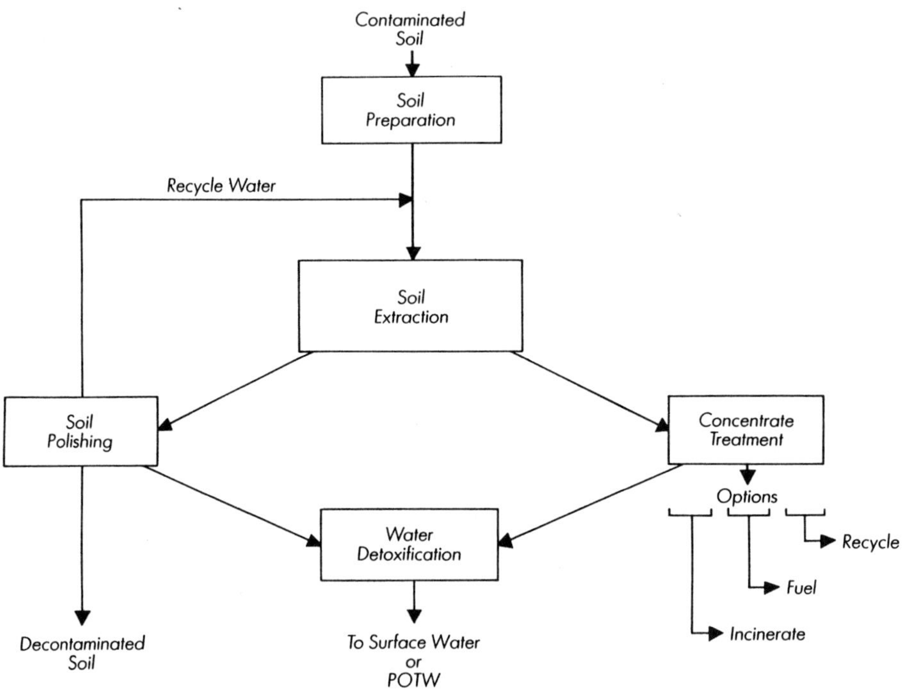

Fig. 4. Diagram of the BioTrol Soil Remediation System.

The advantage of this system is the continuous operation with low energy requirements. It will operate at ambient temperatures. It is a mobile process with a number of flexible features to adapt to site needs; we believe the system also will be adaptable to a wide range of contaminants. Most important, we expect the cost of the system to be most advantageous when compared to the alternative treatment technologies, including incineration and other chemical, biological, and physical treatments.

CONCLUSION

We are convinced that microbiology has much to offer in the elimination of toxic and hazardous contamination. Successful application of this technology will require engineered systems and turn-key services such as BioTrol is uniquely developing and providing today. With the success of this technology, we can achieve the true intent of Congress in providing permanent remedies for the wide-ranging problems of organic contamination that we face in this country today. We look forward with considerable enthusiasm to the coming months and years.

REFERENCES

1. Code of Federal Regulations, 40 CFR Part 261.
2. Pignatello, J.J., L.K. Johnson, M.M. Martinson, R.E. Carlson, and R.L. Crawford (1985) Response of the microflora in outdoor experimental streams to pentachlorophenol: Compartmental contributions. Appl. Environ. Microbiol. Vol. 50:127-132.
3. United States Environmental Protection Agency (1987) National Priority List of Proposed and Final Sites, July, 1987.
4. United States Environmental Protection Agency (1988) Superfund records of decision, EPA Reports, 1985-1987.
5. Wood Preserving Statistics, 1983 and 1984. A Report to the wood Preserving Industry in the United States, January, 1986, pp. 3-7.

PRELIMINARY DEVELOPMENT OF A BENCH-SCALE TREATMENT SYSTEM FOR AEROBIC DEGRADATION OF TRICHLOROETHYLENE

Michael J.K. Nelson,[1]* P.H. Pritchard,[2] and Al W. Bourquin[2]*

[1]Technology Applications, Inc. and
[2]United States Environmental Protection Agency
Environmental Research Laboratory
Gulf Breeze, Florida 32561

INTRODUCTION

Trichloroethylene (TCE) is a volatile, chlorinated, aliphatic compound that has been used extensively as a general cleaning and degreasing solvent by the military and industry. This extensive use has resulted in its becoming one of the most predominant groundwater contaminants, of particular environmental concern because of its potential carcinogenicity and its recalcitrance. Although some decomposition has been reported under anaerobic conditions, the decomposition is slow and results in the production of other hazardous compounds, including dichloroethylenes and vinyl chloride (1,4,5). Aerobic degradation has been reported in soil columns exposed to natural gas, suggesting that methanotrophs may be capable of TCE degradation (6). We recently reported on the isolation of a gram-negative bacterium, designated strain G4 (2), that apparently utilizes the enzymatic machinery of an aromatic biodegradative pathway to detoxify TCE (3).

Use of biological methods for detoxification of hazardous wastes is an expanding technology with great potential as an effective, inexpensive alternative to past methods for cleanup of polluted environments. Isolation of microorganisms with TCE-degrading ability could lead to their eventual use for cleanup of TCE-contaminated areas. Knowledge of the mechanisms by which TCE degradation may occur could lead to development of methods for stimulating naturally occurring microorganisms to detoxify the compound in situ. This chapter summarizes our progress in understanding the aerobic biodegradation of TCE by strain G4 utilizing bench-scale batch cultures. Detailed descriptions of this work have been previously published (2,3).

---

* Present address: ECOVA Corporation, 3820 159th Ave. N.E., Redmond, Washington 98052

RESULTS

A number of soil and water samples from six sites in the Pensacola, Florida, area that had histories of contamination with various synthetic aliphatic and aromatic compounds were screened for biological TCE degradation (2). Trichloroethylene biodegradation was defined as an obvious decrease in TCE concentration in active samples as compared to autoclaved controls. Only one sample from a holding pond at an industrial waste treatment facility was active. During attempts to isolate a pure culture from this sample, certain requirements for maintenance of TCE degradative activity became apparent. Subcultures of the sample retained TCE degradative ability only if sterile water from the original sampling site (sitewater) was used to make up the culture medium. Thus, some abiotic component in the water was required for TCE degradation.

Several isolated colony-types were obtained from the active sample by plating on glucose minimal medium. One isolate, designated strain G4, degraded TCE in the presence of sitewater. Maintenance of the activity also required aerobic conditions, as shown by the lack of degradation in nitrogen-sparged experiments (Tab. 1).

Gas chromatographic analysis (FID detector) of sitewater indicated that some organic compound disappeared during TCE degradation by strain G4 (Fig. 1), suggesting that this compound may be the factor required in the water during TCE degradation. By examining the retention time during gas chromatography as well as mass spectroscopy data, this compound was identified as phenol (3). Phenol replaced the requirement for sitewater and thus was the factor. Other aromatic compounds that stimulated degradation were toluene, o-cresol, and m-cresol (Tab. 2). Inactive compounds included m-xylene, sodium benzoate, and p-cresol. All compounds tested, except m-xylene, were growth substrates for strain G4.

Protein synthesis was required for the stimulation of TCE-degradative activity by the aromatic substrates. Lactate-grown strain G4 degraded

Tab. 1. Effect of O_2 removal on TCE metabolism by strain G4.

Experimental conditions	Inoculum	% Trichloroethylene remaining†
Unsparged	Sterilized	88.0
Unsparged	Active	<2[a]
N_2-sparged	Active	81.1
Air-sparged	Active	<1.8[a]

Note: All experiments employed medium made with sitewater. N_2- and air-sparging was performed for 30 min at 30 ml/min prior to addition of 50 nmol TCE and inoculum. From Nelson et al. (2).

†After 24 hr incubation.

[a]Minimum detectable level.

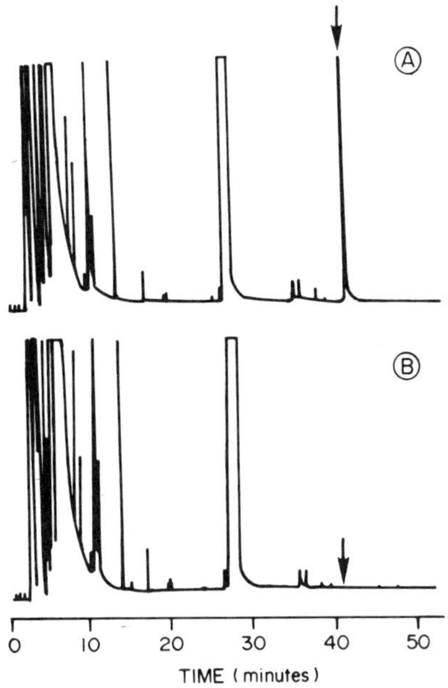

Fig. 1. Gas chromatographic profile of ether extracts of sitewater before (A) and after (B) its use as medium for TCE metabolism by strain G4. The peak of interest is indicated by the arrow. From Nelson et. al. (3).

TCE if phenol was included in the incubation medium (Tab. 3). However, TCE degradation was substantially inhibited if chloramphenicol was also included to prevent protein synthesis. Cells that were preinduced with phenol were not inhibited by chloramphenicol and did not require addition of phenol to the medium.

The results indicate that induction of one or more enzymes involved in degradation of certain aromatic compounds is responsible for the degradation of TCE. Some information has been obtained regarding the type of aromatic degradative pathway induced by the inducers of TCE degradation. Both phenol and toluene induce the production of the <u>meta</u>-fission enzyme catechol-2,3-dioxygenase. The presence of this enzyme was indicated by the production of the <u>meta</u>-fission products from catechol or 3-methylcatechol by cells induced with these compounds. Benzoate did not induce this activity, and was oxidized by an <u>ortho</u>-fission pathway. Attempts to elucidate the earlier steps of phenol and toluene degradation have, to date, been unsuccessful.

The degradation of TCE by strain G4 apparently results in complete conversion of the compound to innocuous products. Experiments conducted with [U-^{14}C]-TCE (Tab. 4) indicated that after TCE degradation occurred, 60% of the carbon in TCE was present as CO_2, whereas no $^{14}CO_2$ was detected in autoclaved controls. Another 35% of the total ^{14}C was present as nonvolatile material. This material is assimilated into cellular carbon, since up to 78% was associated with particulate material. This was indicated by its removal during passage through a 0.22-μm pore filter. The remaining

Tab. 2. Compounds tested for stimulation of TCE degradation by strain G4.

Compound added (1 mM)	Trichloroethylene remaining† (μM)
None	3.35 ± 0.26
Phenol	0.04 ± 0.0
Toluene	<0.02[a]
_m_-Xylene	3.92 ± 0.08
Sodium benzoate	4.17 ± 0.29
_o_-Cresol	<0.02
_m_-Cresol	0.07 ± 0.09
_p_-Cresol	4.04 ± 0.45

Note: Each experiment contained 200 nmol TCE and 1 ml of a late log-phase culture (approximately 3 x 10^9 CFU) of strain G4 grown on 20 mM lactate. Data are the means and standard deviations of triplicate experiments. From Nelson et al. (3).

†Trichloroethylene remaining after 24 hr incubation. A typical control with sterile inoculum contained 3.5 μM TCE.

[a]Minimum detectable level.

Tab. 3. Trichloroethylene degradation in the presence of chloramphenicol by strain G4 with and without preinduction with phenol.

Preinduction with phenol	Added to experiment: Phenol (1 mM)	Added to experiment: Chloramphenicol (0.05 mg/ml)	Trichloroethylene remaining† (μM)
-	+	+	2.62 ± 0.17
-	+	-	0.03 ± 0.01
+	+	-	<0.02[a]
+	-	+	<0.02
+	+	+	1.21 ± 0.70

Note: Each experiment contained 200 nmol TCE and 1 ml (approximately 3 x 10^9 CFU) of a culture of strain G4 grown 20 hr on 20 mM lactate and induced, where indicated, by including 2 mM phenol in the culture medium. The culture and chloramphenicol were added to the indicated tests 15 min prior to addition of TCE. Data are the means and standard deviations of triplicate experiments. From Nelson et al. (3).

†Trichloroethylene remaining after 24 hr. A typical control with sterile inoculum contained 3.5 μM TCE.

[a]Minimum detectable level.

Tab. 4. Distribution of radioactivity after metabolism of [^{14}C]-TCE by strain G4.

Strain G4 inoculum	Nonvolatile products[a]		Carbon dioxide[b]	
	dpm	% of total[c]	dpm	% of total[c]
Active	31,445 ± 1,869	34.7 ± 1.5	53,996 ± 220	59.6 ± 1.0
Autoclaved	1,785 ± 204	3.4 ± 0.5	0	0

Note: Tests were performed in triplicate; data are the means and standard deviations of three replicates. From Nelson et al. (2).

[a]Radioactivity remaining in solution after acidification and sparging to remove CO_2.

[b]Carbon dioxide was determined by sparging the acidified samples with N_2 and trapping the CO_2 by passage of the effluent gas through 1 M NaOH. Trapped radioactivity was confirmed as $^{14}CO_2$ by precipitation with $BaCl_2$.

[c]Based on total ^{14}C present after incubation (active = 90,552 ± 1,759 dpm; sterilized = 53,618 ± 2,329 dpm) as 100%.

^{14}C could be accounted for as unmetabolized TCE. The degradation of TCE resulted in the stoichiometric production of inorganic chloride (Tab. 5). From 2.7 to 2.9 chloride ions were produced per TCE molecule consumed, representing 90% to 98% of the chloride that theoretically could be produced from TCE. Time course studies indicated that chloride was produced at a rate of 3.6 nmol/min in the test system (approximately 10^9 CFU), representing 1.2 nmol TCE/min (assuming three Cl^- per TCE).

Trichloroethylene degradation occurred in environmental water samples with or without the addition of strain G4 (Tab. 6). When conpared to sterile controls, slight degradation apparently occurred even without any additions. However, the addition of phenol or toluene with or without strain G4 stimulated degradation. When strain G4 was added to the sam-

Tab. 5. Stoichiometry of chloride produced from TCE by strain G4.

Experiment	Trichloroethylene added (nmol)	Chloride produced† (nmol)	Chlorides per trichloroethylene	Percent theoretical*
1	200	586 ± 34	2.93 ± 0.17	98
2	200	541 ± 20	2.70 ± 0.10	90

Note: Each experiment contained 1 mM phenol and 7 x 10^9 CFU of strain G4. Data are the means and standard deviations of three replicates. From Nelson et al. (3).

†Corrected for background chloride in controls with no added TCE. Chloride was determined with an Orion 94-17B chloride-specific electrode and a 90-02 reference electrode calibrated with KCl standards.

*Three chlorides per TCE were used as 100%.

Tab. 6. Trichloroethylene degradation in environmental samples with and without the addition of strain G4.

Additions[c]	Percentage trichloroethylene not degraded[a]		
	Environmental samples[b]		
	Estuarine	River	Groundwater
None	87	92	85
Phenol, strain G4	2.1	1.6	<0.7[d]
Phenol, strain G4 (minus basal salts)	4.1	2.2	<0.5[d]
Toluene	73	42	45
Phenol	93	73	30
Phenol, enrichment[e]	-	-	<0.6[d]

[a] Percentage of TCE remaining in active versus identical sterile samples after four days incubation. Data are the means of three replicates.

[b] All samples were water samples collected near Pensacola, Florida. The estuarine sample was from a pristine saltmarsh (Range Point); the river sample was from the Escambia River near the University of West Florida; and the groundwater was from a jet fuel-contaminated area in the Pensacola Naval Air Station.

[c] When included, toluene and phenol were 1 mM and strain G4 was approximately 10^9 CFU. Except where indicated, all samples were supplemented with basal salts. All samples initially contained 200 nmol TCE, and experimental conditions were as previously described (3).

[d] Minimum detectable level.

[e] After addition of 2 mM phenol, samples were preincubated aerobically for four days prior to addition of 10 nmol of phenol and 200 nmol of TCE to initiate the TCE degradation experiments.

ples, from 96 to 100% of the TCE was consumed. In most of the experimental sets, addition of toluene or phenol to the samples in the absence of strain G4 stimulated TCE degradation, as compared to the sets with no additions. Preincubation of one of the water samples with phenol to enrich for phenol degraders resulted in the ability to degrade all detectable TCE.

SUMMARY

Our results have demonstrated that TCE may be biodegraded to nontoxic products under certain conditions. Trichloroethylene is apparently degraded by strain G4 to CO_2, cellular carbon, and inorganic chloride. This activity requires aerobic conditions and exposure of the organism to certain aromatic compounds. This exposure is required to induce the biosynthesis of one or more enzymes that fortuitously degrade TCE. The normal function of the enzyme(s) is the metabolism of the aromatic compounds. Evidence from the work with environmental samples indicates that the natural microflora in a variety of areas is capable of TCE degradation if stimulated by the appropriate aromatic compounds. These results are presently being applied to the development of bench-scale continuous treatment systems for further assessment of biodegradation as a means for detoxification of TCE-contaminated sites.

REFERENCES

1. Kleopfer, R.D., D.M. Easley, B.B. Haas, Jr., T.G. Deihl, D.E. Jackson, and C.J. Wurre (1985) Anaerobic degradation of trichloroethylene in soil. Environ. Sci. Technol. 19:277-280.
2. Nelson, M.J.K., S.O. Montgomery, E.J. O'Neill, and P.H. Pritchard (1986) Aerobic metabolism of trichloroethylene by a bacterial isolate. Appl. Environ. Microbiol. 53:383-384.
3. Nelson, M.J.K., S.O. Montgomery, W.R. Mahaffey, and P.H. Pritchard (1987) Biodegradation of trichloroethylene and the involvement of an aromatic biodegradative pathway. Appl. Environ. Microbiol. 53:949-954.
4. Parsons, F., P.R. Wood, and J. DeMarco (1984) Transformation of tetrachloroethene and trichloroethene in microcosms and groundwater. J. Am. Water Works Assoc. 76:56-59.
5. Vogel, T.M., and P. McCarty (1985) Biotransformation of tetrachloroethylene to Trichloroethylene, dichloroethylene, vinyl chloride and carbon dioxide under methanogenic conditions. Appl. Environ. Microbiol. 49:1080-1083.
6. Wilson, J.T., and B.H. Wilson (1985) Biotransformation of trichloroethylene in soil. Appl. Environ. Microbiol. 49:242-243.

BIODEGRADATION OF HYDROCARBONS IN THE ENVIRONMENT

Ronald M. Atlas

Department of Biology
University of Louisville
Louisville, Kentucky 40292

INTRODUCTION

Although most oil pollutants enter the environment as a result of chronic inputs associated with the extraction, production, transport, utilization, and disposal of this fossil fuel resource, most public concern occurs when there are large accidental oil spills, particularly if these pollute water supplies or recreational areas. Great public attention has been given to oil pollution of the seas, where several major tanker and offshore oil well accidents have released large amounts of oil into the marine environment, visibly contaminating beaches and seabirds. The absence of a "major" spillage in the past few years that could have contaminated recreational beaches, and the scientifically accepted recognition that oiled ecosystems recover as a result of natural processes that remove oil pollutants, have led to diminished public and scientific interest in the environmental fate of oil pollutants. Chronic sources of petroleum pollution, spillages that contaminate groundwater, and land disposal of oily sludges remain problems that may be mitigated by bioremediation and deserve continued study.

When petroleum or petroleum products enter the environment, they are subjected to various weathering processes that alter their chemical and physical nature. Biodegradation of petroleum pollutants by microorganisms is a major process mediating the fate of oil in the environment (2,4,5,8, 15,20,23,40,44,46,53,54). The chemical nature of petroleum and refined oils--which are very complex mixtures of aliphatic, alicyclic, and aromatic hydrocarbons and of some nonhydrocarbon compounds, such as naphthoic acids, phenols, thiols, and heterocyclic nitrogen, sulfur, and oxygen (NSO) compounds, as well as some metalloporphyrins (8)--presents a challenge to microorganisms. No crude oil sample is subject to complete (100%) biodegradation. Extensive degradation of a petroleum mixture generally requires the combined activities of several different microbial populations.

METABOLISM OF PETROLEUM HYDROCARBONS

Hydrocarbons differ in their capacity to serve as microbial substrates (4,5,24,31,33,57,60,61). This is true with respect to single microbial strains and also, in a broader sense, to the microbial community of a given environment.

Some generalizations about the relative susceptibilities of hydrocarbons to microbial attack (16) are:

(a) When hydrocarbons become available to a microbial community in a complex mixture such as petroleum, biodegradation of most of the hydrocarbon compounds occurs simultaneously but at widely differing rates; the biodegradation of petroleum hydrocarbons varies greatly, leading to sequential disappearance of individual components and to successional changes in the degrading microbial community.

(b) The presence of one hydrocarbon substrate may influence the biodegradation of another in either a positive manner (by co-metabolism) or a negative manner (by toxicity or sparing).

(c) Utilization of C_1-C_4 alkanes is restricted to a few specialized species; alkanes in the C_5-C_9 range are toxic to many microorganisms but can be biodegraded by some. _n_-Alkanes of the C_{10}-C_{22} range usually are readily metabolized. Higher-molecular-weight alkanes tend to be solid waxes and are not readily biodegraded; however, slow biodegradation of _n_-alkanes up to C_{44} has been shown (41).

(d) _Iso_-alkanes are less readily utilized as compared to _n_-alkanes; methyl branching can hinder or completely block biodegradation.

(e) Olefins tend to be more toxic and, at least under aerobic conditions, are less readily utilizable than the corresponding alkanes.

(f) Monoaromatic hydrocarbons can be toxic, but in low concentrations many microorganisms rapidly utilize these compounds. Condensed polyaromatics having two to four rings are biodegradable at rates that decrease with the level of condensation. Condensed polyaromatics with five or more rings are poor substrates for microbial attack.

(g) Low-molecular-weight cycloalkanes only rarely serve as growth substrates. At low concentrations, cycloalkanes are degraded at moderate rates; highly condensed cycloalkanes are refractory.

(h) Heterocyclic NSO compounds, when not highly condensed, may be subject to limited microbial metabolism. Highly condensed asphaltenes are very resistant to biodegradation.

HYDROCARBONOCLASTIC MICROORGANISMS

Hydrocarbonoclastic microorganisms (i.e., microorganisms capable of degrading hydrocarbons) are ubiquitously distributed in marine, soil, and freshwater ecosystems. The relative abundances of hydrocarbonoclastic microbes, though, vary considerably according to environmental conditions and hydrocarbon pollution history. Populations of hydrocarbon-degrading microorganisms are substantially higher in areas that have been exposed previously to petroleum pollutants than in pristine areas (4,17,28,49,51, 55). For example, hydrocarbonoclastic microbial populations along the coast of Brittany, a region that had received chronic inputs of oil pollutants from tanker ballast washings, were high and ready to attack the oil spilled from the wreck of the tanker Amoco Cadiz (3,11,38). Thus, microbial communities adapted to hydrocarbons because of prior exposure are better able to attack hydrocarbon pollutants than are microbial communities that lack such adaptation (3,11,43,67).

The recognition that multiple enzymes and metabolic pathways are needed to degrade a significant portion of the hydrocarbons found in petroleum led some to search for mixtures of microorganisms or to create, through recombinant DNA technology, a genetically engineered microorganism (GEM) that would be capable of attacking all or nearly all the components of an oil (2). Several mixtures of microorganisms are produced commercially for biodegrading hydrocarbon pollutants; many of these have been marketed for the biodegradative disposal of oily wastes. In many cases, however, the application of commercial microbial mixtures does not enhance rates of hydrocarbon biodegradation above rates that would be accomplished by the indigenous microbial community (7). In sites used for oily waste disposal, however, application of effective microbial mixtures can be beneficial and necessary.

With respect to using recombinant DNA technology to create an oil-degrading microbial strain with special metabolic properties, a patent was granted to Chakrabarty and General Electric for a hydrocarbon-degrading bacterium formulated by molecular breeding; under current EPA definition, this is a GEM. In a landmark decision, the Supreme Court of the United States ruled that the patent was valid because a novel organism had been created. While of enormous significance for the field of genetic engineering and of scientific importance for the approach used to create a novel organism with specific degradative capabilities, the organism itself has limited metabolic capabilities with respect to oil pollutants of environmental concern. The bacterium patented by Chakrabarty and General Electric degrades a variety of low-molecular-weight hydrocarbons but not the higher-molecular-weight, highly condensed, and highly substituted compounds that tend to persist in the environment because of their resistance to microbial biodegradation. Because of its metabolic limitations, the extensive regulations regarding the deliberate release of GEMs, and the lack of further development at General Electric, this organism has never been applied in an effort to enhance the biodegradative removal of oil pollutants.

Even if a permit for the release of such an organism could be obtained, it is probably unreasonable to expect that genetic engineering can produce a single organism that can be used to enhance the removal of environmental oil pollutants, in large part because hydrocarbon biodegradation is usually limited by abiotic factors and not the availability of diverse hydrocarbonoclastic microorganisms. On the other hand, recombinant DNA technology should be able to produce microorganisms that are capable of aiding in the cleanup of oily wastes before their release to the environment by the selective removal of individual components, such as compounds with extensive branching and/or condensed aromatic ring structures, that otherwise would be difficult to biodegrade.

Specifically, organisms are needed to degrade toxic aromatic components of refinery waste streams before environmental treatment; organisms can be designed with the specific metabolic capabilities for transforming the compounds in the particular waste stream. Additionally, organisms are needed that can be used in bioreactors to degrade oily sludges. Currently, land farms are used for the disposal of many oily sludges; that is, oily sludge wastes are added to soil and fertilizers are applied so that the indigenous soil microorganisms can slowly degrade the complex components of the sludge. Because there are only a limited number of acceptable disposal sites, and also because there is serious concern that groundwater and agriculturally valuable soils will be contaminated with the toxic and carcinogenic pollutants found in oily sludges, alternative means of treating these wastes are needed. Conventional anaerobic sludge digesters are inadequate, because the microbes in such reactors will not attack hydrocarbons. Specially designed reactors with specific microbial populations are needed if oily sludges are to be degraded by biological means, either aerobically or anaerobically, in contained, environmentally safe reactors.

FACTORS INFLUENCING RATES OF PETROLEUM BIODEGRADATION

A number of factors influence the rates of petroleum biodegradation, including the physical state of the oil, environmental temperature, oxygen availability, and concentrations of mineral nutrients--especially nitrogen, phosphorus, and iron (4,5,27). Clearly, the fate of oil in the marine environment will depend on the particular set of abiotic parameters of a given habitat. The interactions of multiple factors will determine the overall rate of biodegradation. Factors such as favorable oxygen concentrations and a large surface area for microbial attack on a surface spill may be offset by low nutrient concentrations; similarly, the favorable nutrient concentrations of benthic sediments may be offset by the concentration of oil in anoxic pockets. While the rate-limiting factors have been elucidated, the interactions of these factors with respect to determining rates of oil biodegradation have not; therefore, it remains difficult to predict the fate of oil in diverse habitats.

Physical State of the Oil

Oil spilled on water may form a thin surface film (slick), may form a stable oil-water emulsion (mousse), may become associated with suspended particles in the water column, and/or may become entrained in sediment as concentrated pockets of oil. The surface areas in these various physical forms of oil that are available for microbial colonization have a marked influence on the fate of the petroleum hydrocarbons. In a study of oil spilled from the IXTOC I well blowout, Atlas et al. (13) concluded that oil was protected within the stable mousse emulsion that formed, because the low surface area of the mousse and the low flux of oxygen and mineral nutrients to the oil-degrading microorganisms at the oil-water interface limited the rate of degradation. Compared to the extremely limited degradation of oil in the mousse, Pfaender and Buckley (59) found that dispersed oil from IXTOC I was rapidly degraded within the water column. Similar limitations occur in sediments and soils where pooling of oil can limit the ability of microorganisms to attack the oil, leading to environmental persistence of hydrocarbon pollutants. The formation of tar balls also restricts biodegradation and weathering for similar reasons. Mousse and tar ball formation have been suggested as major limiting factors of petroleum biodegradation in the IXTOC I (12) and Metula (21) oil spills.

Dispersing oil may cause increased toxicity problems for higher organisms, but it makes the oil more susceptible to microbial attack (7). Hydrocarbonoclastic microorganisms typically position themselves at the hydrocarbon-water interface, a position that assures them maximal access to both the hydrocarbon substrate and the water-soluble mineral nutrients. Many hydrocarbonoclastic microbes produce surface-active agents that emulsify oil, thereby increasing surface area (29,37,73); the increased surface of the finely dispersed oil provides additional interface area for microbial proliferation. Chemical dispersants may also be added to disperse spilled oil, thereby removing the unsightly oil pollutants and simultaneously enhancing rates of microbial hydrocarbon decomposition. Some surface-active agents are produced commercially and, together with the addition of mineral nutrients, are intended to enhance rates of oil biodegradation. Use of dispersants is limited to areas where they will not cause significantly increased toxicity to birds, fish, and other valuable biological resources.

Temperature

As is true for other microbial activities, oil biodegradation is strongly temperature-dependent (32). Although some thermophilic hydrocarbon-degrading bacteria (optimal growth at 55° to 65°C) have been isolated from estuarine sediments (50), it is unlikely that such microorganisms have a significant role in an environment where temperatures range from mesophilic (20° to 40°C) to psychrophilic (<20°C). Hydrocarbon biodegradation has been reported at temperatures below 0°C (76), but the rate and extent of hydrocarbon biodegradation at low temperatures generally is severely restricted (1,39,52,75).

Oxygen

The initial attack on hydrocarbons commonly involves oxygenases, which places a mandatory requirement for molecular oxygen on the process. The question of whether anaerobic hydrocarbon metabolism occurs has been quite controversial. There have been several reports of anaerobic conversion of hydrocarbons (19,45,56,66,68). Recently, Galic and colleagues have isolated anaerobes that are able to degrade aromatic hydrocarbons and have identified the intermediary metabolites of these pathways (35,36,69); anaerobic mineralization of low-molecular-weight aromatics has also been reported by Zeyer et al. (74). These studies establish the potential for anaerobic microbial attack on hydrocarbons.

However, even though such in vitro studies show the potential for such biodegradation, anaerobic hydrocarbon biodegradation in field samples either has not been detected at all or has been detectable only at extremely low rates. For the utilization of hydrocarbons, which are fully reduced substrates, anaerobic conditions offer a rather marginal energy balance. Ward and Brock (71) have reported that alkanes are also subject to slow anaerobic degradation. For this or other as yet undiscovered reasons, anaerobic hydrocarbon biodegradation in the environment is either undetectable or is orders of magnitude slower than aerobic hydrocarbon biodegradation (14,25,71,72). In studies on sediments impacted by the Amoco Cadiz oil spill, it was found that one percent of added oil was biodegraded anaerobically in a year compared with ten percent aerobically in a week (72). Further investigations on anaerobic hydrocarbon metabolism are needed to elucidate why the metabolic potential of some microbes to degrade hydrocarbons anaerobically is not expressed in the environment at rates that would cause the rapid disappearance of hydrocarbon pollutants from anoxic zones.

Presently, the availability of molecular oxygen must be considered as an important limiting factor for petroleum biodegradation. When low concentrations of oil are thoroughly dispersed in an aerobic water column or as a surface slick in direct contact with atmospheric oxygen, it is unlikely that lack of oxygen is a limiting factor. However, in cases where pools of oil accumulate in soils and sediments, in regions of intense decomposition, and in groundwater, the development of anoxic conditions can severely limit the extent of hydrocarbon biodegradation. In sediments, bioturbation (agitation by living organisms) by, for example, polychaete worms has been found to be important in providing oxygen for oil biodegradation (34,42). Such involvement of macroorganisms in determining the rates of microbial oil degradation usually is overlooked and is not included in microcosms aimed at modeling the environmental fate of oil.

By recognizing that oxygen can be the key rate-limiting factor in cases of oil contamination of groundwater supplies, it has been possible to enhance microbial hydrocarbon degradation by providing oxygen, as well as mineral nutrients, to support microbial growth. Initially, oxygen for such stimulated biodegradation of oil-contaminated groundwater was supplied by forced aeration through several holes that were drilled so that air could be introduced to the oil-contaminated groundwater. Supplying oxygen in this manner is difficult and costly. Raymond (62) has developed a very clever alternative method for introducing oxygen into groundwater that involves peroxides at concentrations that are not toxic to microorganisms and in formulations that slowly release oxygen upon decomposition at levels needed to support oil biodegradation. Thus, oxygen becomes available as needed to support microbial biodegradation of hydrocarbon pollutants.

Mineral Nutrients

Petroleum is rich in carbon and energy but deficient in those mineral nutrients--nitrogen, phosphorus, and iron--needed to support microbial growth. Consequently, petroleum-degrading microorganisms must obtain their essential mineral nutrients from the environment. Many ecosystems are mineral nutrient-limited and are unable to support extensive and rapid biodegradation of oil pollutants. In vitro experiments have convincingly demonstrated the phosphorus and nitrogen limitation of petroleum biodegradation in seawater (6). Floodgate (30) also concluded that the scarcity of mineral nutrients in seawater is often limiting for petroleum biodegradation, not only in vitro but also under environmental conditions. Iron limitation was confirmed in clear offshore seawater, but not in sediment-rich coastal seawater (26). Based on the data of Atlas and Bartha (6) and Reisfeld et al. (63), 1.5 to 2.5% N and 0.2% P (w/w) addition (relative to petroleum) allows maximal petroleum biodegradation in marine systems.

RATES OF PETROLEUM BIODEGRADATION

Petroleum is a multicomponent mixture, and each of its components degrades at a different rate--some fast, some slow, and some not at all. It is possible to measure an "overall" biodegradation rate (the average of the individual rates), but it is misleading to use such data for construction of a linear regression curve. As the more biodegradable components are depleted and the more recalcitrant ones remain, the average rate will continuously decline. Nevertheless, we can make some approximations of the rates of environmental biodegradation of petroleum hydrocarbons, as shown in Tab. 1, which summarizes the orders of magnitude of hydrocarbon biodegradation rates obtained in different systems. Optimization of hydrocarbon biodegradation in the environment is difficult, but may be achieved by modification of temperature, aeration, or mineral nutrients, or by all of the above.

SUMMARY

Studies on the environmental fate of petroleum have demonstrated the nearly ubiquitous distribution of microorganisms that can metabolize hydrocarbons. The rates of degradation depend upon the concentrations of such microbes and upon the environmental characteristics of an oil-contaminated ecosystem. Given the appropriate environmental conditions, microorganisms effectively decontaminate, by their biodegradative metabolism, environments that have received petroleum pollutants. Higher-molecular-weight compounds, especially those with multiple condensed ring structures and with highly branched or substituted compounds, are relatively resistant to microbial attack. Despite the fact that a genetically engineered

Tab. 1. The order of magnitude of hydrocarbon biodegradation rates. (From Ref. 16.)

System	Order of magnitude ($gm/m^3/day$)	References*
1. High-density monocultures in fermenters, optimal conditions	10,000-100,000	26, 47
2. Seawater community under partially optimized conditions	5-2,500	12, 18, 22, 70
3. In situ marine enrichment (long incubation times)	0.5-60	9, 10, 11, 12
4. In situ marine potential (short incubation times)	0.001-0.030	48, 64, 65

*The cited references served as basis for calculations; the above listed figures do not necessarily appear in the papers in this particular form.

hydrocarbon degrader was the first organism ever patented and that seed cultures are produced by various commercial firms, enhanced biodegradation as a result of seeding generally has not been shown to be effective. Also, even though some anaerobes have now been demonstrated to be capable of hydrocarbon metabolism, hydrocarbons persist indefinitely in anoxic environments. Environmental modification, on the other hand, such as that achieved by aeration or fertilization with nitrogen and phosphorus, has been shown to enhance biodegradative removal of hydrocarbons.

Having considered the various factors that influence the rates of hydrocarbon biodegradation, we are left with the question of what to do when environmental oil contamination occurs in order to minimize its persistence and thus its long-term effects. Clearly, treatment methods should enhance rather than inhibit the natural rates of oil biodegradation. In some cases, it is possible to modify environmental parameters to enhance rates of hydrocarbon biodegradation, but such methods are rarely undertaken. The translation of our scientific knowledge of hydrocarbon biodegradation into practical applications remains a major challenge. Specifically designed organisms are needed to degrade toxic aromatic components of refinery waste streams before environmental treatment. Specially designed reactors with specific microbial populations are also needed if oily sludges are to be degraded by biological means, either aerobically or anaerobically, in contained, environmentally safe reactors.

REFERENCES

1. Atlas, R.M. (1975) Effects of temperature and crude oil composition on petroleum biodegradation. Appl. Microbiol. 30:396-403.
2. Atlas, R.M. (1977) Stimulated petroleum biodegradation. Crit. Rev. Microbiol. 5:371-386.
3. Atlas, R.M. (1981) Fate of oil from two major oil spills: Role of microbial degradation in removing oil from the Amoco Cadiz and Ixtoc I spills. Environ. Int. 5:33-38.
4. Atlas, R.M. (1981) Microbial degradation of petroleum hydrocarbons: An environmental perspective. Microbiol. Rev. 45:180-209.
5. Atlas, R.M., ed. (1984) Petroleum Microbiology, Macmillan Publishing Company, Inc., New York.
6. Atlas, R.M., and R. Bartha (1972) Degradation and mineralization of petroleum in seawater: Limitation by nitrogen and phosphorus. Biotechnol. Bioeng. 14:309-317.
7. Atlas, R.M., and R. Bartha (1973) Effects of some commercial oil herders, dispersants and bacterial inocula on biodegradation of oil in seawater. In The Microbial Degradation of Oil Pollutants, D.G. Ahearn and S.P. Meyers, eds. Publication No. LSU-SG-73-01, Center for Wetland Resources, Louisiana State University, Baton Rouge, Louisiana, pp. 283-289.
8. Atlas, R.M., and R. Bartha (1973) Fate and effects of oil pollution in the marine environment. Residue Rev. 49:49-85.
9. Atlas, R.M., and R. Bartha (1973) Stimulated biodegradation of oil slicks using oleophilic fertilizers. Environ. Sci. Technol. 7:538-541.
10. Atlas, R.M., and A. Bronner (1981) Microbial hydrocarbon degradation within intertidal zones impacted by the Amoco Cadiz oil spillage. In Proceedings of the International Symposium on the Amoco Cadiz: Fates and Effects of the Oil Spill, Centre National Pour l'Exploitation des Oceans, Paris, pp. 251-256.

11. Atlas, R.M., P.D. Boehm, and J.A. Calder (1981) Chemical and biological weathering of oil from the Amoco Cadiz spillage within the littoral zone. Estuarine, Coastal and Shelf Sci. 12:589-608.
12. Atlas, R.M., G. Roubal, A. Bronner, and J. Haines (1980) Microbial degradation of hydrocarbons in mousse from IXTOC-I. In Proceedings of Conference on Researcher/Pierce IXTOC-I Cruises, National Oceanographic and Atmospheric Administration, Atlantic Oceanographic and Meteorological Laboratories, Miami, Florida, pp. 1-24.
13. Atlas, R.M., G. Roubal, A. Bronner, and J. Haines (1982) Biodegradation of hydrocarbons in mousse from the IXTOC-I well blowout. In Energy and Environmental Chemistry: Volume I - Fossil Fuels, L.H. Keith, ed. Ann Arbor Science Publishers, Ann Arbor, Michigan, pp. 199-217.
14. Bailey, N.J.L., A.M. Jobson, and M.A. Rogers (1973) Bacterial degradation of crude oil: Comparison of field and experimental data. Chem. Geol. 11:203-221.
15. Bartha, R., and R.M. Atlas (1977) The microbiology of aquatic oil spills. Adv. Appl. Microbiol. 22:225-266.
16. Bartha, R., and R.M. Atlas (1987) Transport and transformation of petroleum in biological processes. In Long-Term Environmental Effects of Offshore Oil and Gas Development, D.F. Boesch and N.N. Rabalais, eds. Elsevier Applied Science, London, pp. 287-341.
17. Buckley, E.N., R.B. Jonas, and F.K. Pfaender (1976) Characterization of microbial isolates from an estuarine ecosystem: Relationship of hydrocarbon utilization to ambient hydrocarbon concentrations. Appl. Environ. Microbiol. 32:232-237.
18. Caparello, D.M., and P.A. LaRock (1975) A radioisotope assay for the quantification of hydrocarbon biodegradation potential in environmental samples. Microb. Ecol. 2:28-42.
19. Chouteau, J., E. Azoulay, and J.C. Senez (1962) Anaerobic formation of n-hept-1-ene from n-heptane by resting cells of Pseudomonas aeruginosa. Nature (London) 194:576-578.
20. Colwell, R.R., and J.D. Walker (1977) Ecological aspects of microbial degradation of petroleum in the marine environment. Crit. Rev. Microbiol. 5:423-445.
21. Colwell, R.R., A.L. Mills, J.D. Walker, P. Garcia-Tello, and V. Campose-P (1978) Microbial ecology studies of the Metula spill in the Straits of Magellan. J. Fish. Res. Board Can. 35:573-580.
22. Coty, V.F., and R.I. Leavitt (1971) Microbial protein from hydrocarbons. Dev. Ind. Microbiol. 12:61-71.
23. Crow, S.A., S.P. Meyers, and D.G. Ahearn (1974) Microbiological aspects of petroleum degradation in the aquatic environment. La Mer 12:37-54.
24. Davis, J.B. (1967) Petroleum Microbiology, Elsevier Publishing Company, New York.
25. Delaune, R.D., G.A. Hambrick, and W.H. Patrick, Jr. (1980) Degradation of hydrocarbons in oxidized and reduced sediments. Mar. Pollut. Bull. 11:103-106.
26. Dibble, J.T., and R. Bartha (1976) The effect of iron on the biodegradation of petroleum in seawater. Appl. Environ. Microbiol. 31:544-550.
27. Dibble, J.T., and R. Bartha (1979) Effect of environmental parameters on the biodegradation of oil sludge. Appl. Environ. Microbiol. 37:729-739.
28. Floodgate, G.D. (1976) Oil biodegradation in the oceans. In Proceedings of the Third International Biodegradation Symposium, J.M. Sharpley and A.M. Kaplan, eds. Applied Science Publishers, Ltd., London, pp. 87-92.

29. Floodgate, G.D. (1978) The formation of emulsifying agents in hydrocarbonoclastic bacteria. In Microbial Ecology, M.W. Loutit and J.A.R. Miles, eds. Springer-Verlag, Berlin, pp. 82-85.
30. Floodgate, G.D. (1979) Nutrient limitation. In Microbial Degradation of Pollutants in Marine Environments, A.W. Bourquin and P.H. Pritchard, eds. EPA-66019-79-012, Environmental Research Laboratory, Gulf Breeze, Florida, pp. 107-118.
31. Foster, J.W. (1962) Hydrocarbons as substrates for microorganisms. Antoine van Leeuwenhoek J. Microbiol. Serol. 28:241-274.
32. Gibbs, C.F., K.B. Pugh, and A.R. Andrews (1975) Quantitative studies in marine biodegradation of oil. II. Effects of temperature. Proc. Royal Soc. London Ser. B. 188:83-94.
33. Gibson, D.T. (1968) Microbial degradation of aromatic compounds. Science 161:1093-1097.
34. Gordon, D.C., J. Dale, and P.D. Keizer (1978) Importance of sediment working by the deposit-feeding polychaete Arenicola marina on the weathering rate of sediment bound oil. J. Fish. Res. Bd. Can. 35:591-603.
35. Grbic-Galic, D. (1986) Anaerobic production and transformation of aromatic hydrocarbons and substituted phenols by ferulic acid-degrading BESA-inhibited methanogenic consortia. FEMS Microbiol. Ecol. 38:161-169.
36. Grbic-Galic, D., and T.M. Vogel (1987) Transformation of toluene and benzene by mixed methanogenic cultures. Appl. Environ. Microbiol. 53:254-260.
37. Guire, P.E., J.D. Friede, and R.K. Gholson (1973) Production and characterization of emulsifying factors from hydrocarbonoclastic yeasts and bacteria. In The Microbial Degradation of Oil Pollutants, D.G. Ahearn and S.P. Meyers, eds. Publication No. LSU-SG-73-01, Center for Wetland Resources, Louisiana State University, Baton Rouge, Louisiana, pp. 229-231.
38. Gundlach, E.R., P.D. Boehm, M. Marchand, R.M. Atlas, D.M. Ward, and D.A. Wolfe (1983) The fate of the Amoco Cadiz oil. Science 221:122-129.
39. Gunkel, W. (1968) Bacteriological investigations of oil-polluted sediments from the Cornish coast following the Torrey Canyon disaster. Helgol. Wiss. Meeresunters 17:151-158.
40. Gutnick, D.L., and E. Rosenberg (1977) Oil tankers and pollution: A microbiological approach. Ann. Rev. Microbiol. 31:379-396.
41. Haines, J.K., and M. Alexander (1974) Microbial degradation of high-molecular weight alkanes. Appl. Microbiol. 28:1084-1085.
42. Haines, J.R., and R.M. Atlas (1983) Biodegradation of petroleum hydrocarbons in continental shelf regions of the Bering Sea. Oil and Petrochem. Poll. 1:85-96.
43. Heitkamp, M.A., J.P. Freeman, and C.E. Cerniglia (1987) Naphthalene biodegradation in environmental microcosms: Estimates of degradation rates and characterization of metabolites. Appl. Environ. Microbiol. 53:129-136.
44. Hughes, D.E., and P. McKenzie (1975) The microbial degradation of oil in the sea. Proc. Royal Soc. London Ser. B. 189:375-390.
45. Iizuka, H., M. Iida, and S. Fujita (1969) Formation of n-decene-1 from n-decane by resting cells of C. rugosa. Z. Allg. Mikrobiol. 9:223-226.
46. Jordan, R.E., and J.R. Payne (1980) Fate and Weathering of Petroleum Spills in the Marine Environment, Ann Arbor Science Publishers, Ann Arbor, Michigan.

47. Kanazawa, M. (1975) Production of yeast from n-paraffins. In Single Cell Protein, Vol. 2, S.R. Tannenbaum and D.I.C. Wang, eds. MIT Press, Cambridge, Massachusetts, pp. 438-453.
48. Lee, R.F. (1977) Fate of petroleum components in estuarine waters of the southeastern United States. In Proceedings of the 1977 Oil Spill Conference, American Petroleum Institute, Washington, D.C., pp. 611-616.
49. LePetit, J., M.-H. N'Guyen, and S. Tagger (1977) Quelques donnees sur l'ecologie d'une zone marine littorales recevant les rejets d'une raffinerie de petrole. Environ. Pollut. 13:41-56.
50. Merkel, G.J., S.S. Stapleton, and J.J. Perry (1978) Isolation and peptidoglycan of gram-negative hydrocarbon-utilizing thermophilic bacteria. J. Gen. Microbiol. 109:141-148.
51. Mironov, O.G. (1970) Role of microorganisms growing on oil in the self-purification and indication of oil pollution in the sea. Oceanology 10:650-656.
52. Mulkins-Phillips, G.J., and J.E. Stewart (1974) Effect of environmental parameters on bacterial degradation of Bunker C oil, crude oils and hydrocarbons. Appl. Microbiol. 28:915-922.
53. National Academy of Sciences (1975) Petroleum in the Marine Environment, National Academy of Sciences, Washington, D.C.
54. National Academy of Sciences (1985) Oil in the Sea: Inputs, Fate, and Effects, National Academy of Sciences, Washington, D.C.
55. Oppenheimer, C.H., W. Gunkel, and G. Gassman (1977) Microorganisms and hydrocarbons in the North Sea during July-August 1975. In Proceedings of the 1977 Oil Spill Conference, American Petroleum Institute, Washington, D.C., pp. 593-610.
56. Parekh, V.R., R.W. Traxler, and J.M. Sobek (1977) n-Alkane oxidation enzymes of a Pseudomonad. Appl. Environ. Microbiol. 33:881-884.
57. Perry. J.J. (1977) Microbial metabolism of cyclic hydrocarbons and related compounds. Crit. Rev. Microbiol. 5:387-412.
58. Perry. J.J. (1979) Microbial cooxidations involving hydrocarbons. Microbiol. Rev. 43:59-72.
59. Pfaender, F., and E.N. Buckley (1980) Response of the pelagic community to oil from the IXTOC-I blowout: In situ studies. In Proceedings of Conference on Researcher/Pierce IXTOC-I Cruises, National Oceanographic and Atmospheric Administration, Atlantic Oceanographic and Meteorological Laboratories, Miami, Florida, pp. 545-560.
60. Pirnik, M.P. (1977) Microbial oxidation of methyl branched alkanes. Crit. Rev. Microbiol. 5:413-422.
61. Ratledge, C. (1978) Degradation of aliphatic hydrocarbons. In Developments in Biodegradation of Hydrocarbons-1, J.R. Watkinson, ed. Applied Science Publishers, Ltd., London, pp. 1-46.
62. Raymond, R., and V. Jamison (1987) The microbiology of bioreclamation. In Abstracts of Annual Meeting of the American Society for Microbiology, American Society of Microbiology, Washington, D.C.
63. Reisfeld, A., E. Rosenberg, and D. Gutnick (1972) Microbial degradation of oil: Factors affecting oil dispersion in seawater by mixed and pure cultures. Appl. Environ. Microbiol. 24:363-368.
64. Robertson, B., S. Arhelger, P.J. Kinney, and D.K. Button (1973) Hydrocarbon biodegradation in Alaskan waters. In The Microbial Degradation of Oil Pollutants, D.G. Ahearn and S.P. Meyers, eds. Publication No. LSU-SG-73-01, Center for Wetland Resources, Louisiana State University, Baton Rouge, Louisiana, pp. 171-184.

65. Seki, H. (1976) Method for estimating the decomposition of hexadecane in the marine environment. Appl. Environ. Microbiol. 31:439-441.

66. Senez, J.C., and E. Azoulay (1961) Dehydrogenation of paraffinic hydrocarbons by resting cells and cell free extracts of Pseudomonas aeruginosa. Biochim. Biophys. Acta 47:307-316.

67. Spain, J.C., P.H. Pritchard, and A.W. Bourquin (1980) Effects of adaptation on biodegradation rates in sediment/water cores for estuarine and freshwater environments. Appl. Environ. Microbiol. 40:726-734.

68. Traxler, R.W., and J.M. Bernard (1969) The utilization of n-alkanes by Pseudomonas aeruginosa under conditions of anaerobiosis. Int. Biodeterior. Bull. 5:21-25.

69. Vogel, T.M., and D. Grbic-Galic (1986) Incorporation of oxygen from water into toluene and benzene during anaerobic fermentative transformation. Appl. Environ. Microbiol. 52:200-202.

70. Walker, J.D., R.R. Colwell, and L. Petrakis (1976) Biodegradation rates of components of petroleum. Can. J. Microbiol. 22:1209-1213.

71. Ward, D.M., and T.D. Brock (1978) Anaerobic metabolism of hexadecane in marine sediments. Geomicrobiol. J. 1:1-9.

72. Ward, D.M., R.M. Atlas, P.D. Boehm, and J.A. Calder (1980) Microbial biodegradation and the chemical evolution of Amoco Cadiz oil pollutants. Ambio 9:277-283.

73. Zajic, J.E., and C.J. Panchal (1976) Bio-emulsifiers. CRC Crit. Rev. Microbiol. 5:39-66.

74. Zeyer, J., E.P. Kuhn, and P. Schwarzenbach (1986) Rapid microbial mineralization of toluene and 1,3-dimethylbenzene in the absence of molecular oxygen. Appl. Environ. Microbiol. pp. 944-947.

75. ZoBell, C.E. (1969) Microbial modification of crude oil in the sea. In Proceedings of Joint Conference on Prevention and Control of Oil Spills, American Petroleum Institute, Washington, D.C., pp. 317-326.

76. ZoBell, C.E., and J. Agosti (1972) Bacterial oxidation of mineral oils at sub-zero Celsius. In Abstracts of the 72nd Annual Meeting of the American Society for Microbiology, Philadelphia, April 23-28, Abstr. E11.

DISCUSSION

Michael Weishaar (Chairman), Monsanto Chemical Company

KAUFMAN: Dr. McCarty, could you address the issue of microorganism threshold levels with reference to cleanup standards for drinking water in the aquifer?

McCARTY: We've essentially found that with most contaminants, if we have used another compound as the primary substrate, the contaminants can be transformed down to the lower parts per trillion level. Even when used as a primary substrate, the target chemicals can often be reduced to concentrations that are below the standard. The rate of utilization, however, is proportional to the concentration, so that the effort required to get down to low concentrations may be great.

KAUFMAN: Are you suggesting that, in field implementation, there is a distinct advantage in using the co-metabolic activity rather than the primary substrate type of approach?

McCARTY: Depending upon the compound with primary substrates, you might get into the low parts per billion. Generally, to get the concentration lower, however, the organism needs some other energy source.

CAMPION: What is your feeling as to the locus of biotransformation for contaminants adsorbed on soil? Is it necessary for the contaminant to desorb before it can react, or is there a possibility that the action is actually surficial on soil particles?

McCARTY: There's evidence supporting both possibilities. From the limited literature on this subject, whether sorbed compounds can be transformed directly depends on the compound, the soil matrix, and the particular organisms. I have used as a working hypothesis that the compound has to be in solution to be transformed. Some of our evidence with enphanstrophs is suggestive, but we don't have a definitive study.

BARNTHOUSE: It seems that one of the major problems in trying to degrade toxic contaminants in the subsurface environment is the engineering difficulty of creating the appropriate physical and chemical conditions necessary for the reaction to proceed. For instance, the right amount of methane to promote growth and not inhibit it. Do we have any engineering solutions?

McCARTY: I haven't been able to think of any methods to obtain spectacular improvements. This problem is difficult and certainly needs some good ideas.

IGBENE: We found that the primary contaminants in the Silicon Valley area are trichloroethylene (TCE) and tetrachloroethylene. What is the source of TCE?

McCARTY: Trichloroethylene has been used as a solvent quite widely, so in many cases it has become the primary contaminant in groundwater.

IGBENE: Is there evidence for any transformation reaction of the tri- to the tetrachloroethylene?

McCARTY: No, I haven't seen anything suggesting such transformation. However, one can get the tetra- from hexachloroethane, as I mentioned.

ALEXANDER: If species diversity in aquifers is low, will organisms move with the plume?

McCARTY: There's pretty good evidence that organisms can move within aquifers. They can move with the plume if they are present in the aquifer. However, not much is known about the rates of movement.

LEWIS: Dr. Pflug, in many of these creosote-contaminated soils from the wood-preserving industry, there also occurs a certain amount of dioxin. Have you looked into what would be done if the remaining dioxin level is high enough to prevent delisting?

PFLUG: We're currently conducting two separate investigations into the ultimate fate, using the original reactor we put on the Minnesota site. We have confirmed at this point that transformation does not occur.

LEWIS: I would expect that the dioxin probably would not be a problem with the water pump; but there would be a problem if you're treating soils and were going to delist them.

FOCHT: Dr. Nelson, does strain G4 grow on chlorobenzene? Have you tried that?

NELSON: We did briefly, at first, and didn't get significant growth. However, subsequent work indicated that the concentrations we tried may have been toxic. So I can't say for sure, but the data that we have say that it does not.

LITTLE: Have you made any attempt to obtain constitutive mutants that would be able to degrade TCE without the aromatic inducers?

NELSON: Yes, but at present we haven't had any success. We're continuing to look at the genetics of the system in hopes of being able to clone it.

LITTLE: Do you know what the breakdown products are? In addition to a lot of CO_2, you said there might be about a third that was water-soluble.

NELSON: After CO_2 is extracted, the majority of the aqueous fraction is actually particulate associated with the cell material. We assume that it is cell carbon.

LITTLE: You might be able to develop a screening procedure to look for a mutant like that by reacting something with CO_2 to form a colored colony.

NELSON: Trichloroethylene is not the substrate, but it is fortuitously degraded by the organism. So the selection process of the TCE degradation must be indirect.

QUESTION: Dr. Brubaker, in your emergency response, I was confused as to the mechanisms responsible for obtaining lower concentrations in situ. Additional blotter would be diluting it, and increasing the temperature would increase the volatility. Could you clarify whether the immediate effects obtained were due to microbial degradation or to physical means? Also, being mindful that capillarity and volatility are primary mechanisms for transport in recycling, what precautions are taken to prevent further dispersion of the waste?

BRUBAKER: Obviously, within the field environment, one can't do as much of the kinds of monitoring to look at some of the alternative mechanisms as one does in a laboratory. There have been preliminary studies done in the laboratory with microorganisms in sealed flasks to verify that microbial activity will, in fact, do what we want it to do. No rigorous monitoring controls were done in the field to look at some of the counteracting influences. Volatilization from the aeration tank is always a possibility. We did monitor microbial populations, which increase and decrease, as you'd expect. There were drainage ditches on both sides of the contaminated area which were fed into a sump and recycled into the recirculation system.

QUESTION: Dr. Crawford, in a carbon-limited system with the addition of a secondary nontoxic carbon source, the phenomenon of catabolic repression is mitigated, thereby giving rise to simultaneous degradation rather than preferential sequential degradation. This then increases the biomass, which in turn allows the system to be more efficient and cost-effective, with a decrease in capital investment being paramount. Please comment on the applicability of this method in aquifer and zone reclamation, with mixed microflora, and in above-ground bioreactors with pure cultures.

CRAWFORD: What I refer to as an ancillary carbon source can be useful to build biomass, especially in the early stages of colonization of some of these supports. It does work well, particularly when the contaminants to be treated are at a high concentration so that the toxicity itself is a selective pressure. There is the danger in using an easily metabolized ancillary carbon source, such as acetate for example, that you may tend at some point to select a microbial population that is tolerant of the pollutant and that doesn't degrade it, yet builds biomass at the expense of the acetate. You have to monitor that very carefully. We have used an ancillary carbon source successfully, but we've also had a few problems. With certain pure cultures, such as flavobacteria, an ancillary carbon source does help. It speeds the reaction rate, as well as degradation, and it helps prevent toxicity.

TORPY: How long did it take to form the films on your ceramic saddles before you got into an operational mode?

CRAWFORD: The original biofilm, as I mentioned, came from natural materials in the experimental channels at the Monticello field station. Those channels were subjected to PCP contamination for some months. Stones were collected from the ripples, brought into the laboratory, and then continuously exposed to PCP for more than a year. That was a very mature film. The film was then transferred to the ceramic saddles. That took a relatively short period of time compared to the maturing stage. It's now possible to transfer that film in a few weeks.

TORPY: Did you encounter any problems with plugging or did you go through a backwash?

CRAWFORD: Thus far, we've had no problems with plugging. We can go through a backwash and a slough-off, if that's a problem. Thus far, it hasn't been a significant problem.

LEWIS: First, a comment to Perry McCarty concerning delivery/recovery systems. Our remedial action program at EPA is funding a number of studies in different recovery systems, including electro-osmosis and hydrofracturing. We're also getting information that is being collected in other countries.

Dr. Atlas, we have experienced problems at times, using hydrogen peroxide for the oxygen source in in situ treatment. This is because hydrogen peroxide often must be stabilized with fairly high amounts of phosphate to avoid immediate breakdown. Especially with soils containing a high amount of iron, you can get plugging. Do you have any special recommendations as far as the method of delivery of the hydrogen peroxide?

ATLAS: One has to be very cautious about what else you have to add; in fact, you do have to add something to maintain stability of the hydrogen peroxide. But I don't have any special methods. We've not been working in that area of stabilizing peroxides. There certainly are many site-specific problems that result from using oxygen carriers, such as iron, calcium, and magnesium. Calcium and magnesium are not a problem because of the peroxide, but because of the phosphates. We have formulated nutrient blends which combine polyphosphates as well as monophosphates, thus allowing one to sequester some of the calcium and magnesium while stabilizing the peroxide.

LEWIS: Both the Air Force and the EPA are funding some studies at Virginia Polytechnic Institute to evaluate use of microfine gas bubbles for oxygen delivery systems.

KAUFMAN: Dr. Nelson, in relation to the use of phenol as an inducer to TCE degradation, what was the concentration you used or ratio of inducer to the actual substrate?

NELSON: Substrate concentrations were typically around part per million. Inducer concentrations were usually around 1 to 2 mM. Typically, when we grow the organism, we grow it in an alternate carbon source besides the aromatic inducer, and then add 1 to 2 mM of the inducer to activate the aromatic pathways.

KAUFMAN: Among the various environmental features that have direct or indirect influence on the potential effects of bioremediation in the field vs the effects that we see in the laboratory, pH may be a very significant factor.

ATLAS: In most of our work the pH problem is not great, since we have a well-buffered system in the marine environment. In fact, that's been the case also when we've looked at contaminated soil and water systems. We did not find any natural alteration of more than a 0.1 pH unit in those systems. So far we've not had to consider pH as a problem, nor have we had to formulate anything exotic to maintain a buffering capacity.

McCARTY: Certainly, pH is among the environmental variables that need to be controlled. Although the groundwater is naturally buffered, the pH can be of concern when mixtures of waste and different alkalinities of water are involved.

KAUFMAN: One of our projects involves nitroaromatic treatment in 40 acres of soil. The pH varies considerably from site to site, and the organisms that we find that are capable of degrading various nitroaromatic isomers have well-defined pH optima. These optima were not reflected in the soil in all situations and, therefore, we had to look at some environmental modification to provide the pH optimum. The pH is one of the most significant factors in the overall biodegradative efficiency in this project.

HAGEDORN: Dr. Nelson, I noticed that you were working with moderately high cell densities. What is the ratio of the specific activity of the oxidation of phenol to the oxidation rate toward TCE? What happens to the oxidation of the TCE when you go to lower cell densities?

NELSON: Our cell densities are a 1:10 dilution of the late log cell growth. I can't recall if we've actually operated at lower cell densities. The organism will completely degrade phenol, since it can serve as a carbon source, but we haven't really investigated the kinetics. Phenol oxidation is not required for TCE degradation. In the chloramphenicol experiment I reported, we did not add phenol because the cells that had been preinduced with phenol no longer required that substrate.

HAGEDORN: Dr. McCarty, you had mentioned that anaerobic metabolism of vinyl chloride to carbon dioxide occurs at a very slow rate. Could you say something about the relative rate of anaerobic metabolism of vinyl chloride vs its migration through soil into the groundwater?

McCARTY: Vinyl chloride, which quite frequently is detected in groundwater, is probably a product of the transformation of PCE and TCE. In some cases, it does not move at all; in other cases, it can move fairly fast. As far as transformation is concerned, sometimes it doesn't occur at all and sometimes it does, depending upon the environment and microorganisms present. Thus, transformation and migration are quite different aspects. In the laboratory studies in which we observed transformation of vinyl chloride over long periods of time, the rates of transformation seemed to be a couple of orders of magnitude less than the rates of, say, TCE transformation. In fact, vinyl chloride transformation rates are so low under anaerobic conditions that unless we find some other organism, I don't believe natural mineralization of vinyl chloride under anaerobic conditions will be significant from a practical standpoint.

CAMPION: I'd like Gayle Brubaker and Dale Pflug to comment on two points. First, what considerations do you use in choosing injection and recovery mechanisms for your groundwater recirculation systems? Second, how much of the degradation in the case studies you presented actually occurs in sites below ground and how much occurs in the mixing tanks or pile reactors above ground?

BRUBAKER: The hydrogeological considerations for injection vs recovery are somewhat site-specific. Most important is the depth of the beta zone, the distance between the surface and the water table. One needs to create some kind of mounding in order to induce flow. If the water table is very near the surface, then one is limited to what you can do with wells and use of galleries. Another factor is the depth of the water table. The thickness of the saturated zone is very important in controlling the zone of influence of a pumping well or the construction of a gallery. These considerations, plus physical constraints of where the railroad tracks or the intersection are, usually lead you to determine whether the situation is appropriate for a gallery or a trench. Then one needs to sit down with a good hydrogeological modeler to figure out the specific design of injection and recovery.

With regard to how much of the activity is actually in situ compared to how much is just being flushed and treated on the surface, for gasoline we typically find from material balances that 90 to 95% is in situ and maybe 5% is captured on carbon or captured and dislodged through air-strippers. It varies, of course. With diesel, which has less solubility, you'll find that a higher percentage is in situ. With something like ethylene glycol or MEK, we typically would not do in situ bioreclamation, because these compounds are so soluble and they flush fairly readily. In general, the in situ approach is most attractive for those pollutants which are least soluble.

PFLUG: On the question of the recharge back to the subsurface, the biggest factor usually is the specific permit requirements associated with that recharge. There are specific prohibitions in many individual states. In our experience, the transformation from activity in the surface reactor vs the subsurface is a function also of concentration. As your pumping operation goes on, you may have a transition from the levels at the surface vs the levels in the subsurface.

BULL: I live approximately 600 kilometers from the Amoco Cadiz disaster site. I want to add a tailpiece to Ron Atlas' comment which I think may be a little disturbing. I read in the last few weeks that there are still long-term effects being felt from that oil spill--signs of significant teratological problems appearing in benthic-feeding fish. I think that particular situation does give us pause for longer-term concern about the introduction of hazardous waste into the marine environment.

PRYFOGLE: Problems with the release of genetically engineered microorganisms in the environment seem to relate to whether they are natural and trying to define what is natural and what is unnatural. What about pH and nutrient adjustments? What regulatory constraints can you address in changing what was once a natural environment and is now becoming an unnatural environment?

McCARTY: Any time one proposes to inject something into the groundwater, people do get very concerned. On the other hand, once there is contamination, the natural, background environment no longer exists. So the argument then has to be the one of benefit. One has to look at all the trade-offs. So, beyond the question of adding nutrients, whether one can add an organism and whether or not the organism is natural raises many questions. The adding of engineered organisms is another realm. Now, getting permits for any alterations is always difficult.

SPIKER: Dr. Atlas, how long after a typical oil spill does it take for the oil degraders to appear, and what effect does the concentration of oil have on the rate of degradation or the rate of appearance?

ATLAS: The length of time it takes before the organisms appear varies with the environment. At the Amoco Cadiz spill, we found very rapid colonization and degradation of the oil. Within a couple of days, we were seeing extensive microbial populations degrading the oil. The reason is that the area was well adapted, because the community had been experiencing exposure to petroleum hydrocarbons from ballast discharges for a long time. In the course of collecting samples, we obtained some very unusual results. We were monitoring populations of hydrocarbon degraders; samples in a particular site shot up four orders of magnitude within 24 hours. It made no sense until we picked up the newspaper the next day and found there was yet another spillage. The oil happened to impact very close to one of our sampling sites and not the others. In other settings, such as in arctic systems, we observe much slower responses to spillage of oil.

The main question relating to the concentration of oil is whether toxic components are present initially in the oil that need to dissipate before any extensive colonization occurs. That varies very much with whether it is a refined oil or a crude oil. Crude oils tend not to have such toxic effects. Nor do we find the concentration being at the other extreme, that is, so low that you don't get a threshold for a response. So we've not seen much of a concentration effect.

OMENN: With regard to permits, in these specific projects that have been described, are there EPA regional permits to be obtained, as well as state, and sometimes local, permits? What is the process for coordinating the kinds of information that must be obtained? Are there specific requirements for monitoring the disappearance of the target chemicals and for assuring either that the mineralization is complete or that other potential or predictable compounds have been assayed?

PFLUG: First, as to the types of permits, they are a function of where you're discharging to and the particular state of application. In some cases, it's a state permit versus an EPA permit. It's who has authority in the particular region. In the cases of surface recharge, the principal concern seems to be whether it is a listed or nonlisted waste. But that's just one consideration associated with it.

BRUBAKER: There's sometimes a misconception that agencies permit processes. Our experience is that agencies do not permit processes as a whole, but instead allow you to put in a particular well or to inject a particular nutrient. They do not give a blessing on an overall process, except to say that none of the individual things which you were doing appear to be causing harm anywhere else. The implication is that the contractor and the person who's paying the bill have ultimate responsibility for performance, and the agency and person offering that permit aren't giving their endorsement to the process. We get involved in reinjection permits whose interpretation varies from state to state and EPA region to EPA region. Minnesota, for example, does not allow injection of anything into the ground. So, in situ bioreclamation processes are by law prohibited, unless there is some reinterpretation or redrafting of that law. Minnesota state is the only place in the nation, that I'm aware of, where injection itself is a problem--although one has to be very careful to have approval for the construction of the well and the monitoring routine. It's through the process of monitoring and demonstrating control that the permits are issued.

OMENN: Many of us are interested to know, as experience is gained, how well the analytical results from the lab- or pilot-scale operation can be translated out into the field. In these examples you have presented, how much detailed information was required and how much did you choose to obtain about the pattern of removal of the target chemicals and the demonstration of lack of appearance of any other intermediates that might be of concern, such as the ones that Dr. McCarty illustrated?

BRUBAKER: Obviously, there's a giant leap between a two-liter flask batch reactor and fifty feet between injection well and recovery well. Mass transport becomes very limiting. It's very unusual to see oxygen levels in the groundwater increase over several parts per million, even when you're injecting at 250 parts per million, because the oxygen consumption is so rapid. The microbial processes are much more rapid than mass transport. Thus, one uses a laboratory study only as an indication of what processes one can expect. The real world is often very different from the particular laboratory column studies. There are always gross differences in the field in terms of rate, but we don't see major differences in terms of the efflux and influx compounds.

CRAWFORD: In our studies with the 200-liter bioreactor, we had a commitment to meet a certain effluent standard. This is research and development, of course, so we were doing HPLC analyses daily and guaranteeing that the discharged water met certain standards. We were committed to report the results.

POGEL: If you go several meters into the soil, where the *Streptomyces* dominate, does anyone have information on the kinds of organisms one finds?

McCARTY: There have been several studies showing that there is a great diversity of organisms, primarily bacteria, but even some protozoa, even in deep aquifers. There are many bacteria down there--about 10^4 to 10^7 bacteria per gram of soil in very deep aquifers. There was a misconception for a number of years that such aquifers are sterile; they certainly are not. The extent of diversity is still an open question.

POGEL: Are there both aerobes and anaerobes?

McCARTY: Yes, varying with the particular environment. Of course, there's a whole range of anaerobes. In some aquifers that we studied with high sulfate, some nitrate, and no oxygen, we couldn't find any methanogens. In other situations, methanogens are quite prevalent. We puzzled over why methanogens were in two locations in an aquifer and not in others. It was identified that a sanitary sewer ran over those particular locations, possibly leaking and seeding the aquifer with additional organisms.

LITCHFIELD: Some elegant work done by Dick Raymond back in the early 1970's--and some recent work of ours--has shown quite a diverse population, including Streptomyces. Initially, you will get one or two colony types. If you set the plates aside and come back four weeks later, you'll find a mixture of organisms.

AKALI: Dr. Nelson, what are the possibilities that in your experiment there was a consortium of microorganisms in the groundwater? Is it sufficient to use your G4 strain and your phenol-induced organism to treat TCE, or was there a consortium of microorganisms in the groundwater you were treating that completed the entire degradation process?

NELSON: Initially, we were using the site water, so there was the possibility that some other component might have been necessary. However, a majority of our work has been in a completely defined system using distilled water and basal salts, so that the only addition to the system besides basal salts is the aromatic inducer. There is nothing else besides the aromatic inducer that is necessary for maintaining the activity of the organism.

TAUB: When you are dealing with a historically dirty site or a site that has potential multiple insults, impressive chemical analyses are directed at the target chemical. Are any bioassays routinely used to see whether the material is, in fact, made considerably less toxic, or whether there is any residual toxicity perhaps from side products or from other contaminants? Are bioassays routinely used in addition to the chemistry?

BRUBAKER: Most of our experience to date has been with the petroleum hydrocarbon types of spills where bioassays have not been routinely used. However, as people move toward the more complex Superfund-type, mixed-contaminant spills, I do, in fact, expect to see a lot of bioassay work being done, either prior to reinjection of groundwater or prior to offsite discharge of groundwater.

OMENN: In about 20 oil drilling sumps in Alberta, Canada, that were biologically decontaminated by Ecova Corporation, a 96-hour fish-kill bioassay was used routinely to demonstrate that the clarified aqueous phase could be released for irrigation.

PEELER: My understanding is that the EPA is looking at the standards for bioassays. Dr. Crawford, I wonder whether, through the testing and research and development that are being done, you have included bioassays in the research and development that is being done?

CRAWFORD: They are not included in our work routinely, but we have on occasion done bioassays, such as fish bioassays with fathead minnows, demonstrating that the material to be treated is highly toxic. The fish bioassay involved a pentachlorophenol which is very toxic to minnows. We have demonstrated that inoculation with the degraders completely detoxifies this compound. Actually, inoculation of the bacterium simultaneously with the addition of minnows does protect them, showing by bioassay that there is detoxification occurring in this process.

PEELER: Dr. Brubaker, have you dealt with any contaminated sites where the groundwater is affected by either tidal influence or salt water intrusion, and what procedures have you used to clean up those sites?

BRUBAKER: We're currently looking at several sites that have one to two feet of tidal influence, nothing larger, but we've not implemented those sites yet. Obviously, that affects the smear, the pumping strategy, and the way that you monitor the site within the one- to two-foot zone. It's not a major problem. We saw large fluctuations in a demonstration site that kept us from starting it up during the winter. Instead, we started up injection in the spring as the water came up. I expect these patterns to follow as we get into other kinds of sites that have large seasonal, daily, or weekly influences based on rainfall or tidal influences. We have not at this point done anything in a brackish area, although we are getting ready to do a site. Our assessments in those types of sites suggest that, although the organisms and the environments may be different, hydrocarbon degraders are there. The trick is to make sure that you're not dramatically changing the alkalinity, pH, or whatever is native to a particular environment.

PEELER: I assume that the temperature is also going to change during that process, so there is a certain shock amount of diffusion. And are there problems with diffusion of the groundwater?

BRUBAKER: The temperature, we find, readjusts very quickly because of the thermal mass of the soil compared to the water, so we haven't found major temperature effects. Ideally, we recover groundwater down gradient and, in some cases, run it through activated carbon or surface treatment to cleanse it, if required, then add nutrients to that same water and reinject it. Thus, we are minimizing the change in the groundwater that's moving through the treated environment.

SPAIN: I have a question to all of the speakers from this afternoon. It seems that we have a lot of successful treatments at contaminated sites. I wonder if there are any that were unsuccessful, or if there is anything that you have discovered in the field that you might go back to the lab and learn more about? Is there anything the biochemists or microbiologists could help with, or should we just send them home now and have the engineers take over?

UNTERMAN: I think it's quite obvious that going from the laboratory into the field makes a tremendous difference for us. For example, temperature fluctuations in that tent that we've set up are tremendous, with air temperatures reaching 130°F. We have probes every few centimeters in the soil and up into the air. So, obviously, you have to deal with these kinds of situations. The soil is drying out on us. We have to develop a way to keep the bacteria wet. The surface is drying out so that the bacteria are dying in the top centimeter of that test. They're fine when you get below a centimeter. Then there is the viability of the organism. LB400 is an outstanding polychlorinated biphenyl (PCB)-degrading bug, but it's not very viable. We're now working with a bacterium on soil. Although it's not as good, it is more viable so we're seeing better degradation. These are some of the variables. We have a long way to go. It's just a first shot for us.

O'REILLY: Mr. St. John, in the bioreactor in your first system, do you just have free cells or is there biofilm formation?

ST. JOHN: Free cells.

LITCHFIELD: In your second case study, was the contamination confined to just a few feet of the surface of that site or was it much deeper?

ST. JOHN: It can generally be characterized as surface contamination.

LITCHFIELD: So because of the clay, it hadn't penetrated down to the core and you were able to excavate.

ST. JOHN: Right, it decreased with the depth.

CAMPION: In the last case study, what was the approximate unit cost to your client to treat the soil?

ST. JOHN: About $100 per cubic yard. The cost of this type of system is variable depending on the nature of the treatment process, but it is generally about half the cost of incineration.

TORPY: Dr. Furlong, you touched on the possibility of using proteins for specific identification of organics, but all of your work so far has been with metals. Is there a technical problem in making the transition to the organics, or is this just another simple step that you just haven't had time to address?

FURLONG: For the organics with the biosensors, you need to be able to detect a signal when the immobilized protein interacts with the ligand. We chose phosphate as a convenient model system. For any given organic, a specific binding protein must be identified, and we have several. For example, for glutamine we can at least bind ribose and cysteine, and those are targets for future development. We want to see if we can pick up the signals when those are bound to their proteins and immobilized on the silicon. So we plan to look at the organics. There are a lot of different ways of looking at the signals.

FOCHT: Dr. Isbister, could you tell us something about the organism you isolated that is responsible for organic sulfur removal? How did you find it, and is the released sulfate available for cell growth?

ISBISTER: I isolated it from a number of different soil samples. CB2 came from soil obtained from a coal mine in Pennsylvania. CB2 is the one that works pretty well, we think, with organically bound sulfides. CB1 actually came from the creek bed in my backyard. We took the microbial consortium that we got and looked at the wide spectrum of microbes we had. We tested over a period of time for any, and I mean <u>any</u>, activity with respect to dibenzothiophene (DBT).

Of course, there are a lot of organisms that degrade DBT or at least oxidize it. We found a couple, but activity was minimal: about 7% in seven days. I took that culture and subjected it to selective mutagenesis with rescue in between. Interestingly, I obtained a pure culture that survived mutagenesis, rescue, then high levels of DBT. We used DBT as a sole source of carbon, but DBT itself can be toxic to microorganisms in high concentrations. We had gone from 7% DBT oxidation in seven days to 90% in 18 hours overnight.

FOCHT: And is the sulfate used for cell growth?

ISBISTER: No. The microbes are grown totally on something else before I put them with the DBT. So this is a nongrowing cell. I guess you would call it a "fortuitous occasion."

OMENN: I have a question for Ron Unterman and Joe Watt about industrial hygiene and occupational health. As companies with relatively little experience in dealing with hazardous waste sites and hazardous waste streams are beginning to apply microbiology and engineering techniques, what kinds of guidelines do you have and what kinds of protection do you provide your workers? Bill St. John showed a slide with a man with a respirator and modified moon suit working in a greenhouse on-site in Texas. What kinds of worker protection would be advisable when dealing with priority pollutants in various contaminated soils and groundwater?

UNTERMAN: The guidelines for us were set up by our consulting engineering firm, Woodward Clyde, Inc. All that is required for the study of the toxicity of PCB and for the air sampling, if you don't generate dust, are boots and gloves. For those situations requiring dust protection--for example, when we rototill--we put on Tidex suits and face masks. Those are the guidelines that they set up, and we're following them. The EPA was also at the drag strip conducting the chemical tests, using the same guidelines.

WATT: As far as design of facilities, there are standard OSHA guidelines, FDA requirements, and NIH guidelines. As for sites, there are EPA documents that specify the types of protection required for certain types of materials. In addition, we have a couple of in-house industrial hygienists who are consulted frequently. There are several site-safety industrial hygiene consultants around the country who are used by a lot of the engineers.

SPAIN: Is it safe to say that you're not concerned with protection from the microorganisms that you're using?

UNTERMAN: The microorganism that we're using is a native organism, *Pseudomonas putida*, isolated two miles away at a dump site. One reason we're using it for the site test is that there are really no known pathogenic strains of *P. putida*, except for compromised individuals. One reason we did not do the study with NB1 is that we don't know what species it is; we sent it to American Type Cell Culture and to CDC, but they couldn't identify it. Because it was of a genus that does contain pathogens, we decided, at least at present, not to use it. There are plenty of innocuous bacteria, so we're not concerned with the organism.

QUESTION: Over many years I've seen the problems of medicine and industrial hygiene gradually coming together as they should have been all this time. What are your plans for protection of laboratory workers from the standpoint of occupational health?

SUIDAN: We have tried to educate them to the problems of the compounds we are using. We do most of our analyses under the hood, so as to ventilate gases that are produced. That's about it at most universities.

UNTERMAN: The toxicity of PCBs is somewhat limited, I think, in terms of human exposure, and so the levels that we use in the laboratory--the parts per million range--are not a problem, I believe. We stopped using PCBs more than ten years ago, and so exposures are not much of an issue today. But many workers were exposed over 20 and 30 years of their lives, and we've conducted extensive epidemiological studies along with the New York State Department of Health and other groups. Interestingly, there is very little indication of toxicity. I handle other chemicals in the laboratory, such as toluene, that I would worry about much more than PCBs, and I use them in larger volumes, of course.

I could address a different problem. We have a plant in Indiana that synthesizes several different plastics. They use toxic chemicals, such as phosgene and chlorine gas, and the protective measures there are extreme. For example, when they had a chlorine leak, sirens went off and everybody got to his station. Ironically, there was less chlorine in the atmosphere than in a laundry room with a lot of bleach being used. We do use toxic chemicals--dangerous chemicals. That plant in Indiana has direct phone lines to all of the residents within a certain radius of the plant, so that those people can be called immediately in the event of a spill. There are gas masks throughout the plant and they run drills. They ran a drill for a simultaneous fire, earthquake, and explosion scenario at the phosgene plant. They even had the press there. There are a lot of dangerous chemicals in our society, and we have to protect our workers. Polychlorinated biphenyls are an old problem, not a new one. We've stopped using them, but they're still in the environment.

CHAPMAN: Dr. Unterman, was the failure of your LB400 to continue PCB degradation because of its rather poor viability, or does it also lose enzymic activity?

UNTERMAN: This is a reflection of the fact that a lot of those congeners are double-para-substituted. LB400 is outstanding, but, as I showed, neither LB400 nor HA50 does a very good job on the simple dichlorobiphenyl compared with indigenous cultures from the site, which readily degraded 4,4'-dichlorobiphenyl. That's why the dual culture between LB400 and NB1 was so much better. NB1 readily degrades double-para-substituted biphenyls. At 50% degradation we're getting to congeners that ordinary organisms can do, but that our outstanding ones can't. Thus we need a dual culture to take care of more PCBs.

FOCHT: Were the dual culture experiments done with growing cells?

UNTERMAN: We did them both as growing and as resting cells. The results were for resting cells.

FOCHT: I like your idea about taking two organisms which have different specificity for the different PCB congeners; but I don't think that it will work if they're grown together, because both of them utilize biphenyl as a growth substrate. By invoking the competitive exclusion principle, one can predict that only the better fit will survive. The indigenous one, DS100, grows better on monochlorobiphenyls, so I think that it is a better competitor for the biphenyl and the monochlorobiphenyl. Perhaps this is why you can't get the two of them to grow together.

UNTERMAN: In the study that I showed involving the mixed culture, the LB400 and the NB1 were grown separately. You are right; I don't think they can be grown together. Thus, one option is to grow them separately and then add them to the soil. The other is to go the genetic engineering route. We talked here about not doing so, and, as Jim Tiedje pointed out, each organism may have its specific role. In our case, it may not be possible to combine the existing organisms. We have cloned the genes for PCB degradation from LB400. NB1 will be next. Can we then combine them genetically into a single strain? That is what's on tap in the laboratory. So, I agree with you, Dennis. Incidentally, the indigenous organism didn't grow better on monochlorobiphenyls, but better degraded the 4,4'-dichlorobiphenyls.

STRAUSS: I had a question about the work of the bacterial binding proteins that Dr. Furlong described for phosphate. Is it possible that other compounds besides nutrients have such proteins elaborated that could be used for affinity columns?

FURLONG: I wish to reinforce Dr. Focht's plea yesterday for commercially available metabolic intermediates. Transport systems have been characterized for many substrates that the biochemist and microbiologist can readily obtain for laboratory studies. But where even a single step of enzymology or of chemical synthesis is required, the substrates haven't been made or characterized. For example, a transport system has been characterized for xylose because radiolabeled xylose is readily available. Xylulose, one step removed, hasn't been done. If you have the molecule of interest available in radiolabeled form, you can use a general method of finding its transport or binding protein. Another general point is that if a molecule has a biological effect at a low concentration, surely it is having that effect because it is binding with high affinity to a cellular protein site. Finally, for these kinds of applications you need only a stable binding domain, not the whole protein function.

STRAUSS: Do you have any kinetic information on how the column performed against monoclonals? Have you deduced whether the column transfer has a diffusion limitation or a transfer limitation?

FURLONG: That type of column is diffusion-limited with monoclonals.

COMPLEX INDUSTRIAL WASTE SITES

William D. St. John and David J. Sikes

Ecova Corporation
Redmond, Washington 98052

INTRODUCTION

The biodegradation and transformation of contaminants at complex toxic waste sites represents one of the largest and most challenging problems facing the biotechnology industry. A "complex waste site" is defined here as a site with a large number of diverse toxic compounds contained in a difficult substrate. The diversity of toxic compounds can be illustrated with a portion of the waste manifest listing materials placed in a major Superfund site (Tab. 1). The challenge is to determine how to design and apply effective biological treatments in the field. This chapter presents two case studies where Ecova is incorporating biologicals in a system that also includes chemical and physical treatment strategies. This approach can be best described as a "mixed technology approach."

In case 1, a groundwater plume contaminated with a complex mixture of chlorinated and nonchlorinated solvents stretches across property boundaries and underneath an adjacent building (Fig. 1A).

In case 2, hazardous organic compounds are contained in a clay soil, which is another example of a complex substrate. The contaminants include a diverse mixture of chlorinated and nonchlorinated volatile compounds and a number of polyaromatic hydrocarbons. The concentration of volatile organic compounds (VOCs) varies dramatically in the clay matrix, from less than a few ppb to several thousand ppm.

In each of these studies, we begin with a site assessment to define the types and quantity of contaminants (Fig. 2). Site assessment includes an evaluation of the potential biodegradative activity in contaminated materials taken from the site, with multiple samples to evaluate variability of biological activity at the site. The objective is to determine whether indigenous microorganisms can provide an acceptable level of contaminant destruction. This biotreatability assessment is performed under laboratory conditions optimized to promote biodegradation of the contaminants.

In process design, the treatment system is identified, and then operating parameters for pilot- or full-scale treatment are determined in a bench-scale simulation of the treatment system. Process engineering can

Tab. 1. Sample of waste types at a Superfund NPL site.

Material	Quantity (gallons)
Oyster shells with copper	6,000
Oil and water	58,150
Paint	2,457,904
Perchloroethylene	800
Paint/Formaldehyde	4,250
Paint thinner	5,600
Paint stripper	84,425
Paint sludge	209,885
Polychlorinated biphenyls	14,000
Paint sludge and epoxy	9,740 ± 20 yd^3
Pesticide-affected fabric	500
Perchloroethylene, oil, and alcohol	18,400
Pesticides	7,582
Pesticides with arsenic	1 Box
Phenolic resins	89,362
Phenol-formic acid and methylene	900
Phosphoric acid solution	2,940
Phosphorus	350
Plastic sludge	19,983
Plating sludge	22,505
Potassium cyanide and candy	168
Poisoned cookies	1

be as much a challenge as developing the microorganisms for a treatment system. Toxicity at high levels of contamination and process instability at low or fluctuating contaminant concentrations are typical bioreactor design problems faced at these complex sites.

CASE HISTORY 1

The first case study deals with the development of a treatment system for the remediation of contaminated groundwater at a site in the Silicon Valley. This site can be considered typical for the electronics industry. The process includes bench-scale development of a biological inoculum, field implementation, and a review of field results.

History of the Site

The site, a chemical distribution center with VOCs stored in underground tanks of 6,000- and 10,000-gallon capacity, was contaminated with 11 types of chlorinated hydrocarbon solvents, including toluene, vinyl chloride, acetone, various ketones, alcohols, and ethylene glycol. Hydrogeological investigation of the site resulted in the following four conclusions:

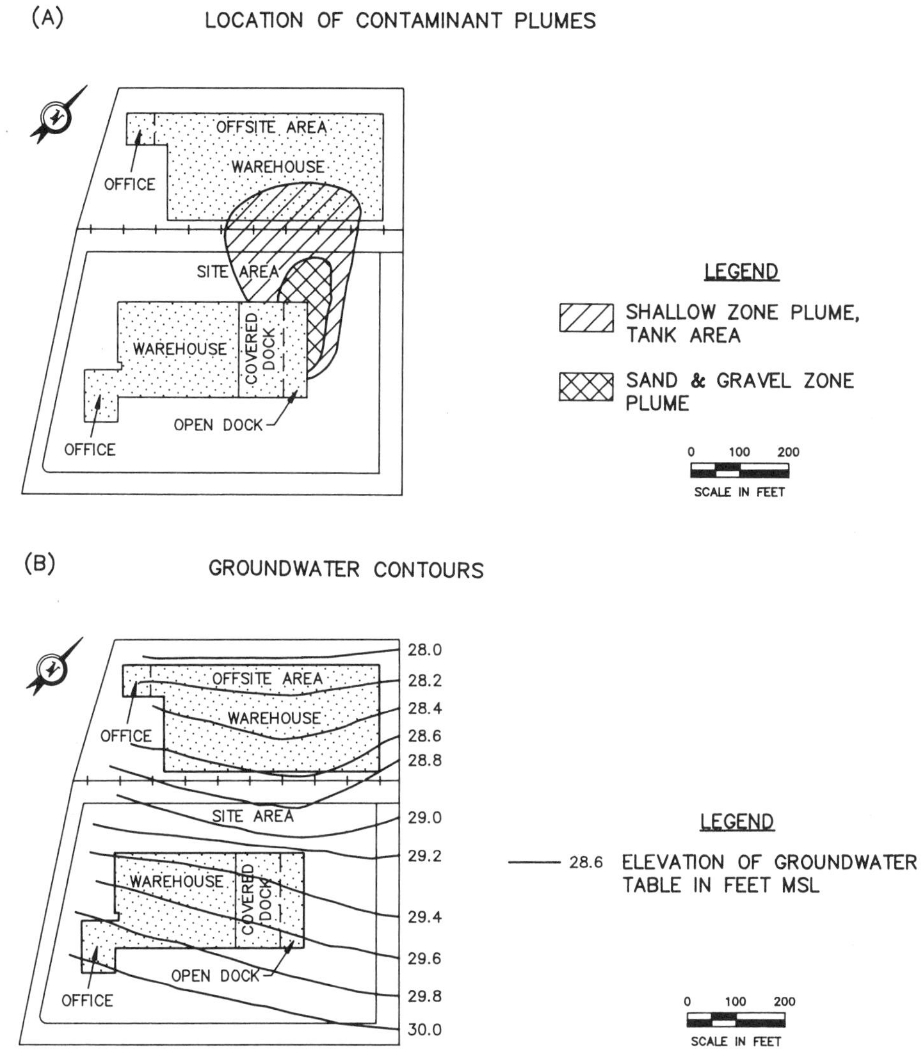

Fig. 1. Approximate location of contaminant plumes (A) and shallow groundwater contours (B).

(i) The site is underlain by three water-bearing zones: silty clay, clayey sand/sandy clay, and sand and gravel.

(ii) Groundwater flow direction is generally toward the northwest, with the average seepage velocity in the shallower zones calculated to be about 20 feet per year (Fig. 1B).

(iii) The extent of on-site groundwater contamination by VOCs is well defined both horizontally and vertically. High concentrations of VOCs and alcohols were found in shallow wells near the underground storage tanks; concentrations near the northwest boundary of the site are low but indicate that off-site migration has occurred in the upper water-bearing zone. The vertical extent of groundwater contamination is more clearly defined near the tank area than at the site boundary.

(iv) The concentrations of alcohols in the wells near the tank area are very high; a comparison between alcohol concentrations in the water and those in the soil suggests that the source could be either surface spills or leaks from underground tanks.

Treatment System Design

A hydraulic system for controlling the movement of groundwater was implemented to contain the existing plumes so that further downgradient migration of chemicals was prevented or reduced to acceptable levels. This was accomplished by pumping a series of wells placed in the water-bearing zones and creating a zone of influence within which groundwater migration would be toward the pumping wells. Capture zone analyses for shallow and deep water-bearing wells indicated that pumping from these two zones could effectively control the migration of the contaminant plumes. Treatment of the water extracted from pumping wells was required before disposal in a storm sewer system.

A thorough evaluation of available treatment technologies resulted in the decision to use an air stripping system to remove the contaminants of low aqueous solubility, followed by a biological treatment unit to treat the organic solvents of high aqueous solubility (Fig. 3). A pilot stripper was designed and constructed with one-tenth the capacity of the field unit, and evaluated with a synthetic groundwater feed whose components covered the range of volatilities and densities present at the site. During the pilot testing, the air stripper operating parameters were determined. The maxi-

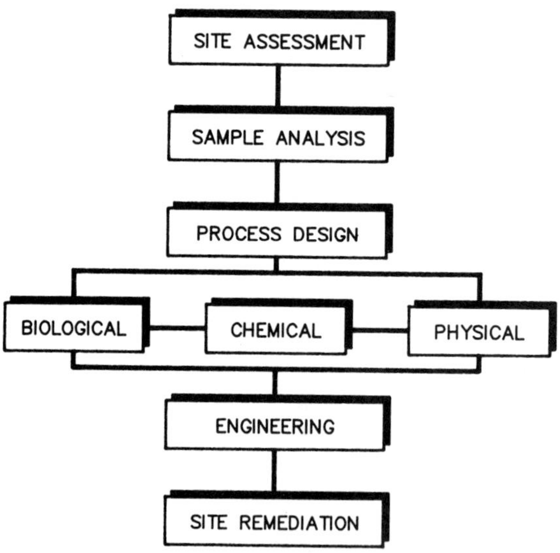

Fig. 2. Flow diagram for the development of a process to treat hazardous waste sites.

mum groundwater and air flow through the stripper was established, thus identifying the optimal volume of water which should be produced from the recovery well field.

Prior to field implementation, microbial enrichment and biotreatability assessment were performed in the laboratory to develop the biological inoculum for the field bioreactor. Contaminated groundwater samples were collected and analyzed for chemical and microbiological characteristics.

Inorganic chemical analyses revealed that the contaminated groundwater was deficient in several key nutrients necessary for effective biodegradation. Heterotrophic plate counts on plate count agar (PCA) indicated an existing microbial population of 4.6×10^3 colony-forming units per ml (cfu/ml) in the contaminated groundwater. A Marubishi bench-top chemostat was inoculated with one liter of contaminated groundwater and supplemented with a concentrated solution of basal salts to provide an excess of inorganic nutrients to stimulate biological metabolic activity. Additionally, several microbial isolates from Ecova's culture collection that were shown to metabolize organic solvents were inoculated into the chemostat. The chemostat was initially operated as a fermentor to enrich for microorganisms capable of biodegrading the target contaminants. At periodic intervals, aliquots were removed from the chemostat and analyzed for target contaminants to monitor the effectiveness of microbial metabolism. Contaminant concentrations were quantified on an HP 5890 gas chromatograph. The gas chromatograph was equipped with a 30-m DB wax or a DB-1 capillary column with a flame ionization detector. The pH was monitored daily and adjustments made to maintain pH near neutrality.

Biotreatability Test

A batch biotreatability test was performed by adding one liter of contaminated groundwater and mineral salts to the fermentor described above. The temperature was maintained at 25°C, and air was sparged through the unit at 0.65 l/volume/min. The effluent air was passed through activated

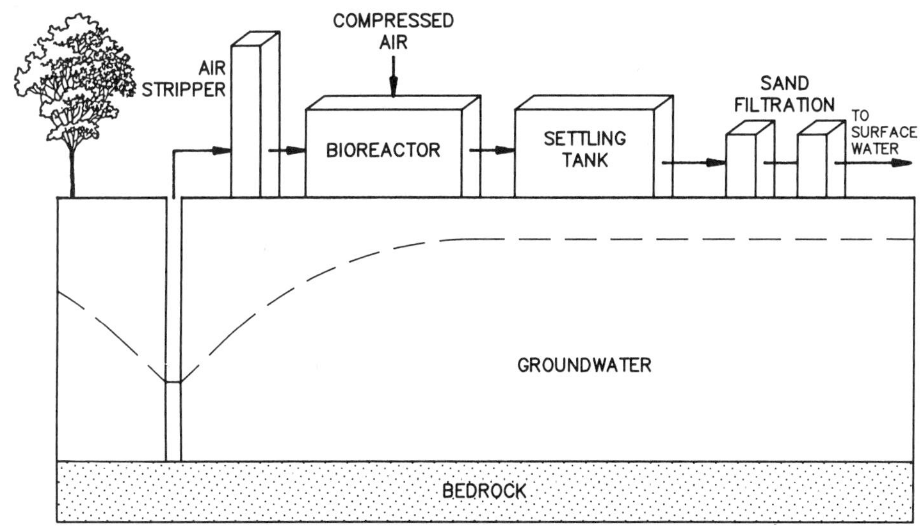

Fig. 3. Schematic diagram for a groundwater treatment.

charcoal, and biodegradation in the reactor was followed by analyzing for reduction of target contaminants and total of bacterial counts. The results indicated that following a 24-hr incubation period, the bacterial population had substantially reduced the concentration of acetone (ACE), methyl ethyl ketone (MEK), methyl isobutyl ketone (MIBK), isopropanol (IPA), and ethylene glycol (EG). Additionally, the total bacterial population density had increased by more than 4 logs. Since these results indicated a high level of microbial metabolic activity, the system was established as a continuous-flow bioreactor.

Continuous-Flow Process Evaluation

The influent for the continuous-flow process was a synthetic mixture that closely simulated the level of contaminants and inorganic ion concentration in the contaminated groundwater. The inorganic ions previously found to be limiting microbial biodegradation were added as a basal salt solution to the reactor. A multichannel peristaltic pump was used to pump the synthetic contaminated water into the reactor at a rate of 1.6 ml/min. This flow rate resulted in a retention time in the reactor of 10.4 hr. The results for the first ten days of operation are presented in Tab. 2. In a ten-day operational period, the microbial consortium in the reactor was able to reduce the contaminants to below detection limits, with the exception of IPA. Continued operation of the unit resulted in the reduction of IPA to below detection limits by day 15. The lowest retention time possible with the contaminants as a sole source of carbon and energy was 5.9 hr.

Tab. 2. Concentration of contaminants in effluent from continuous flow laboratory bioreactor.

	Concentration of organic contaminants (mg/l)									
	Influent contaminant					Effluent contaminant				
Day	ACE	MEK	MIBK	EG	IPA	ACE	MEK	MIBK	EG	IPA
1	159	35	40	103	80	4	4	ND*	32	6
2	159	35	40	103	80	4	ND	ND	69	27
3	151	31	38	97	71	6	ND	ND	70	12
4	156	30	39	99	77	4	ND	ND	75	14
5	155	32	39	101	73	ND	ND	ND	40	18
6	156	31	37	101	74	ND	ND	ND	3	6
7	154	30	35	102	73	ND	ND	ND	4	9
8	151	30	35	97	75	ND	ND	ND	ND	3
9	153	31	36	101	73	ND	ND	ND	ND	5
10	151	31	35	100	74	ND	ND	ND	ND	4

*Not detectable (detection limit: 1 mg/l).

Note: Retention time in reactor was 10.4 hr.

In previous studies (2,5), isopropanol had been stoichiometrically biodegraded to acetone. However, this does not appear to occur in a continuous culture process, as the IPA was the last and most difficult contaminant to biodegrade. The complexity of the contaminant mixture (i.e., five compounds) and the subsequent metabolism of various components in a continuous culture process may account for the variations observed.

In order to address the possibility of large fluctuations in contaminant concentrations that may occur in the field, a series of experiments was undertaken to determine whether the addition of low levels of a supplementary carbon source would stabilize the microbial population against such fluctuations and allow a shorter retention time. The results indicate that the use of a supplementary carbon source in the bioreactor can reduce the retention time by approximately 30%. This reduction in retention time from 5.9-6.1 to 3.8-4.1 hr is maintained over a variety of individual and total contaminant concentration ranges. The microbial population remained constant throughout the experimental period, between 8.0×10^7 to 3.0×10^8 cfu/ml.

Microbial Characterization and Field Scale-up of Inoculum

The members of the microbial consortium were characterized through a series of morphological and biochemical tests resulting in the following identifications (1): *Acinetobacter calcoaceticus* var. *anitratus*; *Pseudomonas putida*; *Pseudomonas paucimobilis*; *Pseudomonas testosteroni*; and *Geotrichum candidum*.

The microbial consortium was then inoculated into a 25-l Microferm fermentor (New Brunswick Scientific) containing the target contaminants in a basal salts medium, and incubated at 25°C for 48 hr. The growth of the microbial consortium was measured by turbidity; the metabolic activity was measured as a reduction in the target contaminant levels in the medium. Following the incubation period, the total microbial count was approximately 2.2×10^8 cfu/ml, with 90 to 99% reduction in the target contaminants. The inoculum was stored at 4°C until transport to the site. After four weeks at 4°C, there was no observable loss in metabolic activity as measured by reduction of target contaminants in a shake flask experiment.

Field Implementation and Process Data

A mobile fermentation unit (capacity 150 gallons), fabricated by Ecova, with temperature, oxygen, agitation, and pH controls, was transported to the site. The fermentation unit was inoculated with the prepared consortium into a basal medium with target contaminants as a source of carbon and energy. Following a five-day fermentation period in which the removal rates of the target organic contaminants were approximately those of laboratory studies, the 150-gallon suspension was inoculated into the bioreactor containing 4,000 gallons of contaminated groundwater.

The bioreactor was monitored daily for removal of the target organic contaminants. As the levels of contaminants approached the discharge limits, the bioreactor was placed in a continuous-flow mode. The initial flow rate was 5 gallons per minute (gpm), which resulted in a retention time of approximately 20 hr. Within a ten-day acclimatization period, the treatment system effluent quality was meeting the discharge requirements for all contaminants: 5.0 µg/l or less for chlorinated compounds; 1.0 mg/l or less for alcohols and glycols; and 0.5 mg/l or less for acetones and ketones.

The treatment system is now processing contaminated groundwater at a flow rate of 15 to 18 gpm and is continuing to meet the above discharge requirements. The average compositions of the process influent and effluent for the first eight weeks of operation are summarized in Tab. 3.

The combination of physical and biological techniques was effective in resolving a complex groundwater contamination problem. The laboratory development of the biological inoculum was effective in reducing the time frame necessary for field implementation and stabilization of the biological system. The advantages of implementing a physical/biological treatment system for groundwater restoration at this site include on-site destruction of contaminants, minimization of liability, and cost-effectiveness.

CASE HISTORY 2

The second case study involves the design, construction, and operation of a pilot-scale, solid-phase soil treatment facility at a Superfund site located in Texas. The site has a large area of soil contaminated with styrene, still bottom tars, and chlorinated hydrocarbon solvents.

History of the Site

Prior to 1956, the 58-acre site was part of an oil field with diked oil stock tanks, lagoons, and other structures associated with crude oil production and storage located on-site.

From 1957 to 1969, the major operating processes at the site were the regeneration of copper catalysts and reprocessing of styrene and vinyl chloride production residues. Copper catalyst regeneration was discontinued in 1970. Between 1969 and 1972, another process operation used hydrogen sulfide blended with spent caustic to produce cresylic acid, sodium sulfide, and sodium cresyllite. Operations during 1972 to 1975 were not well documented; however, styrene tar reprocessing apparently continued throughout this period. From 1975 to 1978, styrene tar, off-spec diesel, ethylbenzene, phenol bottoms, cutter stock, caustic, crude oil, blend oil, polyethylbenzene bottoms, and crankcase oil feedstocks were used to produce aromatic oil, fuel oil, ethylbenzene, toluene, cumeme, sodium sulfide, creosote extender, and 50% caustic.

The recycling and recovery plant at the site was converted to jet fuel production in 1978. Jet fuel, diesel fuel, residual oil, naphtha, kerosene, and fuel gas were produced by distilling crude oil. Jet fuel production continued until 1982, when all operations at the site were discontinued. The site has remained inactive since that time.

The site was nominated for inclusion on the National Priorities List primarily on the basis of incomplete closure of raw material storage pits, 21 pits in all. The first pit closure was completed during 1969-1970. The final on-site pit was closed in 1980. Closures were conducted by mixing soil, sand, or calcined clay with the residues left in place. A soil cover was placed over the stabilized pit material.

Tab. 3. Treatment process summary for field-scale treatment system.

Week	Influent concentration (mg/1)						Effluent concentration (mg/1)					
	ACE	MEK	IPA	MIBK	EG	Chlorinated compound	ACE	MEK	IPA	MIBK	EG	Chlorinated compound
1	5.8	5.5	3.8	2.1	0.7	100-189	0.8	0.8	ND	ND	ND	≤0.005
2	12.0	5.8	10.5	2.0	0.8	68-157	ND	ND	ND	ND	ND	≤0.005
3	9.2	9.4	12.0	2.9	0.5	70-148	0.6	ND	ND	ND	ND	≤0.005
4	8.3	3.7	12.6	3.5	0.6	40-125	ND	ND	ND	ND	ND	≤0.005
5	4.6	4.4	11.5	2.7	0.5	48-115	ND	ND	ND	ND	ND	≤0.005
6	2.1	5.0	1.1	1.3	0.9	45- 79	ND	ND	ND	ND	ND	≤0.005
7	2.9	7.1	1.3	1.4	0.7	52- 74	ND	ND	ND	ND	ND	≤0.005
8	2.7	3.7	5.5	2.0	1.3	48-762	ND	ND	ND	ND	ND	≤0.005

*ND = ≤0.5 ppm.
**ND = ≤1 ppm.

The toxicity of the pit backfill material was determined by Microtox testing. Of the 11 pits tested, two were found to have relatively low toxicity levels. Of these two pits, the one with the lower concentration of VOCs was selected as the source of material for the pilot-scale biodegradation demonstration. Additional samples of the pit backfill were collected in order to conduct a bench-scale evaluation of the biodegradability of the semivolatile organic compounds present in the backfill soil. This testing indicated that the ketones and short-chain chlorinated hydrocarbons found in the pit backfill could be removed by air stripping or could be biologically destroyed by indigenous microorganisms. On this basis, the potentially responsible parties decided to undertake a pilot-scale demonstration of biodegradation of backfill material from the pit.

Treatment System Design and Construction

The pilot-scale, solid-phase biological treatment facility consisted of a plastic film greenhouse enclosure; a lined soil treatment bed with an underdrain; an overhead spray system for distributing water, nutrients, and inocula; an organic vapor control system; and a fermentation vessel for preparing microbial inoculum or treating contaminated leachate from the backfill soil. Two standard plastic film greenhouses were erected side by side at the site of the pilot-scale treatment facility to contain any vapor emissions from the treatment bed and eliminate the need to control or treat rainwater (Fig. 4).

Once the greenhouses were erected, the soil treatment bed was constructed. This bed consisted of a permeable sand underdrain overlying a synthetic membrane liner. The underdrain was a layer of bank sand ranging in thickness from six inches at the center of the treatment bed to 18 inches at its outer edge. The upper surface of the underdrain formed a nearly level surface on which the pit soil was applied. The leachate collection system, designed for a 20-gallon per minute inflow, included two lateral perforated pipe drains and a 140-gallon capacity gravity-flow sump. Leachate was pumped from the collection sump to the on-site fermentor

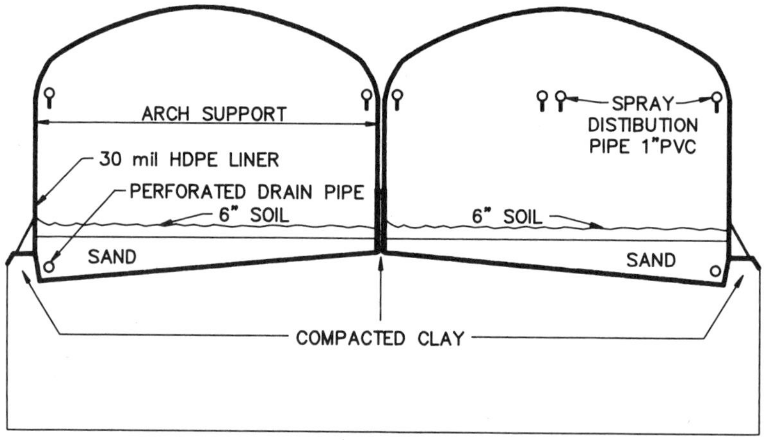

Fig. 4. Plastic film greenhouses. Schematic diagram of the solid-phase treatment facility.

vessel for treatment. The synthetic membrane liner was constructed of 80-mil high-density polyethylene with heat-welded seams. The edges of the liner were secured to the inside of the greenhouse structural arches with self-tapping screws.

The overhead spray system for distributing water, nutrients, and inocula consisted of four individual headers, one for each of the four treatment lanes. The spray pattern was designed to eliminate overspray from one treatment lane to another. Water, nutrient, or inocula mixtures were applied through 32 individual spray heads per treatment lane. Each spray head was capable of supplying 0.79 gallons per minute at 30 psi.

Organic vapor emissions were controlled by adsorption on activated carbon. Air was removed from the treatment facility by three fans [3,000 cubic feet per minute (cfm)] and directed through three activated carbon absorbers. Each of the carbon absorbers was capable of treating an air flow of 3,000 cfm.

Following construction of the treatment facility, approximately 200 cubic yards of soil was removed from the pit and transferred to the treatment facility. Soil was excavated with a 12-foot-reach backhoe and transported to the treatment bed using two tracked front-end loaders. The excavated soil was placed on top of the prepared treatment bed. Due to the cohesiveness of the clay soil, the pit backfill material was allowed to dry before final grading. For several days the tracked front-end loaders were run back and forth over the pit backfill to break up large blocks of clay and distribute material evenly over the treatment bed. After three days of manipulation, the clay was amenable to tillage by a power rototiller attached to a tractor. Soil moisture content was low enough after six days to allow for the addition of nutrients through the spray irrigation equipment.

Experimental Treatment Lanes

The soil treatment bed was divided into four lanes so that different methods of optimizing microbial activity and biodegradation rates could be evaluated (Fig. 5). A control lane, which received only tilling and water additions, was established to provide a baseline for evaluating the effectiveness of the following three treatment processes: (i) nutrient addition, (ii) single microbial inoculation, and (iii) multiple microbial inoculations.

The nutrient-adjusted lane was sprayed with inorganic nutrients (nitrogen and phosphorus) to stimulate the activity of indigenous microorganisms in the affected soil. This treatment was designed to determine the rate at which the indigenous microbial population could degrade organic compounds through the addition of oxygen and inorganic nutrients. The single-inoculation treatment was sprayed once at the start of operations with inorganic nutrients and a high concentration of microorganisms that had been isolated from the pit. This treatment process was designed to determine whether the rate of organic compound degradation could be increased by augmenting the existing microbial activity from pit soil. After an initial inoculation, the multiple-inoculation lane was subsequently sprayed with nutrients and inoculum at approximately ten-day intervals. The inoculum was developed from soil removed from the lane and water from the leachate collection system. This treatment was designed to determine whether the rate of removal of organic constituents could be accelerated by increasing the frequency of application of microorganisms.

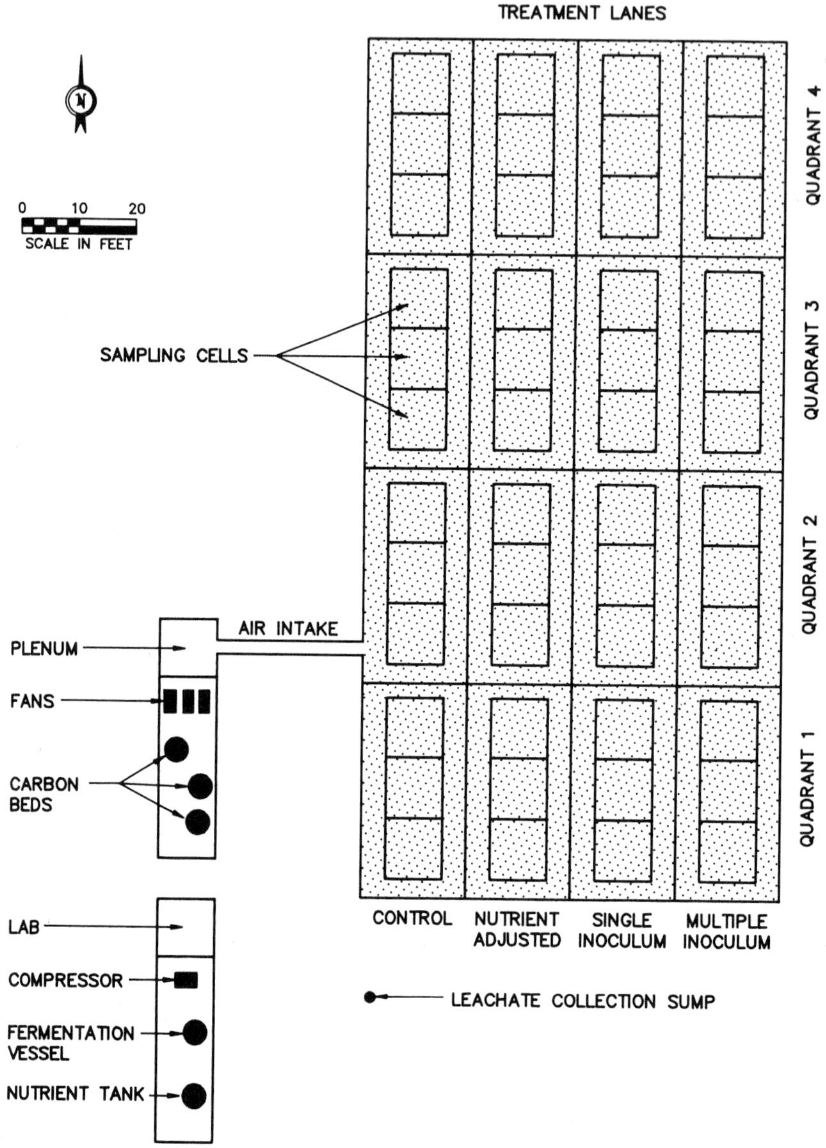

Fig. 5. Solid-phase treatment facility layout.

Treatment Facility Operation

The pilot-scale treatment facility was operated for 94 days. The soil in the treatment facility was tilled daily to optimize contact between microorganisms and the organic constituents present in the pit backfill material, and to ensure adequate aeration for microbial activity. Soil moisture content, soil temperature, and soil pH were monitored to ensure that they remained within ranges conducive to microbial activity. Water, nutrients, and inocula were added as required to the treatment bed through the overhead spray system.

Sample Collection and Analysis

Soil samples were collected 0, 21, 58, and 94 days after initiating operations. Each of the four treatment lanes was divided into four quadrants. These quadrants were in turn subdivided into three ten-foot by ten-foot sampling areas (Fig. 5). During each sampling episode, one sample was collected from each of the 16 quadrants. Samples from each quadrant were a composite of three subsamples. The subsample locations were selected randomly within each quadrant.

Soil samples were analyzed for volatile and semivolatile organic compounds in order to determine the rate of organic compound losses and to measure the effectiveness of the three treatment processes. In addition, the soil samples were analyzed for soluble ammonium, nitrate, and phosphate to determine whether the concentrations of these nutrients were sufficient to ensure maximal microbial growth and organic compound degradation.

Microbial growth was monitored using standard plate count techniques. Radiotracer analyses were done to confirm biological mineralization of organic compounds. Soil samples were spiked with ^{14}C-labeled glucose and phenanthrene, and the amount of these compounds biologically converted to ^{14}C-labeled carbon dioxide was measured. The ^{14}C-labeled glucose mineralization was used to monitor general microbial activity. The formation of $^{14}CO_2$ from radiolabeled phenanthrene was used to prove biological mineralization of the predominant semivolatile contaminant and to demonstrate stimulation of phenanthrene-degrading microorganisms.

Removal of Volatile Organic Compounds

The predominant VOCs detected in the backfill soil that was placed in the treatment facility were ethylbenzene, styrene, and toluene. These compounds were detected at maximal concentrations of 4,400 mg/kg, 240 mg/kg, and 510 mg/kg, respectively. Methylene chloride, 1,1,2-trichloroethane, and several other VOCs were also detected, at lower concentrations.

Tab. 4. Volatile organic compound removal during pilot-scale bioremediation.

Lane	Total volatile organics (μg/kg)				Reduction (after 21 days of operation)
	Day 0	Day 21	Day 58	Day 94	
Control	15,972	81	17	29	99.69%
Nutrient adjusted	39,460	40	14	12	99.90%
Single inoculation	273,184	13	16	25	99.99%
Multiple inoculation	101,868	10	19	27	99.97%

The average concentrations of the VOCs in the treatment facility were reduced by more than 99% over the 94-day period of operation (Tab. 4). Most of this reduction occurred within the first 21 days of operation and was predominantly due to air stripping. Volatile compounds of both high and low volatility were removed with equal efficiency. For example, the concentrations of methylene chloride and 1,1,2-trichloroethane, both highly volatile compounds, were reduced by more than 99%. The concentrations of ethylbenzene and styrene, both low-volatility compounds, were also reduced by more than 99%.

Two methods were used to estimate the amount of VOCs removed from the affected soil by air stripping: (i) concentration of volatile compounds absorbed in the activated carbon units, and (ii) air emission data collected during facility operation. The amount of volatile compounds air-stripped from the affected soil ranged from 137 kg to 159 kg, a removal rate of approximately 7 kg per day.

Biodegradation of Semivolatile Organic Compounds

The initial populations of microorganisms in the backfill soils ranged from 10^4 to 10^6 colony-forming units per gram wet weight of soil (cfu/gm). During the operation of the pilot-scale treatment facility, the numbers of microorganisms increased to 10^7-10^8 cfu/gm (Tab. 5). The increase and eventual stabilization of the numbers of microorganisms in the treatment bed indicated that stable microbial communities were present.

Phenanthrene was the predominant semivolatile organic compound detected in the backfill soil. Phenanthrene concentrations ranged from 440 to 170,000 μg/l, with the average phenanthrene concentration 36,300 μg/l. 2-Methylnaphthalene and naphthalene were also detected in the affected soil. 2-Methylnaphthalene concentrations ranged from 6,200 to 170,000, with an average concentration of 50,700 μg/l. Naphthalene concentrations ranged from 130 to 96,000 μg/l, with the average concentration 19,500 μg/l. Over the 94-day operation of the pilot-scale biological treatment facility, semivolatile organic compound concentrations were reduced an average of 89% (85, 91, 96, and 83%, respectively, in lanes 1-4).

Due to its predominance in the backfill soil, phenanthrene was used to determine the effect of the various treatment processes on the degradation rates of semivolatile organic compounds. Over the 94 days of facility operation, phenanthrene concentrations were reduced an average of 84% (79, 86, 92, and 78%, respectively, in lanes 1-4). During the first 21

Tab. 5. Number of aerobic heterotrophic organisms present in treatment lanes.

Lane	Number of microorganisms (cfu/gm soil)			
	Day 0	Day 21	Day 58	Day 94
Control	1.8×10^6	1.7×10^7	3.0×10^7	9.7×10^5
Nutrient adjusted	7.5×10^5	4.8×10^7	5.5×10^7	8.4×10^6
Single inoculation	2.9×10^4	7.1×10^7	3.3×10^7	4.7×10^6
Multiple inoculation	2.8×10^6	6.9×10^7	4.9×10^7	5.3×10^6

days of operation, phenanthrene degradation occurred at a relatively rapid rate (approximately 1,000 μg/kg/day); for the remainder of the demonstration project, the rate was 124 μg/kg/day. At this degradation rate, approximately 131 days would be required for the phenanthrene concentration to reach 330 μg/l, the analytical detection limit using the EPA-approved procedure.

Phenanthrene half-life values for the control, nutrient-adjusted, single-inoculated, and multiple-inoculated lanes were 40.8, 33.0, 25.7, and 43.3 days, respectively. Thus, all of the data were pooled to determine the average rate of phenanthrene biodegradation in the treatment facility. The average half-life value was 33 days, significantly less than reported half-life values of 69 to 298 days in other solid-phase biodegradation systems. Initial phenanthrene concentration was apparently the parameter controlling the rate of phenanthrene degradation (Tab. 6). There was no statistically significant difference in the rate of phenanthrene degradation in the different treatment lanes. The data collected during this demonstration project suggest that aeration and the amount of contact between the microorganisms and the affected soil were the primary parameters that governed the rate of phenanthrene degradation.

The initial percent mineralization of ^{14}C-labeled glucose to ^{14}C-labeled carbon dioxide within the pilot test area (day 0) was low and extremely variable. During the first 21 days of operation of the treatment facility, the percent mineralization stabilized at approximately 45% and subsequently decreased to approximately 17% by day 94 (Tab. 7). The initial percent mineralization of ^{14}C-labeled phenanthrene to ^{14}C-labeled carbon dioxide ranged from 11 to 22%. During the 94-day operation period, the percent mineralization increased to and stabilized at approximately 44%.

The initial increase and stabilization in the percent mineralization of [^{14}C]glucose and [^{14}C]phenanthrene indicates that the pilot test operations stimulated and promoted an even distribution of phenanthrene biodegradation within the treatment bed. The subsequent decrease in [^{14}C]glucose activity suggests that facility operations promoted the establishment of a microbial population that preferentially degraded phenanthrene, the most predominant semivolatile organic constituent in the soil.

The pilot-scale biological treatment facility constructed at the Superfund site conclusively demonstrated that target compounds such as 1,2-dichloroethane, 1,1,2-trichloroethane, and phenanthrene could be removed

Tab. 6. Effect of initial concentration on phenanthrene biodegradation.

Initial concentration (μg/kg soil)	Average reduction (%)
1,000- 4,999	27.4
5,000- 9,999	33.4
10,000- 49,999	67.2
50,000-100,000	94.0
Greater than 100,000	96.7

Tab. 7. Radiotracer demonstration of phenanthrene biodegradation.

Lane	[^{14}C]Glucose mineralization (percent)			
	Day 0	Day 21	Day 58	Day 94
Control	33.0	45.5	25.5	9.9
Nutrient adjusted	17.7	47.5	30.4	23.0
Single inoculation	10.6	44.9	29.1	17.8
Multiple inoculation	24.1	45.6	35.2	15.9

Lane	[^{14}C]Phenanthrene mineralization (percent)			
	Day 0	Day 21	Day 58	Day 94
Control	19.0	42.0	42.8	42.6
Nutrient adjusted	22.5	51.2	42.4	46.7
Single inoculation	11.6	38.0	43.3	43.8
Multiple inoculation	10.7	46.2	43.3	42.8

Note: All counts corrected for nonbiological mineralization in controls.

effectively from soils using an on-site treatment technology other than incineration. The process removed VOCs by air stripping and destroyed semivolatile organic compounds by biodegradation.

CONCLUSION

Two case studies demonstrate the applicability of bioremediation to complex waste sites when coupled with chemical and physical methods. Such mixed-technology approaches will become increasingly important as the country begins to clean up its growing number of Superfund sites.

REFERENCES

1. (1986) Bergey's Manual of Systemic Bacteriology, Vol. 1 and 2, William and Wilkens, Baltimore, Maryland.
2. Flathman, P.E., and G.D. Githens (1985) In Groundwater Treatment Technology, Evan K. Nyer, ed. Van Nostrand Reinhold Company, New York.
3. Jhaveri, V., and A.J. Mazzacca (1983) Bio-reclamation of ground and groundwater: Case history. Presented at the Fourth National Conference of Management of Uncontrolled Hazardous Waste Sites, October 31-November 2, 1983.
4. Kobayashi, H., and B.E. Tittman (1982) Microbial removal of hazardous organic compounds. Environ. Sci. Technol. 16:170A-183A.
5. Taylor, D.G., R.W. Trudgill, R.E. Cripps, and P.R. Harris (1980) Microbial metabolism of acetone. J. Gen. Microbiol. 118:159-170.

BIOLOGICAL APPROACHES FOR POLYCHLORINATED BIPHENYL DEGRADATION

Ronald Unterman, Donna L. Bedard, Michael J. Brennan, Lawrence H. Bopp, Frank J. Mondello, Ronald E. Brooks, David P. Mobley, John B. McDermott, Charles C. Schwartz, and David K. Dietrich

General Electric Company
Corporate Research and Development
Schenectady, New York 12301

INTRODUCTION

The accumulation of xenobiotic compounds in our environment has had a profound effect on our physical, social, economic, and political well-being. Some of these substances have been found to be harmful to humans, while others, including polychlorinated biphenyls (PCBs), have only been implicated as such. In either case, there has been a growing demand for their safe disposal or, preferably, destruction.

More than 100 forms (congeners) of PCBs, differing in the number and position of chlorine substituents, were commonly used over the last half-century. The chemical and physical properties that formed the basis for their use, thermal and chemical stability, have also contributed to their persistence in the environment. In addition, they are highly insoluble in water and only slightly mobilized by aqueous systems. Thus, PCBs in the environment are generally found in soil and sediments at sites of their production, use, storage, or disposal.

In the last 15 years, there have been numerous laboratory reports of bacterial degradation of biphenyls containing one to three chlorines, but few reports of degradation of the more highly chlorinated congeners (Ref. 4, 5, 10, 12, and 15; reviewed in Ref. 9). The more highly chlorinated PCBs, those with four or more chlorines, have generally been considered resistant to biodegradation in the environment. However, both aerobic and anaerobic biotransformations may be much more prevalent than previously thought (6-8). These observed environmental biotransformations now permit us to reevaluate our options for specific site remediation, and are stimulating an increased interest in the anaerobic degradation of PCBs. Traditional methods for the remediation of PCB-contaminated sites involve incineration of the material or its removal and burial in secure landfills. Both of these techniques are very expensive and often encounter public opposition. Thus, there is great interest in the development of acceptable and economical means for destruction of PCBs.

Several years ago we began a research program designed, first, to isolate and characterize new and better strains of PCB-degrading bacteria and, second, to characterize those biotransformations that are occurring in nature. The results of these studies, and those of others, will help us to better understand the biodegradability of PCBs and the biological approaches potentially available for their destruction.

ISOLATION OF NEW STRAINS

Soil and sediment samples were obtained from PCB-contaminated sites in New York, Massachusetts, and California, and from a non-PCB-containing site in Mississippi. These soils and sediments were added to a phosphate-buffered minimal salts medium (PAS) containing biphenyl or biphenyl plus PCB as sole carbon and energy source. Bacteria were selected for their ability to grow on biphenyl. After several passages on PAS/biphenyl, the mixed bacterial cultures were assayed for their ability to degrade PCBs. One such assay is shown in Fig. 1; peaks D and J (2,5,2',5'-tetrachlorobiphenyl and 2,4,5,2',4',5'-hexachlorobiphenyl, respectively) are generally resistant to bacterial degradation and are only degraded by exceptional bacterial strains.

Prior attempts to assay microorganisms for PCB-degradative competence by measuring disappearance of Aroclors (commercial PCB mixtures containing 40 to 60 congeners) have frequently produced false positive findings because of volatilization or adsorption. Furthermore, these assays have generally left the chemical nature of the competence obscure because of incomplete gas chromatographic (GC) resolution and uncertain identification of the Aroclor peaks. We have avoided these problems by using defined mixtures of PCB congeners and by adopting incubation and extraction techniques that prevent physical loss of the PCBs.

The assay mixtures included PCBs ranging from dichlorobiphenyls to hexachlorobiphenyls representing several structural classes: those chlorinated on a single ring (e.g., 2,3-dichlorobiphenyl); those chlorinated at all possible 2,3 sites (e.g., 2,5,2',5'-tetrachlorobiphenyl); those chlorinated at all possible 3,4 sites (e.g., 4,4'-dichlorobiphenyl); and those lacking adjacent unchlorinated sites (2,4,5,2',4',5'-hexachlorobiphenyl). Packed column GC analysis of these mixtures (e.g., Fig. 1) following a 24-hr incubation with resting cells permits quantitation of differential congener competence. Results for all cultures were calibrated using a nondegradable PCB congener (e.g., 2,4,5,2',4',5'-hexachlorobiphenyl or 2,4,6,2',4'-pentachlorobiphenyl) as internal standard. These studies have allowed us to definitively and unambiguously define the PCB-degradative competence of all new isolates.

Pure cultures were isolated by selection of individual colonies from PAS/agar plates that contained biphenyl as the sole carbon source. These clones were alternately colony-purified on PAS/biphenyl/agar and Luria agar plates. All pure strains were initially characterized using the N/F system (Flow Laboratories, Inc., McLean, Virginia). Selected isolates were sent to the American Type Culture Collection (Rockville, Maryland) for positive identification. The pure cultures were then assayed for their PCB-degradative competence as described above. An example of an Aro-

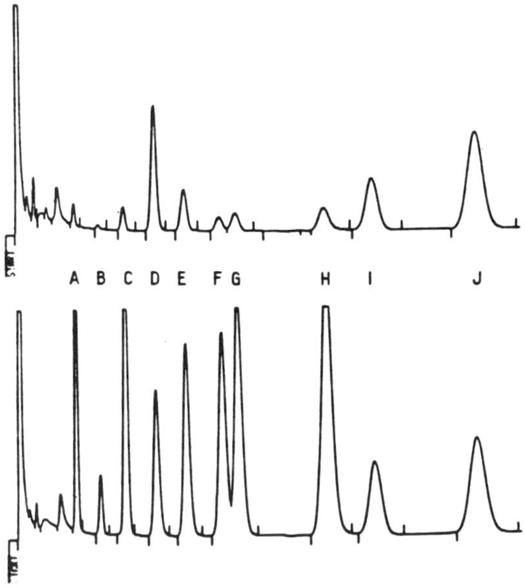

Fig. 1. MB1 degradation of defined PCB mixture. Lower gas chromatogram is control. (See Ref. 4 for details.) A: 2,4'; B: 4,4'; C: 2,4,4'; D: 2,5,2',5'; E: 2,3,2',5'; F: 2,3,2',3'; G: 2,4,3',4'; H: 2,4,5,2',3'; I: 3,4,3',4'; J: 2,4,5,2',4',5'.

clor 1248 (48% chlorine by weight; mostly tri-, tetra-, and pentachlorobiphenyls) depletion assay is shown in Fig. 2. Complete experimental details are described elsewhere (4). As shown previously for defined PCB mixtures (Fig. 1), it is also evident here that MB1 can only degrade specific congeners in this commercial PCB mixture.

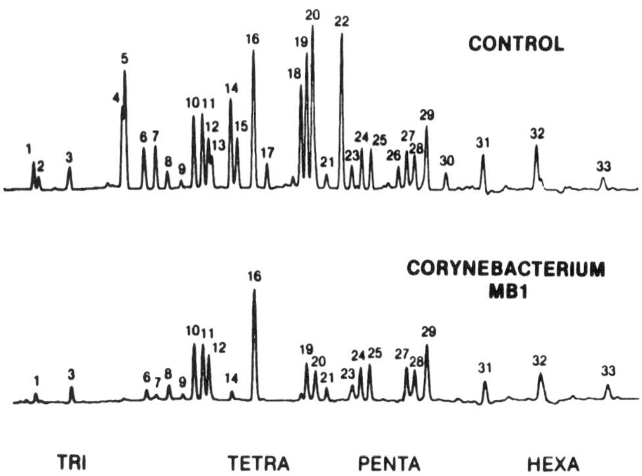

Fig. 2. MB1 degradation of Aroclor 1248 (10 ppm). (See Ref. 4 for details.)

Using this approach, we have isolated a diverse group of PCB-degrading bacteria. Our isolates (Tab. 1) include a gram-positive *Corynebacterium* species; pseudomonads belonging to all three ribosomal RNA homology groups (*Pseudomonas putida*, *P. cepacia*, and *P. testosteroni*); an *Alcaligenes faecalis*; and an *A. eutrophus*. Thus, diverse environmental microorganisms have developed the ability to degrade PCBs. We would like to determine whether the genes specifying the PCB-degradative enzymes in these organisms are related. Such information would give us a better understanding of the natural origin of PCB-degradative competence.

Figure 3 shows the PCB-degradative competence of 26 pure cultures assayed on defined mixtures of 18 PCB congeners (4). As reported with the mixed cultures (7), there is a wide spectrum of degradative competence. This format illustrates a natural progression of degradative capability. The nine strains with the poorest ability to degrade PCBs primarily attacked congeners containing a nonchlorinated ring or a ring with a single chlorine in the *para* position. The remaining strains attacked congeners containing a ring with a single *ortho* chlorine and showed varying abilities to degrade tetra- and pentachlorobiphenyls. These trends in degradative competence may indicate that the PCB-degradative enzymes of these strains are closely related. On the other hand, the differences in congener specificity exhibited by strains MB1 and H850 suggested that there are several distinct pathways of PCB degradation (1-3). This is discussed more fully below.

Several of our pure cultures easily degrade tetra- and pentachlorobiphenyls (2,3,5). Even more significantly, *P. putida* LB400 attacked every congener in the defined congener assay. Other congeners that are attacked by LB400 include 3,5,3',5'-tetrachlorobiphenyl, 2,4,6,3',5'-pentachlorobiphenyl, and 2,4,5,2',4',5'- and 2,4,6,2',4',6'-hexachlorobiphenyls. It is particularly noteworthy that none of these congeners contains adjacent unchlorinated sites. A possible explanation for this activity is the presence of a monooxygenase or a dioxygenase with the ability to attack at a chlorinated position. These would be novel and extremely important findings in terms of basic bacterial biochemistry and potential decontamination of environmental PCBs.

Tab. 1. Identification of PCB-degrading bacterial strains.

Corynebacterium sp.	MB1
Alcaligenes:	
Alcaligenes faecalis	Pi434
Alcaligenes eutrophus	H850
Pseudomonas:	
Pseudomonas putida	LB400, LB410
Pseudomonas cepacia	H201, Pi704, RJB
Pseudomonas testosteroni	H128, H336, H430
Pseudomonas sp.	Pi939, H1130, Pi304
(*Acidovorans* group)	H702, Pi101

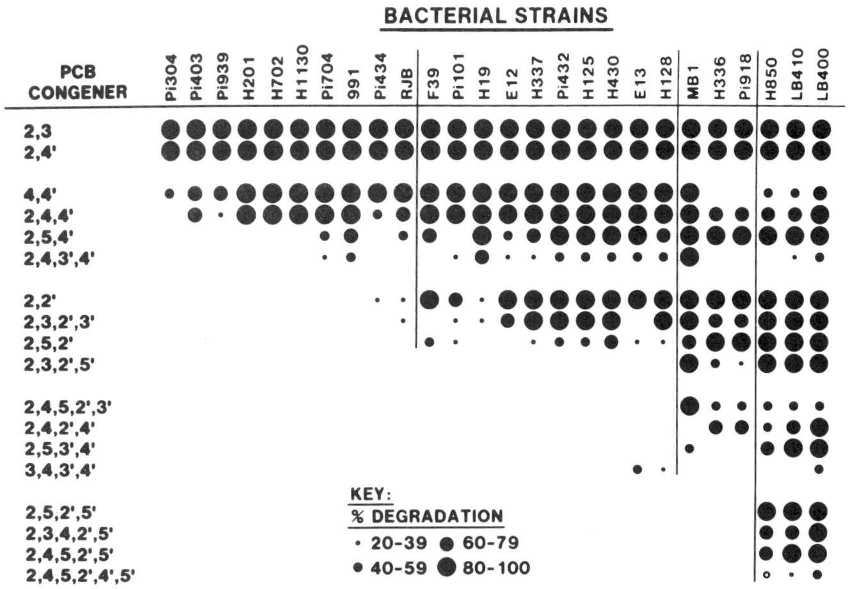

Fig. 3. Polychlorinated biphenyl-degradative competence of environmental isolates. Two different defined-congener assays (as in Fig. 1), including 18 PCB congeners, were used to assess the specific PCB-degradative competence of 26 bacterial strains. (o) H850 degraded less than 20% of 2,4,5,2',4',5'-CB when assayed against the mixture, but clearly degraded this congener and generated a metabolite when the 2,4,5,2',4',5'-CB was assayed separately.

POLYCHLORINATED BIPHENYL DEGRADATIVE PATHWAYS

Our bacterial strains show a varied yet substantial capability for degrading PCBs in our substrate depletion assays. We have now furthered the biochemical analysis of several of these strains by using a metabolite assay to characterize the products of these bacterial oxidations (2,5,14). The demonstration of metabolic products is important for several reasons. First, it unequivocally confirms that the PCB depletion observed in the various depletion assays is due to bacterial oxidation. Second, it allows one to begin to elucidate the metabolic pathway(s) for these oxidations. Third, it shows that more than one route for PCB degradation exists in bacteria, since different organisms can produce different metabolites from the same PCB congener.

A time-course analysis of the metabolites produced from *P. cepacia* H201 incubation with 2,4'-dichlorobiphenyl is depicted in Fig. 4. H201 rapidly oxidizes 50 ppm 2,4'-dichlorobiphenyl (A) to a hydroxylated intermediate (B), which is subsequently degraded to 2-chlorobenzoic acid (C). This is consistent with the general observation that bacteria oxidize biphenyls and PCBs via benzoic acid intermediates, and often accumulate the chlorobenzoic acid products (9). Our GC-mass spectrometry (GC-MS) studies with many other PCBs have shown that the transient intermediates of PCB degradation are hydroxylated-chlorinated biphenyls. With the more highly chlorinated PCBs, these intermediates often accumulate, which has

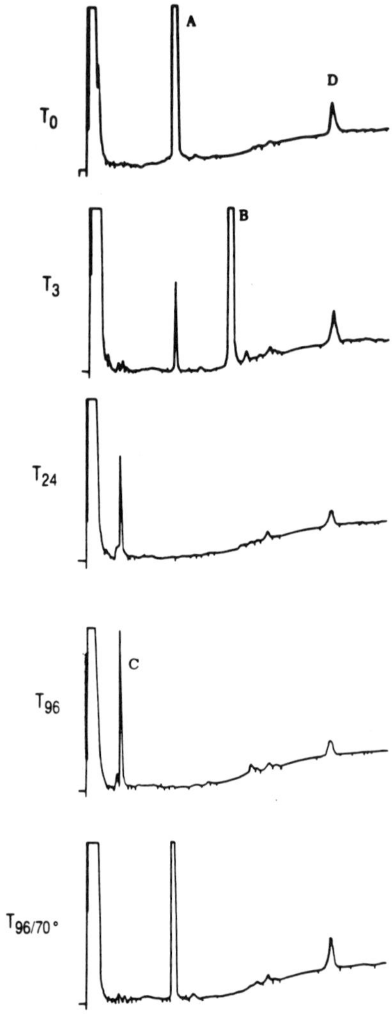

Fig. 4. H201 oxidation of 2,4'-dichlorobiphenyl. Incubation times are indicated at the left. Bottom panel is heat-inactivated control. A: 2,4'-CB; B: transient high molecular weight metabolite; C: 2-chlorobenzoic acid; D: internal standard.

facilitated our analysis of the complete PCB-degradative pathway. An example of such an oxidation is shown in Fig. 5. The parent pentachlorobiphenyl (A) is only partially degraded by LB400; at 20 hr, several metabolites are present (B-D). These compounds have been identified by GC-MS as pentachlorobiphenyl dihydrodiol and dihydroxy pentachlorobiphenyl (5).

A comparison of the PCB competence of *Corynebacterium* sp. MB1 with *A. eutrophus* H850 and *P. putida* LB400 reveals that MB1 is far superior in its ability to degrade congeners substituted in both *para* positions

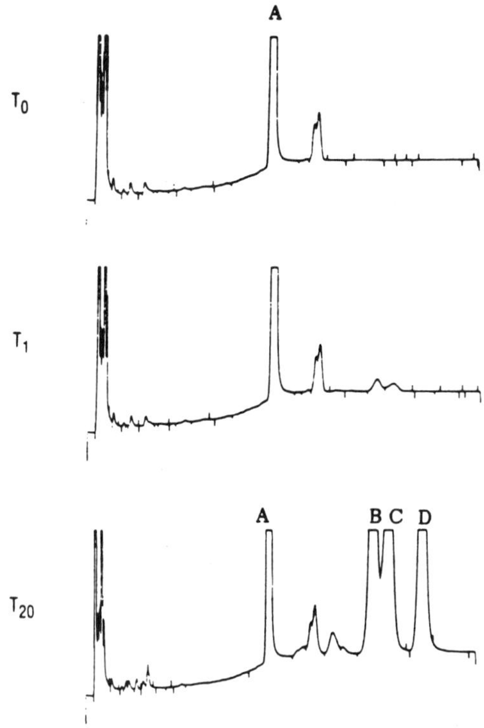

Fig. 5. Oxidation of 2,4,5,2',5'-pentachlorobiphenyl (50 ppm) by LB400 resting cells for 0, 1, and 20 hr. A: 2,4,5,2',5'-CB; B-D: hydroxylated metabolites. (Note: The metabolites have a higher response factor on the electron capture detector, so no direct molar comparison can be made based on peak areas.)

(Fig. 3). On the other hand, congeners that have a 2,5-chlorophenyl ring are preferentially degraded by H850 and LB400, particularly if the opposite ring contains a chlorine at carbon 2'. (MB1 cannot oxidize 2,5,2',5'-tetrachlorobiphenyl to any significant extent.) MB1 degrades 4,4'-dichlorobiphenyl almost completely to the 4-chlorobenzoate, whereas H850 and LB400 show only poor activity on this congener. These differences are most pronounced at higher PCB concentrations. In addition to exhibiting distinct congener preferences, MB1 and H850 produce different metabolites from the same congener.

Based on these data, we proposed that a significant mechanism of PCB metabolism in _A. eutrophus_ H850 and _P. putida_ LB400 involves a dioxygenase that attacks at carbon positions 3,4 (1,2,4,5), whereas _Corynebacterium_ sp. MB1 employs a more common 2,3-dioxygenase mechanism (Fig. 6). The dioxygenase attack of PCBs at positions 3,4 has now been conclusively demonstrated by D.T. Gibson and co-workers (11). Other studies have shown that _A. eutrophus_ H850 degrades some PCB congeners via chloroacetophenone intermediates (2). These novel activities may partially explain the exceptional PCB-degrading activity of H850.

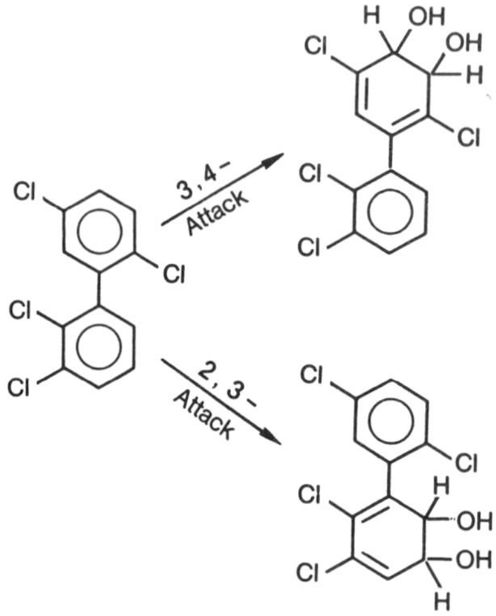

Fig. 6. Initial oxidation of PCBs by 2,3- and 3,4-dioxygenases.

ISOLATION OF THE GENES FOR POLYCHLORINATED BIPHENYL METABOLISM

As part of our investigation of bacterial PCB metabolism, we have begun to examine the regulation and control of the enzymes involved. The ability to isolate and manipulate the genes encoding these enzymes would greatly facilitate these studies and our attempts to maximize the PCB-degradative competence of these organisms.

A genomic library of P. putida LB400 was constructed and placed into an Escherichia coli host. From this library, recombinant strains that contained and expressed the genes for the PCB-degradative enzymes of LB400 were isolated. Several of these E. coli strains have been found to possess all of the enzymes required to metabolize a variety of PCBs to their corresponding chlorobenzoic acids. This result conclusively demonstrates the presence of the four key enzymes of the major PCB/biphenyl metabolic pathway (Fig. 7): biphenyl dioxygenase, biphenyl dihydrodiol dehydrogenase, biphenyl catechol oxygenase, and hydrase.

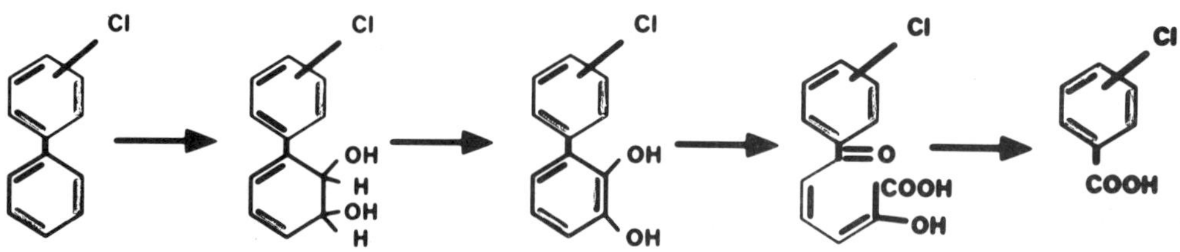

Fig. 7. Polychlorinated biphenyl metabolic pathway.

BIODEGRADATION OF SOIL-BOUND POLYCHLORINATED BIPHENYLS

Recently, we expanded our research program in an effort to explore the possibility of using one or more of our new bacterial isolates for the decontamination of PCB-laden soil. *Corynebacterium* sp. MB1, *A. eutrophus* H850, and *P. putida* LB400 had all been shown to have excellent PCB-degrading capabilities under laboratory assay conditions (1-5,14); however, little was known about their ability to oxidize PCBs adsorbed to soil. Our initial studies using sand as a model demonstrated that PCBs were biodegradable even when bound to a solid substrate.

We have continued with model biodegradation studies using PCB-laden soil containing either Aroclors or pure PCB congeners (14). Experimentally, the approach is identical to our previously described resting cell assays (4), except that the PCBs are bound to soil when mixed with the bacteria. The cultures were grown on PAS buffer with biphenyl as the sole carbon source, then harvested, washed, and resuspended in buffer at a concentration of approximately 10^9 cfu/ml. These washed cells were then incubated with one of several PCB/soil formulations: 50 ppm of Aroclor 1242 spiked onto clean (i.e., non-PCB-containing) soil; 500 ppm of Aroclor 1242 on clean soil; 50 ppm of Aroclor 1254 on clean soil; an environmental sample (South Glens Falls, New York, drag strip site 43C) that contains 525 ppm PCB (very similar in composition to Aroclor 1248); and various concentrations of pure PCB congeners spiked onto clean soil. In addition, some assay variations were conducted using unwashed cells, Luria-grown cultures, cultures actively growing on biphenyl, unstirred samples, and limited amounts of water.

As we observed for free PCBs, each of our strains also exhibits congener-specific degradation of soil-bound PCBs. That is, MB1 is better able to degrade soil-bound, double-*para*-substituted chlorobiphenyls (e.g., 4,4'-dichlorobiphenyl), and H850 and LB400 are better able to degrade soil-bound congeners with 2,5-substituted phenyl rings, such as 2,5,2',5'-tetrachlorobiphenyl (2-5). This specificity is clearly evident in the GC profiles in Fig. 8. LB400 more readily degrades peak 1 (2',2,5- and 2',2,4-trichlorobiphenyls) and the center peak of the triplet designated by three dots (3,4,2',5'-tetrachlorobiphenyl). On the other hand, MB1 more readily degrades peak 4 (2,3,4,4'-tetrachlorobiphenyl) and the first and third peaks of the triplet (2,4,5,4'- and 2,4,3',4'-tetrachlorobiphenyls, respectively). Because the specificities of MB1 and LB400 complement each other, it seemed reasonable to try a biodegradation using a mixed culture consisting of both strains. Figure 8 (third panel) shows the results of such an assay and demonstrates the additive effect of using a mixed culture.

We have extended our Aroclor biodegradation studies to higher concentrations and more difficult Aroclors. LB400 was able to degrade more than 85% of the PCB in a 500-ppm Aroclor 1242 assay incubated for two days. In a 50-ppm Aroclor 1254 assay, LB400 degraded greater than 65% of the PCB in one day.

We have now begun studies with an actual PCB-contaminated soil from New York state. This environmental sample is from the site of a former drag-racing track (drag strip) where PCB oils were used for dust control. The site contains Aroclor 1242. It is depleted in the di- and trichlorobiphenyls and therefore appears similar in composition to Aroclor 1248.

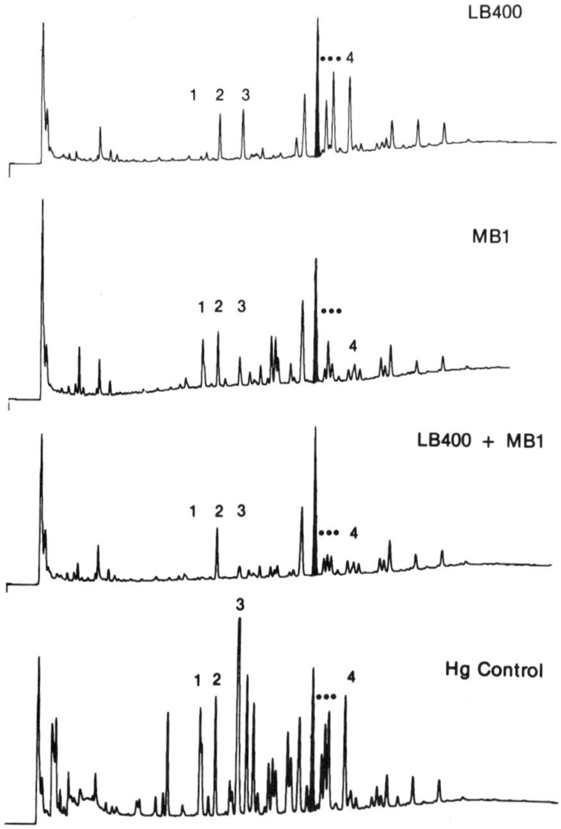

Fig. 8. Biodegradation of soil-bound Aroclor 1242 by MB1 and LB400 applied separately or as a mixture. Soil spiked with 50 ppm of Aroclor 1242 was incubated with resting cells for three days. Cells (2 ml at approximately 10^9 cells/ml) were incubated with 0.4 g of soil. The shaded peak is a nondegradable PCB congener (2,4,6,2',4'-CB) used as an internal standard. The control cells were poisoned with Hg prior to PCB addition.

Resting-cell biodegradation studies using the drag strip soil have shown substantial PCB biodegradation. LB400 degraded 15% of the PCBs in one day and 51% in three days. This compares with 85% degradation of Aroclor 1242 (500 ppm) on laboratory soil. MB1 and H850 also biodegraded the drag strip PCBs, but to a somewhat lesser extent.

In addition to our Aroclor studies we have been conducting resting-cell assays using soil-bound pure PCB congeners in an effort to define the rates of degradation of selected PCB congeners and to determine the metabolites produced from these biotransformations. The steps of PCB oxidation that we have observed in aqueous resting-cell assays (Fig. 7) also occur under soil assay conditions. For example, soil-bound 2,3-dichlorobiphenyl is completely degraded to 2,3-chlorobenzoic acid, and other PCB congeners are degraded to chlorobenzoic acids via transient hydroxylated intermediates.

In the next stage of our soil research we focused our efforts on scaling up the drag strip study in preparation for conducting a site test at the drag strip. We therefore initiated three programs: (i) preparation of the site (water, power, fencing, test plot tents, test wells, equipment, etc.); (ii) scale-up culturing of LB400 to deliver up to 1,000 l cell culture per week; and (iii) determination of conditions and final protocol for conducting in situ biodegradation (in contrast to our previously used resting-cell laboratory protocols). The consulting engineering firm of Woodward-Clyde (Wayne, New Jersey) completed the site preparation in Fall, 1986.

The mass culturing of LB400 has been successfully carried out on a 200-l scale, using biphenyl as sole carbon source and reaching a cell density of greater than 10^9 cells/ml in 24 hr using a 1% inoculum. This rate of cell production is more than sufficient for the required bacterial applications.

The third preparative program required our greatest efforts. A series of experiments was begun to model the conditions, scale, and procedures applicable to the full-scale test. Thus, 1- to 5-kg samples of drag strip soil were incubated with LB400 using experimental conditions more suitable to an in situ application of bacteria. For these simulated in situ conditions, we applied only 3 to 4% of the cell number used in the earlier laboratory resting-cell experiments. In addition, the cells were not washed before use, the amount of water per gram of soil was decreased, the temperature was decreased by 8°C, and there was no shaking or aeration of the reaction mixture. The experiments were designed to test parameters such as: (a) rate of PCB degradation as a function of dosing frequency and bacterial concentration in the soil; (b) effect of soil mixing following bacterial application; (c) degradation rate as a function of temperature; and (d) response of various control applications (e.g., water, buffer, Hg^2 -poisoned LB400, and a mutant strain of LB400 that cannot grow on biphenyl).

Polychlorinated biphenyl biodegradation was first detectable after 30 days, a much slower rate than observed with the resting-cell protocol. The laboratory study with drag strip soil eventually showed significant PCB biodegradation, albeit in weeks rather than days. In an undisturbed soil sample (1 kg) inoculated three times weekly with LB400, 50% of the PCB in the topmost soil was biodegraded in 15 weeks (Fig. 9). However, much less degradation (10%) was seen at depths below 1 cm. Bacterial cell counts indicated at least ten-fold fewer bacteria at these depths. When a duplicate of the undisturbed soil experiment was mixed at three months with continued inoculation, the redistributed soil again exhibited the highest biodegradation rate at the surface. Thus, as expected, mixing is an important variable and must be considered for any site application.

In parallel experiments, 5-kg containers of drag strip soil were inoculated three times weekly with LB400, but then mixed after each application. After 23 weeks, the PCBs in this soil were 35% biodegraded at all depths. This actually represents a greater amount of PCB destruction because the PCBs were degraded throughout the entire sample, not just at the surface.

In an unexpected result, a control sample inoculated three times per week with heat-killed LB400 showed a small amount of PCB biodegradation. Coincident with this PCB depletion we observed the appearance of PCB-de-

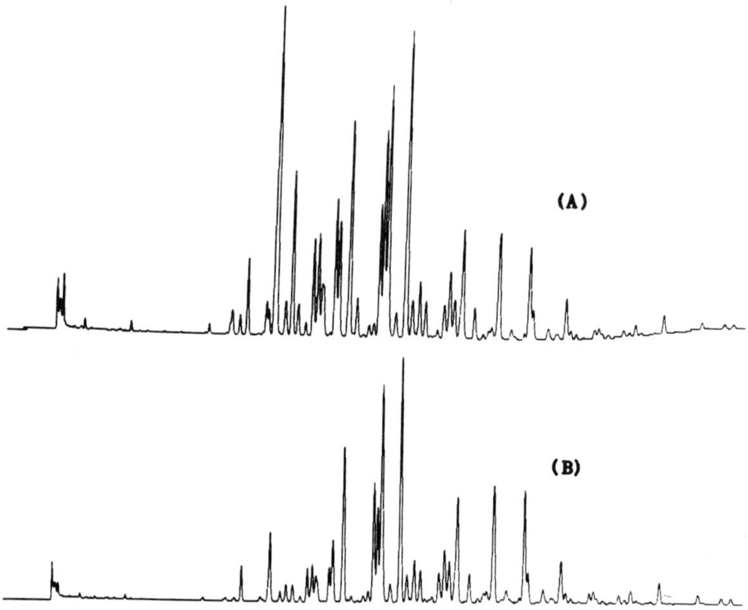

Fig. 9. Gas chromatograms of biodegraded, soil-bound PCBs. (A) Buffer-dosed control, mixed. (B) Inoculated, unmixed, 15 weeks.

grading bacteria in this control soil. Purification and analysis of these bacterial strains has shown that they differ substantially from LB400. Their PCB congener selectivity pattern is indicative of the more common 2,3-dioxygenase type, and matches that of a bacterial isolate (strain DS100) enriched from drag strip soil independently using a standard biphenyl growth selection. It therefore appears that inoculating with cell debris (i.e., nutrients) and residual biphenyl from the culture medium may enrich the population of indigenous PCB-degrading organisms in drag strip soil. These bacteria are vastly inferior to LB400, as determined by both resting-cell assays and drag strip PCB biodegradation.

Our model studies have clearly demonstrated that the rate of PCB biodegradation using in situ conditions is significantly slower than in the resting-cell conditions conducted at higher cell concentration and higher incubation temperatures. We must, therefore, anticipate that it will take weeks instead of hours or days to see the same extent of PCB degradation in the field as that seen under optimized laboratory conditions. In addition, the frequency of dosing is very significant, perhaps due to the inability of LB400 to survive for extended periods in the test soils. Therefore, we chose a thrice-weekly dosing of LB400 for the actual site test.

The test is currently underway (August, 1987). For the past 11 weeks we have been inoculating a small test area (3 m x 3 m) with LB400 and monitoring for PCB biodegradation as compared to an adjacent control plot dosed with buffer. Preliminary results indicate significant PCB degradation after eight to ten weeks.

In addition to the current site test, we have expanded our model laboratory studies in an effort to further define the best conditions for conducting in situ PCB biodegradation. We are attempting to define the limits of PCB biodegradation using LB400 and perhaps to extend these by use of co-cultures with Corynebacterium sp. MB1. Eventually, we would like to conduct laboratory soil decontamination studies using genetically engineered strains with enhanced PCB-degrading competence (described in a later section).

MODELING NONBIOLOGICAL SOURCES OF APPARENT BIODEGRADATION

It has been important throughout our biodegradation process-modeling research to demonstrate that our bacterial soil decontamination results are unequivocally due to biological activity. One pitfall that we, as well as other investigators, must be concerned with is congener depletion in a "biodegradation" process that is actually due to physical loss of the PCB and not to true biological degradation. With Aroclor studies, these processes can easily be distinguished, because biodegradation results in depletion of specific congeners yielding GC profiles that are distinctly different from those of the original Aroclor mixtures. Physical depletion, on the other hand, results in uniform depletion of all congeners (e.g., adsorptive loss) or depletion of lower-chlorinated congeners due to their higher volatility (e.g., evaporative loss). The production of PCB metabolites is of course another unequivocal method for demonstrating the biological basis of PCB depletion.

In order to better evaluate results from open-air, aerated, stirring reactors, we set up a model process that appears to be biologically mediated, but whose conditions were adjusted so as to preclude biodegradation. A sample of Oakland, California, soil contaminated with Aroclor 1260 (7,800 ppm) was air-dried at room temperature, homogenized with a glass mortar and pestle, and then sieved. Argon (instead of air) was bubbled through a water/soil mixture (approximately 125 ml) in a 250-ml round-bottom flask at a flow of approximately 200 ml/min. A florosil sample tube was attached to trap PCBs in the argon as it exited the vessel. Samples (20 ml) were taken periodically while mixing to ensure a homogeneous sample. The volume in the vessel was maintained by adding 20 ml of distilled water after each sampling, and the florosil sample tube was replaced each time a sample was taken. The soil was mixed and purged with argon at room temperature for 19 days, after which the vessel was disassembled.

Each soil sample and florosil tube was extracted for GC analysis. The remaining soil and water were removed and pooled for GC analysis. The entire vessel was washed several times with hexane/acetone (1:1) to remove any PCBs bound to the vessel. These extracts were also pooled for GC analysis. Upon disassembly, we observed a tar-like substance sticking to the Teflon stirrer. This was removed and added to the soil and water fraction before the hexane/acetone extractions.

Neither oxygen nor bacterial inoculum was introduced in this mock process, yet the analytical results (Fig. 10) could be mistaken for biodegradation. Although the time-course analysis indicated greater than 90% PCB depletion, it is clear from the mass balance calculations (Fig. 10) that the aeration and stirring of the soil resulted in the redistribution of PCBs

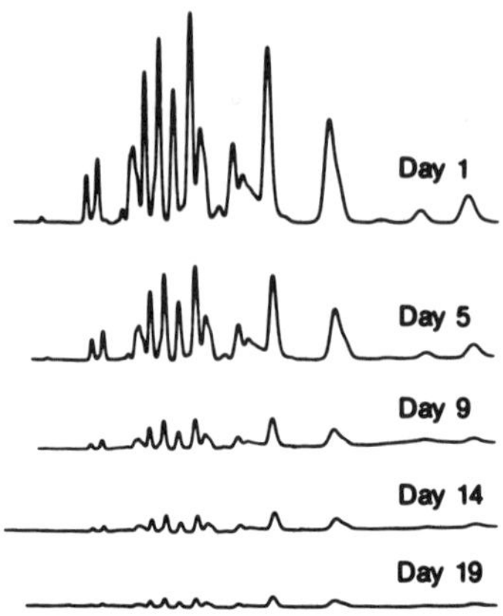

Day	Sample wt.	ppm	mg
1	2.75 g	7800	21
2	2.84 g	5300	15
5	2.75 g	3400	9.4
6	2.06 g	2900	6.0
7	2.16 g	2100	4.5
9	1.73 g	1500	2.5
12	1.57 g	990	1.6
14	1.69 g	830	1.4
16	1.84 g	640	1.2
19	1.53 g	620	0.9

Starting material = 35 g Oakland soil

Total starting PCBs = 273 mg

PCBs Recovered:

Sampling total (for 20.9 g soil) = 64 mg
Glassware = 62 mg
Florosil sample tubes total = 0.4 mg
Final soil + water = 41 mg
Tar balls (in final soil) = 68 mg

Total PCBs recovered = 235 mg (86% recovery)

Fig. 10. Depletion of Aroclor 1260 in an experiment to demonstrate the nonbiological sources of apparent biodegradation.

from the soil to difficult-to-sample locations in the reactor (i.e., glassware, stirrer, and coalesced droplets of PCB). The GC profiles also demonstrate that the observed depletion was not due to a biological process, since all GC peaks were depleted proportionally. Therefore, experiments that purport to show biodegradation of PCBs by quantifying total GC peak areas must be carefully evaluated. It is for this reason that we include nonbiodegradable PCB internal standards wherever possible. If such standards or dead-cell controls cannot be included, then one must rely on differential congener depletion (or metabolite production) as evidence for the biological basis of PCB "biodegradation" processes.

APPROACHES FOR BIODEGRADING SOIL-BOUND POLYCHLORINATED BIPHENYLS BY GENETICALLY ENGINEERED BACTERIA

We are now in a position to explore the potential of using genetically engineered bacteria to biodegrade soil-bound PCBs by integrating our two previously independent lines of investigation: (i) the promising model studies using P. putida LB400 to biodegrade PCBs on soil; and (ii) the recent successful cloning of the PCB-degradative genes from LB400. In the coming months our genetic research will include: (i) studies on the regulation of PCB/biphenyl degradation so that expression can be enhanced and controlled; (ii) use of the LB400 gene clones to isolate and study the PCB/biphenyl genes from other organisms; and (iii) combining of the genes from different organisms to facilitate the metabolism of PCBs and PCB intermediates.

Following attainment of each of these three goals, we will have available new and/or improved recombinant DNA organisms. Thereupon, we can begin to assess the activity of these recombinant bacteria for degrading soil-bound PCBs. The approach would be similar to that used for our laboratory PCB/soil studies using native LB400 (as discussed above). Each new genetically engineered bacterium could be assayed for its ability to degrade PCBs on: (a) Aroclor 1242-spiked clean soil (500 ppm), and (b) drag strip soil (525 ppm Aroclor 1248-like), using both resting-cell and in situ protocols. Specifically, one could assay the soil PCB degradative activity of:

(a) A first-generation, high-expression, genetically engineered bacterium. This organism would be designed to synthesize higher levels of PCB-degradative enzymes from cloned LB400 DNA, and should exhibit increased rates of PCB biodegradation.

(b) A possible second PCB-degrading recombinant strain whose PCB-degradative activity complements that of LB400. If this is achieved, then soil studies with this genetically engineered organism could be conducted using both pure and mixed cultures (with the LB400-derived strain). This may eventually lead to a superior mixed-culture approach to PCB biodegradation (as discussed above and illustrated in Fig. 9).

(c) A recombinant strain designed to combine from different organisms genes for the metabolism of PCBs and/or recalcitrant PCB metabolites. This organism might exhibit novel or improved PCB-degradative activity and would be an excellent candidate for assaying degradation of PCBs bound to soil.

In summary, it will be important in the future to evaluate genetically engineered PCB-degradative strains for their ability to degrade PCBs as actually found in the environment. This integration of genetic and soil studies is a logical and practical extension of important and promising research on PCB biodegradation. Thus, we are now in a position to exploit the potential of modern biotechnology for destroying environmental contaminants such as PCBs.

ACKNOWLEDGEMENTS

We would like to thank Robert Wagner and Ralph May for their expert GC and GC-MS analyses, and Carolyn Meyer, Jennifer Gensler, and Marsha Berry for typing the manuscript. This research was funded in part by the United States Environmental Protection Agency (contract No. CR-812727-02-0), Dr. P.R. Sferra, Project Officer.

REFERENCES

1. Bedard, D.L., M.J. Brennan, and R. Unterman (1984) Bacterial degradation of PCBs: Evidence of distinct pathways in Corynebacterium sp. MB1 and Alcaligenes eutrophus H850. In Proceedings of the 1983 Polychlorinated Biphenyl Seminar, G. Addis and R. Komai, eds. Electric Power Research Institute, Palo Alto, California, pp. 4-1-1 and 4-1-18.
2. Bedard, D.L., M.L. Haberl, R.J. May, and M.J. Brennan (1987) Evidence for novel mechanisms of PCB metabolism in Alcaligenes eutrophus H850. Appl. Environ. Microbiol. 53:1103-1112.
3. Bedard, D.L., R.E. Wagner, M.J. Brennan, M.L. Haberl, and J.F. Brown (1987) Extensive degradation of Aroclors and environmentally transformed PCBs by Alcaligenes eutrophus H850. Appl. Environ. Microbiol. 53:1094-1102.
4. Bedard, D.L., R. Unterman, L.H. Bopp, M.J. Brennan, M.L. Haberl, and C. Johnson (1986) Rapid assay for screening and characterizing microorganisms for the ability to degrade polychlorinated biphenyls. Appl. Environ. Microbiol. 51:761-768.
5. Bopp, L.H. (1986) Degradation of highly chlorinated PCBs by Pseudomonas strain LB400. J. Ind. Microbiol. 1:23-29.
6. Brown, J.F., D.L. Bedard, M.J. Brennan, J.C. Carnahan, H. Feng, and R.E. Wagner (1987) Polychlorinated biphenyl dechlorination in aquatic sediments. Science 236:709-712.
7. Brown, J.F., R.E. Wagner, D.L. Bedard, M.J. Brennan, J.C. Carnahan, R.J. May, and T.J. Tofflemire (1984) PCB transformation in upper Hudson sediments. Northeastern Environ. Sci. 3:167-179.
8. Brown, Jr., J.F., R.E. Wagner, H. Feng, D.L. Bedard, M.J. Brennan, J.C. Carnahan, and R.J. May (1987) Environmental dechlorination of PCBs. Environ. Toxicol. Chem. 6:579-593.
9. Furukawa, K. (1982) Microbial degradation of polychlorinated biphenyls. In Biodegradation and Detoxification of Environmental Pollutants, A.M. Chakrabarty, ed. CRC Press, Inc., Boca Raton, Florida, pp. 33-37.
10. Furukawa, K., N. Tomizuka, and A. Kamibayashi (1979) Effect of chlorine substitution on the bacterial metabolism of various polychlorinated biphenyls. Appl. Environ. Microbiol. 38:301-310.
11. Nadim, L.M., M.J. Schocken, F.K. Higson, D.T. Gibson, D.L. Bedard, L.H. Bopp, and F.J. Mondello (1988) Bacterial oxidation of polychlorinated biphenyls. In Proceedings of the United States Environmental Protection Agency Thirteenth Annual Research Symposium on Land Disposal, Remedial Action, Incineration and Treatment of Hazardous Waste, Cincinnati, Ohio (in press).
12. Sayler, G.S., M. Shon, and R.R. Colwell (1977) Growth of an estuarine Pseudomonas sp. on polychlorinated biphenyl. Microb. Ecol. 3:241-255.

13. Unterman, R., M.J. Brennan, R.E. Brooks, and C. Johnson (1987) Biological degradation of PCBs. In Proceedings of the International Conference on Innovative Biological Treatment of Toxic Wastewaters, U.S. Army, CERL, Champaign, Illinois, pp. 379-389.

14. Unterman, R., D.L. Bedard, L.H. Bopp, M.J. Brennan, C. Johnson, and M.L. Haberl (1985) Microbial degradation of polychlorinated biphenyls. Proceedings of the International Conference on New Frontiers of Hazardous Waste Management, United States Environmental Protection Agency, Cincinnati, Ohio, pp. 481-488.

15. Yagi, O., and R. Sudo (1980) Degradation of polychlorinated biphenyls by microorganisms. J. Water Poll. Control Fed. 52:1035-1043.

EXPLOITATION OF THE HIGH AFFINITY AND SPECIFICITY OF PROTEINS IN WASTE STREAM TREATMENT

Clement E. Furlong,[1] Joseph A. Sundstrom,[1]
Edward B. Weiler,[2] Peter W. Cheung,[2]
John Yin,[3] and Harvey W. Blanch[3]

[1]Departments of Genetics and Medicine
Division of Medical Genetics
Center for Inherited Diseases and
Charles A. Dana Program in Genetics
and Environmental Health
University of Washington
Seattle, Washington 98195

[2]Microsensor Research Program
Washington Technology Center
University of Washington
Seattle, Washington 98195

[3]Department of Chemical Engineering
University of California
Berkeley, California 94720

INTRODUCTION

During the conference on "Genetic Control of Environmental Pollutants" held in Seattle in 1983 (18), it became clear that adsorption systems with high affinities and specificities would be useful in removing specific pollutants from waste streams. Following the conference, we decided to explore the possibility of developing such systems, taking advantage of the high specificity and affinity offered by proteins. The present chapter addresses three areas of research: (a) the development of protein, peptide, and biomimetic adsorbers; (b) the development of microbiosensors and control systems for the continuous monitoring and automation of water treatment procedures; and (c) automation of the production of proteins used in ligand adsorbers and biosensors.

PROTEIN-BASED BIOADSORBERS

The periplasmic transport proteins found in gram-negative bacteria not only offer the specificity and affinity required for specific adsorption, but also are very stable proteins that bind their ligands over broad ranges of pH and ionic strength (5). To investigate the possibility of using these unique proteins to remove pollutants from water, we chose the phosphate-binding protein of Escherichia coli as a model protein (6,14). This protein offers several advantages for studying proteins as specific bioadsorbers. First, two radiolabeled substrates, $^{33}PO_4$ and $^{32}PO_4$, are available for this protein. Binding of the higher-energy isotope ($^{32}PO_4$) may be determined nondestructively. Second, the gene encoding this binding protein has been cloned (13,23), and the protein can be expressed at high levels under conditions of arrested cell growth (15).

The first model system that we developed involved the immobilization of the phosphate-binding protein of E. coli to cyanogen bromide-activated (19) agarose beads (Fig. 1). The derivatized beads were packed into a water-jacketed column which was connected to an in-line scintillation counter (Fig. 2). A valve on the inlet line allowed the feed to be switched from a phosphate-containing feed stream to a phosphate-free rinse stream. The in-line scintillation counter provided both continuous analog and digital analysis of the column exit stream. In operation, the adsorber binds phosphate until all of the sites are filled. The bound phosphate is then removed simply by heating the column. Cooling regenerates the native protein for subsequent cycles of adsorption, washing, release, and renaturation.

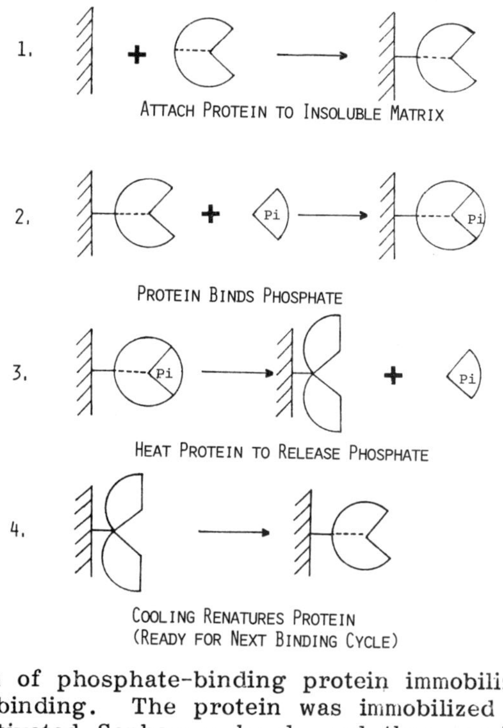

Fig. 1. Diagram of phosphate-binding protein immobilization and cycling ligand binding. The protein was immobilized to cyanogen bromide-activated Sepharose beads and then used to remove phosphate from a feed stream. The immobilized protein functioned through many repeat cycles.

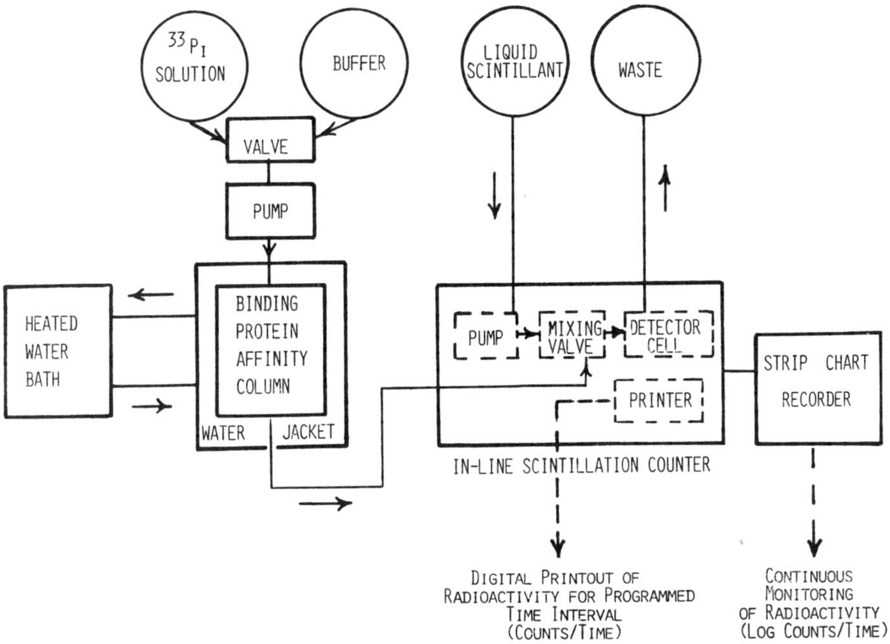

Fig. 2. Diagram of the analytical apparatus for the continuous monitoring of ligand adsorber performance. The apparatus was the same for the phosphate and cadmium ligand adsorbers, except that the bound phosphate was released by heat and the bound cadmium, by a low pH, chelating buffer.

To demonstrate the efficiency of the adsorber when run below maximal capacity, the feed stream was first run directly through the detector, showing that the feed stream contained approximately 2,500 cpm. After the flow cell was washed free of the feed stream, the feed stream was directed through the adsorber. The adsorber effectively scrubbed the phosphate levels down to background.

Next, we developed an adsorber for the removal of cadmium from feed streams. Cadmium is an extremely toxic metal. Exposure to cadmium can result in kidney dysfunction, testicular and liver damage, high blood pressure, and osteomalacia in humans (3,4,11). Land used for sludge disposal has a fixed lifetime cadmium load. Thus, lowering cadmium levels of waste streams extends the amount of sludge that may be loaded on a given disposal site, and decreases toxicity to organisms exposed to the cadmium contained in the outflow.

Organisms have evolved protective proteins and peptides that are expressed when they are exposed to cadmium or other related metals such as zinc (3,4,11,12). The 61-amino acid (in one-letter code) sequence of one of these proteins, metallothionein I from mouse and rat (7), is shown below:

MDPNCSCSTGGSCTCTSSCACKNCKCTSCKKSCCSCCPVGCSKCAQGCVCKGAADKCTCCA

Note the distribution of cysteine residues (C) in this protein. These residues are involved in chelating cadmium, reducing its activity (11,22). Plants (10,21) and yeast (12,16) express small cysteine peptides (relatives of glutathione) that are effective binders of cadmium. These peptides have the general structure (γ-Glu-Cys)n-Gly.

It seemed reasonable to make use of naturally existing cadmium binders in developing cadmium adsorbers. Rabbit metallothionein was immobilized onto cyanogen bromide-activated agarose beads. The same continuous analysis system used for the phosphate adsorber was used to monitor the performance of the cadmium adsorber with ^{109}Cd as the radiotracer. A column of the protein-derivatized beads was used to remove 100 μM cadmium from a feed stream. Figure 3 shows the effective scrubbing of 1 μM cadmium from the feed stream and release of the bound cadmium from the adsorber before adsorber saturation is achieved. These four cycles are from a run of 20 consecutive binding and release cycles, with constant performance of the column. A calculation of the column capacity suggests that the column is effectively binding 0.6 moles of cadmium per mole of immobilized metallothionein. (Maximum binding would be 7 moles Cd/mole protein.)

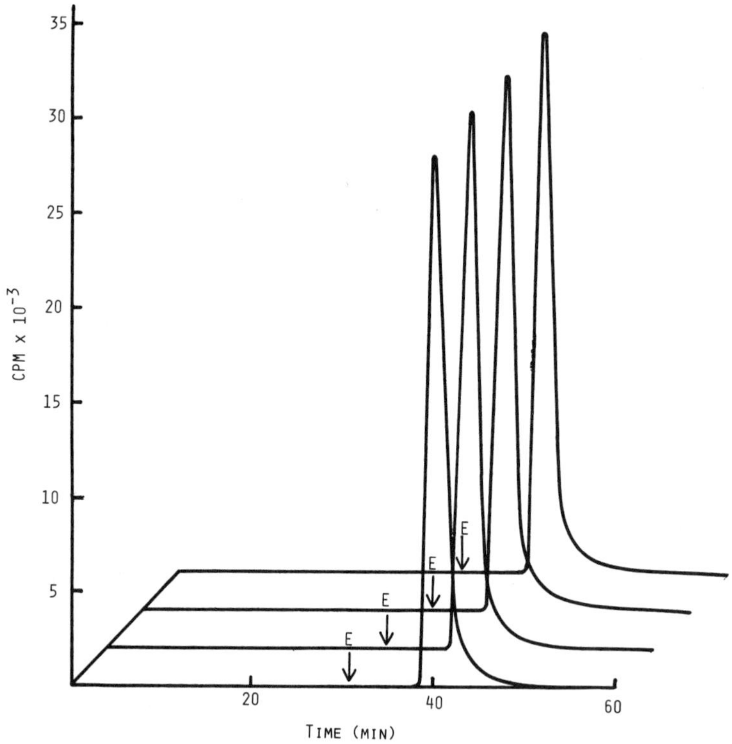

Fig. 3. Experiment illustrating the scrubbing and release of cadmium from a feed stream containing 1 μM ^{109}Cd. The cadmium was released from the adsorber by pH 2.0, 100 mM oxalate buffer at elution time E, following which the column was regenerated with 10 mM Tris-HCl buffer, pH 8. Four sequential cycles are shown in this figure.

To investigate the structural requirements for high affinity-specific cadmium binding, synthetic adsorbants have been designed that mimic the cysteine-rich cadmium-binding peptides synthesized by plant (10,21) and yeast (12,16) cells in response to cadmium stress. Mercapto-ethylamine was attached to the amido dicarboxylic groups of an ion exchange resin (Chelex) through a peptide bond. The disulfide form of mercapto-ethylamine was used to minimize reactions of the sulfhydryls. Subsequent reduction of the resulting products with $NaHSO_3$ at 65°C produces the biomimetic resin termed "Thiolex." An initial characterization of the resin indicated a K_D of 6 x 10^{-7} M and a capacity of 0.05 meq/g dry resin at pH 8.5. The adsorption isotherm is shown in Fig. 4. The cadmium is readily eluted from the resin by treatment with HCl. Preliminary specificity studies indicate a ten- to 20-fold preference of cadmium to zinc.

DEVELOPMENT OF MICROBIOSENSORS

In order to automate such adsorber systems, it would be useful to develop biosensors that would detect saturation of the adsorber and trigger a controller to shunt the flow to a second adsorber while regenerating the first adsorber.

We are presently working on the development of such a biosensor which will take advantage of the high affinity and specificity of the *E. coli* phosphate-binding protein (6,14). This protein is immobilized on the surface of a derivatized silicon chip which allows the binding of phosphate by the protein to be converted into an electronic signal.

We have performed a series of preliminary experiments in which protein was immobilized on the front surface of single crystal (100) p-type silicon squares (0.5 in^2) on which an 800-A-thick layer of silicon dioxide (SiO_2) had been thermally grown in dry oxygen at 1,000°C. The backs of the squares were 5,000 A of Apiezon black wax-coated aluminum.

Silicon dioxide does not naturally lend itself to the immobilization of protein. The surface consists of silanol groups which may be thought of as very small hooks. These silanol groups do not easily form covalent

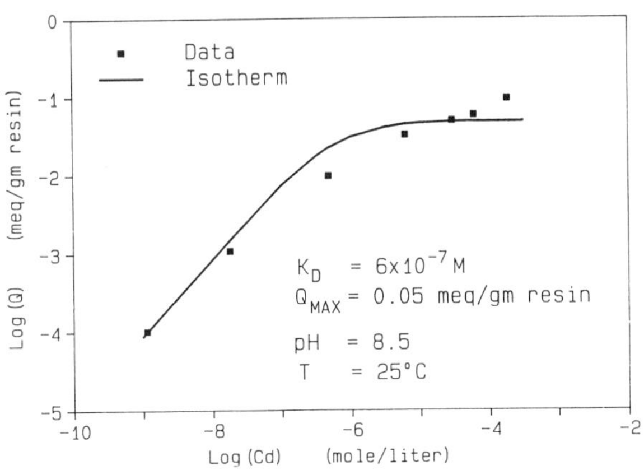

Fig. 4. Adsorption isotherm for the binding of cadmium to Thiolex.

bonds. A derivatization step must first be performed to provide a reactive site that can be used to immobilize the protein. The method we have chosen is based on a procedure described by Haller (9). Samples are placed in a vessel of refluxing vapors of aminopropyl triethoxysilane dissolved in toluene.

The deposition of this silane compound leaves amino-propyl groups extending from the surface. The amino groups are then reacted with a homo-bifunctional cross-linking reagent, dimethyl suberimidate (20), to provide a covalent attachment site for the binding protein. The activated surface is exposed to a solution of the phosphate-binding protein, allowing the protein to be covalently bonded to the surface.

Evidence of the successful attachment of the protein, including its continued viability, was demonstrated by experiments in which protein-modified Si/SiO_2 samples were exposed to a labeled phosphate solution and then quantified by Cerenkov counting. Since this method is completely nondestructive, the samples can be cycled by exposing them to a 75°C solution to cause reversible denaturation of the protein and release of the phosphate. The protein is renatured in room temperature solution. We have demonstrated that the samples can be cycled a number of times and retain functionality even after dry storage.

In order to develop an effective biosensor, the binding of the phosphate by the protein must be converted into an easily manageable electronic signal. We have conducted preliminary experiments to determine the electronic sensitivity of the semiconductor to the phosphate binding. The experiments are based on the use of capacitance-voltage measurements (17) to determine the flat band voltage (the voltage at which the stored charge on the electrolyte-insulator-semiconductor capacitor is reduced to 0) of an electrolyte-insulator-semiconductor (EIS) capacitor. The EIS capacitor is formed by placing an Si/SiO_2 sample with protein immobilized on its surface in an apparatus that allows an electrolyte solution to come in contact with the surface. Contact to the back of the sample is made through aluminum deposited by electron beam evaporation. The circuit is completed by placing a reference electrode in the electrolyte solution, as shown in Fig. 5.

Samples prepared as described above were placed in the EIS apparatus. The phosphate concentration of the electrolyte solution was varied while maintaining a constant pH. The flat band voltage of the EIS capacitor was measured at varying phosphate concentrations. As phosphate was

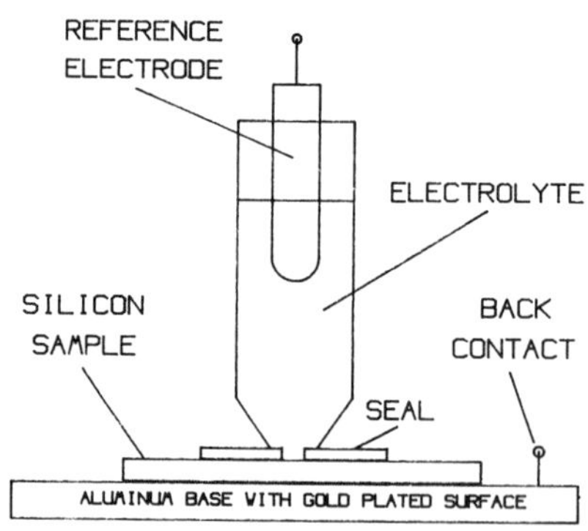

Fig. 5. Diagram of an electrolyte-insulator-semiconductor (EIS) apparatus.

bound at the surface, the change in charge at the surface due to the binding of the divalent anion generated a change in the flat band voltage. Results indicated a sensitivity of approximately 20 mv/decade phosphate concentration, as shown in Tab. 1. Above the K_D of the protein (approximately 1 μM), no consistent EIS results were obtained.

A phosphate-sensitive field-effect transistor can be produced by immobilizing the protein on the exposed gate oxide of a field effect transistor (FET). The resulting CHEMFET can then be used to produce a signal or to control other circuitry in response to phosphate binding or, by analogy, Cd^{2+} binding.

The development of biosensor/controllers should prove useful in automating water treatment. Biosensors will also find important applications in waste stream monitoring, process control, and many other areas.

AUTOMATION OF PROTEIN PRODUCTION

It is necessary to develop procedures for automating protein production to make it cost-effective to use proteins in the construction of ligand adsorbers and biosensor/controllers, and in bioaugmentation. Toward this goal, we have selected strains of E. coli that secrete proteins directly into the medium. In order to generate high purity of the secreted protein product, the gene encoding the desired protein is placed downstream from appropriate regulatory sequences in a high copy number plasmid [pAN92(23)]. The effect of phosphate limitation on the expression of the phosphate-binding protein from the plasmid is shown in Fig. 6.

Encapsulation of this strain in small agarose beads provides the packing for a protein bioreactor that allows for the automation of continuous protein production. The limitation of phosphate provides for both induction of the desired protein and limitation of cell growth to prevent mechanical breakage of the encapsulation matrix. Sterile aerated medium is

Tab. 1. Flat band voltage as a function of phosphate concentration measured in the EIS apparatus.

Phosphate concentration (μM)	Flat band voltage (mv)
0.0	-1030.
0.1	-1010.
1.0	-0990.
10.0	-1010.

Note: Samples minus the wax coating were mounted in the EIS apparatus (see Fig. 5). The flat band voltage was taken from intermediate frequency capacitance/voltage curves generated in the presence of varying concentrations of phosphate.

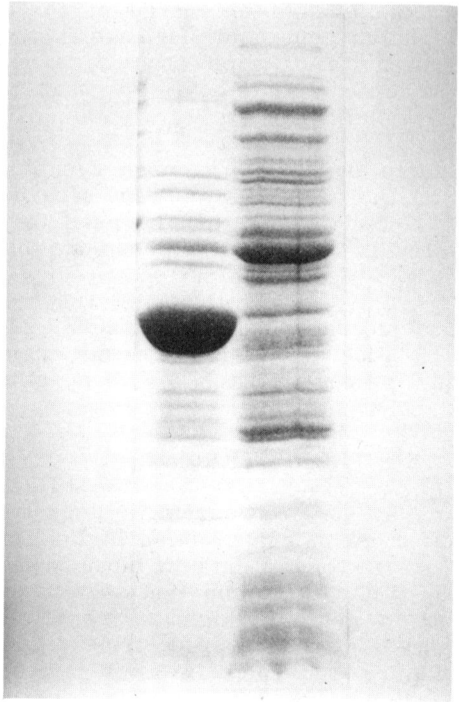

1 2

Fig. 6. Sodium dodecyl sulfate gel electrophoretic analysis of the periplasmic fractions of cells grown on low (lane 1) and high (lane 2) phosphate medium. The cells contained a plasmid directing the synthesis of the binding protein.

pumped through the column of immobilized cells. Secreted protein leaves the cells and is concentrated and dialyzed in a membrane ultrafiltration/-counter-current dialysis apparatus. At this point, many proteins will be suitable for direct coupling to solid phase supports. For sensor applications, where high specificity and lack of interference are required, the apparatus will need an in-line continuous purification step (or steps). The automated protein production technology should be useful in making proteins for many different applications, including those described here. Other applications would include vaccines, industrial enzymes (including those used in bioaugmentation), and growth factors.

IDENTIFICATION OF SPECIFIC BINDING PROTEINS

The development of specific protein-based ligand adsorbers and the development of specific biosensors require procedures for identifying proteins with the appropriate binding characteristics. We have developed a general procedure for identifying both soluble and membrane-bound proteins with high-affinity binding of specific ligands (2). The proteins are first resolved by two-dimensional gel electrophoresis (first dimension, isoelectric focusing; second dimension, SDS gel electrophoresis), then rinsed to remove excess detergent in the gel. The developed gel is then dipped into a solution of radiolabeled ligand, briefly rinsed, and dried at moderate temperature. The gel is exposed to X-ray film and stained for protein. Proteins of interest are identified by superimposing the autoradiogram over the stained gel. Strategies for purification and cloning of the protein can be designed from the information obtained from the two-dimensional gel.

ACKNOWLEDGEMENTS

The authors gratefully acknowledge support provided by the United States Environmental Protection Agency (Cooperative Agreement No. CR811948) and The Microsensor Research Program, Washington Technology Center. We thank Dr. H. Rosenberg for the plasmid used in these studies.

REFERENCES

1. Belcher, E.H. (1953) The luminescence of irradiated transparent media and the Cerenkov effect. I. The luminescence of aqueous solutions of radioactive isotopes. Proc. Royal Soc. London A216:90-102.
2. Copeland, B.R., R.J. Richter, and C.E. Furlong (1982) Renaturation and identification of periplasmic proteins in two dimensional gels of Escherichia coli. J. Biol. Chem. 257:15065-15071.
3. Friberg, L., M. Piscator, G.F. Nordberg, and T. Kjellstrom, eds. (1974) Cadmium in the Environment, CRC Press, Inc., Cleveland, Ohio.
4. Friberg, L., C.-G. Elinder, T. Kjellstrom, and G.F. Norberg, eds. (1985) Cadmium and Health: A Toxicological and Epidemiological Appraisal. Vol. 1. Exposure, Dose, and Metabolism, CRC Press, Inc., Cleveland, Ohio.
5. Furlong, C.E. (1987) Osmotic-shock-sensitive transport systems. In Escherichia coli and Salmonella typhimurium: Cellullar and Molecular Biology, J.L. Ingraham, K.B. Low, B. Magasanik, M. Schaecter, and H.E. Umbarger, eds. American Society for Microbiology, Washington, D.C., pp. 768-796.
6. Gerdes, R.G., K.P. Strickland, and H. Rosenberg (1977) Restoration of phosphate transport by the phosphate-binding protein of Escherichia coli. Biochim. Biophys. Acta 351:77-86.
7. Glanville, N., D.M. Durnham, and R.D. Palmiter (1981) Structure of mouse metallothionein-I gene and its mRNA. Nature 292:267-269.
8. Grill, E., E.-L. Winnacker, and M.H. Zenk (1985) Phytochelatins: The principal heavy-metal complexing peptides of higher plants. Science 230:674.
9. Haller, I. (1978) Covalently attached organic monolayers on semiconductor surfaces. J. Am. Chem. Soc. 100:8050-8055.
10. Jackson, P.J., E.J. Roth, P.R. McClure, and C.M. Naranjo (1984) Selection, isolation and characterization of cadmium-resistant Datura innoxia suspension cultures. Plant Physiol. 75:914-
11. Kagi, J.H.R., and M. Nordberg, eds. (1979) Metallothionein. Experientia Supplement 34, Birkhauser Verlag, Basel.
12. Kondo, N., M. Isobe, K. Imai, T. Goto, A. Murasugi, and Y. Hayashi (1983) Structure of Cadystin, the unit-peptide of cadmium-binding peptides induced in a fission yeast, Schizosaccharomyces pombe. Tetrahedron Lett. 24:925-928.

13. Magota, K., N. Otsuji, T. Miki, T. Horriuchi, S. Tsunasawa, K. Kondo, F. Sakiyama, M. Amemura, T. Morita, H. Shinagawa, and A. Nakata (1984) Nucleotide sequence of the phoS gene, the structural gene for the phosphate-binding protein of Escherichia coli. J. Bacteriol. 157:909-917.
14. Medveczky, N., and H. Rosenberg (1970) The phosphate-binding protein of Escherichia coli. Biophys. Acta 211:158-168.
15. Morita, T., M. Amemura, K. Makino, H. Shinigawa, K. Magota, N. Otsuji, and A. Nakata (1983) Hyperproduction of phosphate-binding protein, phoS, and pre phoS proteins in Escherichia coli carrying a cloned phoS gene. J. Biochem. 130:427-435.
16. Murasugi, A., C. Wada, and Y. Hayashi (1981) Cadmium-binding peptide in fission yeast, Schizosaccharomyces pombe. J. Biochem. 90:1561-1564.
17. Nicollian, E.H., and J.R. Brews (1982) MOS (Metal Oxide Semiconductor) Physics and Technology, John Wiley and Sons, New York.
18. Omenn, G.S., and A. Hollaender, eds. (1984) Genetic Control of Environmental Pollutants, Plenum Press, New York.
19. Porath, J., K. Aspberg, H. Drevine, and R. Axen (1973) Preparation of cyanogen bromide-activated agarose gels. J. Chromatogr. 86:53-56.
20. Pulliam, M.W., L.F. Boyd, N.C. Baglan, and R.A. Bradshaw (1975) Specific binding of covalently cross-linked mouse nerve growth factor to responsive peripheral neurons. Biochem. Biophys. Res. Commun. 67:1281-1289.
21. Rauser, W.E., H.-J. Hartmann, and U. Weser (1983) Cadmium-thiolate protein from the grass Agrostis gigantea. FEBS Lett. 164:102-104.
22. Stillman, M.J., W. Cai, and A.J. Zelazowski (1987) Cadmium binding to metallothioneins. J. Biol. Chem. 262:4538-4548.
23. Surin, B.P., D.A. Jans, A.L. Fimmel, D.C. Shaw, G.B. Cox, and H.R. Rosenberg (1984) Structural gene for the phosphate-repressible phosphate-binding protein of Escherichia coli: Nucleotide sequence and gene-polypeptide relationships. J. Bacteriol. 161:189-198.

BIOPROCESSING OF COAL

Jenefir D. Isbister, Robert E. Wyza, Joseph Lippold,
Alice DeSouza, and George Anspach

Atlantic Research Corporation
Research and Technology Division
Alexandria, Virginia 22313

INTRODUCTION

Bioprocessing of coal is a new and exciting frontier in biotechnology. Biologically mediated processes in industrial microbiology have been successfully applied to high-value, low-volume products in the fields of specialty chemicals, medicine, and agriculture. The potential for high profit margins for such products breeds attractive investment opportunities and justifies funding for research and development in these areas. In contrast, coal is a high-volume, low-value product; thus, any biological process for cleaning or conversion of coal must be relatively inexpensive.

Coal is an abundant natural resource in the United States and will become an increasingly important fuel source in the future as oil prices increase and/or availability decreases. A major disadvantage to the use of coal as a fuel is that pollutants are released into the environment during combustion. The sulfur oxides and nitrogen oxides released are of increasing concern with respect to air quality and acid rain control. Regulations governing sulfur dioxide emissions have created new criteria for determining the degree to which coal must be cleaned before it can be used as a fuel. These regulations are likely to become more stringent as the individual states and the Congress become increasingly concerned about the environmental impact of burning "dirty" coal.

Although bioprocessing of coal is in its infancy, this technology may have potential as a means to give greater flexibility in the use of coal as a fuel. Bioprocessing of coal today focuses on two main areas: (i) beneficiation of coal to remove undesirable contaminants (i.e., sulfur, nitrogen, and trace metals); and (ii) coal conversion (i.e., microbial liquefaction, microbial solubilization, gasification, and pretreatment). The objectives of coal cleaning are the removal of sulfur, nitrogen, metals, and perhaps ash, using mild microbial processes for production of clean coal. Microbial coal conversion is less well defined, although the overall goal of these conversions is the production of a more desirable fuel form. Because biological processes generally do not require extremes of temperature or pressure, they may be more economically feasible than the physical/thermal or chemical processes currently being investigated.

Microorganisms are extremely diverse and can adapt to survive in inhospitable environments. Microbes are capable of breaking down many complex molecules by adaptation of their degradative enzyme systems. Microbial degradation of compounds depends not only on the microbial degradative capacity but on the molecular structure of the target molecule, the types of chemical bonds present in the molecule, and the spatial environment around the bonds. The database available for understanding the macromolecular structure of coal is very limited, and investigations into the biochemical mechanisms responsible for microbial activity on coal have just begun.

Coals from different regions of the United States vary widely in their sulfur content. Western coals may contain an average of 0.68% total sulfur, whereas an average Western Midwest coal may contain an average of 5.25% total sulfur. Conventional physical methods for removal of coal impurities reduce the coal particle size and separate the coal from impurities using differences in specific gravity and/or surface properties of coal and waste minerals. These physical cleaning methods remove ash and inorganic sulfur, leaving the organic sulfur in the coal product. Thus, the inorganic sulfur content of a coal establishes the maximum sulfur removal possible using these methods. Furthermore, maximum inorganic sulfur removal is seldom achieved, because finely dispersed pyritic sulfur is not removed by these processes.

Estimates for reduction of total sulfur by these conventional methods range from 15% for Eastern coals to 44% for Western coals (9). The coal reserves in the United States that are able to meet different long-term average sulfur dioxide (SO_2) emission limits are shown in Tab. 1. It is evident that strict enforcement of long-term average SO_2 emission limits will require advanced cleaning technologies in order for most coals to meet a 1.2 lb/MBTU emission limit. These advanced technologies, which are innovative variations on the conventional processes, remove more ash and inorganic sulfur and give more effective treatment of fine coal particles. At best, the advanced cleaning techniques remove 75% of the pyritic sulfur (47% of the total sulfur) (9).

MICROBIAL REMOVAL OF INORGANIC SULFUR

Certain naturally occurring microorganisms generate ferric iron, one of the most effective and least expensive oxidizing agents capable of transforming solid metal sulfide minerals into water-soluble metallic ions and elemental sulfur. These microorganisms contribute to the leaching of pyrite from coals in the environment, resulting in acid mine run-off. Recently, the use of these microbes for metal extraction and recovery from ores and waste streams has received considerable attention. Extraction and recovery of metals such as lead, copper, zinc, cobalt, mercury, aluminum, radium, molybdenum, selenium, cesium, and uranium are being pursued (13).

Microbial leaching of pyrite from coals has been studied extensively as a means of removing inorganic sulfur from coals. This activity occurs in coal piles at pH 2.5 under ambient conditions. It is generally accepted

Tab. 1. Coal reserves able to meet long-term average sulfur dioxide (SO_2) emission limits. (From Ref. 9.)

SO_2 Emission limit (lb/MBTU)	Region	% Total sulfur†	No emission control (%)	Physical coal cleaning (%)*
1.2	N. Appalachia	3.01	5	11
	S. Appalachia	1.04	50	62
	Alabama	1.33	27	36
	E. Midwest	3.92	1	2
	W. Midwest	5.25	6	6
	Western	0.68	66	86
	TOTAL U.S.	3.02	38	50
2.0	N. Appalachia		15	30
	S. Appalachia		83	88
	Alabama		68	75
	E. Midwest		6	9
	W. Midwest		11	13
	Western		90	96
	TOTAL U.S.		56	62

†Average sulfur values for size United States coal regions.
*Coal cleaning at 1.5, 1.6 specific gravity.

that rates and quantities of pyritic sulfur removed from coal by microbial action are dependent on microbial strain(s), type of coal, surface area, temperature, pH, and microbial numbers present. The time required for this process may be as long as months in piles, but may be decreased to days in stirred or air-agitated vessels containing pulverized coal. This process can be used when space and time are not critical limiting factors.

A number of investigators are evaluating the use of these microbes to modify the surface characteristics of pyrite to aid in its rejection (removal) during coal cleaning.* Attia and Elzeky (3), Atkins et al. (1), and Kempton et al. (10) have reported enhanced sulfur and ash rejection by flotation (or agglomeration) following bacterial pretreatment. Atlantic Research also investigated microbial pretreatment for enhanced sulfur rejection. It was found that optimization of Flotaire column conditions for removal of sulfur resulted in removal of most, if not all, of the liberated pyrite from a number of coals without microbial pretreatment. The cleaned product coal was monitored for liberated pyrite using scanning electron

*Coal is hydrophobic. Pyrite exhibits hydrophilic characteristics and is rejected during flotation, falling to the bottom of the flotation vessel with the ash, which is also hydrophilic.

microscopy. During these tests it was noted that coals pretreated for days with these microorganisms appeared to be reduced in size. Further investigation of this observation showed that one coal was reduced in mean particle size from 1,024 μm to 61 μm over a 240-hr treatment, while a second precleaned coal was reduced from 258 μm to 69 μm. Comparison of pyrite rejection during flotation of mechanically comminuted and microbially comminuted coals indicated that selective liberation of pyrite may occur during the microbial treatment.

Preliminary results of some of these studies are presented in Tab. 2. A pyritic sulfur rejection of 49% was noted following flotation of coal #1 mechanically ground to a D80 of 200 mesh, while 55% pyritic sulfur rejection was observed from microbially comminuted coal at a D80 of 65 mesh and 74% rejection at a D80 of 230 mesh. The precleaned coal #2 showed similar results. Pyritic sulfur rejection from mechanically ground coal at a D80 of 200 mesh was 16%; for microbially comminuted coal, pyrite rejection at a D80 of 100 mesh, 120 mesh, and 230 mesh was 36%, 41%, and 68%, respectively. Such a pretreatment could be carried out in an oxidation pond to provide a useful, economical tool in developing a combined physical-biological sulfur removal process for certain applications.

Tab. 2. Mechanical versus microbial comminution of coal for flotation.

Sample	Comminution	Treatment	% Yield	% Ash	% Total sulfur	Reduction pyritic sulfur (%)
Coal #1*	Mechanical (-200 mesh)	Feed		22.2	2.31	
		Float†	64.5	11.0	1.54	49
	Microbial (-65 mesh)	Float	85.0	9.63	1.46	55
	(-230 mesh)	Float	85.0	8.34	1.18	74
Coal #2**	Mechanical (-200 mesh)	Feed		8.00	2.91	
		Float	84.0	5.06	2.72	16
	Microbial (-100 mesh)	Float	64	4.94	2.49	36
	(-120 mesh)	Float	77	5.04	2.44	41
	(-230 mesh)	Float	84	4.46	2.12	72

†Wemco flotation cell.
*Pyritic sulfur, 1.45%; SO_4, 0.05%.
**Pyritic sulfur, 1.16%; SO_4, 0.0%.

ORGANIC SULFUR REMOVAL FROM COAL

Organic sulfur is more difficult to remove from coal than pyritic sulfur, as it is covalently bonded to the carbon in the coal matrix. This type of sulfur is likely to be present as aliphatic or aromatic thiols, sulfides, and disulfides, or as heterocyclic compounds of the thiophene type. Atlantic Research began work on microbial organic sulfur removal using a model organic sulfur compound, dibenzothiophene (DBT). This model compound was selected based on limited available analyses of several coals for organic sulfur, forms which indicated that thiophenes were likely the most prevalent organic sulfur form in the coals tested (2).

A naturally occurring microbe isolated by Atlantic Research demonstrated minimal DBT oxidation. This microbe was subjected to mutation, adaptation, and selection to produce a microbe called CB1 (Coal Bug 1). This microbe tolerated fairly high concentrations of DBT and did not use DBT as a sole carbon source, but did alter the molecule to render it water-soluble. Two possible DBT oxidation pathways are illustrated in Fig. 1. In one pathway the ring is hydroxylated and ring cleavage subsequently occurs. Sulfur is not released from the molecule in this reaction, although the product is water-soluble. Increases in sulfate have been detected in the aqueous phase by several investigators, indicating the release of sulfur from the molecule as sulfate by an alternate pathway which may proceed through the sulfoxide/sulfone intermediate.

Atlantic Research demonstrated that CB1 removed organic sulfur from some coals by oxidizing the sulfur from thiophenes, releasing the sulfur as water-soluble sulfate. The microbe does not break the carbon ring structure of the thiophene molecule. Radiotracer studies using ^{14}C and ^{35}S labels in the model organic sulfur compound, DBT, were conducted to determine the mechanism of action of CB1 on DBT. $^{14}CO_2$ was not detected during the treatment period, and no incorporation of ^{14}C into the biomass was detected. ^{35}S was found in the aqueous phase of the medium following treatment of [^{35}S]-DBT with CB1. Data indicated that the ^{35}S in the aqueous phase was in the form of sulfate ion. Production of sulfate from DBT was confirmed by ion chromatography. Gas chromatography/mass spectra analysis identified the organic product as 2,2'-dihydroxybiphenyl.

The first coals tested were low sulfur coals in which more than 90% of the sulfur present was reported as organic sulfur. These coals were selected to reduce the analytical problems associated with quantifying organic sulfur. Organic sulfur in coal is calculated as the difference between total sulfur and combined pyritic and sulfatic sulfur. Coal containing a high percentage of impurities is difficult to analyze accurately for pyrite, and inaccuracies in the pyrite determination lead to inaccurate organic sulfur values.

Our goal was to demonstrate sulfur removal in excess of the reported pyritic sulfur content. Results of laboratory studies on these coals are presented in Tab. 3. Organic sulfur reductions in these coals ranged from 18% to 35%, and are greater than the reported pyritic sulfur content of some of the coals. Treatment of run-of-mine (ROM) coals (coal as mined; no washing or cleaning to remove contaminants) showed that CB1 removed from 25 to 34% of the organic sulfur from these coals, as shown in Tab. 4.

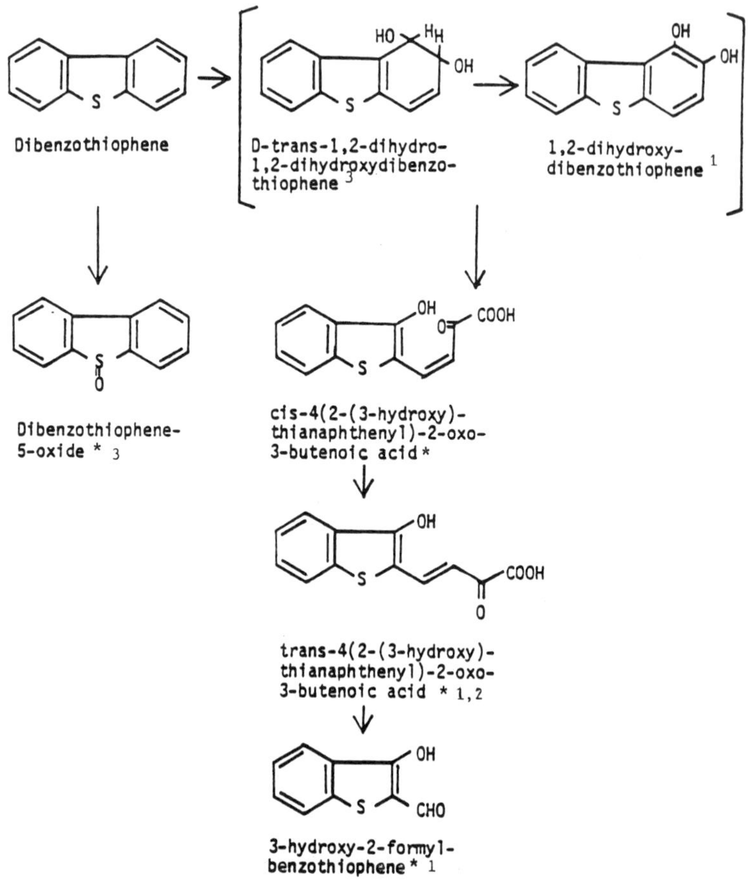

Fig. 1. Proposed oxidative pathway of DBT oxidation. (From Ref. 7, 8, 11, and 12.)

Because of the analytical problems associated with conventional sulfur analyses and quantification of organic sulfur, tests were conducted to confirm oxidation of organic sulfur in coal by CB1. A ROM Upper Freeport coal (D80 of 100 mesh) was analyzed for total, pyritic, and sulfate sulfur content before and after microbial desulfurization. The total sulfur content of the coal was reduced from 2.3 to 1.8% by the activity of CB1. Pyrite analysis of the samples before and after treatment did not show significant differences. Sulfate, however, was reduced by 0.2%. By calculation, organic sulfur was reduced from 0.79% to 0.53%, a reduction of 33%. Energy dispersive X-ray microanalysis combined with scanning electron microscopy (EDS-SEM) for direct determination of organic sulfur (15) indicated a 34% reduction in organic sulfur, confirming calculated values. Scans representative of average values obtained for each coal sample are presented in Fig. 2 and 3.

A continuous bench-scale unit for microbial desulfurization has been tested at Atlantic Research using five coals. Organic sulfur reduction ranging from 19 to 36% was demonstrated, as shown in Tab. 5.

Tab. 3. Microbial desulfurization of low-sulfur coals.

Coal	Before treatment		After treatment	
	Total sulfur (%)*	Organic sulfur (%)	Total sulfur (%)	% Reduction total sulfur
Wentz	0.67	0.60	0.53	20.9
Montcoal	0.80	0.72	0.62	22.5
Black Mesa	0.61	0.49	0.56	8.2
Upper Freeport	1.12	0.70	0.91	18.8

*Washed control values. Soluble sulfate washed out.

Tab. 4. Microbial desulfurization of run-of-mine (ROM) coals.

Coal	Before treatment		After treatment		
	% Total sulfur	% Organic sulfur	% Total sulfur	% Organic sulfur*	% Reduction organic sulfur
Consolidation	3.8	2.0	3.1	1.5	25
Pittston	4.0	1.9	3.3	1.3	32
Sewickeley	5.0	1.9	3.6	1.3	34
Upper Freeport	2.3	0.8	1.8	0.5	33
Kentucky #9	4.2	1.2	3.7	0.9	26
Amax Illinois #6 and #7	4.2	1.7	3.5	1.3	24
Illinois #6	3.5	2.1	2.9	1.6	24

*Total sulfur reduction includes sulfate washed out during treatment.

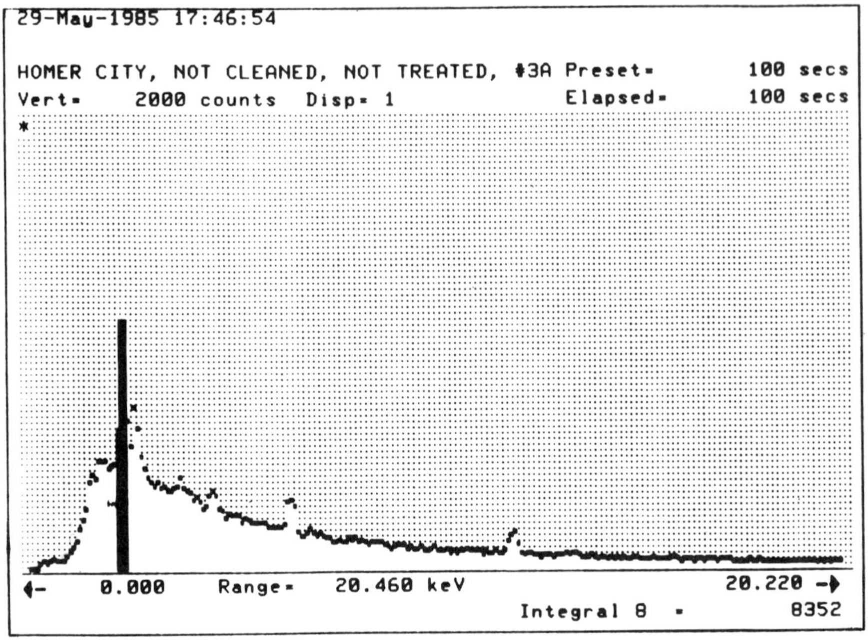

Fig. 2. EDS-SEM analysis for organic sulfur in untreated run-of-mine coal.

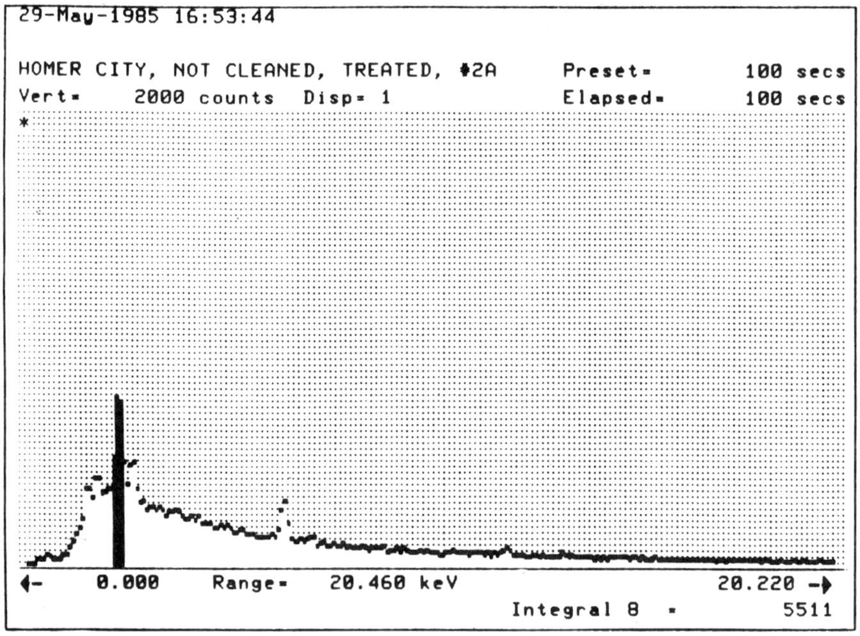

Fig. 3. EDS-SEM analysis for organic sulfur in microbially treated run-of-mine coal.

Tab. 5. Bench-scale microbial coal desulfurization.

Coal	Continuous operation	Retention time (hr)	% Total sulfur Before	After	% Reduction sulfur
Illinois #6[a]	5 weeks	48	3.5	2.9	24
		24	3.5	2.8	29
Illinois #6[b]	1 week	12	3.6	2.9	19
Consolidation[c]	2 weeks	12	3.8	3.1	25
Upper Freeport[d]	1 week	9	1.1	0.9	30
Kentucky #9[e]	1 week	9	4.2	3.4	36

[a]Total sulfur, 3.5%; organic sulfur, 2.1%; pyritic sulfur, 1.3%; SO_4, 0.1%.
[b]TS, 3.7%; OS, 2.2%; PS, 1.2%; SO_4, 0.3%.
[c]TS, 3.8%; OS, 2.0%; PS, 1.6%; SO_4, 0.2%.
[d]TS, 1.12%; OS, 0.68%; PS, 0.44%; SO_4, 0.0%.
[e]TS, 4.2%; OS, 1.4%; PS, 2.5%; SO_4, 0.3%.

A second microorganism (CB2) capable of removing sulfur from a different model organic compound, phenyl sulfide or diphenyl sulfide (DPS), is being evaluated for additional organic sulfur removal from coals. Efforts are focused on enhancing organic sulfur removal to achieve a target removal of at least 60%. Preliminary results indicate cumulative removal of sulfur when CB1 is used followed by CB2 for treatment of some coals. Efforts to culture the microbes together were unsuccessful, and treatment of coals using combined cultures (50% CB1 and 50% CB2) was not reproducible.

COAL CONVERSION

In the area of coal conversion, two main objectives are being pursued: (a) biological solubilization or depolymerization of low-rank coals, and (b) biological gasification of coals and/or coal-derived products. A number of investigators are pursuing studies on degradation of coal by oxygen-requiring reactions. Fakoussa (6) isolated a pseudomonad capable of growing on coal-agar medium. Cohen and Gabriele (4) and Wilson et al. (16) reported coal degradation by the white rot fungus *Polyporus versicolor* and *Poria* sp. Since coal (especially the low-rank coals) is related chemically to the lignin fraction of the plant cell-wall material incorporated into the coal, it may be possible that the enzymes produced by *Polyporus* sp., *Poria* sp., and *Pseudomonas* sp. for lignin degradation may also degrade coal. It has also been reported that *Streptomyces* sp. are active in lignin (5) as well as low-rank coal degradation (14). Products from microbial solubilization of lignites and sub-bituminous coal are water-soluble at high pH and may be recovered as a solid material by lowering the pH to approximately 1.5.

Atlantic Research has isolated a unique microbial consortium that is capable of depolymerizing lignites and sub-bituminous coals. Leonardite, a highly oxidized lignite coal, is depolymerized by this microbial consortium with no pretreatment. Pretreatment is required for solubilization of several South Dakota lignites and a Wyodak sub-bituminous coal. Pretreatment using 8 N HNO_3 for a period of 1 hr was best, although treatment with 6 N HNO_3 was acceptable, and pretreatment of the coals with H_2O_2 was also effective. Solubilization data with product recoveries and mass balances for two low-rank coals are presented in Tab. 6. The average product recovery for the Beulah lignite pretreated with 8 N HNO_3 followed by solubilization with the ARC consortium was 78.6%, with a total mass recovery of 98.9%. Analysis of the lignite prior to microbial solubilization showed that the ash content of this coal sample was 20.49%. Recovery of 78.6% as the dry solid product indicates that nearly all of the organic portion of the coal was solubilized and recovered. Wyodak sub-bituminous coal was also pretreated with 8 N HNO_3, followed by primary and secondary digestion of the coal using the cell-free microbial culture broth containing the enzymes. This resulted in 70.1% of the coal being recovered as dry solid product, with a mass recovery of 88%. Ash in this sub-bituminous coal accounts for 10.2% of the total mass, indicating that the two solubilization treatments were successful in solubilizing most of the organic portion of the coal.

Low-rank coals are thought to consist of aromatic clusters which are cross-linked by ether, ester, methylene, and alkyl linkages. Oxygen may be found in carboxyl groups as well as in other functional groups. Structural units such as these should be amenable to microbial attack using enzymes such as esterases, etherases, and perhaps carboxylases. Such attack could depolymerize the macromolecular structure of the low-rank coals, resulting in more reactive fragments amenable to further degradation.

Atlantic Research has begun studies to determine which of these linkages are cleaved by the enzyme systems produced by our microbial consortium. Preliminary results indicate the presence of esterases and etherases and the cleavage of alkyl bonds. Further studies are under way to confirm these results. Investigation of decarboxylation and cleavage of polymethylene linkages found in sub-bituminous coals will be initiated. Information about biochemical bonds cleaved by the enzyme systems of the microbial consortium will allow us to postulate pathways by which microorganisms depolymerize coal and predict the structure(s) of oligomers produced during microbial treatment. Products from depolymerization of lignite and sub-bituminous coals will be studied to characterize and identify as many of the molecular structures as possible.

Products of microbial solubilization of low-rank coals, as well as products of chemical and/or thermal solubilization of similar coals, are currently being evaluated as a feedstock for microorganisms. Aerobic populations may be able to convert these products to short-chain acids and alcohols. Anaerobic microbial populations are capable of converting these low molecular weight substrates to methane. Evaluation of microbial products along the path to methane may indicate potential for microbial production of alcohols or other chemicals of economic value.

The area of coal bioconversion discussed above is the result of aerobic conversion of coals in which oxygen is added to the coal during conversion. An alternative approach that has received limited attention is anaerobic conversion of coal in which hydrogen is added to the coal. Such enrichment in the coal product could result in a more desirable fuel.

Tab. 6. Microbial solubilization of lignite and sub-bituminous coals.

Coal*	Culture	Average % solubilized	Average total recovery (%)
Beulah lignite	ARC-CP1+2	78.6	98.9
Wyodak sub-bituminous	ARC-cell free	70.1	88

*Coals pretreated with 8 N HNO_3 for 1 hr, washed, and dried before microbial solubilization.

In order for bioprocessing of coal to become a reality on a commercial scale, bioengineering must be an integral part of the effort. Bioreactor design using whole-cell, fixed-cell and/or enzyme systems, nutrient recycle, and waste stream treatment is only one of the total engineering challenges to be addressed.

There are many additional challenges to be faced in bioprocessing of coal. The first is the heterogeneity of coal. Each coal is different, so a single biological process may never be found that is capable of treating every coal. Thus, the microbes or their enzyme systems may have to be tailored to bioprocess coal for beneficiation or for bioconversion. The next area for consideration is microbial stability. Genetically engineered microbes may not be stable for extended periods of time, indicating that enzyme systems may have significant advantages. The time required for bioprocessing must be reduced substantially for this technology to become economically viable. Process efficiency and consistency of clean coal product must be addressed during process optimization.

Market and public perceptions of bioprocessing and what it can accomplish are critical for the further development of this technology. Most people do not understand biology; even more important, they do not understand what microorganisms can do, how they can do it, and, equally important, what they cannot do. Fear of a new technology such as bioprocessing of coal usually stems from misconceptions and lack of understanding. Much effort must be made to provide easily understandable information to allay some of the fears, to show our concern by precautions we take in our work, and to address environmental concerns and potential solutions with respect to waste streams from such processes.

This technology faces significant challenges before it can become a commercial reality. Efforts must be increased to ensure that bioprocessing is a reality when the market is ready, when people are willing to try the technology because oil prices increase, or when more stringent regulations mandate the reduction of sulfur emissions from coal combustion.

SUMMARY

Bioprocessing of coal is a new and exciting frontier in biotechnology. Coal bioprocessing today focuses on two main areas: beneficiation of coal to remove undesirable contaminants (i.e., sulfur, nitrogen, and trace metals) and coal conversion (i.e., microbial solubilization, gasification, and pretreatment). The objectives of microbial coal cleaning are the removal of sulfur, nitrogen, and perhaps ash, using mild microbial processes for production of clean coal. Microbial coal conversion is less well defined, although the overall goal of these conversions is the production of a more desirable fuel form. The low-severity approaches of these biological processes may render them economically more feasible than physical/thermal processes.

Atlantic Research is currently investigating two approaches to microbial beneficiation of coals. A microorganism capable of removing a portion of the organic sulfur (thiophenic sulfur) from coals has been developed and patented. A second microorganism is being evaluated for removal of an additional organic sulfur form (aromatic sulfides) from coals. As much as 40% of the organic sulfur has been removed in a bench-scale process. A second approach uses naturally occurring pyrite-oxidizing microorganisms to liberate pyritic sulfur from coals through selective comminution before beneficiation of the coal by physical processes. Preliminary feasibility studies indicate reduction in particle size from approximately 1,000 μm to 60 μm using microbial comminution. Physical beneficiation of mechanically ground coal was compared with that of microbially comminuted coal. Data indicate that pyrite may be selectively liberated during the microbial pretreatment, with resultant increase in pyritic sulfur rejection during physical beneficiation.

Also under study at Atlantic Research are the biochemical mechanisms by which a microbial consortium depolymerizes coals (lignites, sub-bituminous, and perhaps bituminous coal). A unique microbial consortium has been isolated which is capable of depolymerizing Leonardite, a highly oxidized lignite coal, with no pretreatment. These microorganisms are also capable of depolymerizing other lignites and a sub-bituminous coal, although pretreatment of these coals is required. Relatively rapid and efficient microbial depolymerization has been demonstrated with pretreated lignite and sub-bituminous coals, with production of an alkali-soluble product. Model compound studies are under way to determine the types of bonds that are cleaved by the microorganisms. Size exclusion chromatography is being used to determine the molecular weight of depolymerization products, and a variety of chemical analyses will be evaluated for identification of products from the coals. Future work will address conversion of these products to methane and the potential for creation of alternate fuel forms from the microbially produced products.

ACKNOWLEDGEMENTS

Atlantic Research Corporation's research was sponsored by in-house funds and by United States Department of Energy contracts from the Pittsburgh Energy Technology Center.

REFERENCES

1. Atkins, A.O., A.J. Davis, C.C. Townsley, E.W. Bridgwood, and F.D. Pooley (1985) Production of sulphur concentrates from the bioflotation of high pyritic coals. In Proceedings of the International Conference Sulphur 85, London, United Kingdom, pp. 83-104.

2. Attar, A. (1983) An analytical method for the evaluation of sulfur functionalities in American coals. United States National Technical Information Service, COE/PC/30145-Tl.
3. Attia, Y., and M. Elzeky (1985) Biosurface modification in the separation of pyrite from coal by froth flotation. In Proceedings of the First International Conference on Processing and Utilization of High Sulphur Coals, Columbus, Ohio, pp. 673-682.
4. Cohen, M.S., and P.D. Gabriele (1982) Degradation of coal by the fungi Polyporus versicolor and Poria monticola. Appl. Environ. Microbiol. 44:23-27.
5. Crawford, D.L., A.I. Pometto, and R.L. Crawford (1983) Lignin degradation by Streptomyces viridosporus: Isolation and characterization of a new polymeric lignin degradation intermediate. Appl. Environ. Microbiol. 45:898-904.
6. Fakoussa, R.M. (1981) Coal as a substrate for microorganisms: Investigation with microbial conversion of national coals. Ph.D. Thesis, Friedrich-Wilhelms University, Bonn, Federal Republic of Germany.
7. Finnerty, W.R. (1980) Microbial desulfurization and denitrogenation. Symposium on Microbial Process for Metals Accumulation and Sulfur/-Nitrogen Removal from Fuels, Second Chemical Congress of the North American Continent, San Francisco, California.
8. Hou, C.T., and A.I. Laskin (1976) Microbial conversion of dibenzothiophene. Develop. Industr. Microbiol. 17:351-362.
9. Hutton, C.A., and R.N. Gould (1982) Cleaning Up Coal: A Study of Coal Cleaning and the Use of Cleaned Coal, Ballinger Publishing Company, Cambridge, Massachusetts.
10. Kempton, A.G., N. Moneib, R.G.L. McCready, and C.E. Capes (1980) Removal of pyrite from coal by conditioning with Thiobacillus ferrooxidans followed by oil agglomeration. Hydrometallurgy 5:117-125.
11. Kodama, K., K. Umehara, K. Shimizu, S. Nakatani, Y. Minoda, and K. Yamada (1973) Identification of microbial products from dibenzothiophene and its proposed oxidation pathway. Agric. Biol. Chem. 37:47-50.
12. Laborde, A.L., and D.T. Gibson (1977) Metabolism of dibenzothiophene by a Beijerinckia species. Appl. Environ. Microbiol. 34:783-790.
13. Monroe, D. (1985) Microbial metal mining. Am. Biotechnol. Lab. 3:10-19.
14. Scott, C.D., G.W. Strandberg, and S.N. Lewis (1986) Microbial solubilization of coal. Biotechnol. Progr. 2:131-139.
15. Straszheim, W.E., R.T. Greer, and R. Markuszewski (1983) Direct determination of organic sulphur in raw and chemically desulphurized coals. Fuel 62:1070-1075.
16. Wilson, B., R. Bean, J. Franz, and B. Thomas (1987) Microbial conversion of low-rank coal: Characterization of biodegraded product. Energy and Fuels 1:80-84.

ANAEROBIC FLUIDIZED-BED TREATMENT OF HAZARDOUS WASTES

Makram T. Suidan and Girgis F. Nakhla

Department of Civil Engineering
University of Illinois
Urbana, Illinois 61801

INTRODUCTION

The anaerobic attached film process consists of a columnar reactor containing a packing medium that serves as an attachment surface for the retention of anaerobic microorganisms that metabolize the organic constituents of a waste. The wastewater is usually applied in an upflow mode in order to submerge completely the unit and maintain an anaerobic environment (21). Anaerobic reactors have been operated in both the packed-bed and the expanded-bed modes. When operated in the expanded-bed or fluidized-bed mode, effluent recycle is often practiced to fluidize the bed and thereby establish close to completely mixed-flow conditions. Although the kinetics of substrate removal are generally lower in this type of reactor than in a "once-flow-through" reactor, dilution of the incoming waste can often enhance treatment by reducing inhibition. Dilution of the concentration of degradable inhibitory compounds in the feed, and the prevention of pH depression by suppressing the rapid buildup of volatile fatty acids at the inlet sections of the reactor, are two advantages attributed to effluent recycle (4). In addition, recycle minimizes problems with bed-plugging, leads to improved flow distribution, and allows for the entrapment of small particles within the bed, thus enhancing their degradation. Fluidization also reduces the thickness of the hydrodynamic boundary layer, thus decreasing resistance to mass transport.

EFFECTS OF MEDIA CHARACTERISTICS ON REACTOR PERFORMANCE

The medium in expanded-bed anaerobic reactors provides an attachment surface for the microorganisms that are needed for waste treatment. Media characteristics that impact the performance of anaerobic reactors include particle size and density, surface roughness, and resistance to abrasion.

Particle size affects the attachment surface area, mass transfer resistance, the recycle ratio needed for bed-expansion, cost for effluent recycle, and biomass shear losses. Small particles provide a high specific surface area, thereby increasing the opportunity for biomass attachment. In addition, the availability of surface area for attachment and the high shear stresses in such reactors promote the formation of thin but dense biofilms (17). This, in turn, permits the attainment of a higher concentration of biomass in the reactor; and, since the biofilm is thin, mass transfer is improved and substrate diffusion limitations are minimized.

The recycle ratio, R, needed for bed expansion can be calculated from the following equations (1):

$$R = \left(\frac{V_a}{H}\right)\theta - 1 \tag{1}$$

$$V_a = V_s\,\varepsilon_e^{\,n} \tag{2}$$

$$\varepsilon_e = 1 - (1 - \varepsilon)\frac{L}{L_e} \tag{3}$$

$$n = 4.4\,R_e^{-0.1}, \quad 1 < R_e < 500 \tag{4}$$

$$n = 4.4\,R_e^{-0.03}, \quad 0.2 < R_e < 1 \tag{5}$$

where:

ε = unexpanded bed porosity

ε_e = expanded bed porosity

L = unexpanded bed height, L

L_e = expanded bed height, L

V_s = unhindered settling velocity, L/T

V_a = upflow superficial velocity, L/T

H = height of reactor, L

θ = empty bed hydraulic detention time, T

R_e = Reynolds No. based on unhindered settling velocity

It is apparent from the above equations that a small particle has a low settling velocity; therefore, a low recycle ratio is needed to maintain the desired level of bed expansion. Figure 1 shows the relationship between the empty-bed hydraulic detention time and the recycle ratio needed to maintain a 20% expansion of a fluidized bed packed with 0.5 mm granular activated carbon (GAC), having a density, ρ_s, of 1.35 g/cm^3. The recycle ratio increases linearly with hydraulic detention time for a constant expanded medium height, while the required recycle ratio decreases with increasing medium height when the empty-bed hydraulic detention time is held constant. This is due to the fact that the upflow velocity, and not the volumetric hydraulic loading rate, controls bed expansion. Particle density is also found to influence the recycle ratio and the ensuing rate of biomass loss due to shear stresses. A heavy particle will tend to settle rapidly, thereby increasing the upflow superficial velocity, V_a, needed to

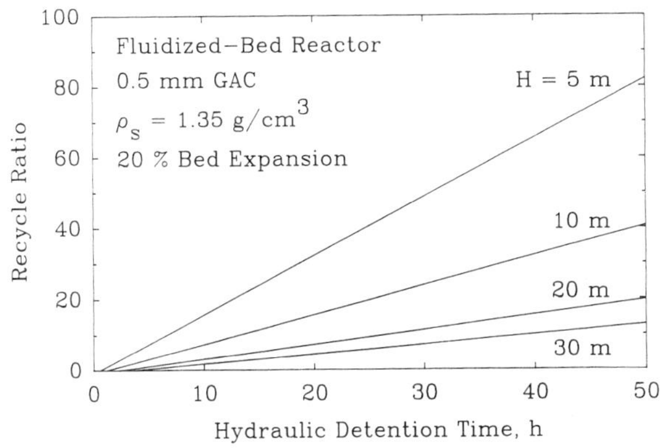

Fig. 1. Effect of bed height and hydraulic detention time on effluent recycle.

maintain bed expansion. The increased upflow velocity results in increased shear stresses on the medium particles and, consequently, increased biomass attrition rates. The data in Fig. 2 illustrate the sensitivity of the recycle ratio that is needed to maintain bed expansion to the empty-bed hydraulic detention time in a reactor, and to the density of the attachment medium. The data in Fig. 1 and 2 seem to suggest that the ideal medium for an expanded-bed reactor should consist of the smallest available particle size that is separable from water, should have a density just slightly greater than that of water, and should be placed in a reactor that has the greatest feasible ratio of height to diameter.

Resistance to abrasion is a desirable characteristic of the attachment medium of expanded-bed anaerobic reactors in order to minimize medium loss and deterioration. Surface roughness appears to be a medium characteristic that exerts a significant impact on biomass accumulation. A smooth surface has been observed to support only a thin biofilm under both high and low chemical oxygen demand (COD) volumetric loading rates. How-

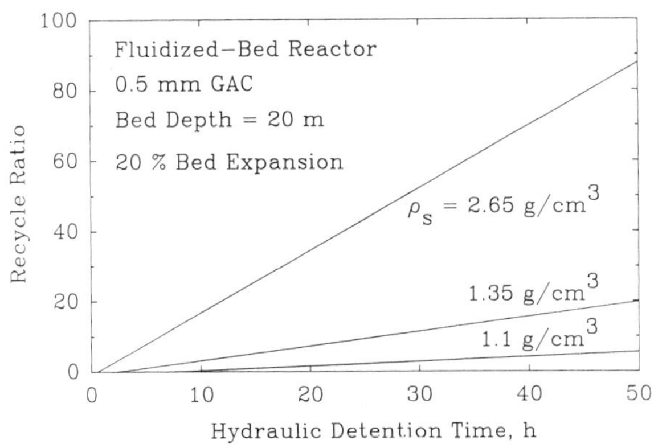

Fig. 2. Effect of particle density and hydraulic detention time on effluent recycle.

ever, a rough surface has been observed to permit the development of thicker attached films on the particle surface (12).

EXPANDED-BED ANAEROBIC GRANULAR ACTIVATED CARBON REACTOR

Granular activated carbon possesses many of the characteristics that are desirable in an attachment medium for expanded-bed anaerobic reactors. Some of the advantages of GAC over other media are its low wetted material density, which is approximately 1.35 g/cm^3, its availability in a wide range of particle sizes, and its relative hardness and resistance to abrasion in the anaerobic environment. Perhaps one of the most important characteristics of GAC as an attachment medium is its exterior surface roughness, which renders it superior to most other media in microbial sheltering and attachment. Several hypotheses have been proposed to clarify the role of activated carbon during biological treatment. Weber et al. (20) observed that the adsorptive properties of activated carbon increase the concentration of soluble organic matter at the interface, thereby stimulating biogrowth and assimilation. Characklis (3) pointed out that the biofilm associated with carbon adhered firmly to the particle, where it is sheltered against high velocities by the pits and crevices on the carbon surface. In side by side comparisons of two expanded-bed reactors treating a prepared solution of 5,000 mg/l acetic acid, Suidan et al. (12) reported that more than 20 times as much biomass accumulated on activated carbon as on sand when equal-sized particles of both media were employed for microbial attachment. The effluent concentrations of volatile suspended solids and acetic acid from the sand reactor were, respectively, nine and 100 times higher than in the effluent from the GAC reactor.

Another advantage of using activated carbon as an attachment medium in expanded-bed anaerobic reactors is its ability to adsorb a wide range of organic compounds from the aqueous phase. This property of activated carbon may be particularly important in the operation of full-scale systems, where the GAC will serve to store excess substrate and resist excursions in performance efficiency caused by perturbations in feed strength, composition, and flow rate. This is clearly evident from Fig. 3 (13), which shows the performance of a completely mixed, expanded-bed GAC anaerobic reactor treating a synthetic wastewater containing 200 mg/l catechol. At the beginning of operation, the main removal mechanism in the reactor was adsorption, as evidenced by the low effluent substrate concentration and the absence of methane production. When the adsorptive capacity of the GAC medium for catechol was partially exhausted, the effluent concentration of that compound increased gradually. This slow rate of increase in the concentration of catechol in the effluent seemed to promote the development of an acclimated culture which, by day 170 of continuous feeding, was capable of handling the incoming substrate as well as the substrate that was previously adsorbed on the carbon. As the concentration of catechol in the liquid-phase effluent started to decrease, the previously adsorbed substrate desorbed off the GAC medium and was metabolized by the microorganisms. This biologically mediated desorption and the ensuing bioregeneration of the carbon were characterized by a methane production rate that exceeded that theoretically anticipated from the feed.

The ability of GAC to buffer fluctuations in the influent load is demonstrated in Fig. 4 (7). The data show the effect of perturbations in the influent phenol concentration on the performance of two completely mixed anaerobic reactors packed with 16 x 20 U.S. Mesh GAC and anthracite, re-

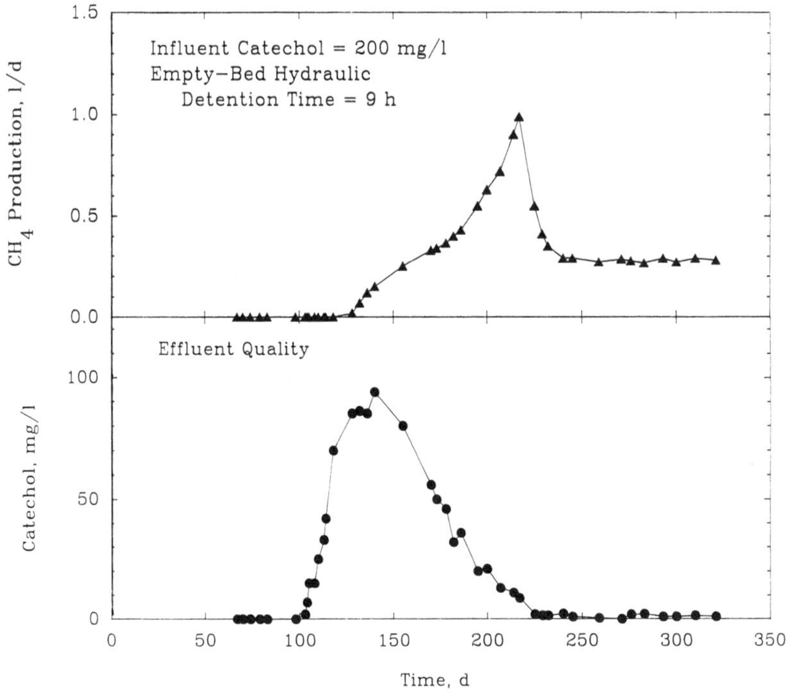

Fig. 3. Performance of expanded-bed granular activated carbon (GAC) anaerobic reactor in treating a catechol-bearing wastewater.

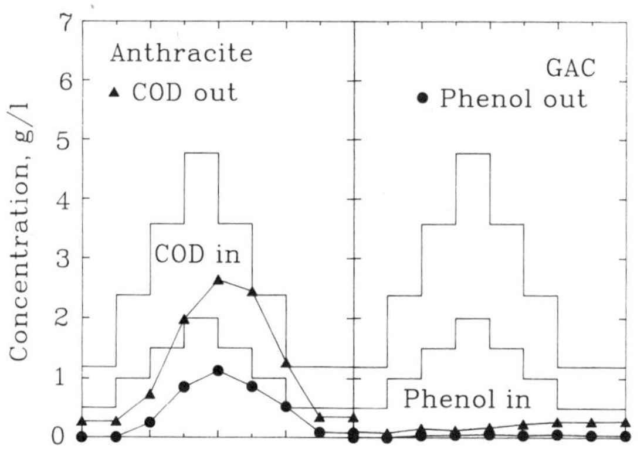

Fig. 4. Effect of perturbation in organic loading of phenol on the performance of granular activated carbon (GAC) and anthracite anaerobic reactors.

spectively. The two reactors were operated in parallel until steady-state conditions were established on a synthetic substrate containing 500 mg/l of phenol. After 274 days of continuous feeding, the concentration of phenol in the feed was increased to 2,000 mg/l in steps of 500 mg/l-day. This peak phenol concentration was maintained for one day, after which the concentration of phenol in the feed was decreased in steps of 500 mg/l-day back to 500 mg/l. The results obtained from this dynamic loading study indicate that, even after 274 days of continuous operation, the adsorptive capacity of the GAC medium was still sufficient to absorb the shock load of phenol. It is worth noting that the combined function of anaerobic biodegradation and physical adsorption enabled the anaerobic GAC reactor to continue delivering a high removal efficiency of phenol, whereas the anthracite-pack reactor failed to handle the increased load.

In another study, a step increase in the feed phenol concentration from 1,500 mg/l to 3,000 mg/l led to substrate inhibition and eventual failure of an anaerobic reactor packed with a nonactive carbon, while an identical increase in loading to a parallel reactor packed with GAC did not adversely affect the performance of the reactor. Methane production from the GAC reactor doubled within a period of 14 days after the step increase in the feed phenol concentration. During the 14-day period, the removal mechanism of the increased phenol load shifted from physical adsorption to biodegradation after sufficient increased growth in the microbial culture had occurred.

The adsorptive capacity of activated carbon also renders it the ideal medium for the expanded-bed anaerobic reactor treatment of wastewaters containing some refractory adsorbable organic compounds that are toxic to microbial cultures that are otherwise capable of biodegrading a major fraction of the organic content of the wastewater. The ability of the GAC medium to adsorb the toxic constituents of the waste casts activated carbon in a new role of enhancing the biotreatability of wastewater.

GRANULAR ACTIVATED CARBON ANAEROBIC REACTOR FOR THE TREATMENT OF HAZARDOUS WASTES

The expanded-bed, GAC anaerobic reactor has been successfully used for the treatment of various types of toxic wastes, such as those produced during coal gasification (9-11,15), coal coking (2), paint stripping operations (7), and refinery sour water stripper bottoms (6). In this section, the role of surface active media in rendering toxic wastes biologically treatable is examined using data from two case studies relating to the treatment of coal gasification and refinery sour water stripper bottoms, respectively.

Case Study 1: Treatment of Coal Gasification Wastewater

Associated with the various coal distillation processes is the production of wastewaters that are highly contaminated with phenolics, ammonia, cyanide, thiocyanate, and a variety of suspected mutagenic substances. The data in Tab. 1 represent chemical characteristics of two batches of wastewater produced from a low BTU elevated pressure, stirred fixed-bed gasifier operated by the Morgantown Energy Technology Center (METC), United States Department of Energy, Morgantown, West Virginia (9,10). Prior to treatment in an expanded-bed anaerobic reactor, the coal gasification wastewaters were ammonia-stripped, since ammonia does not biodegrade

Tab. 1. Chemical characteristics of coal gasification wastewaters.

Parameter	Concentration Batch 1	Batch 2
pH	8.52-8.53	8.48
Alkalinity, mg/l as $CaCO_3$	16,043 ± 190	27,065
COD, mg/l	24,693 ± 750	14,658
DOC, mg/l	7,023 ± 73	4,138
NH_3-N, mg/l	4,612 ± 8	10,400
SCN^-, mg/l	91 ± 13	218
CN^-, mg/l	3 ± 2	275
Phenol, mg/l	2,493 ± 132	1,957
o-Cresol, mg/l	378 ± 4	663
m- and p-Cresol, mg/l	1,139 ± 83	1,473
2,4-Dimethylphenol, mg/l	131 ± 3	67
3,5-Dimethylphenol, mg/l	255 ± 6	233

anaerobically and ammonia concentrations of 3,200 mg/l as nitrogen can cause inhibition to microbial cultures and induce process failure (18). Therefore, the concentration of ammonia was reduced to 1,100 mg/l as nitrogen prior to feeding the wastewaters into expanded-bed anaerobic reactors.

Initial attempts to treat ammonia-stripped coal gasification wastewater in anaerobic reactors packed with nonsurface-active polypropylene berl-saddles resulted in no measurable contaminant removal through the reactor (15). Glucose added to the feed was converted to methane, indicating inhibition to the primary organisms responsible for ring cleavage and not to the methanogenic bacteria. The data in Fig. 5 (5) reflect the performance

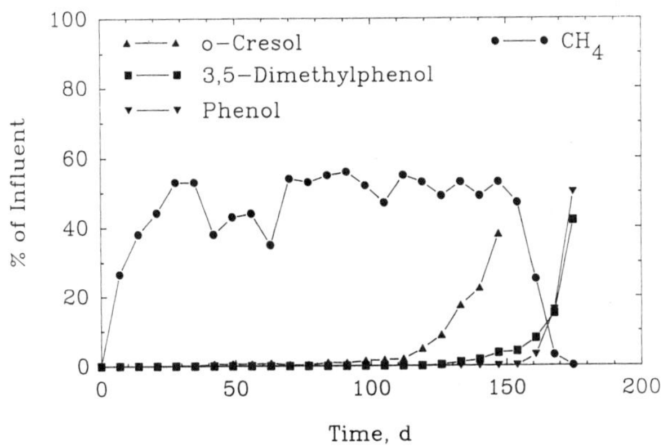

Fig. 5. Performance of granular activated carbon (GAC) anaerobic reactor in treating full-strength coal gasification wastewater.

of an expanded-bed GAC anaerobic reactor in treating the first batch of coal gasification wastewater. This wastewater, which is characterized in Tab. 1, was fed at full strength to an 11-l volreactor (shown schematically in Fig. 6). This volreactor was charged with 1,200 g of virgin 16 x 20 U.S. Mesh F-400 GAC, and 300 g of an identical carbon that was withdrawn from a biologically active reactor treating a similar wastewater. The COD loading rate on this reactor was 13 g per kg of GAC per day. The experimental data presented in Fig. 5 reflect the time-varying fraction of the feed COD that was recovered as methane gas, and the corresponding fractional breakthrough of phenol, o-cresol, and 3,5-dimethylphenol (5). It is apparent from Fig. 5 that the gradual breakthrough of the anaerobically nonbiodegradable o-cresol and 3,5-dimethylphenol (8) was rapidly followed by a breakthrough of the readily degradable phenol.

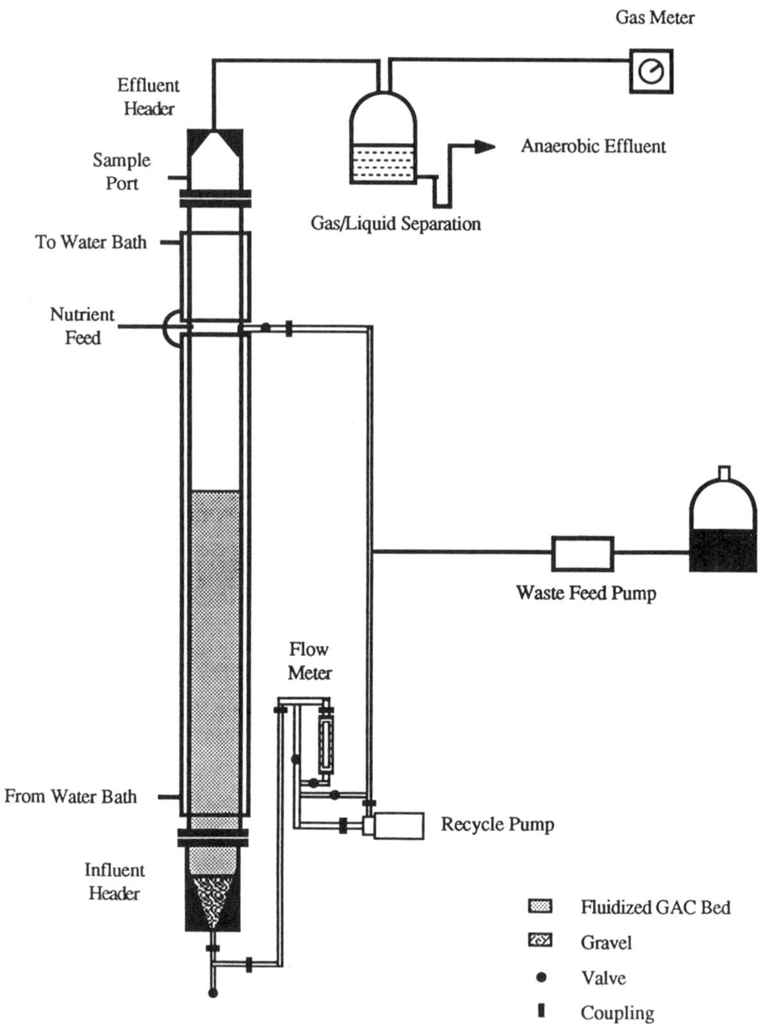

Fig. 6. Schematic of expanded-bed granular activated carbon (GAC) anaerobic reactor.

These results indicate that as long as the biologically inhibiting components (e.g., o-cresol and 3,5-dimethylphenol) of the coal gasification wastewater were sequestered out of the aqueous phase by the GAC, the anaerobic culture within the reactor was capable of degrading over 50% of the organic constituents of the wastewater (i.e., phenol). However, saturation of the adsorptive capacity of the carbon and the breakthrough of toxicity, represented here by the appearance in the reactor effluent of several substituted phenols, resulted in inhibition to the biological activity in the column, accumulation of biodegradable organic compounds on the GAC surface and in the effluent, and, finally, system failure.

Case Study 2: Treatment of Refinery Sour Water Stripper Bottoms

The role of surface activity and particle size of the attachment medium in the expanded bed was investigated by Gardner et al. (6) during the anaerobic treatment of full-strength refinery sour water stripper bottoms. The characteristics of the wastewater were similar to those given in Tab. 1 for coal gasification wastewater, except that in this case the sour water had approximately one-tenth the contaminant concentrations shown in Tab. 1. Three reactor systems were operated in parallel for this evaluation. The reactors respectively contained equal volumes of 16 x 20 U.S. Mesh nonactivated carbon, and 16 x 20 and 30 x 40 U.S. Mesh activated carbon. The effect on reactor performance of the adsorptive capacity of the media was determined by examining data from the reactors containing the 16 x 20 U.S. Mesh media. The effect of particle size, on the other hand, was determined by comparison of data from the two reactors containing the active media.

The nonactivated carbon was obtained from the same process train as the GAC, except it was withdrawn from the manufacturing process train just prior to the final activation step. The nonactivated carbon was determined to have an appreciably lower adsorptive capacity than activated carbon. The reactor that was charged with the nonactivated carbon did not perform as well as the activated carbon reactors, in terms of the reduction in organic contaminants and their conversion to methane. The inferior performance of this reactor led to process failure when methane production ceased. The addition of 100 g of GAC to the failed reactor stimulated significant methane production, thus confirming the hypothesis that the wastewater contained some toxic and adsorbable components and that no biological activity was possible if the toxic components were not removed from the aqueous phase by adsorption (6).

The anaerobic expanded-bed reactor that contained the finer 30 x 40 U.S. Mesh GAC outperformed the reactor charged with the coarser carbon only when higher organic loading conditions were maintained. This superiority of the finer GAC medium was attributed to the increased surface area for microbial attachment and the faster adsorption kinetics associated with the smaller particle size (6).

TOXICITY CONTROL AND CONTINUOUS OPERATION OF GRANULAR ACTIVATED CARBON ANAEROBIC REACTORS

It is apparent from the data presented in Fig. 5 and from several other studies (14,16,19) that breakthrough of specific compounds resulted in system failure. In all cases, the indicator compounds were adsorbable, nonbiodegradable under anaerobic conditions, and biologically inhibiting.

The solution to this shortcoming of the system is to control toxicity by maintaining the concentrations of toxic inhibitory compounds below threshold levels, thus allowing for continuous operation of the process. Realizing that the breakthrough of an adsorbable, nonbiodegradable compound indicates an exhaustion of the adsorptive capacity of the bed, replenishment of this adsorptive capacity provides the means for averting system failure. This is achievable by adopting a strategy for partially replacing the reactor medium with virgin or regenerated GAC. The data presented below will elaborate on a case study where carbon replacement was demonstrated as a viable approach for control of biological inhibition.

The second batch of coal gasification wastewater, the characteristics of which are given in Tab. 1, was fed to an expanded-bed GAC anaerobic reactor at a daily COD loading rate of 11 g per kg GAC. A COD material balance on this reactor is shown in Fig. 7 (16). The data show the time-varying mass of COD applied to the reactor and the portion of that mass that persisted in the effluent or was converted to methane gas. At the start-up of the experiment, the wastewater was applied at half strength and no GAC replacement was practiced. Very little methane was produced during the first 70 days of operation, and COD removal was primarily by adsorption (Fig. 7). On day 70 of continuous operation, half of the reactor GAC was replaced with virgin GAC, and a strategy for a weekly replacement of 10% of the medium was initiated. The treatment system responded positively to the adopted changes, and methane production from the reactor increased rapidly to a rate that was equivalent to the total influent COD, thus suggesting bioregeneration of previously adsorbed degradable organic compounds. As a result of the positive response of the system to the weekly GAC replacement schedule, the feed to the reactor was increased to full strength, which resulted in a two-fold increase in the COD loading rate on the carbon. This mode of operation was continued for 200 additional days, demonstrating that a stable and continuous operation of the system was possible during the treatment of toxic wastewaters.

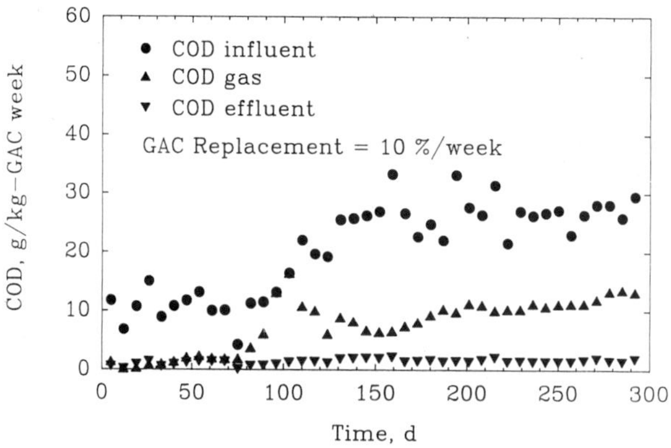

Fig. 7. Continuous operation of expanded-bed anaerobic reactor operating under a granular activated carbon (GAC) replacement strategy.

CONCLUSIONS

The use of GAC as a medium in expanded-bed anaerobic reactors results in a process capable of effecting excellent treatment efficiencies, while at the same time providing resistance to fluctuations in feed composition, flow rate, and strength. The expanded-bed GAC anaerobic reactor has been demonstrated to be a viable process for the treatment of a variety of wastewater, ranging from prepared solutions of phenols, volatile fatty acids, and mixtures of polycyclic nitrogen-containing aromatic compounds, to industrial wastewaters generated during coal coking and gasification, petroleum refining, and paint stripping operations. Throughout these studies, good contaminant removal efficiencies were maintained while high organic loading rates and low empty-bed hydraulic detention times were achieved. Granular activated carbon appears to be superior to other reactor media in terms of microbial attachment and sheltering properties. Lower contaminant concentrations and fewer suspended solids were found in effluents from reactors in which GAC was used as compared to units utilizing sand, anthracite, or nonactivated carbon.

The adoption of a schedule for the partial replacement of reactor GAC with virgin or regenerated carbon provides a means for controlling toxicity and maintaining stable and continuous operation of the GAC anaerobic reactor during the treatment of wastewaters containing biological inhibitors. However, it must be realized that the biomass attached to the removed reactor medium will be lost in the process. The mass of reactor medium that needs to be replaced by fresh GAC depends on the loading rate of the inhibitory compounds, which is directly proportional to the empty-bed hydraulic detention time maintained in the reactor, and the capacity of the carbon for adsorbing these compounds under the environmental conditions maintained within the reactor. Assuming that the biomass lost in the effluent is insignificant compared to that lost during GAC replacement, sludge age becomes a function of only the fraction of medium replaced by virgin or regenerated carbon. Therefore, a close relationship exists between the empty-bed hydraulic detention time, sludge age, and influent toxicity level.

ACKNOWLEDGEMENTS

The material presented in this chapter is a summary of research projects that were funded by The Standard Oil Company, Ohio; the United States Department of Energy, Morgantown Energy Technology Center, Morgantown, West Virginia; and The Advanced Environmental Control Technology Research Center, University of Illinois, Urbana, Illinois.

REFERENCES

1. Andrews, G.F., and J. Przezdziecki (1986) Design of fluidized-bed fermentors. Biotechnol. Bioengineer. 28:802-810.
2. CEBEDEAU (1985) Epuration des eaux de cokerie par voie anaerobie. Rapport Terminal, Recherche CECA No. 7257-15/393/02.
3. Characklis, W.G. (1973) Attached microbial growth, attachment and growth. Water Res. 2:1113-1118.
4. Chian, E.S.K., and F.B. DeWalle (1977) Treatment of high strength acidic wastewater with a completely mixed anaerobic filter. Water Res. 11:295-304.

5. Fraser, J. (1986) Series reactor configuration, and thermophilic temperatures applied to the treatment of coal gasification wastewater using anaerobic, fluidized-bed, granular activated carbon filters. M.S. Thesis, University of Illinois, Urbana, Illinois.
6. Gardner, D.A., M.T. Suidan, and H.A. Kobayashi (1987) Role of GAC activity and particle size during the fluidized-bed anaerobic treatment of refinery sour water stripper bottoms. J. Water Poll. Control Fed. (submitted for publication).
7. Khan, K.A., M.T. Suidan, and W.H. Cross (1982) Role of surface active media in anaerobic filters. J. Environ. Eng. Div., ASCE Vol. 108, EE2, pp. 269-285.
8. Najm, I.N. (1987) Kinetics of anaerobic biodegradation of phenolic compounds. M.S. Thesis, University of Illinois, Urbana.
9. Nakhla, G.F. (1987) Effect of process variables on the treatment of coal gasification wastewater using anaerobic GAC reactors. M.S. Thesis, University of Illinois, Urbana.
10. Pfeffer, J.T., and M.T. Suidan (1985) Anaerobic-aerobic process for treating coal gasification wastewater. In Proceedings of the Industrial Waste Symposium, 58th Annual Water Pollution Control Federation Conference, Kansas City, Missouri, pp. 235-249.
11. Pfeffer, J.T., and M.T. Suidan (1986) Role of carbon replacement in the biological processing of toxic organics in the anaerobic fluidized-bed reactor. In Proceedings of the Industrial Waste Symposium, 59th Annual Water Pollution Control Federation Conference, Los Angeles, California, pp. 349-365.
12. Suidan, M.T., and P. Fox (1987) Treatment of hazardous landfill leachate with fluid-bed anaerobic filter technology. Periodic Progress Report, CERL-EN-5-413A, TASK #7.
13. Suidan, M.T., W.H. Cross, and M. Fong (1980) Continuous bioregeneration of granular activated carbon during the anaerobic degradation of catechol. Progr. Water Technol. 12:203-214.
14. Suidan, M.T., G.L. Siekerka, S.W. Kao, and J.T. Pfeffer (1983) Anaerobic filters for the treatment of coal gasification wastewater. Biotechnol. Bioengineer. 25:1581-1596.
15. Suidan, M.T., C.E. Strubler, S.W. Kao, and J.T. Pfeffer (1983) Treatment of coal gasification wastewater with anaerobic filter technology. J. Water Poll. Control Fed. 55:1263-1270.
16. Suidan, M.T., J.T. Pfeffer, Y.T. Wang, M.A. Deady, P. Fox, S.W. Kao, and M.A. Johnson (1985) Anaerobic treatment of gasifier effluents. Final Report, DOE/MC/19352-1763.
17. Switzenbaum, M.S., and W.J. Jewell (1980) Anaerobic attached-film expanded bed reactor treatment. J. Water Poll. Control Fed. 52: 1953-1965.
18. Taylor, D.W. (1972) Full-scale anaerobic trickling filter evaluation. Technology Science Report EPA-R2-72-018, Washington, D.C.
19. Wang, Y.T., M.T. Suidan, and J.T. Pfeffer (1984) Anaerobic activated carbon filter for the degradation of polycyclic N-aromatic compounds. J. Water Poll. Control Fed. 56:1247-1253.
20. Weber, Jr., W.J., L.D. Friedman, and R. Bloom, Jr. (1972) Biologically extended physicochemical treatment. In Proceedings of the Sixth International Conference on Water Pollution Research, Jerusalem.
21. Young, J.C., and P.L. McCarty (1967) The anaerobic filter for waste treatment. In Proceedings of the 22nd Purdue Industrial Waste Conference, West Lafayette, pp. 559-574.

ENVIRONMENTAL CONCERNS ASSOCIATED WITH THE DESIGN OF GENETIC ENGINEERING FACILITIES

Joe C. Watt, Vincent S. Wroniewicz, and Daniel F. Ioli

Stearns Catalytic Division
United Engineers and Constructors
Philadelphia, Pennsylvania 19101

INTRODUCTION

With the advent of genetic engineering processes, there is much concern regarding wastes generated by the facilities incorporating these processes and the potential effects of these wastes on the environment. The primary process employed in these facilities is fermentation utilizing recombinant DNA organisms. Subsequent processing to extract and purify products can incorporate many different types of unit operations, such as ultrafiltration, ion exchange, extraction, dialysis, electrophoresis, high pressure liquid chromatography, and ion chromatography. While production rates of products produced by genetic engineering facilities are small, their environmental problems can be relatively large and can often have major impacts on these plants.

One of the critical considerations in the design of commercial-scale fermentation facilities (volume >10 liter) is to prevent the release of the recombinant DNA organisms into the environment. Within the United States, the National Institutes of Health (NIH) has issued guidelines for large-scale recombinant experiments (2).

Another area of critical importance is the treatment of wastewaters originating from the process steps. Generally, the yields from these processes are relatively low, but are justified because they produce high-priced products. Because of the low-yield nature of the processes, large quantities of raw materials are used. The added cost to treat these large quantities of discarded chemicals can significantly impact the economics of production. There is, therefore, a need and advantage to maintain an unbroken interface between waste treatment engineers and the process development staff from the development stage, through pilot operations, to detailed design.

THE PROCESS

The fermentation step is usually a batch or semibatch operation. It can result in an intracellular product (contained in the cells' bodies), requiring disruption of the cell structure before recovery, or an extracellular product (the product is produced outside the cell walls). Both generally require cell harvest from the fermenter broth to achieve product recovery.

Containment must be maintained until the organisms are inactivated (i.e., killed). Various means of inactivation (e.g., the use of temperature or chemicals) are employed at various stages of the process prior to purification, depending primarily on the stability of the product to the inactivation process. It is advantageous to kill the batch as early in the process as possible, ideally in the fermenter, but this is not always possible.

Following the fermentation and cell harvest are several, often complex, steps to recover and purify the product. These steps are usually batch or semicontinuous processes.

CONTAINMENT

Containment of recombinant DNA organisms is difficult in a commercial-scale fermentation process because of the large number of potential discharge points, such as vents, valves, and seals. The design of the containment system requires specialized features and depends on a detailed evaluation of the potential discharge of microorganisms in both normal and emergency operations. The degree of physical containment required depends on the potential lethality of the host-vector system. As shown in Tab. 1, the NIH guidelines specify three different levels of physical containment: BL_1-LS, BL_2-LS, and BL_3-LS (with BL_1-LS being the least stringent and BL_3-LS being the most stringent) (1,3,4,5). These containment measures can be augmented by regulated control procedures to insure that containment has been achieved.

For most systems of commercial interest, BL_1-LS and BL_2-LS are suitable. The BL_1-LS and BL_2-LS levels both provide primary process containment: that is, the process is contained, but the physical facility in which the process is located is not. The BL_3-LS level, however, provides primary containment as in BL_2-LS plus secondary facility containment: that is, the physical facility which houses the process is designed to prevent release of viable organisms to the environment should the primary containment system fail. The cost of BL_3-LS facilities is, consequently, significantly higher.

The engineering requirements for adherence to the NIH guidelines fall into three categories: the basic primary safety measures for all three containment levels; the additional primary containment measures for levels BL_2-LS and BL_3-LS; and the secondary facility containment measures required for containment level BL_3-LS.

Basic Primary Containment Measures (BL_1-LS, BL_2-LS, and BL_3-LS)

(1) Operations must be conducted in a closed vessel for organisms containing recombinant DNA molecules. Strict adherence to this

Tab. 1. Physical containment requirements for large-scale fermentations using organisms containing recombinant DNA molecules. The levels of containment required are compared to the NIH physical containment guidelines (2).

Item No.	Description	BL1-LS	BL2-LS	BL3-LS
1	Closed vessel	•	•	•
2	Inactivation of cultures by validated procedure before removing from the closed system	•	•	•
3	Sample collection and addition of material in a closed system	•	•	•
4	Exhaust gases sterilized by filters before leaving the closed system	•	•	•
5	Sterilization by validated procedures before opening for maintenance or other purposes	•	•	•
6	Emergency plans and procedures for handling large losses	•	•	•
7	No leakage of viable organisms from rotating seals and other mechanical devices		•	•
8	Integrity evaluation procedures—monitors and sensors		•	•
9	Containment evaluation with the host organism before introduction of viable organism		•	•
10	Permanent identification of closed system (fermentor) and identification to be used in all records		•	•
11	Posting of universal biohazard sign on each closed system and containment equipment when working with viable organism		•	•
12	Posting of universal biohazard sign on entry doors when work is in progress			•
13	Operations to be in a controlled area:			•
	Separate specified entry			•
	Double doors—air locks			•
	Walls, ceiling and floors to permit ready cleaning and decontamination			•
	Utilities and services process piping and wiring) to be protected against contamination			•
	Handwashing facilities			•
	Shower facilities in close proximity			•
	Area designed to preclude release of culture fluids outside the controlled area in the event of an accident			•
	Ventilation—movement of air, filtration of air, etc.			•

is very impractical, since vessels containing rDNA organisms are often involved in processes that require addition of material to the vessel and subsequent venting. An example is the fermenter, where a large volume of air is sparged into the broth. The spirit of the guidelines, however, can be satisfied. The most efficient way to do this is by providing coalescing and sterile filters on vent lines from the vessel, thereby preventing the release of organisms to the atmosphere.

(2) Inactivation of culture by validated procedure before moving from a closed system. In selecting from the various methods for inactivating cultures (e.g., the use of temperature, chemicals, or antibiotics), the most important consideration is the degree to which the technique accomplishes the inactivation with the least impact on the quality and yield of the product. This decision can have a profound impact on the character of the waste stream and subsequent treatment and disposal. The next question is where the culture inactivation should be performed. In the fermenter? In a

downstream holding tank? After the cell harvesting? From a containment viewpoint, it should be performed in the fermenter, thereby avoiding the need to design containment into cell harvesting or downstream processes. While desirable, it may not be possible to find a technique that can kill the organisms without affecting the product quality. Figures 1 and 2 are conceptual sketches which illustrate containment systems for an extracellular product (Fig. 1) and an intracellular product (Fig. 2).

(3) Closed sample collection and transfer of cultures in a closed system, preventing releases of aerosols or contamination of exposed surfaces. To meet these requirements, the following are required:

(a) Closed sample collection systems. Each sample station should be provided with steam to sterilize the connection before and after sample collection.

(b) All vessels holding active cultures should be equipped with sterile vent filters to prevent release of organisms during transfer.

(c) Equipment that can generate aerosols, such as centrifuges, should be enclosed in biological safety cabinets.

(4) Sterilization of the exhaust gas from the fermenters and all vessels containing viable organisms. This can be accomplished by using sterile filters on the gas outlets, which is, among other considerations, a more economical approach than the use of other sterilization techniques such as incineration. However, the question that must be addressed here is one of practical design considerations. A considerable amount of aerosol formation can occur in the fermenter, and foam can be carried over with the gas; if the sterile vent filters are not protected properly, they

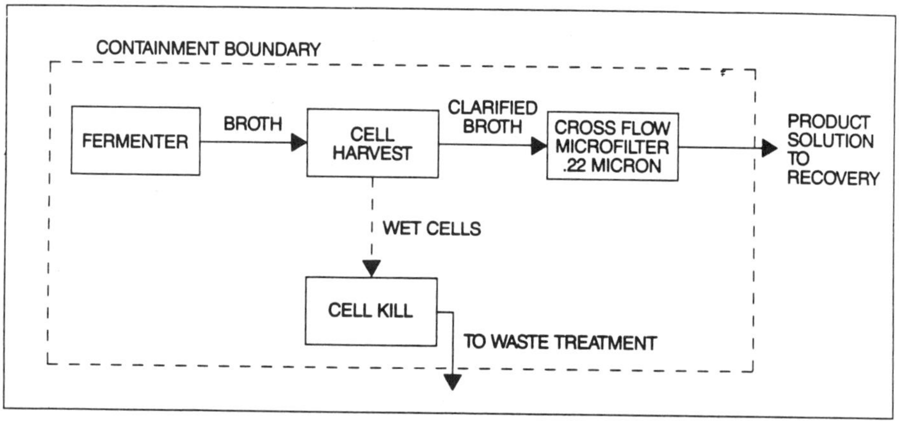

Fig. 1. Containment for an extracellular product. Cells are killed by direct steam, acid, or chemicals.

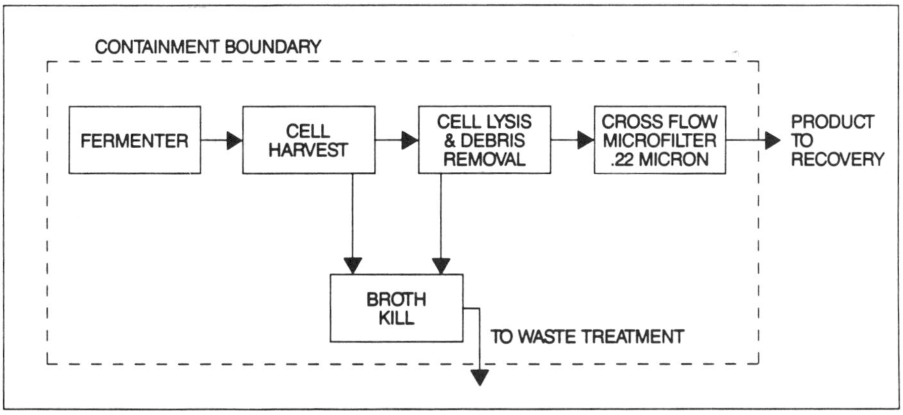

Fig. 2. Containment for an intracellular product. Broth is killed by direct steam injection or by chemicals.

will plug very quickly. Therefore, adequate separation equipment, such as cyclones and coalescing filters, must be provided upstream of the sterile filters to remove liquids and solid particles. Also, there is the important question of reliability of the sterile vent filters: if proper upstream precautions are taken, then the sterile filters would be expected to fail only under extraordinary conditions. The chance of failure can be reduced by providing two sterile filters in series and then testing these filters after each batch to validate the integrity of the filter. Figure 3 provides a sketch of a typical system.

(5) <u>Sterilization by a validated procedure, before opening for maintenance, of any equipment handling recombinant DNA organ-</u>

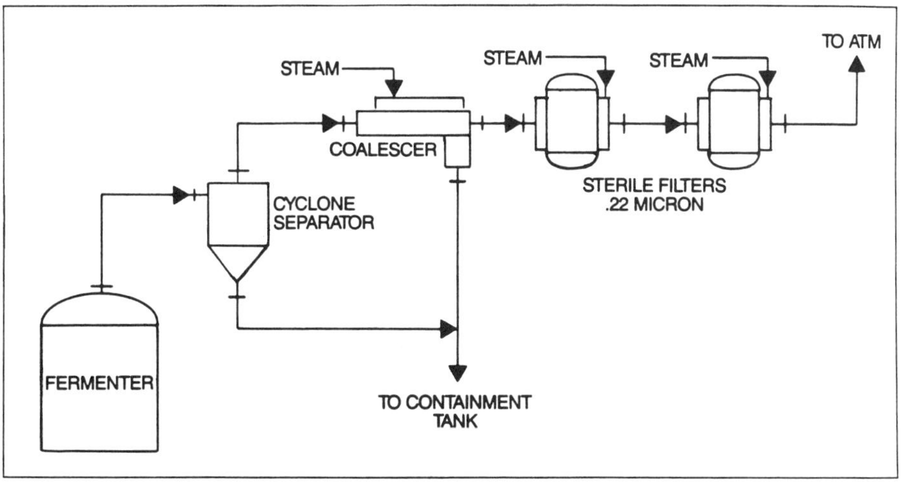

Fig. 3. Fermenter exhaust gas containment system.

isms. This can be accomplished using 20 psig steam, which is the normal procedure for steam sterilization. During the sterilization process, all condensate from the lines, vessels, and vents is collected in a central tank, and the contents of that tank are tested for any viable organisms before discharge to the sewer. A sketch of a typical containment tank is shown in Fig. 4.

(6) Emergency plans and procedures for handling large losses in the fermentation area. To address this requirement, a detailed hazards analysis of the system must be performed to determine the risk of release of viable organisms from the system, based on worst-case scenarios. For each of these worst-case scenarios, a plan and system must be provided to contain the release of the organisms. Examples include pressure relief valve discharge from fermenters, and a plan for containing large spills.

Additional Primary Containment Measures (BL_2- and BL_3-LS)

(7) No leakage of viable organisms from rotating seals and other mechanical devices. This can be achieved by providing double mechanical seals for rotating shafts (such as agitators and pumps) and then by providing an additional steam seal for this equipment. It is difficult to provide containment for some mechanical equipment (e.g., rotary vacuum filters). Area containment may be required.

(8) Integrity evaluation procedures, monitors and sensors. During normal operation, instrumentation must be provided to monitor the area for release of organisms. For example, within diked areas, level-detection monitors can trigger alarms.

(9) Containment evaluation with the host organism, before introduction of viable organisms into the system. This involves testing the system prior to use to make sure that all containment systems are functional; inoculating the fermenter with the host organism; validating the gas outlet filters; validating the inactivation of the culture at the end of the fermentation; and simulating spills in the fermenter area to validate the efficacy of the containment systems.

(10) Permanent identification for the closed system, and use of this identification for all records. In other words, all inlets, outlets, and equipment that comprise the closed system must be labeled as such, and all recordkeeping must make reference to the labeling. Conventional practice is to do this whether or not the fermenter contains recombinant DNA organisms.

(11) Posting of universal biohazard signs on each closed system and on containment equipment when working with viable organisms.

Secondary Containment (BL_3-LS)

(12) Posting of universal biohazard signs on entry doors when work is in progress.

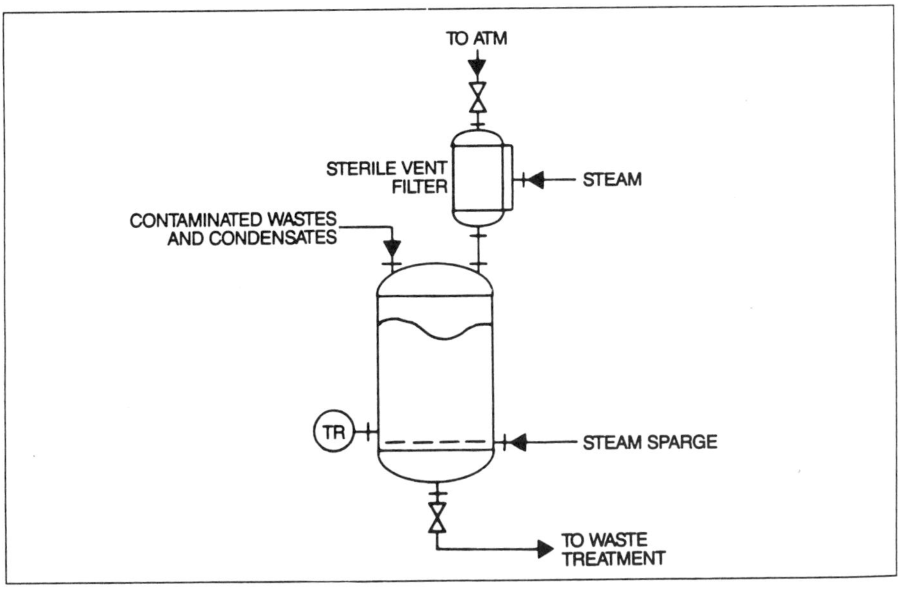

Fig. 4. A typical containment tank.

(13) Where viable organisms are being processed, operations should be conducted in a controlled area. This requirement mandates the use of secondary containment to prevent the accidental release of organisms to the environment. Compliance with the guidelines requires that the following features be built into the facility:

(a) Entrance into the controlled environment area should be through air locks. Separate gowning and washing facilities must be provided at each entrance. By code, emergency exits must be available, and these must also be provided with air locks.

(b) The interior space must be provided with easily cleanable walls, ceilings, and floor. The floor should be coved for cleaning. Typical materials of construction should be epoxy terrazzo floors and PVC covering on the walls and ceiling.

(c) All penetrations into the enclosed environment (electricals, water pipes, medium supply lines, telephone cables, etc.) must be sealed.

(d) All utilities and services are to be protected against contamination. This means that the amount of exposed piping is to be kept to a minimum and should be run inside the walls or above the ceiling.

(e) All drains for the controlled area should be connected to a closed containment tank in case of spills.

(f) The controlled environment area should be at a negative pressure with respect to the adjacent areas.

(g) All the exhaust air from the controlled environment area should be passed through high-efficiency particular air (HEPA) filters.

WASTE CHARACTERISTICS

Commercial-scale fermentation facilities of any size generate significant wastewater discharges. Solid waste generation can also be significant. Disposal of large volumes of sludge and the possible presence of rDNA organisms are of particular concern. Air emissions, although not large in volume, can present problems in odor control and containment of the rDNA organisms.

Sources and Characteristics

Process descriptions, process design, laboratory and pilot plant samples, and material balances can all be used to help define those wastewaters requiring treatment. While there may be many specific discharges from the full-scale process, they fall into four general categories: (a) spent fermenter broth; (b) product recovery wastes; (c) purification wastes containing high concentrations of dissolved organics; and (d) dilute purification wastes, often having large flows resulting from washes, etc. In addition to process wastes, utility services will also generate wastewaters that must be addressed, such as boiler or cooling tower blowdown and ion exchange regeneration streams.

The first step in waste characterization is to establish the waste sources and quantities. These sources must then be individually examined to ascertain the impact they may have on the overall facility and their possible incompatibility, potential for toxicity, and need for pretreatment. The most efficient way to accomplish this is to prepare a simple computer model using a PC spreadsheet program and the process material balance. Figure 5 shows a typical output of such a spread sheet. Each source can be tallied individually. Calculations can be programmed for individual constituents (materials) to convert them into wastewater parameters (such as TOC, COD, TKN, phosphorus, and TDS). These calculations are based on the chemical formula, literature data, and prior experience. Laboratory analysis of pure constituents and/or samples from bench-scale or pilot facilities can be used to fill in "holes" and verify the "model."

The computer model is normally set up on a "per batch" basis. Batch sizes and batches per day are adjusted to derive daily loads at different production rates. Figure 6 is a typical summary of the resulting waste loads, and would be part of the same spreadsheet file as shown in Fig. 5.

As the process develops and changes, or as production projections change, the impact on raw waste load can be easily derived using this spreadsheet model. Having this information available during the process

Fig. 5. Typical spreadsheet output (simplified).

"XYZ" PROCESS—12% YIELD
PER BATCH RAW WASTELOAD

Component (kg/batch)	MW	Specific Gravity	BOD Ratio	COD Ratio	Fermenter: Cell Harvest Centrate	Product Recovery: Centrifuge Centrate	Purification: Displacement Waste	Purification: Backflush Wastewater	Purification: Water Rinse Waste	Vessel Washes	Total
Water	18.0	1	0	0	33213	6661	5670	284000	3976000	20000	4,596,600
Fermenter substrate	92.1	1.26	0.62	0.997	132	19					
Protein substrate		1.46	0.93	0.39	102	16					
Product		1.46	0.8	1.3		19.2					
Soluble by-products		1.46	0.8	1.3		218					
Other protein		1.46	0.8	1.3		29.4					
Cells		1.46	0.8	1.3	20						
Antibiotic A		1	0.8	1.221	1.2						
Antibiotic B		1	0.8	1.221	1.2						
Phenol	94.1	1.07	0.785	2.379	71	9.6					
Toluene	92.1	0.807	0.86	1.41	58	7.8					
Complexing agent	128.0	0.6	0.01	0.93		4.2	15				
Organic ammine	61.1	1.02	0.81	1.14							
Detergent		1									
Other solubles		1.46	0.8	1.3							
Sulfonated organic salt	288.0	1.25	1.32	1.77			20		256		
Sodium phosphate	120.0	2.04	0	0		20.4					
Sodium hydroxide	40.0	2.13	0	0							
Sodium hypochlorite		2									
Sodium chloride	58.5	2.165	0	0							
Sodium carbonate (monohydrate)	124	2.509	0	0			15				
Sodium bicarbonate	84	2.159	0	0			12				
etc.											
....											
....											
Total (kg/batch)					33598	7005	5717	284000	3976256	20000	4,969,400
Flow (liter/batch)					33661	6922	5676	284000	397205	20000	4,632,120
BOD5 (kg/batch)					4683	254.2	75	0.0	337.9		3,361
COD (kg/batch)					2639	425.5	31	0.0	453.1		5,743
TKN (kg/batch)					19.4	40.8					421
Phosphorus (kg/batch)					29.0	5.3					182
etc.											
....											
....											
....											

"XYZ" PROCESS—12% YIELD
DAILY WASTELOAD

Stream Desc / Component (kg)	Combined Fermentor Wastewaters	Product Recovery	Purification: 1st Separator	2nd Separator	Concentration	Final Purification	Final Wash	Total Combined Wastewaters
No. batches/day (avg)	0.5	0.5	0.5	0.5	0.5	0.5	0.5	
No. batches/day (max)	1	1	1	1	1	1	1	
No. batches/day (min)	0	0	0	0	0	0	0	
Flow L/day (avg)	24,684	314	1,988,102	286,287	123,694	192,410	284,823	2,820,314
Flow L/day (max)	49,368	629	3,976,205	572,575	247,387	384,819	409,645	5,640,628
Flow L/day (min)	0	0	0	0	0	0	0	0
COD kg/day (avg)	1671	116	227	0	0	658	157	2830
COD kg/day (max)	3343	232	453	0	0	1317	315	5659
COD kg/day (min)	0	0	0	0	0	0	0	0
BOD kg/day (avg)	1049	86	169	0	0	467	109	1881
BOD kg/day (max)	2098	173	338	0	0	934	219	3762
BOD kg/day (min)	0	0	0	0	0	0	0	0
NH3-N kg/day (avg)	NA	NA	0	NA	NA	NA	NA	0
NH3-N kg/day (max)	NA	NA	0	NA	NA	NA	NA	0
TKN kg/day (avg)	48	0	0	0	0	132	30	210
TKN kg/day (max)	97	0	0	0	0	263	60	421
Total P kg/day (avg)	29	0	0	0	0	0	0	29
Total P kg/day (max)	57	0	0	0	0	0	0	57
TSS kg/day (avg)	25	0	0	0	0	0	0	25
TDS kg/day (avg)	NA	NA	0	NA	NA	NA	NA	3452
COD mg/L	67,712	368,750	114	0	0	3,421	768	1,003
BOD mg/L	42,505	275,000	85	0	0	2,428	534	667
NH3-N mg/L	NA	NA	0	NA	NA	NA	NA	ND 2
TKN mg/L	1,965	0	0	0	0	684	147	75
Phosphorus mg/L	1,164	0	0	0	0	0	00	10
TSS mg/L	1,013	0	0	0	0	0	0	9
TDS mg/L	NA	NA	0	NA	NA	NA	NA	1,224
GPD								

Fig. 6. Typical spreadsheet output (simplified).

development stage will minimize waste treatment problems. Such an approach can also be used to evaluate the environmental effects of process options that might be considered at the start of process development.

Table 2 shows the typical characteristics of wastewaters discharged from the various process areas. The range of values shown is based on a variety of studies and engineering evaluations performed for various bioengineering and pharmaceutical companies. The following section discusses each of these areas and provides information on sources and relative strengths of wastewaters generated by the various processes employed.

Fermentation wastewaters. Wastes from the fermentation steps (spent fermenter broth) are generally very concentrated, relatively low-volume batches of wastewater. They can contain high levels of suspended solids (biomass). The initial cell harvest can thus generate a substantial quantity of biological sludge for disposal. Nutrients (nitrogen and phosphorus) will be present in the broth, sometimes at very high concentrations. While much of the nutrient content is utilized in downstream biological treatment processes, it can still contain sufficient quantities to create a concern in final discharge limitations. The waste can also contain emulsifiers, antibiotics, and other compounds which are added to the fermenter broth. Following fermentation, the inactivation of the batch can result in other waste characteristics, such as a low pH if acid is used or toxicity if organic chemicals are used. In some cases, these organic chemicals can be priority pollutants. The concentrations required for inactivation are usually low enough that, after dilution with the remaining wastes, the compounds that are biodegradable can be handled by aerobic biological

Tab. 2. Typical raw waste characteristics of process wastewaters.

	Fermenter	Product Recovery	Purification		Total Combined
			Strong[A]	Weak[B]	
Volume	0.5 to 4.0 batches/day Up to 75,000 liters/batch	0 to 10% of fermenter batch	15,000-115,000 liters	15,000-3,785,000 liters	1,100-5,700 M^3
COD mg/l	30,000-70,000 (40-60% of total load)	High if soluble organic solvent is used	2,000-70,000	10-500	2,200-9,000 kg/day
BOD_5 mg/l	15,000-50,000		0.1 to 0.8 of COD	0.1 to 0.8 of COD	0.5-0.8 of COD
TKN mg/l	700-2,500		None or 500-5,000	<10-20	9-360 kg/day
PO_4-P mg/l	700-2,500[C]		None or 500-5,000[C]	<10-20	9-270[C] kg/day
TSS mg/l	1,000-20,000[D]	Biomass[D] sludge	Precipitates possibly	<10	45-680 kg/day
TDS mg/l	20,000-40,000		<100-15,000	<100-5,000	—
pH units	3.0[E]-8.0		2.0-12.0	3.0-11.0	—
Other pollutants possible	■ Organic priority pollutants ■ Antibiotics		■ Heavy metals ■ Organic priority pollutants (solvents) ■ NH_3-N ■ Surfactants		—

Notes: [A] High organic concentration, low volume
[B] Low organic concentration (often high volume) washes, etc.
[C] Sometimes as high as 2,600 kg/day (due to phosphoric acid addition)
[D] Depends on handling of biomass in process as to where the TSS comes out
[E] Acid addition after fermentation complete

treatment. However, some minor, residual constituents can exhibit chronic toxicity to a biological treatment system. Treatability studies, therefore, are needed to address this possibility. It can take as long as two to three months to see the effects, but they can be dramatic.

Product recovery wastewaters. Initial product recovery can be achieved by operations ranging from simple physical separation to exotic liquid/liquid extractions. Quite often, though, most of the pollutants are those remaining from the spent fermenter broth. At this point, most of the biomass will be discharged as a sludge, which must be handled and treated. Extraction processes may result in spent organic solvents. These are generally recovered for reuse when large volumes are involved, or incinerated when volumes are relatively small.

Purification wastewaters. Purification wastes vary widely. There may be pure organic streams, wastewaters with high concentrations of organics (usually at relatively low volumes), high-volume (relatively dilute) washes containing organic and/or inorganic salts, acid or caustic streams, or other high-salt streams. Typical waste generators are: (a) absorption media regenerations and washes; (b) ion exchange regenerations and washes; (c) diafiltration wastes (ultrafilter permeate, etc., and washes); (d) distillation overheads/bottoms; (e) spent organic solvents; and (f) acid or caustic treatment wastes. Some streams may contain heavy metals that were used as catalysts or reactants. Sludges may also be generated, often the result of precoat filtration. These sludges can contain rDNA organisms. In these cases, the sludges must be inactivated before they can be discharged from the containment area.

Utility wastewaters. Utility wastewaters from boilers and cooling towers are usually not significant in comparison to the process wastes. Deionized water usage in the process can be high, and wastes from the preparation of DI water, such as reverse osmosis reject brine and/or ion exchange resin regeneration wastes, can be of sufficient volume to be of concern. Separate treatment (equalization/neutralization) and discharge, bypassing the treatment system for the process wastes, is usually recommended.

WASTEWATER TREATMENT METHODS

In treatment of wastewaters from genetic engineering facilities, a full gamut of treatment steps could be required. This could include pretreatment or primary treatment for recovery and recycle; pH adjustment to remove heavy metals and/or suspended solids; secondary treatment to remove organic constituents; tertiary treatment to reduce nutrient concentrations; and sludge handling for treatment and disposal of sludges produced by the facilities. Treatment systems could include all or any combination of these steps.

Pretreatment

Some of the concentrated streams, or those containing unusual pollutants, may be candidates for pretreatment, recovery, or recycle. Table 3 shows some of the unit operations, applicable sources, and pollutants removed. Also shown is a range of typical removals achieved in studies that we have performed. Most of these options were studied in our laboratory to some extent before completing the engineering evaluation, but not all were selected for design.

Fermentation product recovery wastewaters. Pretreatments that might be utilized, separately or in combination, are demulsification, coagulation, solids separation, and dewatering for the suspended biomass. Phosphate removal may be required if large amounts of phosphate buffers are used. Neutralization may be required prior to discharge if acid is used for inactivation. Neutralization may generate significant sludge.

Purification wastes. Pretreatment operations that may be applicable are neutralization, metals precipitation, steam stripping/distillation, and phosphate precipitation. Organic streams would generally go directly to solvent recovery or incineration.

Combined wastes. Equalization will generally be critical in the design of treatment facilities due to concentrated loadings from batch processes. Certain low-volume concentrated sources (mainly fermenter broth) may be best handled by collecting the batch in a separate holding tank and bleeding the wastewaters into the biological system. Some end-of-pipe equalization may still be necessary, however, for waste flows from other operations in the production facilities.

Tab. 3. Pretreatment of wastewater from genetic engineering facilities.

Unit Operations	Sources	Pollutants	Typical Removals
Waste treatment:			
■ Equalization	■ Fermentation wastewaters ■ Strong purification wastewaters	■ Organic load to treatment	—
■ Coag./flocc./demulsifucation	■ Fermenter wastewaters ■ Product recovery wastewaters	■ TSS ■ TSS	— —
■ Physical separation (gravity/ centrifuge, filter, etc.)	■ Fermenter wastewaters ■ Product recovery wastewaters ■ Chilled process steps	■ TSS ■ TSS ■ Inorganic salts organics	96-99% — —
■ Precipitation/clarification	■ Fermenter wastes ■ Extraction buffer ■ Purification reactant	■ PO_4-P ■ PO_4-P ■ Heavy metals	80-95% 85-98% 70-98%
■ Distillation/stripping	■ Solvent recovery	■ Organic chemicals	80-95%
■ Carbon adsorption	■ Variable	■ Toxic organics (low levels)	—
By-product recovery			
All the above plus:			
■ Solvent extraction, wet air oxidation, ion exchange, ultrafiltration, evaporation, drying, etc.	■ Sources, pollutants and removals are variable depending on value of recoverables, treatment required and other requirements		

Secondary Treatment

Generally, the wastewaters are readily biodegradable, although low levels of toxic inhibitors may exhibit chronic toxicity in the system. The potential for this condition is not always evident from the characterization data. Laboratory treatability studies should be run, if possible, to determine if chronic toxicity will occur, and also to ascertain actual effluent quality. These studies will also identify other operating problems that may be present, such as odor.

If a toxicity problem does become apparent, the constituent can often be isolated by further laboratory testing. Possible remedies are removal at the source or pretreatment, such as with activated carbon. Another approach which may be successful is the addition of minimal doses of powdered carbon to the biological system; quite often the toxic organics are preferentially adsorbed over the readily biodegradable portion of the wastewaters.

Biological treatment will almost always be applicable, with discharge to existing systems (in-plant or POTW) a possibility. Anaerobic treatment, particularly of the fermentation wastes, should be reviewed and investigated. This could be a pretreatment step for POTW discharge and could provide a means for energy recovery through the burning of off-gases. For NPDES discharge, multiple-stage systems may be required to attain necessary effluent levels. Anaerobic treatment of concentrated streams, followed by aerobic treatment of the anaerobic effluent plus all other wastes, could prove to be an economical treatment method.

Of the aerobic processes, activated sludge provides the most flexibility in operation and has the ability to fully treat variable wasteloads. Fixed film reactors, which may appear more attractive economically, are more susceptible to intermittent overloads and possible upsets. Two-stage aerobic systems may also be in order if nitrification is required. Some typical design parameters for the activated sludge process are shown in Tab. 4.

Tertiary Treatment

Requirements for nutrient removal are generally addressed as a tertiary process. Ammonia stripping or phosphate precipitation would fall into this category.

Large quantities of ammonia nitrogen can be present in final effluents for a number of reasons. Ammonia or ammonia salts may be significant raw materials used in the process, or organic nitrogen compounds may be biodegraded in the secondary treatment system, releasing ammonia nitrogen as a by-product. Ammonia concentrations might be reduced sufficiently by pretreatment steps or by biological nitrification in the secondary system. However, ammonia nitrogen concentrations may still exceed discharge limits. This may require ammonia stripping as a tertiary removal step.

Phosphate concentrations may also exceed effluent requirements, even after pretreatment. Addition of coagulants to the secondary clarifier may be sufficient to achieve discharge limitations. Separate tertiary precipitation/coagulation of phosphorus is another alternative. If phosphoric acid usage is significant in the process, large quantities of sludge can result.

Effluent-suspended solids levels from biological treatment units may also exceed discharge limitations. This can be caused by such factors as the low F/M ratios necessary to achieve nitrification or high TDS levels due to inorganic salt concentrations. If effluent limitations are strict, media filtration may be required.

Sludge Handling

Sludge will be produced from many areas of the wastewater treatment facilities, as well as from areas of the production facilities. All these sludges will require treatment and handling.

Digestion of waste biological sludge from treatment units will generally be required, based on the typically large organic loads encountered. Waste biological sludges from the manufacturing process must also be considered for inclusion in any sludge handling system. These can be handled separately or in combination with the wastewater biological sludges.

Other sludges that may require disposal are precipitation sludges (heavy metals, sulfates, and phosphates) from pretreatment, solvent recovery still bottoms, filter cakes, and spent ion exchange resins. If acid is used for inactivation, precipitate sludges may be calcium sulfate (sulfuric acid) or calcium phosphate (phosphoric acid), and the sludge quantities might be very significant.

Unit operations that may be required for pretreating inorganic sludges are thickening, dewatering, and disposal (landfill or sale as a by-product). Unit operations for secondary sludges could include thickening,

Tab. 4. Typical design parameters for the activated sludge process.

Hydraulic retention time (HRT)	1- 5 days
Sludge retention time (SRT)	10-25 days
BOD_5 removed	>98%
COD removed	75-90%
Apparent sludge yield	0.3-0.4 mass/mass of BOD removed
F/M ratio	0.15-0.3
O_2 utilized (includes endogenous O_2)	1.35-1.9 mass/mass of BOD removed

digestion, dewatering, and disposal. Disposal may be by landfill, landfarming, or possibly incineration. Landfarming of rDNA organisms, even deactivated, is a difficult and complex issue. Other than pure organic streams, or organic still bottoms, there are generally no RCRA hazardous wastes generated.

SUMMARY

Recombinant DNA technology is being used to produce a wide spectrum of products, such as vaccines, interferon, insulin, and growth hormones. In the design of facilities employing this technology, critical consideration must be given to the protection of the environment, both in the prevention of releases of recombinant DNA organisms into the environment and in the treatment of wastes originating from the production facilities.

The design requirements for containment of large-scale systems are complex and require detailed analysis to insure that the containment system can handle both the normal and emergency releases of recombinant DNA organisms. This must include the prevention of releases through either liquid discharges or air emissions. The "killing" method used in the process for either the cells (extracellular product) or the broth (intracellular product) is an important step and can have significant implications in downstream treatment of wastewaters.

Since fermentation is the primary process used in the production of recombinant DNA products, wastewater characteristics from this area of the process are basically similar to those of other fermentation processes. They differ, however, because of the "killing" step in the process, which can introduce compounds not normally found in fermentation wastewaters. This can complicate the treatment process by requiring additional treatment operations.

Characteristics of wastewaters from other areas of the process can be very diverse, and no general characterization can be made. Techniques for recovery and purification can vary from product to product or even from plant to plant, making characterization difficult. It is important, therefore, that each process be examined in detail so that waste characterization is meaningful and useful in the design of treatment facilities.

Because of the complex nature of the processes involved in the production of recombinant DNA products, wastewater treatment can also become a very complex problem. Systems to treat these wastewaters can include many diverse unit operations, from pretreatment of selected streams to tertiary treatment of the combined streams to meet stringent effluent criteria. While biological treatment is almost always applicable, waste loads are very high, and multiple-stage systems could be required.

Early and ongoing interface between the process development scientists and engineers and the environmental disciplines allows for the early recognition of potential environmental problems. With early recognition, many of these problems can be economically and efficiently addressed in the design of the facility. Environmental evaluation throughout the process development will help direct the design to avoid or minimize difficult discharge problems. An important part of this evaluation is the laboratory and pilot-scale evaluation of waste treatment processes in conjunction with the pilot development of the production process itself. Computer modeling of the waste load can also be a useful tool, projecting the effects of process changes on the waste treatment facilities.

If the process is developed with minimal environmental input, waste treatment may still be successfully designed with a total end-of-pipe approach, but the costs associated with this type of approach could be significantly higher and could have a major impact on the economic justifications for the overall project. An integrated approach allows for continuous technical and economic evaluation of the options.

REFERENCES

1. Giorgio, R., and J. Wu (1986) TIBTECH.
2. National Institutes of Health (1984) Guidelines for research involving recombinant DNA molecules. Federal Register 49(227):46266-46291.
3. Schenke, E., and R. Giorgio (1984) Process Eng. 65(4):27.
4. Seckhar, C. (1985) Chem. Eng. 92(9):57.
5. Wu, G., S.E. Lee, and E. Jenkins (1983) Paper presented at the 186th American Chemical Society National Meeting, Washington, D.C., August 1983.

DISCUSSION ON APPLICATIONS TO CHEMICAL AND ENVIRONMENTAL TARGETS, COMBINING MICROBIOLOGY AND ENGINEERING

James Spain (Chairman), Tyndall Air Force Base
Alan T. Bull, The University of Canterbury
Arnold Demain, Massachusetts Institute of Technology
Perry L. McCarty, Stanford University
James Tiedje, Michigan State University
Michael Torpy, ECOVA Corporation

SPAIN: How can microbiology improve the engineering results that you've seen today? What kind of information do we need from basic research to augment the information that we've obtained from the last two or three years of field work?

TIEDJE: There seems to me to be a difficulty between discovering interesting biodegradative activities and then putting them into new biotechnological practices. The practices that we're using now are logical extensions from existing ones, but there haven't been any major breakthroughs. I think laboratory research can bridge the gap between the techniques of modern biotechnology and their application. It is going to require good protein chemists and genetic engineers, as well as people who can bring the information together and make the application work.

SPAIN: It's going to require people to ask us the right questions.

TORPY: The engineer and the microbiologist have to learn from each other about the opportunities available. An environment in which such questions are being asked and answered is probably the key to the success of biotechnology in hazardous waste treatment.

BULL: I've been extremely impressed by what I've heard today. I think I've taken more notes today at the conference than I have for a long time. I've been impressed by the translation of basic microbiology into field situations which, before coming to this meeting, I would have been extremely pessimistic about. In terms of interaction between microbiologists and engineers, I think the engineers have a pretty good grasp of what is required, judging by the contributions at today's session.

As a microbiologist, I always focus on the first step in developing a process--whether it is for a new pharmaceutical or for an environmental problem--and that is the selection of the organism or organisms. I'm delighted, as a proponent of continuous culture and continuous enrichment isolation, to see how much this selection approach is now being exploited. With regard to Dennis Focht's question earlier about competitive exclusion, one has a better chance of isolating stable communities if one goes about selection initially, using continuous enrichment techniques for the PCB-degrading organisms. In our early studies, we were amazed by the number of organisms that were supposedly competing for the primary substrate that remained in a stable association for years. We can be innovative in our thinking and in our experimentation, with the expectation that we may be able to isolate and mix stable communities of organisms, perhaps avoiding the need for genetic engineering.

McCARTY: Bridging of efforts between microbiologists and engineers has long been apparent in biological treatment processes widely used throughout the United States and the world for treatment of very complex waste mixtures. Many of the hazardous chemicals that we're dealing with today are readily treated. There are both aerobic and anaerobic systems with which to treat such materials. Little more than 20 years ago, phenol was looked upon as something very toxic to organisms and not treatable. However, we know today that even at high concentrations, it is relatively easy to handle aerobically. Ten years ago it was thought that phenol couldn't be broken down by methane fermentation. However, as reported here today, there are now working treatment systems that can do this.

Methane fermentation is a complex ecological system. When one starts looking at it and taking it apart, one wonders how a system like that even works. But this system has been used for a hundred years in engineering applications, even though we don't fully understand the complex microbiology involved. As in nature, however, it tends to run by itself. For this reason, I think one can be very optimistic about engineering applications of new biological approaches. We know quite a bit about applications with fixed-film systems, dispersed-growth systems, and so forth.

One phenomenon that hasn't been taken advantage of, however, is co-metabolism or fortuitous transformations. In most treatment systems the organic material destroyed is used as a primary substrate for energy and growth. In this case natural biological systems respond and mixtures of organisms that can oxidize the organics just naturally develop. But it is in the area of co-metabolism that we need a lot more help from the microbiologists. There is a great diversity of natural organisms in the environment with many uncaptured capabilities. It is encouraging that the microbiological community is becomming much more aware of the importance of this. More support is needed for greater knowledge of microbial ecology, an important field that should be much more actively developed because of the great potential for applications of all kinds.

There has been much interest in recombinant DNA applications, but a much greater need at present is to better understand the diverse abilities of natural organisms. Through better understanding of natural organisms, we should be in a better position to take advantage of recombinant DNA techniques.

DEMAIN: There's no doubt that microbial ecologists can help the engineers in putting together consortia of microorganisms to carry out certain tasks. In the area of breaking down cellulose and converting it to liquid fuel, mixtures of microorganisms can do the job better than any one microorganism. When it actually becomes a commercial process, I'm rather sure it's going to be done with co-cultures. The fermentation industry has been successful because of the cooperation of the engineer and the microbiologist. Those of us who have been involved with the pharmaceutical and fermentation industry can appreciate and speak the language of our opposite number. I can see a number of applications to environmental biotechnology. Microorganisms can carry out certain types of degradation, but certain compounds reach dead-ends and cannot be converted any further by one organism. By putting together sequentially the abilities of anaerobes and aerobes, one might develop a procedure in which compounds are degraded completely.

The fermentation industry might be helpful with the co-metabolism effect, the oxidation or conversion of a compound which cannot be used as a substrate. As was just said by Dr. McCarty, this is an area in which the engineers can use some help. One of the problems, of course, is that no selective advantage results from oxidation of this compound. Industrial microbiologists have worked for many years on fermentations in which there is no advantage to the culture. Why would Penicillium crysogynum make 50 grams of penicillin per liter while isolated in a fermenter? Much of the strain improvement and genetic application in fermentation has been carried out without the benefits of selective procedures. We've developed large numbers of screening procedures, in many cases tedious, by using automation and miniaturization. I think this experience could be useful in developing systems for biodegradation.

Certainly, the modern geneticists have something to contribute. If engineers find that certain steps are extremely difficult, they could ask the genetic engineer to conceive of some construction, some way in which a genetic manipulation could help. Those of us who are here to learn the latest developments in regulations are eager to see genetically engineered organisms being put to work in the environment.

We've also been dealing with the problem of stability in the fermentation field for many years. When making an antibiotic, for example, in a 50,000-gallon fermenter, there's nothing to help the organism to maintain itself. As Dr. Cohen spoke about the stability of plasmids, if plasmids are unstable, the geneticists can tell us how to get desired characteristics integrated into the chromosomes and how to amplify the expression of those characteristics.

Today Bill St. John mentioned fungi. I think it's the first time I've heard fungus mentioned at this meeting. The antibiotic industry has had great success in using organisms other than unicellular bacteria. Fungi, particularly Actinomyces, have been extremely useful, even though in some cases the pathways are also found in unicellular bacteria. Although there's no biological selection in the fermenter, there's tremendous economic selection for fermentation processes. Bioconversions originally discovered in the acetic acid bacteria, such as the conversion of glucose to gluconic acid, are not done with bacteria, but with fungi, which for some reason provide the least expensive process. I wonder what possibilities might evolve with use of filamentous organisms, along with unicellular organisms, to effect the treatment of waste.

TIEDJE: There is a significant group of engineers and others in the hazardous waste treatment industry who believe that microbiology doesn't offer a major contribution to the solution to hazardous waste treatment. Part of the problem is that, in the past, the microbial processes, particularly for the most difficult compounds, haven't been that good. So, there is a credibility issue. We have, in a way, a second chance. The improvements in biological processes and biodegradative organisms must really work and must lead to demonstrably better processes, so that those nonbelievers can become believers.

At this meeting we've heard a number of recent discoveries that are quite exciting. For example, the G4 organism that degrades TCE, I think, is quite a significant advance,which an engineer or chemical treater probably didn't think would be possible. Ron Crawford's Flavobacterium seems to do quite a good job on pentachlorophenol. Fungal lignase, which hasn't been touched on at this meeting [see Omenn and Hollaender, 1984], will certainly treat some compounds. There are the various reductive dechlorinations of chlorinated solvents and of chlorinated aromatic compounds, including PCBs. All of these are quite new, useful discoveries. Now, we've got to make sure that these discoveries work when they're put into practice.

BULL: Jim Tiedje's description of the pessimism about developing biotreatment processes for hazardous waste makes me think that in the past perhaps we've expected one type of biotreatment process to do everything. One of the things that's emerged from this meeting is that we are developing processes for specific types of problems.

OMENN: In a number of the presentations today, the microorganisms have been used inside various kinds of bioreactors. This is true for wastewater treatment, for the wood treatment facility that Ron Crawford described, for the big site down in Texas that Bill St. John described, and for coal and coal gasification wastewater treatment, and so forth. Do the panelists see this practice as a general strategy? Is this simply a result of making effective and efficient use of the combined microbiology and engineering? Or is this strategy explicitly or fortuitously responsive to the concern about release of organisms into the general environment?

BULL: I'll put my neck on the block and say I think it's not only desirable, but also essential. It's the only way we're going to come to grips with hazardous waste destruction. I'm quite impressed with the great degree of optimism that today's speakers have in undertaking in situ bioreclamation projects. Many are sites that contain only one type or one chemical family of waste, where there's a reasonable expectation that the technology will work. We have not heard, for example, of treating environmental pollutant situations involving PCBs or other chlorinated compounds co-present with heavy metals. I know of only one situation in the United Kingdom where a waste site that contains both phenolic compounds and heavy metals is being treated. This is being done by separating the two. Anything that contains heavy metals is put aside and goes to landfill. We need to develop processes at the source--either at the end of the pipe or preferably somewhere down in one of the unit stages where those particular wastes are emerging. That is particularly true with heavy metals.

McCARTY: I would disagree that we should be looking primarily at bioreactors above ground or those that can be better controlled. When the waste or contaminants can be readily extracted from water, then use of controlled reactors is appropriate. But with underground contamination, it is often difficult to get the contaminant into water to extract it and bring it to the surface. Many contaminants are very tightly held to soil. Unless the soil can be removed, then in situ approaches are needed. And this method does work, particularly for hydrocarbons. Thus, we need both strategies.

As for above-ground bioreactors and the great diversity of contaminants often present in a wastewater, a wastewater treatment system must be developed. One part of such a system may be a biological process, but other process components may be chemical or physical. One has to address a particular problem with all the available tools. Let me give an example of heavy metals. Hexavalent chromium is very toxic to microorganisms. This metal could be removed, however, by first using a complete-mix anaerobic reactor where the chromium would be reduced to the trivalent state. The trivalent chromium is not toxic and also could then be removed by the physical process of settling. It would precipitate out, and then all the rest of the biological process would go on. For other metals, one can change the valence and solubility biologically as part of an overall removal scheme.

DEMAIN: There is a famous quote of Salvadore Luria at MIT, who said (paraphrased), "When you have mutants, you're better off than when you don't." By analogy, if you have a bioreactor, you're better off than when you don't.

TIEDJE: Part of Dr. Omenn's question related to whether people were not releasing organisms because of the current political and regulatory climate. My assessment is that right now there are no well-developed genetically engineered organisms that anyone is waiting to test in the environment. If there were, EPA has a process in place to handle any proposal for field testing, if the organism and process are well-tested and effective in the laboratory.

What might be the environmental risks of releasing an improved degrader? At the recent meeting of the American Chemical Society Agricultural Chemicals Division, potential concern surfaced. Improved biodegrading organisms might lead to enhanced degradation of some valuable pesticides, thereby destroying their effectiveness and commercial value. Enhanced degradation is already a problem for some pesticides. Agriculturists and pesticide-producing companies can legitimately expect some oversight over the potential of a GEM contributing to enhanced degradation of some valuable pesticides.

GLASER: I would like to pick up on comments of Drs. Demain and Tiedje about fungal activities. We are exploring the use of the white rot fungus *Phanerochaere chrysosporium* in at least three laboratories to which we are giving support. Researchers are looking at specific extracellular enzymatic activities expressed by the fungus. We see also a potential application of the fungus to water treatment as a potential RBC, and we intend to explore this with researchers at North Carolina State University who have used a bench-scale RBC.

DIBBLE: There seems to be an overemphasis on spills and catastrophes. Our chemical plants and refineries each process millions of gallons of water per day. Our Department of Solids [Texaco Research] processes tens of thousands of tons of hazardous solid waste in our land treatment units every year. Increasingly, there are stricter requirements on what can come out of a discharge pipe in wastewater treatment. There is also regulatory pressure on our land treatment units. I think there should be a greater emphasis on the daily garbage that is produced rather than on spills and catastrophes. There is a lot of material presently being processed, and we must make improvements in these processes.

BULL: I very much agree with you, but one can't disregard spillages and accidents. I'm impressed by the technology that is being developed to contend with those rarer situations. I would be interested to know some of the costs involved, as well as who is going to pay. What type of accident will be deemed sufficiently severe so that this technology that is being developed will have to be applied?

McCARTY: The normal flow of wastewaters from industry into streams is fairly strongly regulated. One can clean up such waste streams from industry to produce distilled water. If the regulatory requirement came to that, I'm sure that it could be done. But the cost would be very high. The main factor that makes possible some of the technologies that we are considering here today is that the cost of alternatives is often enormous--$1,000 to $10,000 per pound of waste chemical removed, in some cases of dilute groundwater contamination. Such high costs make many things possible. On the other hand, when dealing with your normal wastewater treatments for BOD removal, the cost may be on the order of $1 per pound. Economics, regulations, and desires of society all influence what is and is not possible.

BOUMA: Concerning mixed substrates and metals, I think that the DuPont Company developed a packed process about 15 years ago. This is a process that was touched on, in part, by Dr. Suidan when he was discussing the fluidized bed reactor with carbon [Suidan, this Volume]. It is a process that absorbs metals. The reason that the carbon is lost at some point is that the metals become toxic on the carbon surfaces. The packed process was not an economically useful process because carbon regeneration in multiple-hearth furnaces was not a feasible option. DuPont found that wet carbon in multiple-hearth furnaces would simply blow those furnaces to pieces. So it never became a viable option. Two years ago, another company bought the packed process and developed a high-pressure regeneration process in which a catalyst was used to regenerate the carbon. It removes the metals. So presently in this country there is a process that combines bacterial degradation on carbon surfaces with metal removal.

UNTERMAN: Polychlorinated biphenyls are a vestige for us. We don't handle them like we used to; I don't think any industry does today. But we have current waste problems, and we have a bad reputation. Eight months ago, a nitroaromatic problem in one of our plants was posed to us. Eight months later we returned with an organism that can degrade nitroaromatics. They were somewhat overwhelmed; they never expected us to come back. Indeed, we may have a solution for them. It's a complex waste. It's not metals; but it's higher than 10% nitrite, which is quite toxic. By using a denitrifying culture followed by our aerobic culture, we are addressing their waste problem. So our credibility is enhanced. We are trying today in industry to provide the most economic means of dealing with an on-line problem. I think biology, and not carbon or incineration, is going to be the solution. I find that very promising, and I hope our credibility continues to go up.

MECHALAS: Southern California Edison is a pretty big utility with some pretty big problems. We are in a fairly arid area. We have waste sites with very low dilutions of material. In other words, the material is scattered through the soil in low concentrations but the quantities are huge. I am a microbiologist by training, so I have a lot of faith in the microbes. But an axiom I have is that the environment selects the bug, the bug doesn't select the environment. If you place a pure culture or a bioengineered bug into the environment, it probably will disappear overnight because the monsters--the protozoa and the fungi--will eat it all up. I think we need help from engineers or others to produce an environment at a waste site that will simulate a bioreactor. I think this is a missing link. Before we can ever hope to use some of these tailored microorganisms, we need better ways of getting oxygen down to the bugs; we need better ways to maintain moisture content and humidity. There has to be a better approach than excavation of soil. In some of these sites there might be hundreds of thousands of cubic feet of soil to move. Since there is no feasible way to move it, how do you clean up these sites? We have talked about the water being circulated in the site, but many of these sites are on aquifers where the water is flowing through. We face the problem of maintaining the water on the site as we are cleaning it up while there is some more waiting to pass through it. We need help from the engineers to reproduce the environments necessary to the survival of our bugs.

DEMAIN: If you have bacteria that are being chewed up by protozoa or fungi, then we have to learn how to engineer the protozoa or fungi to do the job. Also, I don't think that one discipline can wait for the other. Going back again to the success of industrial fermentations, the development of engineering and microbiology went on for years, with each helping the other. They have to go hand in hand, and they need people from both sides who can talk with each other.

LITCHFIELD: I totally agree with Dr. Demain. In fact, packed process is alive and well at DuPont because engineers and microbiologists have worked together. We are currently dealing with a four million-gallon per day system that is taking hazardous waste from throughout the United States, including California, Florida, and New Jersey. It is a very successful ongoing process.

TAUB: It would do wonders for the field if there could be some ribbon-cutting ceremonies at a previously highly polluted area that is now open--whether it is for a public site, factory, or home sites. This would point out to people that the area has been changed from a highly toxic site to some place that is now safe. But I wonder if that is ever going to happen. I wonder if people are going to want to admit how bad things were. Will companies want to say how much effluent had been released and how much cleaner it is now? They may do so at a stockholders' meeting, but I wonder if they will hesitate to show the public the differential of how bad it was for fear that there may be legal implications.

BULL: There are parts of the United Kingdom that have actually been reclaimed. South Wales is a good example. It has a history of one hundred years of refining heavy metals, and I imagine the levels of copper, zinc, and lead were horrendous. But these areas have been successfully reclaimed for recreational purposes. I agree with Dr. Litchfield that more than anything else, demonstration--and not necessarily on a large scale--will be important to show the general public that we are doing things and getting results.

PEELER: Being in the regulatory business, I would like to comment on Dr. Taub's question. I think the basic problem is that the public has very little understanding of what the process of cleanup involves. They know that something is being cleaned up; they know that we are working in Bunker Hill, Idaho, or some other area in the United States. But they know very little about how it's being done. Is it being done by picking up dirt, or chemically? Very few people understand that the natural order of things for now is we pick up the work and take it to some other place. They are shocked when they find out. I think we need to do a better job of public information.

THE ROLE OF BIOTECHNOLOGY IN POLLUTION CONTROL AND THE ROLE OF GENETIC ENGINEERING IN SUCH BIOTECHNOLOGY APPLICATIONS

ASSESSMENT, COMMUNICATION, AND MANAGEMENT OF ENVIRONMENTAL RISKS

William D. Ruckelshaus

William D. Ruckelshaus Associates
Seattle, Washington 98101

INTRODUCTION

I am delighted to be able to address this conference, drawing upon my experiences at the U.S. Environmental Protection Agency (EPA). Some have called it the "finest environmental protection agency in the world." A theme I have heard repeatedly from environmental leaders in other countries is that what the EPA does in this country largely dictates what happens in many other parts of the world. It strikes me as ironic that an agency with so much prestige abroad has trouble here at home convincing our own people of the wisdom of its actions. Of course, its image has changed over time. I remember speaking to the Economic Club in Detroit in 1970 to some 1,100 businessmen; I was introduced by the Executive Vice President of General Motors as "the greatest friend of American industry since Karl Marx." I know that many of you here in your hearts probably lust after the job of Administrator of the Environmental Protection Agency. If it's ever offered to you, I have one piece of advice: don't take it unless it's offered by a Republican president! As I have now discovered for the second time, if you're the EPA Administrator in a Republican administration and you leave before you're indicted, everybody tells you what a good job you did, proving you can succeed in any job if only you set your goals low enough!

This conference is focused on a specific use of biotechnology, namely, the reduction of environmental contamination through the use of biotechnology. As all of you--many of whom are experts in the field--know, besides technical questions about the application of bugs that eat toxic waste, there are many broader issues relating to the application of biotechnology. Ethical questions about the genetic engineering of human cells, questions about the use of drugs, and environmental questions about the use of genetically engineered plants or the release of genetically engineered microorganisms are certainly serious in the minds of many people, and will be politically debated into the foreseeable future. The application of biotechnology to contaminated Superfund sites also will raise questions. This particularly would be true, I expect, for genetically engineered microorganisms, which might, ironically, be more effective at decontamination than some of the naturally occurring organisms.

I will try to set a framework for some of the discussions that will follow. Representatives of the Administration--David Kingsbury from the White House Office of Science and Technology Policy see Kingsbury, this Volume and Marcia Williams, who has been managing the hazardous waste programs at EPA for the last couple of years see Williams, this Volume--will undoubtedly get more into the details.

THE ENVIRONMENTAL DECADE OF THE 1970s

First, to grapple with the public issues associated with the application of biotechnology, we have to think about where all of this environmental concern came from, and how the topics gaining attention have shifted from those in the explosion of environmental activism 20 years ago. Laws, rules and regulations in our society come from public demand for both national and state actions. In the late 1960s and early 1970s we were dealing with gross kinds of pollution problems--the problems you could smell, touch, and feel. The Cuyahoga River that goes through Cleveland burst into flames when somebody threw a match in it. Lake Erie was declared "dead" back in the 1960s. Air pollution problems were quite visible, rendering other things invisible. There was a desire of people living in Denver to see the mountains again, of people in Los Angeles to see one another. Public demand that something be done about these gross pollution problems led to the Clean Air Act of 1970, the Clean Water Act of 1972, and a host of laws involving solid waste, pesticides, radiation, and noise.

Toward the middle of the last decade, attention began to shift toward less visible problems--problems that really were more scary to the public, relating to dread diseases like cancer. Starting in 1976, there was the Toxic Substances Control Act to achieve premanufacture control of new chemicals, and the Resource Conservation and Recovery Act, which dealt with the existing regulation of hazardous waste facilities. Toward the end of the decade, we discovered Love Canal and many other abandoned hazardous waste sites, which led to the passage in 1980 of Superfund.

SUPERFUND

Superfund, of course, symbolizes the neglect of our society in wisely managing wastes that our industrial processes and other processes were generating, as well as our effort now to deal with these wastes. In the passage of Superfund we transformed what had been essentially a local contamination problem, usually dealt with by the local public health official, who had to balance the cost of alleviating that public health problem against the benefits to the local community. You had that almost constant tension between alleviating a potential health hazard and what it was going to cost the local community. What we did with Superfund was to elevate that problem to the national level. We bypassed the state governments and gave the federal government the responsibility, delegating back to the states a portion of the responsibility, consistent with the modus operandi of the Clean Air Act, the Clean Water Act, and other statutes.

In addition, in the early years of this Administration we were not wise in administering that law. Rita Lavelle, who was put in charge of the Superfund law, wasn't by background or otherwise suited to manage that new responsibility. However, even if we had put Peter Drucker in charge of the Superfund law in 1981, he would have had difficulties taking what had essentially been a local issue, creating out of whole cloth a new federal program, and making it work under the terms that were passed by the Congress. So we had a complicated program in its very early stages, and then the public trust was destroyed. Since then a mix of politics, science, and recrimination has produced a newly reauthorized Superfund law. During the course of that reauthorization fight, there was a loss of momentum. We're still struggling, both with defining what it is we want to do and attempting to implement an ever-changing set of requirements.

We still have to resolve politically such issues as how clean is clean? To what extent should we clean up the Superfund sites? Regarding zero risk, to what extent should we eliminate risk associated with exposure to these sites? Who is going to pay for it and should that payment be allocated? To what extent should the citizen be invited to participate in all of this? We've not finally decided any of these questions. In some respects, reauthorization has complicated further these questions and created other process questions, like what level of government has responsibility for taking the lead. Since we've given citizens additional rights to participate in the form of citizen-right-to-sue, how do we implement all of these provisions?

We have continued a process which, in my view, is inherently flawed. We are going down two tracks to try to cleanup these sites. On the one hand, we raised \$1.6 billion and now will raise \$8.5 billion through a tax on those whose waste is found in the sites, the so-called generators or potentially responsible parties. We've taxed those people, in effect, taxing ourselves as consumers. At the same time we have an enforcement track, through which we are trying to recover funds from the people whose waste is still in the sites. In most instances, the people who actually managed the sites are long gone; now it is only those whose waste is left there that can be held responsible for paying for the cleanup. This dual approach creates incredible complications in terms of trying to administer the laws, both at the federal level and the state level.

Shortly after returning to the EPA, I felt that we should seriously examine whether we need both of these tracks to get at this problem. One thing that makes the Superfund and these abandoned hazardous waste sites at least possible to deal with is that everybody agrees we ought to clean them up, whatever cleanup means. Everybody agrees that we ought to decontaminate these areas, eliminate the environmental and public health risks, and move on. The real questions,again, are: How do you do it? Who pays? How should we use federal or state enforcement responsibilities to achieve an ultimate cleanup?

There are other issues involved, however. Certainly, this conference is addressing a shift in public policy. We originally sought to dig all of this contaminated soil up and move it someplace else. Well, we found there is no place else that wants it. Nobody is clamoring for the next designation as a Superfund site. Therefore, it is very difficult (short of putting it on barges and floating it around the world) to find any place to put it. So we began to focus on triyng to clean it up in situ, where it sits. There are a lot of ways of doing that. You can use chemical treatment; you can excavate and can put liners underneath and on top; or you can biologically treat soil, using organisms that will destroy these toxic substances. Many technologies are emerging, presently at very exciting periods in terms of their development, like the technology you are addressing at this conference. If we give ourselves enough time, we should have some tremendous opportunities for a more cost-effective and really effective way of dealing with this problem.

However, there are tremendous political pressures to clean the sites up immediately. These issues haven't been solved under the reauthorized Superfund. Congress and the public have given this complicated problem to the federal Environmental Protection Agency and said, "Get on with the job. Clean it up. Do it under the precise terms that we have set out for you under statute. If you don't accomplish our goals of the elimination of several of these sites and the risks associated with them within a reasonable time, we'll have hearings every thirty days to find out how you're doing and let you know what we think about your failure to achieve our lofty goals."

PUBLIC PARTICIPATION

Let's take one issue, an important aspect of all this, the issue of public participation. On the 200th anniversary of our Constitution, it's an interesting time to think about this matter. If you go back about twenty years, when I believe all of this started, essentially what happened was that the American people, who had granted the power to govern, took some of that power back. The first three words of the Preamble of our Constitution says, "We the people . . ." The source of power in this society is very clearly the public. The government simply derives its power to act from the consent of the governed. And when some of that consent is taken back, government is inhibited from carrying out its delegated power. What we've seen, at first in foreign policies, starting with the Vietnam War, has been a take-back of some of those powers, a demand on the part of the public to become more involved in the decision-making and implementation processes. In foreign policy, we've seen a forced sharing of the conduct of foreign policy with the Congress, reflected currently in the Iran/Contra hearing.

On the domestic side, exactly the same thing has happened, starting with Watergate and then exacerbated, at least in the area of the societal regulation of health, safety and the environment, by what happened in the early years of this Administration. Again, an erosion of trust. The public said, "I don't trust you birds to continue to do what you're supposed to do. I want to share with you the power, the decision-making responsibilities, and the implementation of those decisions in a much more direct way than I have in the past." Access to the decision-makers is now guaranteed through administrative process. The public has an opportunity to participate at several steps along the way, the right to sue, to go to court to challenge what the Executive Branch is doing, to take relevant information out of the Executive Branch under the Freedom of Information Act, and, most recently, the proliferation of Right-to-Know laws in many states and the reauthorization of environmental laws at the federal level. The public can obtain more information directly from the generators of waste, for example, so that they can determine what, if anything, they want to do in asserting their rights to participate.

Dr. Bull will describe the British and European approaches. If there's one big difference that I've noticed between the way we manage risk in this country and the way in which it is managed in most developed societies (and even more so in the developing societies), it is public participation--the public demand to be involved. Many of my international counterparts, while I was at the EPA, would say, "What in heaven's name are you doing, letting the public into these issues? Don't you realize they don't understand them, that they act very emotionally? You are going to get yourself and ultimately us in real trouble by letting the public into this." Well, maybe they're right, and maybe they're not. I don't believe they're right. In any case, that argument is over in this country. It isn't a question of whether the public is going to get in, it's how. How do we make their participation more meaningful? How do we improve the decision-making process and include the public in a meaningful way in its development? Everybody recently has been citing that sign on Harry Truman's desk, "The buck stops here." The problem in this area, of course, where we're feeling our way as to how to deal with public participation, is that the buck doesn't stop anywhere. It just bounces back and forth from one branch of government to the other, one level of government to the other. We have more hearings, and we sue one another a little more, but we never decide anything. Ultimately, we've got to develop a process that makes the decision somewhere. We haven't figured out how to do that yet.

There are several definable elements of successful public participation. First, we have to have access to the decision-maker, which can be guaranteed by statute. Next, we must have sufficient information flowing from the decision-maker to the public, so that they can respond in a meaningful way, so that they can use the access guaranteed by the Congress and by state legislatures in ways that will increase the likelihood of wise decisions. Most important, we need to help or force the participating public to be responsive to the true societal issue. Addressing just half the equation is not useful. We have asked people whose property values are destroyed as a result of proximity to the Superfund sites, "What level of risk are you willing to accept?" Their answer is quite simple, "None"--the same as yours would be. In contrast, the societal question is, "What level of risk reduction do we want to undertake, at what cost?" Unless the public has to deal with the same issue that the society has to deal with, that the ultimate decision-maker has to deal with, we have a flawed public participation process. Such a process is doomed to failure, and the public is more likely than not to be frustrated because they can't get the right kind of response from their participation.

Where we have tried to involve the public in a substantial way, given them access, given them adequate information to make the access meaningful, and given them the whole equation, "What do you want to do in the form of risk reduction at what cost, part of which will be shared by you, making the demands?" we've gotten very sensible kinds of responses. I frankly believe that the public is quite sensible about hazardous waste issues, much more sensible often than those who speak for them as to what should be done. It isn't true in every case, of course; sometimes the issues get too emotional.

COMMUNITY-LEVEL PUBLIC PARTICIPATION

There's an element of responsibility involved in all this participation. The public itself has to accept responsibility for the decisions it has made, an ultimate obligation of citizenship. We cannot confuse the right to be heard with the right to be heeded. They are not the same. When the public gets into all of the elements of the decision-making process, they, in the last analysis, have to accept the decision. We'll never reach that level of abstraction and acceptance of citizenship unless we have processes that are responsive to the real needs.

We have other kinds of problems that we need to deal with in terms of public participation. For example, I am involved with a group here in the Northwest trying to site a hazardous waste facility. By choice, we've started without a site. We've gone around the state saying that we're searching for a hazardous waste facility, and we don't have a site. Once we go through the site review process and use the state's criteria to eliminate perhaps 95 percent of the state, we may have about 5 percent of the land area left. We've invited communities to talk with us. We say, "If you don't want the site in your community, it won't be there. If you are interested, we will share the benefits with you." We've given them a choice. Of course, we may end up with no communities desiring to be so honored with this site! That will remain to be seen. In fact, we've gotten much favorable reaction by the approach that we've taken.

RISK COMMUNICATION

A lot of people think, "If only we could get the public to see these problems the way we do, then their fears would be eliminated, and we'd be okay. Therefore, we need to educate the public." What is really needed is to listen to the public. What is it they think about these problems? How do they view them? They don't necessarily view them in the same way that scientists do. People talking about 10 to the minus 6 often elicit nothing but glazed eyes from the public. We have enough survey data now that we can assess rather clearly what the public thinks about these problems, especially problems with unknown risk, high dread, high uncertainty. Whether it's nuclear power or toxic substances that cause cancer in animals or man, the public demands very high levels of protection. And the political system will respond to that demand. If it is seen as responding responsibly and in ways that ought to be trusted, then some of the public hysteria will diminish. But the public will never respond precisely as the risk assessment experts would rank comparative risks. Let me give you just one example of a well-known survey.

An Oregon group recently took 30 risks that people are faced with on an almost daily basis, ranging across automobile driving, smoking, drunks (running over them), getting mugged on the street, slipping in the bathtub, exposure to pesticides, and nuclear power. They asked three groups of people to rank those risks on a scale of 1 to 30, 1 being the most risky; they asked risk assessment experts, League of Women Voters members, and recent college graduates. The risk assessment experts ranked nuclear power 24th out of 30 (this was before Chernobyl); the League of Women Voters members and the recent college graduates ranked nuclear power number 1. Nothing was even close. It's the high dread, high uncertainty, and cataclysmic nature of the risk as these groups perceive it, and the involuntary nature of this risk, as opposed to the smoking or other high-risk voluntary activities that rank considerably lower on the uncertainty-dread scale. That's a political reality. If, in the development of biotechnology, as an example, we don't learn from the mistakes made in the development of nuclear power, if we don't let the public all the way in and discuss with them in very frank, open terms not just the benefits associated with its use but also the risks, the unknown nature of some of these risks--and, nevertheless, how we think, in fact, they can be reasonably controlled--we're headed for real trouble. If all of these risks or precautions are discovered through nasty little surprises on the part of the public, they will conclude that the experts have not completely leveled with them as to the nature of these problems. The history of the development of nuclear power in this country is all too good an example.

I'll never forget the first time I testified before the Joint Committee on Atomic Energy, when I was first at EPA. I went through six doors just to get into the hearing room. It was easier to get into the Situation Room in the basement of the White House than into the open testimony of this Congressional committee. I had been given the responsibility then, as the newly installed Administrator of the EPA, to regulate radiation outside nuclear power plants. I had no idea what that meant, and neither did anybody else. The chairman of the Joint Committee on Atomic Energy and all of the other members berated me for four hours, asking me, "What in heaven's name do you know about the regulation of radiation?" I confessed I knew nothing; however, I said, "I didn't ask for the authority to regulate radiation. You gave it to me--you, the Congress. Therefore, I will do the best I can to understand it and try to responsibly regulate it." This experience seemed to me to illustrate the problem in the way in which we developed nuclear power. It was so secret. The problems were downplayed and the benefits up-played. By the time the problems bubbled up, the public got the impression the problems had been hidden from them. They, therefore, conjured up levels of risk several orders of magnitude greater than the risk assessment experts even today would give them. In this country, at least for the near term, nuclear power seems doomed as a source of additional generation of electricity. The same could happen in biotechnology if we repeat those mistakes.

A GLOBAL PERSPECTIVE

In the big picture, we might ask, "Well, what difference does it make? So we don't handle all of these problems just right; democracy doesn't work very well, government doesn't work; who cares? It's been ever thus." People don't trust government; they never have. And when they look at it in retrospect, they find their lack of trust is often warranted.

Well, one of my most recent incarnations has been as a member of an organization called the World Commission on Environment and Development, created by the United Nations. We were given the charge of looking at the issues of pollution, population, resource depletion, energy, and food supply globally over the next 25 years. Not overwhelmed by that modest agenda, we also picked up the issues of Third World debt and nuclear disarmament! It was a very interesting experience for me. The chairman was a prime minister of Norway, a woman named Gru Brutlin; 14 Third World nations were represented on the 22-member commission. We all served in an individual capacity, not as representatives of our countries. I was the only American. We spent two weeks every quarter in some region of the world. We went to Jakarta, Indonesia, and Borneo; to South America, and into the Amazon; to Zimbabwe and southern Africa; to India; and in the Northern Hemisphere, to Ottawa, Geneva, Tokyo, Oslo. We even spent two weeks in Moscow in December of last year--showing how little we know about the environment! [Laughter] We had public hearings in all of these places.

I mention this commission to underscore the importance of your work in developing new technologies for dealing with the unwanted side effects of industrialization. We must show that in a free society we can, in fact, cope with our own complexities. If we look at the issues of pollution in this country and compare them to those in developing countries or even in other developed countries, frankly, ours pale into insignificance. Some of the problems associated with pollution in the rest of the world really are mind-boggling.

The commission's title was not accidental. It was called the World Commission on Environment and Development. Our report this spring emphasized the connection between economic development and the environment in the developing world. There is simply no way that the massive environmental problems associated with the developing world can be addressed without economic growth, without development. Such development must be equitable, sharing broadly the proceeds of that development with the people in those societies, and it must be sustainable, environmentally sustainable as well as economically sustainable. There is tremendous understanding of that connection in the developing world, more so than in our own country.

We know how to develop economically in this country and protect the environment. We know how to clean up these abandoned hazardous waste sites and other contaminated areas to adequate levels of protection. We know how to develop new technologies to do it more efficiently. We can act sensibly if we lower our voices a little bit, if we begin to bring into the public dialogue people like yourselves who have studied these problems over your lifetime and understand them, and who can bring to the public debate a good deal of perspective and wisdom. It is terribly important that the scientists get into the public debate. Now is not the time to stand on the shore and jeer as the ship of state floats by. Now is the time to get in the boat and row. The debate has been left too often to those who represent extremes of points of view. It is terribly important that knowledgeable, sensible people get into the public debate and offer knowledgeable, sensible solutions to these problems. We can grapple with these problems and solve them. By example, we can lead the rest of the world. Some very fragile adherences to free institutions that are taking root in the developing countries of the world can be fostered to flourish if we in our system are capable of successfully dealing with these problems.

What you're about is important, and it is important that all of you raise your voices and be heard, and listen, too, so that your knowledge can be transferred into the formation of public policy.

BALANCING REGULATORY CONTROL, SCIENTIFIC KNOWLEDGE, AND PUBLIC UNDERSTANDING

David T. Kingsbury

Directorate for Biological, Behavioral, and Social Sciences
National Science Foundation
Washington, D.C. 20550

INTRODUCTION

The commercial application of the modern biotechnologies or "New Biotechnology" (recombinant DNA, monoclonal antibodies, and advanced microchemical analysis and synthesis) has been in progress for slightly more than a decade. The initial products that resulted were monoclonal antibodies for disease diagnosis and new drugs resulting from the large-scale fermentation of microorganisms that had been genetically modified. In each case, the living cells were not the products themselves, but were simply the means by which the end product was produced. Production was done in carefully controlled, contained facilities, and product regulation was almost independent of the mode of manufacture.

As the "biotechnology industry" continued to grow and expand into nonmedical applications, it became clear that many of the second-generation products were going to be the living organisms themselves and that most would be applied in the environment. This recognition led to a wave of concern regarding the environmental consequences of these products. Those expressing the most concern have focused on the possible negative environmental impact of the new products and have generally disregarded the potential positive environmental impact (for example, the replacement of chemical fertilizers and pesticides with equally effective microbial products). The government has a heavy stake in this debate. Modern biotechnology is a direct result of many years of government investment in biological and biomedical research, including molecular biology and microbial genetics. The promise of biotechnology is starting to be realized. However, we must be vigilant to ensure that our past investment continues to pay dividends.

One fact is clear, however. A continued commitment to basic research--in biological processes, such as gene regulation, genetics, plant physiology, and cell biology, and in the environmental sciences, such as ecology, population biology, and systematics--will be necessary for many years to come. We are entering an era when the pharmaceutical and chemical industries are calling for a broader research agenda in generic scale-up production technology to assist them in more quickly and economically producing their products. Our continued commitment to funding research in the environmental sciences is equally important in order to continue to build the broad database necessary to ensure the intelligent and scientifically based regulatory decisions needed to enhance progress in the future. Likewise, much of our support for systematic biology benefits work in biological diversity and germplasm preservation, issues which are critical to the future of a well-balanced plant biotechnology program and which, in many ways, are important to the penetration of foreign markets.

THE ROLE OF SOUND REGULATORY POLICY

In the past three years, various government agencies have focused on two principal areas related to biotechnology: (i) fundraising to provide adequate support for continued development, and (ii) establishment of a regulatory policy within which the appropriate commercialization of new biological products is possible. On the research side of the equation, we are all accustomed to the shortage of funds for research. As a result, we have focused heavily on the provision of resources, especially in the areas of environmental biology and the plant sciences. On the regulatory side, a Working Group of the White House Domestic Policy Council has labored for almost three years on the generation of an appropriate regulatory policy.

The process of development of the Administration's policy began in April of 1984 with the formation of the Working Group. In a December 1984 Federal Register Notice, the Working Group published a proposed coordinated framework (1). Each agency attempted to define what it saw as the regulatory challenges of this technology and how the agency might respond. The Environmental Protection Agency's (EPA) most critical element was the definition of a "new" organism to be regulated under the Toxic Substances Control Act (TSCA). The proposed definition was "process" based. A microorganism would be considered "new" if significant human intervention had been used to develop it. Microorganisms developed by recombinant DNA and cell fusion techniques were presumed to be new because they involved significant human intervention. Under these proposed definitions, all organisms that were not "naturally occurring" or whose production required more than "limited human intervention" (such as isolation from nature) would be considered "new" and subject to a Premanufacturing Notice (PMN).

The December 1984 Notice also included a statement from the United States Department of Agriculture (USDA) regarding what constituted a "regulated article" under USDA statutes. In addition, the Working Group proposed a strong central "Biotechnology Science Board" to oversee (and second-guess) all of the government's regulatory activity.

ESTABLISHMENT OF THE BIOTECHNOLOGY SCIENCE COORDINATING COMMITTEE

In response to the December 1984 Notice, the Working Group received hundreds of comments about all aspects of the policy. While each agency was responsible for dealing with its own policy statement, the Working Group served as a central policy-coordinating forum. The Working Group eventually voted to establish the Biotechnology Science Coordinating Committee (BSCC). The BSCC consists of the Commissioner of the Food and Drug Administration (FDA), the Director of the National Institutes of Health (NIH), the Assistant Secretary of Agriculture for Marketing and Inspection Services, the Assistant Secretary of Agriculture for Science and Education, the Assistant Administrator of the EPA for Pesticides and Toxic Substances, the Assistant Administrator of the EPA for Research and Development, and the Assistant Director of the National Science Foundation (NSF) for Biological, Behavioral, and Social Sciences.

The Committee operates as a scientific coordinating committee between the agencies within the context of its charter and a Memorandum of Understanding signed by each of the members. Two factors about the BSCC are important to recognize. Its first responsibility is to focus on scientific, not policy, questions; its second, to act as the vehicle to coordinate agency interaction and to help establish a rational basis for assigning lead agency responsibility in those areas where possible overlap of jurisdiction exists.

THE FINAL COORDINATED FRAMEWORK FOR THE REGULATION OF BIOTECHNOLOGY

On June 26, 1986, the final part of the "Coordinated Framework for the Regulation of Biotechnology" appeared in the Federal Register (2). At the time of publication of this Notice, we felt that we had finally achieved the appropriate regulatory balance necessary to deal with a new and rapidly developing industry. Furthermore, we hoped that the principles established in this policy would be suitable as a model in other countries.

The Coordinated Framework is a broad and complex policy that explains the application of existing statutes to the regulation of biotechnology and outlines the approach to interagency coordination that is so vital to this field.

The policy consists of six elements: the preamble and statements of policy from the FDA, the EPA, the USDA, the NIH, and the Occupational Safety and Health Administration. Proposed USDA rules governing "Introduction of Organisms and Products Produced Through Genetic Engineering Which Are Plant Pests" and an "Advance Notice of Proposed Guidelines for Biotechnology Research," also from the USDA, were published at the same time.

The Federal Register policy statement shows how the policies of the separate agencies form a coordinated framework; yet, each of the agency statements stands alone and each describes the agency's separate and respective policy. The preamble is essentially an executive summary, providing an overview of what constitutes biotechnology, the agencies' separate policies, and how these agencies work together. Biotechnology is not a unitary entity, but an enabling technology; it has broad applications in many diverse aspects of industry and commerce. The sharing of different agency viewpoints and approaches is essential in this new field, where generic review standards are not well established and each product must be reviewed on a case-by-case basis.

The preamble describes efforts of the BSCC to coordinate the scientific aspects of agency activities in biotechnology. One approach is to coordinate, as appropriate, the definitions used by each of the agencies of the items to be regulated. The BSCC section mentions concerns regarding "release" and requests comments on ways to define it. The preamble also mentions international policy and expresses the principles of the Organisation for Economic Cooperation and Development (OECD) relating to recombinant DNA safety and regulation.

Government regulation is not organized around technological processes, but rather around products developed by various technologies for specific intended purposes. Therefore, one critical element in a coordinated regulatory framework is a common definition of the nature of the products subject to particular types of regulatory oversight.

The BSCC, in its role as a science coordinating body, recognized that the EPA's definitions of "intergeneric organism" and "pathogen" would take effect immediately upon publication of the 1986 Federal Register Notice, as it would be the EPA's final notice. This immediate effect was important to ensure that "new" microorganisms would become subject to review on the date of the notice. The Committee worked with EPA scientists and their advisers, as well as with the other agencies, to ensure that the definition of a regulated article would be as consistent as allowed by statute. The BSCC believed that it would be useful to specifically solicit the views of the public, and thus asked for comments. After reviewing the comments received, the BSCC is considering what further steps, if any, should be taken to refine the BSCC definitions. It will be up to each agency to make any changes in its own definitions. The BSCC will continue to work to promote consistency and coordination.

The principal focus of the policy is the environmental release of new organisms. There has been general acceptance of the regulation of nonliving products of biotechnology; and, in fact, the new policy reiterates that the past regulatory practices will be maintained. The new policy explains the application of certain statutes to genetically modified organisms and, in some cases, imposes an abbreviated review over unmodified organisms when applied to environmental uses. A clear policy is established requiring review of genetically engineered microorganisms prior to release into the environment, with some organisms subject to an abbreviated review. This is outlined in Tab. 1, which is abstracted from the Summary Table in the EPA policy section (Ref. 2, p. 23319). For example, all organisms classified as pathogens will be subject to the notification requirements, with only an abbreviated review necessary for small-scale trials (less than ten acres) and a full-scale review required for larger applications. Nonpathogens applied to the environment will receive an abbreviated review under both small- and large-scale conditions.

This policy is now in effect; however, additional rulemaking is needed for full implementation. The first publication of these rules is expected in November 1987, followed by a 90- to 120-day comment period before any possible final publication. Until the rules are final, the EPA expects that persons introducing such organisms into the environment for nonagricultural uses will report to the EPA voluntarily. In the unlikely event of a problem arising in this period of time, the EPA could use its authority under Section 7 of TSCA to immediately limit or prohibit the manufacture, processing, distribution, or use of the product. In addition to the EPA

Tab. 1. Prior notification and review of microorganisms applied in the environment.

Type of microbial product	Coverage by notification and review policy* FIFRA		TSCA	
	<10 acres	>10 acres	<10 acres	>10 acres
1. Genetically engineered microorganisms				
a. Formed by deliberate combinations of genetic material from dissimilar source organisms (intergeneric combinations)	X	X	X	X
b. Formed by genetic engineering other than intergeneric combinations:				
i. Pathogenic source organisms†	X	X	X	X
ii. Nonpathogenic source organisms	0	X	0	0
2. Nonengineered microorganisms				
a. Nonindigenous pathogens†	X	X	0	X
b. Nonindigenous nonpathogens	0	X	0	0
c. Indigenous pathogens†		X	0	X
d. Indigenous nonpathogens		X	0	0

*"X" designates that the microorganism will be subject to EPA review prior to small-scale (ten acres or less) or large-scale (greater than ten acres) environmental applications, as indicated. Under TSCA, submitters would only notify the Agency once (at the first appropriate time), unless EPA specifies during the original review that further reporting is required.
"0" designates that the microorganism will be subject to abbreviated review, prior to small-scale (ten acres or less) or large-scale (greater than ten acres) environmental applications, as indicated. Under FIFRA, this provision is effective immediately. Under TSCA, the abbreviated notification will be implemented through rulemaking.
†Pathogens in this category used solely for nonpesticidal agricultural purposes will not be subject to EPA notification requirements. They will be subject only to USDA review. See Unit IV for a definition of "agricultural uses" and "pathogens."

Source: Federal Register 51:23319, June 26, 1986.

activity, the USDA will review all genetically engineered plant pests and animal pathogens.

FUTURE IMPLEMENTATION OF POLICY

The future implementation of the current policy is of equal significance, as is the continued coordination of agency activity. While each agency has a statutory responsibility to carry out its function, the role of the BSCC is also to help in the coordination and development of full implementation. The BSCC has planned a full and active course for the future. The Committee will carry out the majority of its work through a subcommittee structure designed to provide a broad multiagency and, in some cases, nongovernmental representation. At present, the BSCC has established the following subcommittees: (i) risk assessment; (ii) greenhouse containment standards and definitions of "containment" and "release"; (iii) education and training; (iv) research and research needs; (v) public information and education; and (vi) international coordination. The current subcommittee structure is by no means final.

RESEARCH AND RESEARCH NEEDS

Three of the six subcommittees have been especially active. The Subcommittee on Research and Research Needs has compiled agency statistics regarding funding levels and areas of research support. The goal was to identify any obvious omissions or shortfalls in the federal funding picture. While all areas of biological and biomedical research are yielding unprecedented results, the BSCC feels that our commitment to an adequate level of research funding of the environmental sciences is particularly important. Such funding is needed to continue building the broad database necessary to ensure the intelligent and scientifically based regulatory decisions needed to enhance progress in the future.

GREENHOUSE STANDARDS AND DEFINITIONS

The subcommittee reviewing the definitions of "containment" and "release" has been working diligently. The results of that effort will appear in several forms. The final rule-making from the Animal and Plant Health Inspection Service (APHIS) was published in June 1987, and an addendum is expected in July or August 1987. A combined publication from the USDA and the NIH is expected in August 1987. Each of these publications will reflect the work of this subcommittee.

INTERNATIONAL COORDINATION

The Subcommittee on International Coordination has been working closely with the Department of State, which chairs an international subgroup of the Domestic Policy Council Biotechnology Working Group. This international group has been the focal point of coordination with the European Economic Community (EEC) and the OECD.

In addition, planning for another part of the international effort has begun. We have proposed the establishment of an international database related to the introduction of new organisms into the environment. The initial planning has been a joint activity of the U.S. and representatives of the Commission and member states of the European Community. It is our intention to extend the activity outside of this group to include, at least, the member countries of the OECD. During the initial meeting, we developed a working model that consisted of a database of seven interlinked database files: taxonomy, literature, organisms, release events, guidelines, a directory of related information sources, and an "electronic bulletin board" through which investigators and regulators could share ideas and data.

Within the taxonomy file, we propose to provide the latest in taxonomic data for all species that are included in some form in any of the other database files. In the literature file, we propose a bibliographic listing containing all of the references cited in any part of the database. This file will be searchable by conventional means, such as author, title, journal, and any textword present in the abstract. In the organism file, we propose extensive synopses of known data, extracted from the literature by a technical staff and reviewed and scored for quality by an expert panel. Where appropriate, synopses are to be referenced with pointers to the literature file. A wide array of data about the biology, genetic description, uses, and known genetic modifications will be included, as well as comments about the availability and source of the organism.

The release event file will contain synopses of results of previously conducted prerelease and release experiments extracted from the literature and from nonpublished technical reports. This file will include the known reports on the introductions of new organisms that have occurred over the past 100 years.

The guideline file is intended to assist in the design of future experiments, providing information on relevant procedures, technical considerations, and regulatory guidelines in effect in various countries or locales. The directory of related sources will reference such sources as GENBANK and the Protein Sequence Data Bank.

In discussing this concept, we agreed that the volume of available data that has resulted from the past century of research on microbial ecology, soil microbiology and chemistry, and plant and animal breeding is enormous. This volume has, in itself, been a barrier to obtaining comprehensive reviews, thereby giving the impression that data do not exist. We hope that compiling this large body of existing data into one central collection will address the problem of scientific access and simultaneously will assist both those responsible for funding basic research and those responsible for research in support of regulation to focus on areas of greatest need.

A survey of the world's literature on the introduction of new plant varieties (such as specific corn hybrids, strawberry varieties, or varieties of tomatoes) and the use of microorganisms in the environment (for example, <u>Rhizobium</u> sp. for nitrogen enrichment or <u>Bacillus thuringensis</u> as a pesticide) immediately identifies the need for broad international participation in this data sharing. Many countries of the world have had a much broader experience with the agricultural applications of microorganisms than we have had in the United States, and those experiences will aid us greatly as we try to examine future research directions. Likewise, such long-term ecological studies as the Long Term Ecological Research (LTER) sites (3) supported by the NSF will be critical to the provision of baseline information upon which we can build interpretations of the possible environmental perturbations associated with the large-scale release of a new organism into the environment. We must remember that understanding ecological balance is based on longitudinal observation and cannot be represented by cross-sectional, "snap-shot" data collection.

INFORMATION REQUIREMENTS OF THE PUBLIC

In addition to the information needs of the scientific and regulatory community, we must pay attention to the information requirements of the public we serve. Public apprehension of some of the products of biotechnology has surfaced as a problem in many countries of the world. This public apprehension has been responsible for considerable delays in the field testing of new products, despite the fact that these small-scale tests have been approved by the regulatory agencies. We must do something to assuage the fears of that small number of persons who feel that this technology is a threat to the environment.

One critical element is public confidence that we are not trading one environmental problem for another. This is why the U.S. has worked diligently to put in place a sound regulatory policy and is constantly reviewing that policy in the context of new scientific findings. The challenge is to be always on the boundary between the over-regulation of new products, with the accompanying repression of innovation, and under-regulation, with its potential for environmental harm, without stepping significantly onto either side of that boundary.

A broad range of research activities related to assessing the environmental impacts of new products is essential. In the U.S., we attempt to achieve that broad range principally through the research programs of two government agencies. The combined programs of the NSF and the EPA cover the full spectrum from theoretical ecology through specific ecosystem models for laboratory assessment of environmental effects. Broadly speaking, the NSF programs focus on the very basic side of this research spectrum, with little direct attention paid to questions related to regulation, while the EPA programs focus on many aspects of regulatory needs, in addition to some fundamental work. In the middle area, there are NSF grantees working together with EPA scientists on common approaches. In addition to this work, the USDA supports a significant amount of work addressing the environmental effects of new crop plants and new crop varieties.

THE ROLE OF THE GOVERNMENT IN THE IMPLEMENTATION OF THE NEW BIOTECHNOLOGY

Government is clearly playing a dual role in the implementation of the new biology. On the one hand, government agencies are imposing restrictive conditions on research and the introduction of new products and ideas; on the other hand, these agencies are responsible for much of the underlying research and have taken the responsibility for generating the data necessary to cope with the regulations. This is not a new role for the government, especially in health-related issues; it is a long-awaited and much-needed move in the environmental and agricultural sciences.

Government agencies and their employees have an additional responsibility, one that can easily be satisfied. A recent study commissioned by the Congressional Office of Technology Assessment (OTA) (4) examined public attitudes toward biotechnology and its products. The results were both encouraging and sobering. First, the public generally felt that biotechnology products were safe and beneficial. This positive attitude should be reinforced whenever possible. On the other side, the polls showed that the public did not trust the government for information about biotechnology. They did trust university professors as a source of information. The challenge for government researchers and regulators is to reaffirm their competence and judgment, while allowing the emerging biotechnology industry to progress quickly in a safe and rational fashion. This industry has not had a safety problem in its more than ten-year history--a point worth noting and reiterating often.

While it is not possible to predict precisely what impact biotechnology will have on the future of the world economy, most observers feel it will be substantial. Recent predictions state that the current situation of relatively few U.S. sales will change to several billion dollars in the 1990s and as much as 40 billion dollars by the year 2000. The product mix by that time is expected to be pharmaceuticals, agricultural chemicals and pesticides, growth-promoting hormones for animals and plants, and a line of very high value-added industrial chemicals. Many observers predict that special microorganisms engineered to degrade pollutants and chemicals, as well as modern biological waste-treatment processes, are also highly probable products for the near future.

Clearly, many investigators at present are using enrichment culture techniques to isolate microbes capable of degrading many petroleum-based pollutants. These organisms will form the starting place for the development of newly "custom designed" organisms that will become the next generation of these products. The development of this technology has a long way to go. Likewise, the acceptance and understanding of the technology by the public and the regulators also needs to be developed.

To the concern of many observers, it appears that two nonscientific issues may play a dominant role in the future of biotechnology development and implementation in the U.S. First, if not based on sound scientific principles, the regulatory climate could present U.S. industry with an insurmountable problem for the eventual introduction of products now being developed, leading to a future withdrawal of activities in certain areas. Second, within the financial community, concerns about the long-term stability of companies doing business in environmentally regulated fields will lead to decreased company values and lack of capital for the continued development and testing that are necessary to overcome the regulatory barriers of safety and effectiveness.

All of us associated with regulatory questions related to the issues of biotechnology have a major responsibility to act quickly and definitively to balance the various opposing forces facing this technology. We cannot accept the notion that all biotechnology-derived products are too dangerous to introduce into the environment; yet we must acknowledge the legitimate concerns of the public and work to establish principles that govern the safe environmental use of these products and thus allow the underlying research to proceed. In such a fast-moving technological environment, it is necessary to review regularly the appropriateness of the scientific basis of existing regulations and to make any required adjustments in either the technology of regulation or the statutory basis for regulation.

The interagency committees that worked on the current policy are very confident that existing regulations cover the current and near-term biotechnology industry. As the technology changes, we must continue to monitor those changes and work together to keep our regulatory "coordinated framework" modern and effective for fulfilling our responsibilities to the American public.

SUMMARY

In summary, I would like to emphasize the continued need for broad and vigorous basic research, with a balance between the fundamental work that may eventually lead to commercial products and the fundamental work that is necessary for an understanding of the interaction of many types of organisms within the environment. I would like also to reiterate the need for balance in the regulatory approach so that we do not repress innovation in research and development. Over-regulation has many side effects. In addition to repressing innovation and not taking advantage of our research base, over-regulation leads to reluctance by the capital markets to invest in the future of our new industries, thereby halting their development at an early stage. At the same time, under-regulation leads to lack of confidence by the public and paralysis of the industry based on public outcry and legal proceedings.

It is my personal belief that the combination of a sound approach to regulatory practice, based on current scientific knowledge, combined with appropriate communication with the public regarding the new products, will lead to an exciting future for all sectors of industry that use the new biotechnology.

REFERENCES

1. Federal Register, FR 49, 50856-50907, December 31, 1984.
2. Federal Register, FR 51, 23302-23393, June 26, 1986.
3. Halfpenny, J.C., and K.P. Ingraham, eds. (1984) Long Term Ecological Research in the United States, Long-Term Ecological Research Network, Corvallis, Oregon.
4. Office of Technology Assessment (1987) Public Perceptions of Biotechnology (New Developments in Biotechnology, Vol. 2), Washington, D.C.

ENVIRONMENTAL POLLUTION POLICIES IN LIGHT OF BIOTECHNOLOGICAL ASSESSMENT: ORGANISATION FOR ECONOMIC COOPERATION, UNITED KINGDOM, AND EUROPEAN ECONOMIC COUNCIL PERSPECTIVES

Alan T. Bull, Geoffrey Holt, and David J. Hardman

Institute for Biotechnological Studies
University of Kent
Canterbury, Kent, United Kingdom CT2 7NJ

> I've always admired those reporters who can descend on an area, talk to key people, ask key questions, take samplings of opinions, and then set down an orderly report very like a road map. I envy this technique and at the same time do not trust it as a mirror of reality. I feel that there are too many realities.
>
> John Steinbeck, Travels with Charley

INTRODUCTION

In preparing this chapter, we have been struck by the correspondence of events in the 1960s and 1970s, when strategies for controlling pollution were first being developed in the United States and in other industrialized countries, with those in the 1980s, when the technological discontinuity of genetic engineering has forced itself upon pollution control as it has on so many other sectors of our everyday existence. The sequence of events, contradictions, and uncertainties that was driven by the adoption of technology-based standards for environmental management has its parallel now in the debate on deliberate release of genetically engineered microorganisms (GEMs) and related biotechnology issues. The problems include: inadequate databases; a general underestimate of time, costs, and acceptance; the required listing of priority pollutants; and ill-considered questions of analysis, culminating in guidelines and regulations that mirror only a portion of reality (30).

The destruction of environmental chemical pollutants poses an immense challenge in both technological and legislative terms. Few, if any, would take issue with this assertion. This chapter will focus attention on aspects of European environmental policy, having in mind the new biotechnology and the pervasive influence of genetic engineering. Is the technosphere--the "built world" of Max Nicholson (33)--altering the biosphere to such an extent that its self-regulating capacity is being put in jeopardy? We will not indulge in scenarios in this discussion, popular as they are in environment-pollution debates. Scenarios for the most part are misleading and, in this context, often unjustifiably alarmist, based as they are on linear extrapolations of present trends. As Sydney Brenner has noted: "It is no use asking whether one leaf is higher than another one on a tree, if you do not even know which branch you are on" (5).

OECD, UNITED KINGDOM, AND EUROPEAN PERSPECTIVES

The OECD Recommendations

The Organisation for Economic Cooperation and Development (OECD) published its first consideration of biotechnology in 1982 (12). At that time we drew attention to a number of interrelated matters pertaining to the emerging technology of genetic manipulation:

(a) Public safety must be of prime concern in designing and operating biotechnological processes.

(b) Unified standards for good laboratory practice and manufacturing procedures should be encouraged.

(c) The benefits of the technology far outweighed conjectural risks, provided that suitable precautions were taken.

(d) At a time when pressures on governments to legislate safety measures for genetic engineering were at their greatest, many countries had no regulations for the handling of known dangerous natural pathogens.

(e) Different countries prescribed different regulations or guidelines for genetic engineering.

(f) Excessively restrictive regulations must be avoided, as these would impose major constraints on biotechnological innovation.

The OECD Committee for Scientific and Technological Policy made a commitment to undertake further work on biotechnology, including safety and regulations. Four years later the OECD published its follow-up report, "Recombinant DNA Safety Considerations: Safety Considerations for Industrial, Agricultural, and Environmental Applications of Organisms Derived by Recombinant DNA Techniques" (34). This study, made by an ad hoc group of government experts from 22 of the 25 member countries, unanimously supported the following:

(a) National positions on safety of use of GEMs against the background of extant or planned legislation.

(b) Identity of criteria for the monitoring or authorization for production of GEMs in industry, agriculture, and the environment.

(c) Exploration of possible ways and means for monitoring future production and use of GEMs in industry, agriculture, and the environment.

The OECD report's first reference to recombinant DNA techniques in the control of environmental pollution is revealing: "Here organisms must be found or modified to meet demanding requirements of extreme environments, and they must survive and proliferate in competition with the existing flora, predators, and sometimes extreme or widely fluctuating chemical and physical conditions" (34). The earliest developments are perceived to be for the enhanced degradation of toxic wastes in well-controlled environments (bioreactors) at or near the point of generation; but other major targets include reducing the volume and bulk of municipal wastes, biohydrometallurgy, and enhanced oil recovery.

The OECD study did not address risk assessment techniques but referred to the earlier (1981) Office of Technology Assessment (OTA) descriptive model as a suitable point of departure for biotechnology applications (Tab. 1). Whereas there are opportunities for analyzing stages 1 and 2 by quantifying the probabilities associated with the magnitudes of consequences, and stage 5 by adapting epidemiological and toxicological procedures, it is the analysis of stages 3 and 4 that presents the greatest difficulties. In this context the development of appropriate methodologies and of realistic analytical protocols is a sine qua non (see below for further consideration of this point).

We share the view expressed in the OTA report that potential environmental effects of GEMs (stage 5) can be expected to mirror those observed following deliberate release of natural organisms, that is: (a) direct, unexpected effects on nontarget species; (b) disturbance of direct species interactions; (c) disturbance of indirect relationships between species; (d) disturbance of biochemical processes that sustain the ecosystem (photosynthesis, mineral cycling, nitrogen fixation); and (e) changes in the rate and direction of evolutionary responses of species to each other and to their abiotic environment.

Tab. 1. Risk assessment: sequential stages of a descriptive model applicable to environmental deployment of genetically engineered microorganisms. (After Ref. 34 and 35.)

1. Formation:	Deliberate or accidental production of the GEM.
2. Release:	Deliberate release or accidental escape of the GEM.
3. Proliferation:	Multiplication, growth, transport, modification, and die-off of the GEM in the environment; gene transfer.
4. Establishment:	Within an ecological niche including other biota.
5. Effects:	Perturbation of the ecosystem by interaction of the GEM with some host or environmental factor.

The main OECD recommendations can be considered to be those relating specifically to environmental applications and those with a more general reference:

Environmental

(a) The considerable existing data on environmental and human health effects of organisms should be used to guide risk assessments.

(b) Genetically engineered microorganisms should be evaluated for potential risk prior to application. Until such time as general international guidelines for such applications are formulated and agreed upon (if ever), independent reviews of potential risks should be made on a case-by-case basis.

(c) Development of GEMs for environmental application should proceed in a step-by-step manner from the laboratory, through microcosms, to limited field testing, and, ultimately, to full-scale testing.

(d) Research to improve the prediction, evaluation, and monitoring of the results of GEM applications should be encouraged.

(e) The large-scale production of GEMs for environmental applications should, wherever possible, use low-risk microorganisms and operate under conditions of Good Industrial Large-Scale Practice (GILSP).

General

(f) National guidelines and regulations should be harmonized.

(g) Guidelines and regulations should not impede future development in recombinant DNA techniques.

(h) Development of testing methods and related matters should be considered on a national and international basis in order to minimize trade barriers.

(i) Special efforts should be made to improve public understanding of recombinant DNA technology.

(j) For certain environmental applications of GEMs, notification schemes may be desirable.

(k) Protection of intellectual property and confidentiality interests, consistent with safety, needs to be assured.

Implementation: The United Kingdom Guidelines

The OECD Council Recommendations are acts of mutual agreement, submitted for the consideration of the member countries, that are considered to be opportune for implementation. Several OECD member countries are considering the introduction of guidelines and regulations for the release of genetically altered organisms: those for the United Kingdom (UK) were ratified early in 1987.

The UK "Guidelines for Risk Assessment and for the Notification of Proposals for Such Work" (1) were drafted by the Advisory Committee on Genetic Manipulation (ACGM), an arm of the Health and Safety Commission. These guidelines are in broad agreement with the OECD proposals. The ACGM concluded that it is not yet possible to devise a single, broadly applicable risk assessment scheme. Nevertheless, it believes that factors relevant to the majority of foreseeable release applications can be identified and that these should be used to develop a consistent approach to risk assessment. The ACGM intends to keep the guidelines under review, but it must be notified of all proposals for releasing GEMs. The notifier has to (a) address points listed under "Risk Assessment Factors" (Tab. 2); (b) give details of the aims and rationale of the work; and (c) be advised by an "appropriately constituted local body." The notifier can be protected by a "commercial in-confidence" agreement and can expect to receive a response from the ACGM within 30 days. At present, although there is no statutory requirement to do so, the ACGM considers that all proposals must be notified. It seems likely that the notification requirements of the Health and Safety (Genetic Manipulation) Regulations of 1978 will be extended to include the use of GEMs as defined by the ACGM guidelines.

Assertions of environmental policy often imply a strategy for everyone about everything and therein reflect the fact that most policies carry some environmental implications. One practical consequence of such horizontal or universal influence of environmental matters (20) is the potentially large number of government departments and laws applicable to the release of biological agents with which an applicant or notifier has to contend. The UK guidelines were prepared under the auspices of the Health and Safety Commission in collaboration with the Departments of the Environment and of Health and Social Security and the Ministry of Agriculture, Fisheries and Food; in addition, these and other departments of state operate with and through sectors of local government and intermediary agencies.

A similar situation exists in other OECD countries. Illustrative of the range of legislation that impinges on the release of biological agents is that enacted by the United States Congress (Tab. 3). The first applications to release GEMs reveal the sort of inconsistencies and difficulties that can be expected, at least during this early phase of development. The United States Environmental Protection Agency (EPA) uses two acts, the Federal Insecticide, Fungicide and Rodenticide Act (FIFRA) and the Toxic Substances Control Act (TSCA), to monitor genetically altered organisms: "ice-minus" bacteria have been approved under the FIFRA on the grounds that ice on plants is a pest and that such bacteria are inter alia pesticides; "fortified" nitrogen-fixing bacteria, on the other hand, do not kill anything and are best considered under the TSCA [unlike chemical substances, however, they are capable of proliferation and establishment (Tab. 1)]. "Since most European countries follow the regulatory actions of the United States or the United Kingdom, and since Europe typically strives for uniformity ..." (28), these test cases clearly show that there is already a regulatory tangle.

Tab. 2. Risk assessment factors to be addressed by notifiers of proposals to release genetically engineered microorganisms into the environment in the UK (1).

1. General

   1.1. Nature of organism to be released.

   1.2. Release procedure.

   1.3. Nature and source of altered nucleic acid.

   1.4. Verification of the genetic structure of the GEM.

   1.5. Genetic stability of the GEM.

   1.6. Predicted effects of manipulation on behavior in the organism's ecosystem.

   1.7. Survival ability of the GEM.

2. Release

   2.1. Geographical location, size, and nature of site.

   2.2. Details of target ecosystem; predicted effects of release.

   2.3. Method, amount, frequency, and duration of application.

   2.4. Monitoring capabilities and intentions.

   2.5. On-site worker safety.

   2.6. Contingency plans in event of unanticipated effects of the GEM.

3. Survival and dissemination

   3.1. Growth and survival of host organism and effect of genetic manipulation.

   3.2. Susceptibilities to environmental stress.

   3.3. Ability of the GEM to survive and transfer genetic material.

   3.4. Potential for rDNA transfer.

   3.5. Control of superfluous organisms or nucleic acid.

The EEC Position

The question of deliberate release of genetically engineered organisms in the context of waste disposal and agriculture has been under active consideration by the European Communities for some time, with a report to the Council of the European Communities scheduled for September 1987. The establishment of Communities policy in this area involves Directorate-General (DG) III Internal Market and Industrial Affairs, DG V Employment, Social Affairs and Education (including Directorate E for Health and Safety), DG VI Agriculture, DG XI Environment, Consumer Protection and

Tab. 3. United States legislation applicable to a greater or lesser degree to the release of biological agents.

Federal Food, Drug and Cosmetic Act (FFDCA)
Public Health Service Act (PHSA)
Resource Conservation and Recovery Act (RCRA)
Clean Air Act (CAA)
Federal Water Pollution Control Act (FWPCA) amended by Clean Water Act (CWA)
Toxic Substances Control Act (TSCA)
Marine Protection, Research and Sanctuaries Act (MPRSA)
Federal Insecticide, Fungicide and Rodenticide Act (FIFRA)
Comprehensive Environmental Response, Compensation, and Liability Act (CERCLA)*

*Popularly known as "Superfund."

Note: See Krimsky and Fraenkel (25) for an overview of United States environmental laws. See, also, Kingsbury, this Volume.

Nuclear Safety, and DG XII Science, Research and Development. Four directives, covering deliberate release of GEMs, waste disposal, processes based on "generally-regarded-as-safe" (GRASS) organisms, and water protection, will form the foundation of the Communities policy (EEC Comm. Doc. 57386).

The European Communities have set great store on standardization and harmonization in the formulation of policy, not least to remove technical barriers to trade and cooperation in research and development. However, it appears that in this situation considerable discordance may be the outcome of the Commission of the European Communities' desire to formulate acceptable guidelines. Most noticeable is the position adopted in the Federal Republic of Germany, where the Parliamentary Commission has recommended a five-year moratorium on the release of GEMs into the environment. Such a decision clearly reflects a strong ecology presence in West German politics and the need to convince the public that new technology is acceptable.

The European Federation of Biotechnology (EFB) has produced a document intended to encourage safe biotechnology (26), but no consideration was given to deliberate release either of GEMs or of natural microorganisms. Microorganisms are classified on the basis of pathogenicity and are placed in four risk classes along lines determined earlier in the United States and the United Kingdom. A fifth class (Group E) contains organisms that present "a more severe threat to the environment than to man," especially to animals and plants. The EFB has announced that it will keep a watch on safety developments and produce periodic updates of regulations and guidelines.

Perception of Risk

So much has been written about risk, not least in the forum of environmental pollution, that it may seem to be redundant to devote more space to it in this discussion. Nevertheless, because of the concern that release proposals have aroused in the United States and the strength of feeling in certain parts of Europe, it is more important than ever that scientists assist the public in gaining a better understanding of relative risk.

At the Royal Society of London in early 1987, William Waldegrave, Minister for the Environment in the UK government, delivered a forceful lecture on science and policy-making in which he developed the theme of the perception of risk and the part that scientific and other considerations play in changing that perception. Waldegrave contended that public acceptability of an activity--be it a new technology or motorcycling--will be conditioned by four factors: (i) the probability of injury resulting from the activity (safety); (ii) the perceived importance of the activity; (iii) the exercise of free will (hence, greater public tolerance of motorcycling than of nuclear waste reprocessing); and (iv) confidence in the advice of experts and confidence that even unanticipated, serious consequences of activities will be containable temporally as well as spatially. Taking motorcycling (a Pirsigian metaphor beautifully apposite), nuclear reprocessing, and genetic engineering, Waldegrave attributed scores of up to 10 to each of the above factors in terms of hypothetical acceptability, a high score signifying high acceptability (Tab. 4). The score for genetic engineering was pitched low deliberately.

How can one change the public perception of something like genetic engineering, especially in regard to its application to environmental management? It is necessary (a) to make better known the benefits of the technology, while not concealing risks, and (b) to consult public opinion, thereby raising the level of public confidence. Of course, these simplistic statements totally belie the difficulty of the tasks. Confidence, for example, is much harder to instill when a technology is totally novel, as is the case with genetic engineering, and when the environmental risks are conjectural. The contrast with the development of nuclear power in the 1950s is clear: during the decision-making process there was abundant evidence that radioactivity was hazardous, but the level of public debate was rather

Tab. 4. A scorecard of perceived risks.

	Motorcycling	Reprocessing	Genetic engineering
Safety	1	8	9
Importance	0	7	1
Free will	10	1	1
Confidence	9	2	1
	20	18	12

low; thirty years later, the situation with environmental release of GEMs is quite the reverse--conjectural hazards and wide public debate. In our opinion, confidence in the release of GEMs for agricultural and pollution control purposes will rise appreciably in the wake of sound practical demonstrations based on good environmental science and technology. The case-by-case review advocated at this time by the OECD and others is very much in tune with this thinking.

Experimental Evaluation: Mastering the Probable and Managing the Unpredictable

Description of containment criteria and facilities for the large-scale (greater than 10 liters) production of GEMs is somewhat outside the scope of this chapter. It can be expected that most, if not all, microorganisms designed for environmental use will be those considered to be of low risk vis-a-vis exposure of plant workers and other persons, and that Good Industrial Large-Scale Practice (GILSP) will be applicable. Criteria for GILSP are given in the OECD Report (34), while Giorgio and Wu (18) have reviewed the design of containment into bioreactors and downstream operations. In the UK, the Department of Trade and Industry and the Public Health Laboratory Service have recently established an Industrial Biosafety Club whose task it will be to provide much-needed quantitative research and to apply risk assessment methods to bioprocessing with GEMs. There are suggestions that current practices (e.g., with regard to air sampling) may provide misleading information (3) and that revisions to monitoring procedures may be necessary (13,47).

The problem of risk assessment for GEMs released into the environment is much greater because the relevant knowledge base is so low. Johnston and Robinson's criticisms (23) remain valid: microbial ecology has produced a great amount of descriptive information on the environmental establishment of introduced organisms, but little of this information enables the successful prediction of the establishment of any given species in any given ecosystem; molecular biology has provided very detailed knowledge of gene exchange, but in a very limited range of microorganisms having only small roles in the environment. In other words, we have little more than conjecture upon which to base risk assessments.

The acquisition of relevant knowledge and the development of appropriate test systems ideally would be done on the basis of defining the probability of a deleterious effect resulting from introducing GEMs into the environment. The factors upon which such probabilities could be adduced are well known (Ref. 2, 14, and 48; Tab. 1 and 2). Given the general lack of relevant knowledge, there has been a mistaken tendency to place undue faith in simplistic, inappropriate tests and data, the "something is better than nothing" philosophy. At times in such discussions the wish for a tabula rasa is almost overwhelming. An equally mistaken position, in our opinion, is that which leads to protracted approval procedures, or even the imposition of moratoria, in the comforting expectation that all the "relevant knowledge" will become available as a consequence of more research. The very complexity and idiosyncrasy of different natural and man-made environments makes this impossible. Furthermore, it is unnecessary.

What is required is a relatively small number of guiding principles based on accepted data, complemented by a portfolio of sound practical demonstrations at the "field" scale. It has been argued that commercial confidentiality will limit the general usefulness of case-by-case review by hindering the dissemination of accumulated wisdom. The OECD report provides a useful and realistic checklist of required information together with the following caveat: "Although the list attempts to be comprehensive as far as present knowledge allows, not all the points included will apply to every case. It is to be expected therefore that individual proposals will address only the particular subset of issues that are appropriate to individual situations. The level of detail required in response to each subset of considerations is also likely to vary according to the nature of the proposed activity" (34).

We have already referred to the desirability of developing organisms and processes for treating environmental chemicals in an incremental, step-by-step fashion. It is important to emphasize that laboratory-scale exploratory research must be based on holistic principles and must be focused on some type of microcosm system. This latter point has been made repeatedly in discussions of biodegradation and biodetoxification (7,11, 38,40,49), but only slowly have experimental approaches addressed this requirement.

We can refer briefly to the development of appropriate methodologies and of realistic analytical protocols (cf. Tab. 1 and related text). The detection of organisms (and genes) introduced into a pollution treatment situation is absolutely necessary; new technology has aided this objective in dramatic fashion. Until quite recently, spontaneous antibiotic-resistant mutants were used widely for monitoring and enumerating bacteria introduced into soils and other habitats. Stability of antibiotic resistance has been improved greatly by producing appropriate strains via plasmid transformation and transposon mutagenesis. Some of these early studies followed the prudent dictum of "the best strategy for efficacy and safety is to isolate organisms from environments of interest ... and then return the organism to its habitat," having made the genetic alteration of choice (36). For example, in Dupler and Baker's study (17) of *Fusarium* wilt-suppressive *Pseudomonas putida*, a strain of *P. putida* was isolated from soil in which it suppressed the wilt fungus and a rifampicin-resistant mutant was selected. The Rfm strain enabled the survival characteristics of the bacterial control agent to be determined throughout field trials and under a variety of soil environments (clay and sandy loams, water potential, season).

The use of transposon mutagenesis and transformation for studying population dynamics under environmental conditions is illustrated by the work of van Elsas et al. (51). These authors reported that antibiotic resistances introduced in these ways remained stable for more than 50 generations without selective pressure. They argued that the use of these markers was preferable because (a) the genetic changes were precisely defined, (b) specific gene recognition is possible in field experiments via colony hybridization and plasmid patterns, and (c) no substantial reduction in growth rate was caused. These advantages notwithstanding, many microbiologists have argued that antibiotic resistance markers are not ideal. Two attractive alternatives which offer sensitive, accurate, and highly specific detection and monitoring opportunities are immunofluorescence microscopy and DNA probing (in vitro and colony hybridization); the latter is proving useful in biotreatment situations (50). Gene probes have been developed and used to detect TOL plasmids, for example, with a sensitivity of detection of one colony to 10^6 in environmental samples (44).

Moreover, such DNA-DNA hybridization provides a powerful means of revealing cryptic (unexpressed) catabolic functions in pollutant-receiving situations.

Given these techniques, it is now possible to evaluate gene transfer (homologous and heterologous) within test microcosms and biotreatment sites. Stotzky and Babich's (48) recommendations should be heeded in this context: (a) potential recipient species need to be identified among the indigenous microflora; (b) survival and growth characteristics of parental recipient strains, as well as those of recombinants, need to be determined in order to understand the population dynamics governing the persistence of heterologous genes in the environment; (c) the influence of abiotic factors on gene transfer under field conditions needs to be evaluated; and (d) the influence of indigenous microorganisms on gene transfer needs to be evaluated.

There exists quite a body of information on genetic interactions among microbial communities. Reanney and his colleagues (41) reviewed the known routes of intergeneric genetic exchange among bacteria in a group of interactive ecosystems: rhizosphere, rhizoplane, nonroot soil, soil-faceces, and gastrointestinal tract. Potential genetic, entry, temperature, and ecological barriers to gene flow in microbial communities are very important to risk assessment. Very few studies have demonstrated gene transfer in soils per se. The advent of immunofluorescence and DNA probes will enable more rapid progress to be made and, most interestingly, will facilitate the detection of organisms, carrying target genes, that are not culturable. Compared with soil ecosystems, there has been a more thorough examination of gene transfer in aquatic environments. Trevors (50) has calculated the following transfer frequencies: conjugation, the most common mechanism, 5 x 10^{-5} to 5 x 10^{-8}; unknown, but probably conjugation, 2 x 10^{-3} to 1 x 10^{-6}; and transduction, 8.8 x 10^{-5} (a single report).

Two recent reports illustrate the way forward. First, plasmid mobilization has been demonstrated under a range of operating conditions in a laboratory-scale waste treatment facility. Mancini et al. (27) constructed a *recA* transformant of *Escherichia coli* HB101 carrying plasmid pHSV106 (herpes simplex thymidine kinase marker) for this study and showed mobilization to laboratory and to wastewater strains. Second, Saye et al. (43) have examined plasmid transduction in a natural freshwater habitat using *Pseudomonas aeruginosa* plasmid Rms149 and a generalized transducing bacteriophage. Transduction occurred in the absence and presence of the natural microbial community; however, under the latter condition, both the numbers of introduced donors and recipients and the number of recovered transductants decreased rapidly.

These investigations still beg the question as to what effects on the ecosystem are created by the introduction of GEMs. A first level of investigation might attempt to discern structural and metabolic changes within a microcosm community. Thus, a change in the microbial diversity index might signal an imposed stress on the ecosystem as a result of GEM loading. Metabolic and biochemical quantities which change in response to perturbations to the ecosystem include nutrient fluxes, photosynthesis (and chlorophyll content), ATP, lipid profiles, and rRNA sequences. The changes, in many if not all cases, are likely to be minor, and to this extent there may be a case for setting up selected "worst case" experiments in order to assess the probability and severity of hypothetical environmental disasters.

To date, only a few reports have appeared on the genetic manipulation of microorganisms in order to improve their biodegradative activities, and even fewer have transferred the investigation to environmentally valid models of the type mentioned above. However, the recent work from Sayler's laboratory on in situ treatment of contaminated aerobic groundwater aquifers (21,22) has very positive implications. Microcosms were made aseptically from pristine sample cores from the saturated zone of the aquifer. The microcosms were contaminated experimentally with a toluene-chlorobenzene-styrene mixture and inoculated separately either with indigenous bacteria or with *P. putida* carrying TOL plasmid pWWO. The latter was stably maintained in the aquifer microcosm for at least eight weeks, during which time its population size was monitored by using a gene probe, and it removed the added chemical pollutants.

Similarly, observations of natural "evolution" within communities of pollutant-degrading microorganisms are uncommon. Nevertheless, the known instances of this phenomenon have implications for the introduction of GEMs into the environment, as brief reference to two cases will show: (a) alteration in a regulatory mechanism, and (b) direct gene transfer. In the first case, our analysis of Dalapon (Na_2-2,2-dichloroproprionic acid) biodegradation revealed that a community member, in this instance a strain of *P. putida*, acquired the ability to utilize the haloalkanoic acid as a result of selection of a spontaneous mutation, effecting a dehalogenase induction mechanism (46). The mutant strain thereafter competed successfully with other dehalogenase-producing members of the community for the growth-limiting substrate Dalapon. The second case came from Knackmuss' laboratory (45) during a study of chlorophenol degradation by defined mixed bacterial cultures. Synthetic effluent containing phenol, acetone, alkanols, 4-chlorophenol, and isomeric chlorophenols was completely degraded by the community. A strain of *Pseudomonas*, B13, was responsible for metabolizing the chlorophenols. Subsequently, the chlorocatechol genes from B13 were transferred to another member of the community, *Alcaligenes* A7-2, which was much more competitive with respect to phenol and chlorophenol utilization than *Pseudomonas* B13.

BIOTECHNOLOGY AND POLLUTION CONTROL

General Considerations

"On scientific grounds, waste treatment should be a field particularly open to the application of advanced biotechnology" (16). Table 5 summarizes the opportunities. Unfortunately, waste treatment usually has to contend with large quantities and extremely low unit value; consequently, processing needs to be simple and cheap. The marketplace will be a major test for any innovation in waste treatment, not just for recombinant DNA techniques, and the real challenge is to develop strategies for commercializing such innovation. We are of the opinion that public pressure and legislation will be the chief stimuli for forcing advanced biotechnology into the waste treatment industries and, perhaps, for orchestrating the necessary cost-benefit incentives. Some industrial spokesmen have stated that the pollution control laws simply do not exert enough pressure on major corporations to stimulate investment in better technologies to control pollution.

Tab. 5. Applications of biotechnology to waste management. (From Ref. 16.)

1. Improvement of existing processes by application of bioscience and biochemical engineering (e.g., knowledge of mixed cultures and flocculation).

2. Use of genetically improved organisms, including GEMs (e.g., treatment of toxic wastes and correction of a poorly performing plant).

3. New catalysts and novel bioreactors (e.g., immobilized organisms and fluidized beds).

4. New sensors and automation (e.g., trace amounts of toxic organics and heavy metals).

5. Recovery of products from wastes (e.g., methane and metals).

It must be stating the obvious to say that the starting point for considering biotechnological innovation is the precise definition of the pollution problem. Nevertheless, this is crucial if the problem is to be addressed effectively and competitively. In the remainder of this discussion we will consider the application of modern biotechnological principles to processing wastes that have been considered extremely difficult to treat due to their toxicity, high BOD, and/or recalcitrance. A rational approach to the disposal of such wastes will include:

(a) Analysis of the problem: is it a single component of the waste that is difficult to treat, or is it a multicomponent problem?

(b) Can the manufacturing plant be adapted to remove the problem component at the source?

(c) Is the waste amenable to treatment by topical application of specialized microorganisms, or by specific enzymes?

(d) Is biorecovery of materials from the waste an option?

(e) Does the pollution arise as a result of (i) deliberate discharge; (ii) a side effect of application (e.g., pesticides); (iii) mining, dumped-container, or landfill leachates; or (iv) accidental spillage?

Unfortunately, information of this type is frequently not available, and the situation is exacerbated by certain features of the NIMBY ("not in my backyard") syndrome with regard to waste and waste processing. In Europe, for example, there is no present harmonization of regulations on toxic waste disposal, transportation, or notification. Thus, the more "relaxed" regulations in the UK have led to the deplorable practice of importing large quantities of various and often unidentified toxic wastes for disposal by landfill with or without pretreatment. It has taken the tragic circumstance of the Townsend ferry disaster to alert public attention to such matters.

It is also worth pondering that the OTA has estimated that the cost of remedial cleanup of, for example, 600 hazardous waste sites in the United States could be as high as $100 billion. Again, unfortunately, the drafting of regulations and guidelines in this area is notoriously difficult [cf. the 1972 revisions to the United States Federal Water Pollution Control Act covering municipal, industrial, and toxic discharges (30)].

Existing Biotreatment Processes

It is not our intention to survey the range of existing technologies for waste treatment and pollution control or the applications to which they have been directed. Simply, we wish to state, as have others, that considerable improvement can be made to current practices by more intelligent application of microbiological and process engineering principles; perhaps only then should we think of further improvements via recombinant DNA technology. The waste management biotechnologist will need to be just as knowledgeable about phenotypic variability of microorganisms as he or she is about genetic engineering. Thus, increasingly complex microbiological problems associated with the treatment of xenobiotic pollutants have prompted modifications in reactor design and operation. Control of solids retention time (SRT) is one of the most important introductions (29), together with the ability to control it independently of the hydraulic retention time (HRT). In this way, biotreatment can be optimized by minimizing HRT (to reduce the reactor volume) and maximizing SRT (to increase pollutant degradation or detoxification rates and prevent washout). In comparison with other industries, the waste treatment industry has lagged behind in the use of instrumentation, control, and automation; major control requirements are for aeration and solids inventory.

The practice of bioaugmentation (adding competent wild-type or mutant organisms to treatment sites) has become an established waste treatment procedure. It is a flexible operation that can be applied to reactors, waste collection lines, sumps, and other contained areas, and to hazardous dump sites. Although landfill and related systems (entombments and lagoons) have been the customary means of disposing of hazardous wastes [over 210 te per annum in the United States, according to Nicholas (32)], the contamination of groundwater by leachates is forcing the authorities to investigate and invest in biodetoxification (and nonbiological destruction) technologies.

The Impact of Biotechnology

In this section the concern is largely with impacts of genetic engineering. We open with some cautionary comments. At the previous meeting on this subject, Owings (37) remarked, "We have [genetic engineering] tools coming along; how are we going to apply them? Right now we are dealing mostly with concepts." Thus, attempting to apply such technology when it is inappropriate or unnecessary is likely to be unsuccessful, leading to frustration and a squandering of resources. The appropriateness and necessity of using recombinant DNA technology in environmental biotechnology are crucial points to consider. Are the long-term objectives likely to be attainable? Will GEMs function in the target environment? The state of the art is still predominantly in the laboratory. It is incumbent upon us to address Nicholas' challenge that biotechnology in hazardous waste disposal is "an unfulfilled promise" (32).

Just as gene cloning has become the prosaic part of genetic engineering, the identification of appropriate microorganisms for developing pollution control processes is not particularly difficult. The needs are obvious to all, and the means of selecting an organism or organisms carrying the necessary catabolic genes are well documented. However, one point above all others must be emphasized: it is not sufficient to select organisms possessing the required biodegradative activity; the selection must be made with reference to the design of a process, a criterion not always appreciated by or of interest to the academic microbiologist. Here the concern has to be for: (a) the rate at which biotransformation or mineralization takes place; (b) the physiology of the organism, especially with respect to the utilization of mixed substrates (most waste streams contain a complex of toxic, recalcitrant, and/or isomeric chemical types); (c) the resistance(s) of microorganisms or enzymes to waste stream or environmental components other than the target pollutant (e.g., acidity, alkalinity, heavy metals, and ionic strength); and (d) the competitive advantage that the biodegradative agent may have or need to have for in situ operations. There is nothing particularly novel in reciting these requirements; it is simply depressing that the required research remains to be done in so many instances. [We refer the reader here to the approaches, and prejudices, of our own laboratory: biodegradation (7); process optimization (8); continuous enrichment selection (9); and mixed substrate utilization and mixed cultures (10).] To conclude, therefore, as in so many areas of contemporary biotechnology, there exists a crucial gap between innovative research in the laboratory and the working technology.

Goals and strategies. It is helpful to separate pollution problems and their treatment into (i) predischarge and (ii) in situ categories, and to examine if and how biotechnology might have a role in their resolution. The greatest opportunities for exploiting the biotechnologies in pollution control come with predischarge treatment in the factory, where the full range of sophisticated fermentation technology can be considered. Moreover, the options increase if we move upstream from the complex, time-variant end-of-pipe waste streams to particular manufacturing unit processes that produce simpler, more consistent and concentrated waste streams. Patterson (39) suggests that such an approach could be viewed as a "manufacturing process modification" rather than pollution control in the conventional sense: the end result is the same, but the efficiency, cost, and potential to recover by-products could be much improved. We present one example of such thinking below. It is in this context that we believe that applications of genetic engineering to pollution control will be the most effective and acceptable. The point should also be emphasized that, in most circumstances, it will be important to effect mineralization or conversion to readily metabolizable compounds, rather than to simply detoxify the pollutant chemical. Incomplete biotransformation of the pollutant can generate equal or greater problems of toxicity, recalcitrance, and pollution via the formation of secondary materials (7,45).

The legacy of leaking landfills and a similar "throw it away" philosophy is a horrendous, world-wide problem; "a national disaster," in Nicholas' terms (32), for the United States. The role of biotechnology in the remedial treatment of very extensive environmental pollution is far from clear; for the most part, we tend to take a somewhat pessimistic view. The recent amendments to Superfund (Tab. 3) call for, among other things, the demonstration of alternative or innovative technologies to be used at Superfund sites. We are forced to conclude that, faced with the scale and urgency of these problems, nonbiotechnological remedies, such as incineration, may be called for in many of these situations. The development of plasma arc pyrolysis is leading to mobile treatment units that can achieve "nine nines" efficiency (99.9999999) at significantly lower costs than conventional incinerators. In countries such as The Netherlands, where landfill operations are precluded because of geography and topography, the implementation of these alternative means of hazardous waste disposal is well-established.

It is important, but often avoided, to note the use to which the site will be put after reclamation. The Dutch, for example, have insisted on maintaining the multifunctionality of the soil, so that contaminated land has to be cleaned to enable unrestricted future use. Contrast this with the UK policy (guidelines set by an Interdepartmental Committee on the Reclamation of Contaminated Land), in which the intended use is largely dictated by the amount and nature of any decontamination action. The efficacy of biotechnological methods for decontaminating major toxic waste sites remains to be proved. The first attempt at such processing in Europe is being made by BioTechnica on a ten-hectare gas works site in Blackburn. The approach being taken is that advocated by most microbiologists working in pollution control: isolate from the site, analyze, select, and return to the site, augmented with microbial nutrients. The problem is one of treating cyanides, phenols, oil tars, and heavy metals. The heavy metal-contaminated soil is too toxic for microbial treatment and will be confined to sealed landfills. The remaining soil (approximately 26,000 cubic meters) is put into mounds and sprayed with bacteria. The project was expected to be completed in two years and is ahead of schedule (4). In situ biodegradation is in its infancy as a technology to reclaim most hazardous sites. Among the problems are the extent of waste removal in relation to EPA and other regulatory standards, and the fact that rather permeable soils are a prerequisite for its application.

Incentives. We have concluded above that only the combination of public pressure and government legislation can provide the impetus for safe, _effective_ hazardous waste _destruction_. Commitment to this end will require: (a) long-term research and development, particularly technology transfer; (b) prompt resolution of regulatory uncertainty and controversy; (c) enactment of punitive legislation on defaulters, and an acceptance by the consumer of certain knock-on costs; (d) government determination and harmonization of international action; and (e) entrepreneurial investment (given the above, this should be guaranteed).

Case histories. In this final section we draw upon the work and interests of our own laboratories to illustrate some aspects of biotechnological innovation in waste management.

Large volumes of cyanide-rich wastes result from the production of electroplated objects, plastics, and steel, and are often mixed to various degrees with phenol, cyanate, thiocyanate, heavy metals, and organic nitriles. Knowles and his colleagues have developed effective biotreatment technology for cyanide degradation (19,24). Several microbial enzymes are potentially suited for cyanide degradation: β-cyanoalanine synthase, rhodanese, cyanide hydratase, nitrogenase, and cyanide "oxidase."

The Kent group have focused on the cyanide hydratase of the phytopathogenic fungus _Gloeocercospora sorghi_ (leaf spot of sorghum), which

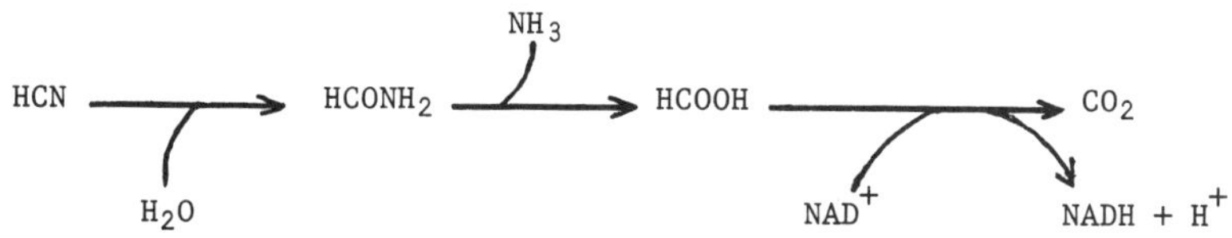

Fig. 1. Cyanide hydratase and formamidase catabolism of cyanide.

converts cyanide quantitatively to formamide; the latter is readily metabolized to ammonia and carbon dioxide by formamidase (Fig. 1). The *G. sorghi* enzyme has a K_m of 25 mM, a broad pH optimum of 7 to 9, and a half-life of three to five days. However, stability can be extended to several months by simple immobilization of the mycelium with polyelectrolytic flocculating agents. The mycelium is freeze-dried and extruded in such a way that it packs well into column reactors. These reactors convert cyanide, up to 100 mM, to formamide with an efficiency of 99.9999% at a flow rate of 0.5 column volumes h^{-1}. This process has been developed by ICI Biological Products, the product being known as Cyclear. Fungal cyanide hydratases are very susceptible to heavy metals, and they do not hydrolyze metal cyanides, which are common components of several industrial wastes. A second generation product is under development that is less sensitive to heavy metals, and the activities of bacteria that can utilize metal cyanides are also being optimized (42).

Our interest in metal-accumulating bacteria has led to the examination of strain development for the simultaneous degradation of cyanide and the accumulation of metals, particularly precious metals. It is known that some bacteria effectively degrade cyanide in the presence of heavy metals and that bioaugmentation of local activated sludge plants with *Pseudomonas paucimobilis* has proved successful in the treatment of large-volume gold mine wastewaters (875 $m^3 \cdot h^{-1}$) (31).

The hazards of heavy metal pollution are well known in terms of human health. Adverse effects of heavy metals on crop yield and crop contamination result from applying contaminated sewage sludge; the potential dangers are significant because in the UK alone about 10^6 tons of sludge are used annually as organic manure. Guidelines controlling the build-up of metals in soils are based on the impairment of crop yield. Only recently (6) have effects on the soil microflora been determined. Even at or below current maximal permissible levels, heavy metals depressed soil microbial biomass, ATP content, and nitrogen fixation. Contamination of aquatic environments occurs variously from agricultural, urban, and road run-off. Considerable understanding exists of the accumulation of metals by microorganisms, and, quite frequently, treatments have been devised whereby the metal can be released nondestructively and the biomass reused several times (45a). Biotechnology provides high affinity, sometimes selective accumulation, and frequently high loading of the biomass. We have selected mixed bacterial populations that are capable of accumulating copper to the extent of 35% of the dry weight (15), while a strain of *Citrobacter intermedius* has been developed to selectively accumulate silver or gold from liquors containing other metals such as copper and cadmium (M.P. Meadows and A.T. Bull, unpubl. data). To date, we remain unaware of any commercial developments of sophisticated metal bioaccumulation processes. However, we believe that these will come, again in response to legislative pressure, as a cost-effective means of secondary recovery and separation of valuable metals.

Finally, we have been working on the detoxification of mycotoxins, a project which in part has led us to a preliminary appraisal of in situ treatment of a natural product (S. Beeton, J. Lomas, and A.T. Bull, unpubl. data). Our particular concern is for trichothecene mycotoxins of the type produced during grain spoilage by *Fusarium*, *Myrothecium*, *Stachybotrys*, and *Trichothecium* species. Chemically, the trichothecenes are very stable and offer a serious health hazard. In surveying a broad range of geographical sites from Antarctica, typical rain forests, and temperate soils, we detected T-2 toxin-metabolizing capacities in 85% of the samples. All of the active isolates were complex bacterial communities, often containing up to nine species; invariably the rate of T-2 toxin disappearance was much greater by the community than by any of the competent individual species. Freeze-dried communities maintained their species diversity and activity for more than 18 months. Detoxification, as monitored by a *Rhodotorula rubra* bioassay, accompanied degradation of T-2 toxin via HT-2 toxin, the triol and the tetraol, with subsequent disappearance of the 12,13-epoxy group. The T-2 toxin (1 mM) was capable of being degraded within 60 minutes by some communities. The possibility of deploying such populations for localized mycotoxin detoxification is under consideration, but this method is very much in its infancy.

THE FUTURE

We have attempted to review and comment upon some major policy issues that will determine the pace and direction of environmental pollution control and the impact of biotechnology. As the conference upon which the present volume is based was held in the Pacific Northwest, it is appropriate to quote from Chief Seattle, who in a letter to President Franklin Pierce wrote, "If all the beasts were gone, man would die from a great loneliness of spirit. For whatever happens to the beasts soon happens to man. All things are connected."

REFERENCES

1. ACGM (1987) *The Planned Release of Genetically Manipulated Organisms for Agricultural and Environmental Purposes*, Guidelines for risk assessment and for the notification of proposals for such work, issued by the Advisory Committee on Genetic Engineering, Health and Safety Executive, London.
2. Alexander, M. (1986) In *Microbial Communities in Soil, FEMS Symposium No. 33*, V. Jensen, A. Kjoller, and L.H. Sorensen, eds. Elsevier, Amsterdam.
3. Anonymous (1987) Industrial safety club born. *Chem. Industry* 1 June, p. 373.
4. Anonymous (1987) "Magic" microbes eat the dirt. *New Scientist* 113 (No. 1550):30.
5. Brenner, S. (1984) Discussion to present state and future of life sciences. In *Conference on the Life Sciences and Mankind*, Hakone, Japan, p. 21.
6. Brooks, P.C., and S.P. McGrath (1986) Effects of heavy metal accumulation in field soils treated with sewage sludge on microbial processes and soil fertility. In *Microbial Communities in Soil*, V. Jensen,

A. Kjoller, and L.H. Sorensen, eds. Elsevier, Amsterdam, pp. 327-343.

7. Bull, A.T. (1980) Biodegradation: Some attitudes and strategies of microorganisms and microbiologists. In Contemporary Microbial Ecology, D.C. Ellwood et al., eds. Academic Press, Inc., London, pp. 107-136.
8. Bull, A.T. (1981) Strategies in microbial process optimization. In Global Impacts of Applied Microbiology Sixth International Conference, S.O. Emejuaime, O. Ogumbi, and S.O. Sani, eds. Academic Press, Inc., London, pp. 623-645.
9. Bull, A.T. (1983) Continuous culture for production. In Basic Biology of New Developments in Biotechnology, A. Hollaender, A.I. Laskin, and P. Rogers, eds. Plenum Press, New York, pp. 405-437.
10. Bull, A.T. (1985) Mixed culture and mixed substrate systems. In Comprehensive Biotechnology, Vol. 1, A.T. Bull and H. Dalton, eds. Pergamon Press, Oxford, pp. 281-299.
11. Bull, A.T., and J.H. Slater (1982) Microbial Interactions and Communities, Vol. 1. Academic Press, Inc., London.
12. Bull, A.T., G. Holt, and M.D. Lilly (1982) Biotechnology: International Trends and Perspectives, Organisation for Economic Cooperation and Development, Paris.
13. Cameron, R., P. Hambleton, and J. Melling (1987) Assessing microbiological safety of bioprocessing equipment. In Proceedings of the Fourth European Congress on Biotechnology, Vol. 1. Elsevier, Amsterdam, pp. 139-142.
14. Colwell, R.R. (1986) Release of genetically engineered microorganisms into the environment. MIRCEN J. 2:41-49.
15. Dunn, G.M., and A.T. Bull (1983) Bioaccumulation of copper by a defined community of activated sludge bacteria. Eur. J. Appl. Microbiol. Biotechnol. 17:30-34.
16. Dunnill, P., and M. Rudd (1984) Biotechnology and British Industry, Science & Engineering Research Council, Swindon.
17. Dupler, M., and R. Baker (1984) Survival of *Pseudomonas putida*, a biological control agent, in soil. Phytopathology 74:195-200.
18. Giorgio, R.J., and J.J. Wu (1986) Design of large scale containment facilities for recombinant DNA fermentations. TIBTECH 4:60-65.
19. Harris, R.E., A.W. Bunch, and C.J. Knowles (1987) Microbial cyanide and nitrile metabolism. Sci. Progr., Oxford 71:293-304.
20. Holdgate, M. (1987) The reality of environmental policy. Royal Soc. Arts J. CXXXV:310-327.
21. Jain, R.K., and G.S. Sayler (1987) Problems and potential for *in situ* treatment of environmental pollutants by engineered microorganisms. Microbiol. Sci. 4:59-63.
22. Jain, R.K., et al. (1987) Maintenance and stability of introduced genotypes in ground water aquifer material. Appl. Environ. Microbiol. 53:996-1002.
23. Johnson, J.B., and S.G. Robinson (1984) Release to the environment of engineered microorganisms. In Genetic Engineering and New Pollution Control Technologies. Noyes Publications, Park Ridge, New Jersey, pp. 73-82.
24. Knowles, C.J., and A.W. Bunch (1986) Microbial cyanide metabolism. Adv. Microbial Physiol. 27:73-111.
25. Krimsky, S., and A.H. Fraenkel (1985) United States and Canadian governmental regulations concerning biohazardous effluents. In Comprehensive Biotechnology, Vol. 4, C.W. Campbell and J.A. Howell, eds. Pergamon Press, Oxford, pp. 609-632.

26. Kuenzi, C.M., et al. (1985) Safe biotechnology. General considerations. Appl. Microbiol. Biotechnol. 21:1-6.
27. Mancini, P., et al. (1987) Mobilization of plasmid pH5V106 from Escherichia coli HB101 in a laboratory-scale waste treatment facility. Appl. Environ. Microbiol. 53:665-671.
28. McGarity, T.O., and K.O. Bayer (1985) Government regulations of recombinant DNA research and manufacturing processes. In Comprehensive Biotechnology, Vol. 4, C.W. Robinson and J.A. Howell, eds. Pergamon Press, Oxford, pp. 587-608.
29. Melcer, H. (1985) Control of toxic and inhibitory contaminants in biological wastewater treatment systems. In Comprehensive Biotechnology, Vol. 4, C.W. Robinson and J.A. Howell, eds. Pergamon Press, Oxford, pp. 535-541.
30. Melcer, H. (1985) Management of toxic pollutants: Technical constraints versus legal requirements. In Comprehensive Biotechnology, Vol. 4, C.W. Robinson and J.A. Howell, eds. Pergamon Press, Oxford, pp. 939-944.
31. Mudder, T.I., and J.L. Whitlock (1984) Biological treatment of cyanidation waste waters. Miner. Metall. Process., 1:161-165.
32. Nicholas, R.B. (1987) Biotechnology in hazardous waste disposal: An unfulfilled promise. ASM News 53:138-142.
33. Nicholson, E.M. (1970) The Environmental Revolution: A Guide for the New Masters of the World. Hodder and Staughton, Sevenoaks.
34. Organisation for Economic Cooperation and Development (1986) Recombinant DNA Safety Considerations, OECD, Paris.
35. Office of Technology Assessment (1981) Impacts of Applied Genetics. Microorganisms, Plants, and Animals. OTA, United States Congress, Washington, D.C.
36. Omenn, G.S. (1986) Genetic control of environmental pollutants: A conference review. Microbial Ecol. 12:129-134.
37. Owings, D. (1984) Discussion: Directions for the future. In Genetic Control of Environmental Pollutants, G.S. Omenn and A. Hollaender, eds. Plenum Press, New York, pp. 363-364.
38. Parkes, R.J. (1982) Methods for enriching, isolating, and analyzing microbial communities in laboratory systems. In Microbial Interactions and Communities, Vol. 1, A.T. Bull and J.H. Slater, eds. Academic Press, Inc., London, pp. 45-102.
39. Patterson, J.W. (1984) Perspective on opportunities for genetic engineering applications in industrial pollution control. In Genetic Control of Environmental Pollutants, G.S. Omenn and A. Hollaender, eds. Plenum Press, New York, pp. 187-193.
40. Pritchard, P.H., and A.W. Bourquin (1984) The use of microcosms for evaluation of interactions between pollutants and microorganisms. Adv. Microbiol. Ecol. 7:133-217.
41. Reanney, D.C., W.P. Roberts, and W.J. Kelly (1982) Genetic interactions among microbial communities. In Microbial Interactions and Communities, Vol. 1, A.T. Bull and J.H. Slater, eds. Academic Press, Inc., London, pp. 287-322.
42. Rollinson, G., et al. (1987) The growth of a cyanide-utilizing strain of Pseudomonas fluorescens in liquid culture on nickel cyanide as a source of nitrogen. FEMS Microbiol. Lett. 40:199-205.
43. Saye, D.J., et al. (1987) Potential for transduction of plasmids in a natural freshwater environment: Effect of plasmid donor concentration and a natural microbial community on transduction of Pseudomonas aeruginosa. Appl. Environ. Microbiol. 53:987-995.

44. Sayler, G.S., et al. (1985) Application of DNA-DNA colony hybridization to the detection of catabolic genotypes in environmental samples. Appl. Environ. Microbiol. 49:1295-1303.
45. Schmidt, E., M. Hellwig, and H.-J. Knackmuss (1983) Degradation of chlorophenols by a defined mixed microbial community. Appl. Environ. Microbiol. 46:1038-1044.
45a. Schumate, II, S.E., and G.W. Strandberg (1985) Accumulation of metals by microbial cells. In Comprehensive Biotechnology, Vol. 4, C.W. Robinson and J.A. Howell, eds. Pergamon Press, Oxford, pp. 235-247.
46. Senior, E., A.T. Bull, and J.H. Slater (1976) Enzyme evolution in a microbial community growing on the herbicide Dalapon. Nature (London) 263:476-479.
47. Steward, I.W., R.H. Watson, and G. Leaver (1987) Aerobiological monitoring for containment validation in bioprocessing. In Proceedings of the Fourth European Congress on Biotechnology, Vol. 1. Elsevier, Amsterdam, pp. 163-166.
48. Stotzky, G., and H. Babich (1984) Fate of genetically-engineered microbes in natural environments. Recombinant DNA Tech. Bull. 7:163-188.
49. Strauss, H.S., et al. (1986) Genetically-engineered microorganisms: II. Survival multiplication and genetic transfer. Recombinant DNA Tech. Bull. 9:69-88.
50. Trevors, J.T. (1987) Gene transfer among bacteria in soil and aquatic environments: A review. Can. J. Microbiol. 33:191-198.
51. Van Elsas, J.D., et al. (1986) Survival of Pseudomonas fluorescens and Bacillus subtilis introduced into two soils of different texture in field microplots. FEMS Microbiol. Ecol. 38:151-160.

ENVIRONMENTAL PROTECTION AGENCY ACTIONS TO STIMULATE USE OF BIOTECHNOLOGY FOR POLLUTION CONTROL AND CLEANUP

Marcia E. Williams

Office of Solid Waste
United States Environmental Protection Agency
Washington, D.C. 20460

In the period 1982 to 1983, when Rita Lavelle and Anne Gorsuch were in charge, the EPA lost a lot of credibility. We may never regain the credibility that everybody believed we had in 1970 when the Agency started. I think one of the reasons that we had that credibility in the early years, as Bill Ruckelshaus noted, was that the problem was a lot simpler back then. Everybody knew what they wanted the EPA to do: clean up the worst air and water problems. The EPA had a single identity and problems that it could do something about, and people could see the results. It could be done in a way that was overburdened cost-wise, yet no terrible uproar was heard over the economic side of things during that period.

I remember doing a three-month assignment on the Senate side of Congress in 1975. When I was trying to provide input to the Senate Public Works Committee on Clean Air Act issues, they said, "We need to balance everything that you say against what industry says because you are speaking on behalf of the environmental people." This struck me as very odd since at that time I did not see myself in that light. I saw myself as a person who was trying to balance issues.

Things have changed since then. Today, the EPA has a real schizophrenic set of personalities. It depends on whom you ask how they view the Agency. There is a large part of the Congress that sees us as very much pro-industry and pro-business. Given that we have over fifty committees that we have to address, some members and staff think we're too pro-environmental. Most industry folks think we're too pro-environmental. Probably the nicest thing the environmentalists have to say is that the Agency is a "shield for industry." The states generally see us as a nuisance and wish we weren't around at all. We view ourselves as trying to take laws that Congress has passed and doing the best job that we can of implementing them. It is not surprising that the public is pretty confused about the Agency.

As with so many other areas, biotechnology puts EPA at the cutting edge of science and information. We're dealing with a great sea of uncertainty surrounding us with regard to the benefits and the risks. Obviously, one would like to be able to take more time, gather more information, and think it out better before doing anything. Typically, however, the public, the Congress, and the political system won't or can't let the EPA take much time. As a result, regulatory decision-making and policies get set in place before we have answers to many of the questions.

Hazardous waste is a prominent example of this predicament of uncertainty about risks and benefits. Everybody would agree that there is no particular benefit to producing hazardous waste; but there are benefits from the products made in this country that are generating the wastes. The public now finds it very difficult to differentiate the actual environmental problem from the solutions to that problem. Often the uproar and concern are as great over the solutions as over the problems. It's very important that we not get ourselves in that situation with biotechnology. I think that there are potentially very large benefits from biotechnology; so we need time to make sure that we can evaluate the risks and the benefits carefully.

Until recently, the EPA's main focus in the biotechnology area has been in taking a look at new products--microbial pesticides or new chemicals--that a company is attempting to introduce into the marketplace. But the whole issue that we're addressing at this conference is the viability of biotechnology to be one of the solutions to our hazardous waste contamination problems. It is important for the EPA, Congress, the states, and other key players to assess the viability of this technology and to put into place a regulatory structure that, while protecting health and the environment, will encourage the development of such a set of technologies. One of the key objectives is regulatory stability. It's extremely hard for anybody to operate in an ever-changing regulatory arena.

THE HAZARDOUS WASTE STATUTES

Where does biotechnology fit into today's activities within the RCRA and Superfund programs? I want to describe how the two hazardous waste statutes deal with new, innovative technology. Then I will note the constraints to biotechnology in the current regulatory structure. Finally, I will discuss some of the research and policy issues that the EPA is dealing with today.

The current concept, shared by the U.S. Congress, the scientific community, and the public, is that it is definitely time to move away from land disposal as a viable solution for the hazardous waste problem. Land disposal is essentially a way of putting off today's problems until tomorrow. Congressional reauthorization of both of the major hazardous waste statutes in the last couple of years emphasized provisions responsive to these views.

The Resource Recovery and Conservation Act Program (RRCAP) has several provisions that try to push one toward more innovative technology. First, the core of that statute, reauthorized in late 1984, was a ban on land disposal of hazardous wastes without prior treatment of those wastes.

That requirement kicks in gradually between 1986 and 1990. To implement this requirement, the EPA needs to take virtually every kind of waste stream, group these streams by the type of waste matrix, develop a set of treatment technologies appropriate for that waste, and then set the levels to which one could expect that waste to be treated using the best means available. Once the waste is treated in that way, it can go into land disposal. For example, if incineration turns out to be the technology that's acceptable, one can incinerate and then put the residue from incineration in the ground. A whole set of treatment technologies that haven't been available in the past should be stimulated.

Second was the concept of waste minimization, as an explicit national policy, and the requirement that begin reporting on waste minimization progress.

The third big change that was put into the RRCAP was what we call corrective action, which is essentially the same as Superfund cleanup. If you want to get a permit to manage hazardous waste on your facility, you need to agree to clean up all the contamination on that facility. This program is massive and has the potential to be even larger than the Superfund program. There are today about 5,000 sites in the United States that manage hazardous waste, as a disposal facility, an incineration facility, or a storage or treatment facility. This requirement applies to every one of those sites. So the amount of cleanup that is going to be effected under the RRCAP is tremendous. Also, unlike the Superfund program, which has a scoring mechanism and attempts to pick the worst sites to be cleaned up, the RRCAP requires cleanup of all sites, regardless of whether they do or do not score above certain levels. The federal government or the state government that is running the RRCAP provides the permits, sets the acceptable health standards to which a particular facility has to clean up, and oversees the cleanup. It also decides whether or not the technology that's proposed for that site will or won't be able to reach the levels that need to be reached. There is still a critical role on the part of the owners of that site in making the decisions and recommendations as to what technologies they want to use. While the RRCAP does not have a specific requirement to develop innovative technologies, it puts in place a set of structures that will force the development and use of more sophisticated technologies.

The Superfund statute, which was just reauthorized in the Fall of 1986, has more specific requirements with regard to innovative technologies. First, it establishes a strong and clear preference for the use of cost-effective treatment remedies that permanently and significantly reduce the toxicity, mobility, or volume of waste at the site to the maximum extent practicable. That's pretty ambitious language. Before any decision gets made at any Superfund site, a full analysis has to be done to decide whether a particular treatment is the best treatment in terms of balancing those factors. There are a lot of issues in this Congressional mandate. What is permanent? Is incineration permanent? What's happening to substances potentially going out the stack of an incinerator? What's going on with air pollution? What's going on with any kind of soil pollution that comes from settling of those substances, crop uptake, etc.? If you do chemical solidification, is that permanent? What happens over 30 years? Over a hundred years? It's not like technology X is permanent and technology Y is not. But, clearly, many technologies can be more nearly permanent than land disposal of untreated waste.

Second, the Superfund amendments of 1986 set up a specific program of research, evaluation, and testing which is essentially a demonstration and development program of alternative technologies. That program is being carried out jointly between the Solid Waste and Emergency Response Office and the Research and Development Office. It's called the Superfund Innovative Technology Evaluation Program or the SITE Program. The Agency expects to fund about ten innovative technology sites per year between now and 1991, which is when the current Superfund authorization runs out. In the process, the Agency will be trying to gather cost and effectiveness information on new technologies. I'll give you a couple of examples of some of the projects that are being considered and worked on at present. There is a biodegradation technology project where naturally-occurring microorganisms are being adapted and then inoculated back into the soil. There is a basic extraction sludge treatment project which is being used on difficult-to-handle oil sludges. Part of what the process does is separate the different phases of the waste. There is an infrared incineration system. Between 12 and 15 projects are actively being looked at right now.

While both the RRCAP and the Superfund statutes are in effect, at EPA we've committed ourselves to an integrated, consistent implementation of these two programs, to avoid creating interests or incentives for people to clean up a location under Superfund versus RRCAP, or vice versa. We ought to get the same kind of cleanup out of both programs. This means that any requirements that the RRCAP sets up, such as land disposal ban restrictions that require treatment of these wastes in advance, will apply equally at Superfund sites, and vice versa. The search for permanent technologies mandated by Superfund will also be backed and applied in the RRCAP. This effort should be helpful to all parties. As far as the Agency is concerned, there are not going to be two entirely different sets of ways to clean up.

CONSTRAINTS TO ENVIRONMENTAL BIOTECHNOLOGY

I think there are two general kinds of regulatory barriers to biotechnology. The first is the complexity of the regulatory system. Nobody wants to get into a game with rules they can't understand. The second barrier is the uncertainty over future federal statutes and regulations, over which we don't have a lot of control, and over state regulation. In 42 states the hazardous waste program is state-delegated. The state may be more stringent than the federal requirements. So you're dealing with a two-tiered set of regulatory requirements.

In addition to these general situations that are hard to quantify, there are other, much more specific constraints. Under the Toxic Substances Control Act (TSCA), EPA is in the process of reviewing new chemicals before they can go to the market through its premanufacturer notice program. As regards biotechnology, they have identified which of the biotechnology organisms they need to see before these come onto the market as a new product, specifically those where intergeneric transfer has taken place. But they are also looking at three additional regulatory requirements.

One is called a "significant new use rule." Even if approval has been granted or the organism is on the market now, if your organism is one of those where intergeneric transfer has taken place and if you want to change the use of it and apply it to a new area, you would need to undergo another review for the new area of use.

The second kind of requirement is called an "8A information reporting rule." This rule would ask or require reporting of general information on environmental uses of microorganisms that are broader than the ones I just mentioned. In other words, with natural organisms, we want to know what's going on in the field. That particular EPA office is today asking people to submit information voluntarily. I think it is in the interest of everybody in the field that EPA have such information. But it looks like the Agency is heading toward a decision to make the reporting of this information mandatory.

The last requirement, which perhaps is the most important, is a rule that is going to define research and development for the purposes of TSCA. Research does not have to come in for premanufacturing notice review. But just as the Office of Pesticides had to define what an experimental use for biotechnology is--and maybe it's not the same for biotechnology as it is for other types of chemical situations--the same kind of discussion is going on with regard to new products. There will probably be a definitional rule which could be extremely important if it narrows the definition of research and development with respect to biotechnology.

Now let's return to RCRA for regulatory constraints. The 1984 reauthorization clearly specified that land treatment--normal land treatment such as oil refineries that are doing soil biodegradation--is considered to be land disposal for the definitional purposes of the Hazardous Waste Act. Thus, land treatment falls under the requirement to pretreat waste. You cannot put waste on the land before it is treated. The only way to get out of this box--of not having to pretreat the waste before you put it on the land--is to be able to make what is called a "no migration" finding. That is, you must persuade us at EPA that there will be no migration of the wastes from the particular unit in which they are being managed for as long as the wastes remain hazardous.

This is a very hard finding to make for a current land treatment unit. The unit is defined as being the confining area, which is probably three feet down in terms of ground, and the surface at the top of the unit in terms of any releases to air. I will touch briefly in a moment on whether this restriction is good or bad, but it is presently the law.

An additional constraint with regard to RCRA is that, if you are going to treat waste under RCRA, you have to get an RCRA permit. An RCRA permit is not the easiest or fastest thing to get. I'm sure that if any of you have had this experience, you have less nice things to say about it! Because Superfund needs to cover the RCRA requirements, if you're treating waste at a Superfund site, and if the waste is in the ground and is not picked up, and if it stays in the ground in its original position, and if one wants to inoculate the soil with certain kinds of organisms, then the land ban is not kicked in because you haven't moved anything. If that waste is picked up to be moved to a separate location for the purpose of treatment, it kicks in the treatment ban requirements. This situation is obviously problematic, because there are many situations that people have pointed out to us where it's not practical to treat the waste in exactly the location where it is.

The other question with regard to Superfund and the issue of regulatory barrier constraints is the whole interpretation of "permanence to the maximum extent practicable." How will biotechnology stack up against some of the other technologies that are out there in regard to these issues?

EPA INITIATIVES IN ENVIRONMENTAL BIOTECHNOLOGY

First, we have "research and development permits." Fortunately, R&D permits take less time to obtain than regular RCRA permits. We've been able to issue a fair number of them. The average time to obtain one is about six to eight months. So that's good to know if one wants to get into the game quickly. Clearly, there are opportunities to use biotechnology in some of the RCRA or Superfund cleanups, and in some of the RCRA closures of old facilities. (Obviously, they are subject to the kinds of requirements I've just laid out.)

There is the possibility that one could get a normal RCRA operating permit for a facility that would basically use a biotechnology process to degrade newly generated waste. Again, the land ban would kick in only if it is a land treatment approach. If it involves biotechnology in tanks or in closed containers, the land ban doesn't kick in.

From a research perspective, the EPA is a drop in the bucket. There is no way that we can be the leaders in research in this field. What we hope to do is to stimulate good research and be a part of the game, but it's going to take a lot of interaction among government, academia, and industry to get the necessary research done.

We are right now primarily focusing our research toward demonstration of the viability of the technology and the degree to which we do or don't have concerns with its impact on health and the environment. We are also looking at some of the approaches that one might be able to use for risk assessment of this technology. Projects that are underway include the study of genetic transfer and survivability, research on biological self-containment, and predictive studies on biodegradation.

What I want to make sure I share with you is some of the policy thinking that is going on, which is likely to result in some future decisions in this area. We are putting a fair amount of time into several issues. One is whether or not land treatment should or should not be considered in the same way as land disposal is. Once the EPA gets its thinking together, we will probably have some extensive dialogues with Congress on this particular issue. They are interested. This issue transcends biotechnology.

Another question is how we would permit a biotechnology unit if someone came in and wanted a permit. That hasn't happened yet, particularly for some of the more potentially sophisticated biotechnology applications. Should we put out particular sets of standards? We have a recent general provision for permitting kinds of facilities that don't neatly fit under any other category. So the question is, "Should we be fairly specific, or should we leave it general and wait until the first case comes in and see if we can come up with a good set of permit requirements?"

We are looking at the degree to which we might be able to use biotechnologies in helping to establish treatment levels for the waste prior to land disposal. An area that I would like to see us explore is the role that biotechnology might potentially play in waste minimization. Are there alternative ways to make chemicals that would reduce the amount of hazardous waste that is being generated?

Finally, one of the big policy issues is whether or not the waste from the development of biotechnology itself needs to be regulated. At present, clearly, there is a fair amount of biotechnology being practiced in the pharmaceutical and agricultural areas. Is that of concern? Do we need to look at the waste products and see whether certain management requirements are necessary?

Clearly, the EPA has a dual role with regard to biotechnology. Our overall role is to protect health and the environment, and that is something that we have to do with all areas in which we are involved. On the other hand, we need to stimulate good, innovative approaches to the cleanup. It certainly seems as if biotechnology needs to be fully explored to see whether it can fit the bill. We would like to help make sure that exploration happens. At this point, the areas that seem to have promise are the cleanup of existing sites and the treatment of new wastes, so long as that cleanup is being done in closed systems.

COMMENTARY: SOME ISSUES IN BIODEGRADATION

Martin Alexander

Department of Agronomy
Cornell University
Ithaca, New York 14853

As an individual, not on behalf of the EPA Science Advisory Board, I would like to share with you some of the problems I have observed with reviews of a variety of technologies. I have been with the Science Advisory Board since the marquee at EPA was first lit up, except for a brief period in 1981 when there was an enforced sabbatical leave for some of us. I would like to indicate some of the issues that I see in the area of biodegradation in the broader sense, and in applying genetic engineering to biodegradation in a narrow sense.

First, how can one most rapidly enhance biodegradation? An obvious need is to add more oxygen, nitrogen, and phosphate. There are some very obvious difficulties. For example, Perry McCarty mentioned all of the water containing oxygen that has to be moved into the system in order to have rapid degradation. How can one get phosphate to move through many environments, when phosphates form insoluble salts? In an anaerobic environment, nitrate as a nitrogen source would be denitrified, and ammonia moves slowly. What can one do in a very practical sense?

A second issue is anaerobiosis. We have discussed the virtues of anaerobic treatment and, to a lesser extent, of aerobic treatment. Where, in fact, would an anaerobic process be better under realistic conditions than an aerobic process, and vice versa? The operational criteria are not clear at this time.

Issue three is inoculation. Inoculation might involve adding discrete existing organisms, the establishment of a remediation column, a soil perfusion apparatus, or adding a genetically engineered organism. I have reviewed the published literature on inoculation and found that there are essentially the same number of failures as successes. Why do the failures not work, and why do the successes work?

It is naive to assume that the possession of one trait---that of biodegradation---is sufficient in an inoculum strain to esure its success. As an agronomist, I have worked with rice. The International Rice Research Institute developed wonderful varieties of rice with high yield. That is what was wanted, more rice production. Unfortunately, when the rice was fertilized, the plant kept growing and then fell over into the water. It could not stand up straight. In addition, predators, parasites, a variety of pests, and ecological constraints affected the plant. The one trait--high yield--was not enough. The same is true of biodegrading microorganisms.

Consider toxins. Waste disposal sites, if they are treated with microorganisms, will pickle those organisms. If any of you have taken the mixture of chemicals in many of these disposal sites and inoculated it with some species, you have found that the outcome is a sterile solution. Under these conditions, biological treatment may not be the way to go, or it may have to be combined with absorption or stripping.

There are other issues. For example, acclimation. In some cases, acclimation is not evident. But suppose a chemical is released episodically into the Mississippi River near St. Louis; by the time the organisms gird their loins, the chemical will be consumed by the people drinking the water of New Orleans. Can one enhance acclimation? Under these conditions, what are some of the research needs in this area?

When I read through the documents sent to me by governmental or private agencies for review, I find problems or potential problems with the movement of organisms. The public health literature says that bacteria do not move through solid matrices. But the public health literature is concerned with the organisms not appropriate for biodegradation concerns. If we consider *E. coli* or *Streptococcus faecalis* as the indicator species, will we conclude that organisms will not move with the chemical? In fact, one of the characteristics for success in biodegradation is mobility in the environment. I know of no researchers studying organism traits who are concerned with mobility over considerable distances.

Consider sorption. Chemicals often are not in the free aqueous phase. They are sorbed to soils, sediments, organic detritus, or aquifer material. Do organisms really use many sorbed chemicals? Will they have to wait for the chemical to go into solution, which in many cases will be very slow? Can one develop organisms that, in fact, enhance removal of the chemical from surfaces? Simply getting an organism to degrade a soluble compound is not enough. It has to work on the compound as it exists in the environment.

The same questions apply to many insoluble chemicals. The literature on the mechanisms of utilization of water-insoluble compounds deals with chemicals that are not toxic to humans at ambient concentrations. Much information exists on nonchlorinated alkanes but there is little concern with nonchlorinated alkanes! One may be able to engineer many traits into a bacterium, but can one engineer it to use the water-insoluble chemicals?

Consider kinetics. Some investigators, my laboratory included, have investigated rates of disappearance of chemicals in order eventually to predict how much will be present in nature at some future time. However, none of the kinetic models is particularly convincing when only a small percentage of the parent chemical remains. First-order kinetics, zero-order kinetics, three half-order model kinetics are very good, and may allow us to publish papers. But how sound are they for determining the kinetics under natural conditions?

A final issue is the matter of risk. What is good for the goose is good for the gander. While sitting on the EPA Science Advisory Board or on other committees examining risk, I have generally evaluated the risks of someone else's technology. In this case, I am concerned about the risks in my own field.

Let me cite just two risks. First, biodegradation often does not remove all of a chemical. Some may be totally removed as we learn more through research. Some will likely not be removed by any biological treatment technology, at for least not in the foreseeable future. Chemicals will come out of the effluent. Other things will remain at the waste disposal site. What can one do to remove the remainder?

Second, organic products accumulate in many biotransformations, and those products may be more toxic than the precursors. I would be happier drinking water containing perchlorethylene than the product of microbial degradation, vinyl chloride. I would be happier having in my agricultural system pentachlorobenzyl alcohol than the microbial product formed from it, trichlorobenzoic acid.

Let me conclude on a philosophical note. Academic scientists ask many questions, and we on the EPA Science Advisory Board can be very critical. Yet the problems and issues remain. Relative to trying to solve important problems, recall the last sentence in Voltaire's Candide. After hearing all of Pangloss' comments, Candide says: "'Tis well said, but let us cultivate our gardens." Let us thus go ahead and do the job.

DISCUSSION

Gilbert S. Omenn (Chairman), University of Washington

BAKST: Dr. Kingsbury, the original coordinated policy statement of December 1984 was a process-based framework. In response to comments received on that policy statement, the second policy statement, which you have described, was organized around an intrageneric approach. Are some of the agencies now reconsidering the process-based approach?

KINGSBURY: Let me provide some context for this question. The December 1984 notice, especially from EPA, involved a regulatory net that would catch everything manufactured by human intervention. Many respondents recommended that the agencies regulate products, not processes. The 1986 notice from the EPA moved toward a much more product-based approach. But I think that's an oversimplified notion. The intergeneric criterion defined both the product and certain aspects of the way the product was derived. Over time, I think there has been wider understanding of the definitions. Recently, the Animal Plant Inspection Service (APIS), the regulatory side of the U.S. Department of Agriculture, issued its final notice about July 1 in the Federal Register; the definitions that were originally agreed on were maintained. The APIS did drop self-fusion as a regulated activity. Each of the agencies is responsible for how it defines a product. Advisory committees, such as the one that Dr. Alexander sits on for EPA, have been very much involved in the decisions. I don't know how it will finally work out. But I don't expect a major change in the approach. There's going to be a combination of process and product bases because it's inherent in the definition of biotechnology, and we've decided that biotechnology is something that we're going to regulate.

FOCHT: One of the problems that scientists have in trying to apply the results of basic research to the field, which is becoming quite acute now, is the issue of liability. Experiments fail. They even fail in the laboratory under ideal conditions. In some cases, we simply don't know why. They fail in cases where we're not familiar with the environment--the Stringfellow acid pits near my home, for example. Under no circumstances would I, as a university professor, touch those pits with a ten-foot pole. There's mercury, cadmium, God knows what in there. Yet, my role as a university professor is to be responsible to the community. How do we address the issue of liability? How are we going to handle the situation where we think we have a good plan, where we think we know the environment, and where we're fairly certain that it should work; yet we go out and do a field test, and it fails, and somebody gets sued for failure? I think this is a situation that one must deal with if one is going to apply this technology. Maybe we need a lawyer on the panel.

ALEXANDER: As a scientist, I usually put in my final statements the words "We suggest," "We propose"; I rarely write "We conclude." When we have field tests that don't work, I think most of us do not even think about the liability question.

WILLIAMS: I don't have an answer. However, I personally have been spending a fair amount of time trying to deal with the liability issue on hazardous waste sites. As many of you know, liability insurance is almost nonexistent. Insurance companies and reinsurers keep saying, "We're only willing to get into the game if we understand the risks that come from that kind of situation."

I think that if an experiment is designed in such a way that what is going on during the experiment can be measured, it may be possible to educate the insurance companies and get limited insurance. The fact is that you're basically liable in this country for anything that you do. There's no way around that. So the challenge is to get somebody, presumably through insurance, to protect you. I'm not necessarily hopeful. I'm only more hopeful than I am in the hazardous waste situation, where the reason they won't play in the game anymore has to do with the fact that they aren't willing to accept all the problems that have come in the past which they're likely to be liable for as soon as they walk into the hazardous waste site. Biotechnology won't have that issue going against it.

OMENN: Dr. Kingsbury, is liability with regard to research and development becoming a concern in the policy discussions with which you're involved?

KINGSBURY: No, it isn't. Dr. Focht's inference was that a research program would generate expectations that you were going to achieve something, thereby leaving liability when the results aren't achieved. I think that most people don't go into research programs that way. If you don't infer that there's more to an exploration than is there, liability should not be a particular problem--assuming that one follows the principle of "do no harm."

LEWIS: At the NPL treatment sites, an important step is the feasibility study. A variety of methods may be tested--chemical, physical, or biological, or combinations. Often just a small segment of the site is tested to determine whether the approach is viable and preferable. Last year, John Wilson of the EPA/ADA Lab and I were asked by the Dutch government to participate in meetings on biodegradation, to see a number of their sites, and to advise them on some of their research. In Holland they often do land treatment now by spreading a synthetic liner out, then putting the soil on it, and providing run-off to sampling areas before release. They aren't just driving the wastes down deeper into the soil, but instead are facilitating and monitoring biodegradation.

KOBAYASHI: I appreciate Dr. Alexander's comments about biotreatment of various kinds of waste, including difficulties from toxicity. Our diversified industry generates many different kinds of waste, so we encounter many of these problems. Regarding land treatment of oil waste, Ms. Williams, would you say that this approach is out of the picture? Would it be allowed in a contained site? Is there hope that it would be considered a treatment technology for the future?

WILLIAMS: I probably get more Congressional questions on land treatment than any other individual subject. In order to continue land treatment without any pretreatment of the waste, the no-migration finding will be required. We believe that some land treatment facilities will be able to make that finding. But many won't--for example, land treatment facilities that are located in groundwater, in the saturated zone. It's going to be case-specific. Air migration may be a bigger issue than groundwater migration, if there are any volatiles in the waste. Some companies are exploring the possibility of pretreatment of oily wastes to strip off some of the volatiles and then going ahead with land treatment as we currently know it. So there are options, but our practices will have to change significantly.

COX: As I understand what you've said about the creation of public confidence in biotechnology, Dr. Bull, and from my own knowledge of the United Kingdom, the academic and the policy-making interests are very much concerned with the regulation of organisms result from recombinant DNA technology. In contrast, the industry, which, though small, is developing very fast, particularly in Europe, is moving to apply microbiology using naturally occurring and selected microorganisms. I think this divergence can cause tremendous confusion in public perception and public opinion of what's going on. I'd like your comments, Alan, and then I'd like the panel to indicate more generally what lessons we Europeans might learn from the United States.

BULL: I'm not sure that I agree with your premise about public concern. I've always been struck by the fact that with new technologies, the acceptance may come from other areas rather than the particular area one is addressing. I'm really quite encouraged, actually quite amazed, by the degree of acceptability of aspects of biotechnology that people relate in public opinion soundings, particularly in the health-care sector. I realize that there are certain ethical and emotive issues that trigger opposition, but I feel that acceptance through that route will make the general public more receptive to some of the approaches we've been talking about at this conference. I don't perceive a big distinction between the academic and the industrial approaches.

KINGSBURY: I've been particularly concerned about the use of the word "biotechnology," which I would like to have removed from the lexicon. This term has become increasingly imprecise. When we first started focusing on it, we thought of the biotechnology industries as the companies that were commercializing monoclonal antibodies and recombinant DNA. Over time, we have been subjected to "biotechnology creep," partly to generate investor interest and extraordinary corporate valuations. Naturally, other companies, that are not using recombinant DNA, but that are using biological approaches, seek to ride that wave to the capital markets. Thus, the boundary conditions around what biotechnology is continue to expand, now including the application of selected organisms. Environmental applications of microorganisms for nitrogen fixation in leguminous and nonleguminous plants have been going on for well over a hundred years. This has been well accepted. Now we've redefined it as biotechnology and have raised a red flag.

FURLONG: I'd like to applaud the efforts to generate an information database. I urge early planning of ways to distribute the database. With the advent of inexpensive CD ROM drives and modern indexing algorithms, it's possible for all of us to access the pertinent literature on our desk tops within seconds.

We are using the distributed mode, as in Medline and Toxline, as a paradigm. We hope to develop a peer-reviewed database, with an expert board, instead of just loading everything in there. We're struggling with getting the funding.

BULL: May I make a heretical comment? I'd just be a little cautious about databases. The feeling that any information is better than none in guiding legislators and ourselves in the scientific community in producing guidelines could be counterproductive, simply because we may not have the necessary information from all the relevant domains. I do agree that ongoing update and review of that information is absolutely crucial.

KINGSBURY: We also recognized that what we call the "gray literature" is as important as the published literature, namely, the technical reports that have come into the agencies. That's where we get the negative data, as long as they are appropriately peer-reviewed, because journals don't publish negative results.

ALEXANDER: We have a vast literature about inoculation of organisms that are not in the computerized system. Beware; people who start with databases end with the same databases.

KINGSBURY: According to a paper in 1926, every organism that was in pure culture in the United States had been added to a field with some kind of a leguminous or nonleguminous plant to see if it would fix nitrogen. It was an unbelievable list, including some human pathogens to see if they would colonize corn plants.

CORPORATE PANEL

USERS AND PROVIDERS OF THE NEW TECHNOLOGIES

Gilbert S. Omenn (Chairman), University of Washington
Christopher Bayley, Burlington Northern Railroad
Gaylen R. Brubaker, International Technology Corporation
Dick A. Davis, Univar Corporation
Dale Pflug, BioTrol, Incorporated
David Sikes, ECOVA Corporation
Ronald Unterman, General Electric Company
Michael Weishaar, Monsanto Chemical Company

OMENN: I've organized our panel into two groups: the users and the providers of these new technologies for pollution control. We would like to know how the users and the corporations they represent look at the emerging technologies. What kinds of strategies are attractive in dealing with these serious chemical hazards, as perceived by the public and by our regulatory system and statutes? What are the relative positions of legal tactics, landfills, incineration, biotechnology, and other combinations of technology? How do they size up what we scientists seem to offer? Then the biotechnology providers will have an opportunity to respond.

DAVIS: It is timely to consider both the innovative technologies and the practical ramifications of reducing environmental risks associated with soil and groundwater contamination. I represent Univar Corporation, the world's largest industrial chemical distribution company, with headquarters here in Seattle. We distribute over 3,500 different chemicals, both hazardous and nonhazardous, from over 100 facilities located throughout the United States and Canada. Our role is to provide the customer with chemical products for a full range of manufacturing industries. We provide delivery quantities ranging from full tanker or railroad car loads to such small containers as 55-gallon drums or even five-gallon pails. We also issue our customers health and safety information through comprehensive labeling and material safety data sheets. Finally, we provide our customers custom packaging and blending services to meet their particular needs. Thus, in any of our operations we may inventory, repackage, and ship hundreds of chemicals daily. These products range from chlorine gas for water purification to perchloroethylene for dry cleaning. No doubt some of you may consider us part of the environmental problem; we can be, and we are in some locations.

But we also think we're part of the solution. Not all of our operations are environmental state-of-the-art distribution facilities. We have operated at some places for over forty years. In other cases, we have acquired pre-existing distribution businesses. The result is the same. Past practices have at times produced detectable levels of environmental contamination in some of our facilities. These are normally not Superfund sites, although I personally am involved with one or two of our sites where we've been named as a PRP [potentially responsible party]. These are not horrifying cases of stockpiled corroding drums. They are simply working chemical distribution facilities that over the years have experienced incidental spillage which resulted in detectable levels of contamination.

Since they are working facilities, we are faced with several options. These options include: (a) taking no action; (b) closing down the site or selling it to someone else; (c) removing contaminated soils and water to an off-site treatment or disposal area; or (d) applying in situ treatment technology. As my presence here indicates, we've got a bias toward in situ treatment, since it becomes a permanent solution to a contamination problem, rather than a transfer of the problem to another location.

Conceptually, these technologies represent economical and safe mechanisms for effectively managing contaminants, often with minimal disruption to our day-to-day distribution operations. Whether in situ technologies involve physical, chemical, or biological treatment, however, we've observed certain disadvantages in their implementation. First, because each facility has its own unique geography, field applications of these technologies can fail unless designed for specific sites. For example, climates vary, and the success of bioremediation is strongly influenced by the local climate in certain applications. An approach that worked in Tampa, Florida, may not be directly applicable in Minneapolis, Minnesota.

The work required to design the site-specific remediation plan is termed research by some. Research takes time; it costs money; it connotes experimentation and also connotes some risk of failure; and, in many cases, it requires a pilot study. All of the tasks require time, and time means delay. Delay means that the plume is moving. Often the easy way out of this dilemma seems to be off-site treatment and disposal.

Another disadvantage we face relates to regulatory affairs. There are substantial costs involved in gaining acceptance of an innovative technology by a regulatory agency. If your site is not a Superfund site, and if your contaminant levels present are on the borderline of environmental or public health risks, you'll often be working with state and local authorities. These individuals are not prepared to act quickly, due to a number of constraints. We often put these regulators under a great deal of pressure by asking them to establish "How clean is clean?" The result can be indecision, unnecessary expense, and excruciatingly slow response to the question "How can we move ahead?"

From this very brief introduction, one might think that we do not seek out innovative approaches. However, we do. We continue to enhance our business capabilities in designing, negotiating, and implementing effective cleanups. After all, we drink the same water that you do, we eat produce grown on the same soil. We're committed to providing an environmentally responsible future for everyone.

We currently are treating one site in San Jose, California, using bioremediation. Several of the gentlemen at this table participated along the way. I'm happy to report that we have been very successful. We have contained a plume. It's not moving away any farther in the groundwater. Next we're going to see whether we can tackle the soil issue, which is basically the source of contamination. Our success in this area will depend to a great extent on our ability to convince regulators to try new technology.

BAYLEY: I'd like to give you an understanding of the Burlington Northern companies and how we are interested in the subject of this conference. Burlington Northern, Inc., is a holding company that includes a railroad company, which itself is made up of eight different railroads, some of which have been in business for 120 years. The holding company also includes a real estate subsidiary, the Glacier Park Company (GPC), which I head. The real estate subsidiary has recently inherited from the railroad the management of all real property that is not actually used for railroad operations; we've become one of the largest land holders in the United States.

Also within Burlington Northern are other companies that have problems of remediation and other environmental problems. We have a timber company and a mineral and coal exploration company. We have a natural gas company, the El Paso Natural Gas Company, which is the major supplier to the state of California; and we have one of the largest independent oil and gas exploration companies. In all these activities, we have a lot of environmental problems and concerns. Since GPC is now the manager of most of the land that isn't useful for one of these other resource purposes or that doesn't have railroad tracks on it, we are the primary company within Burlington Northern concerned about environmental liabilities. Like most companies, we've been caught up in a rapidly changing regulatory environment. We're trying to manage these problems as responsibly as we can, seeking opportunities to reduce our liabilities and costs. Biotechnology is one of the approaches that we have turned to.

I will organize my remarks around three areas. First, a brief description of our environmental liabilities. Second, a statement about the priorities we have in managing them. Finally, a brief look at a case study which exemplifies the way we approach these problems.

Ten years ago the managers of the Burlington Northern companies dealing with environmental problems were in a totally different environment than we are today. As you all know, the regulatory framework in which we operate is shifting. Things you could do to dispose of wastes in the past, you no longer can do; things you can do today, we may not be able to do tomorrow. This requires a great deal of flexibility, not only in terms of the techniques and technologies, but in relations with government officials at all levels. Some disposal sites that we may have thought could be forgotten are now problems again. We have to go back to these sites and control past problems under today's new sets of rules.

In addition, we are concerned not only with property that we have operated, but also with property that we have leased. As you know, the liability for pollution problems cannot be avoided by leasing or by any other legal technique. It attaches to everyone who has anything to do with the land. So, as a large private land owner, we are also a large lessor. We're responsible for our practices, past and present, and for those of our lessees.

All of these problems have received the close attention of upper management at GPC. We're committed to cleaning up problem sites in a responsible way and insuring that our current practices do not create future problems. We have developed and promoted within GPC an internal environmental staff of seasoned professionals. We've charged them with managing these problems, using their own resources and those of outside vendors in a sound and cost-effective manner.

In terms of our priorities in dealing with these environmental problems, we favor the active approach, rather than waiting for something to be discovered. As such, environmental issues have become an important part of our strategic planning effort. We don't want to wait until something is placed on the Superfund list before dealing with it. Nor can we wait for others to develop new solutions for these problems. We have the responsibility to solve these problems in a timely and environmentally sound way; and we have the fiduciary responsibility to minimize our cleanup costs. Our objectives, therefore, are quite straightforward: (a) to identify our environmental liabilities and to implement a plan for remedying them; (b) to implement an environmental audit program so that we're in compliance as we proceed with our activities; (c) to work cooperatively with the regulatory agencies so that the solutions that we come up with are acceptable to those agencies; and (d) to sponsor research and development work into new, cost-effective solutions to these problems.

Let's now look at a case study that shows you the kind of problems we have--a former railroad tie treatment plant in Brainerd, Minnesota. All of you know what a railroad tie looks like. You probably have them in your garden holding up soil. If you take a close look, you'll notice that it's impregnated with creosote. This process historically has produced some significant environmental problems and some Superfund sites in the areas where tie treatment plants were located. Brainerd was listed in 1983, due to soil and groundwater contamination caused by leakage from the lagoons where the residue of this process had been placed.

We entered into a cooperative agreement back in 1983 with the state and with EPA to investigate and remedy the contamination. As part of the effort, we proposed to complete bench-scale and pilot-scale studies to evaluate the feasibility of biological treatment of the waters. Frankly, the agencies were skeptical that such an approach could be successful. But we convinced them that on-site treatment would minimize our long-term liabilities and that the data developed from these studies could be useful, not only to Burlington Northern but to the public at large. After these initial negotiations and studies, we implemented a two-year program demonstrating that the technology can be effective in treating these contaminants.

During that period, we worked closely with these agencies, bringing them in as partners. They approved the results of the studies, and we constructed a full-scale treatment facility in 1985. This system has now been in operation for two years. It has exceeded our expectations. We estimate that this method of treatment has saved Burlington Northern somewhere between $3 and $6 million at the outset, compared to what would have been spent with conventional methods. More importantly, this is the first Superfund site that has been cleaned up using this particular technique. It's paved the way for other cleanups, both on our sites and other sites, and I think this site can be held up as an industry model for cooperation and for the use of this technology.

Since the Brainerd experience, we have continued the use of these techniques at Glacier Park sites. Our friends at Ecova are now helping us in the management of three Glacier Park fueling-facility sites in the state of Montana. They bring to us more advanced techniques than we were able to do ourselves. I'm confident that this particular use of biotechnology is a permanent part of our attempt to solve these problems and a way to deal with them cost-effectively. This is certainly true with tie treatment plants, fueling facilities, and other sites that have been contaminated as a result of decades of use by the railroad.

In closing, it's our belief that biotechnology will play an important role in developing environmentally sound and cost-effective solutions for our hazardous waste problems. We also recognize that there are scientific and administrative obstacles to all forms of this technology. But we'd like to be the ones that help with the innovation, working with the vendors. We feel that there are many opportunities to apply this technology in order to effectively manage the country's environmental problems.

UNTERMAN: I'm representing Steve Hamilton, Manager of Technical Assessment for General Electric's (GE) environmental problems, who is based at our corporate headquarters in Connecticut.

General Electric can act as a model for getting more research or research funding. We have many PCB-contaminated sites. We began a research effort six or seven years ago in response to those sites. The technologies available then were truck-and-dump. I think most of us would agree that's not much of a solution. So we began research programs to find alternative technology. I spoke yesterday about biological treatment of PCBs; we also have explored chemical and physical extraction techniques. Let's not use old or expensive technologies that really are no solution.

It's possible for state agencies to enter into agreements with companies held responsible for industrial pollution problems. Agencies can apply the money paid as penalties to a fund that supports research and development in academia or elsewhere on hazardous waste technologies. Alternatively, they can accept compulsory research in lieu of a penalty. The problem is still there, and we must come back to that.

Our PCB research, about $10 million over six to seven years, has gone beyond some of our own problems. For example, bioremediation is a viable option for wood-preservative sites with pentachlorophenol or phenol. The problem is contamination with 60 parts per million Aroclor 1242. How do you biotreat that site? People familiar with our research call us, asking "Can you help us?" I say, "Sure, let's try. Send us some soil. Let's see what we can do. Maybe our organisms can deal with Aroclor 1242." In one case, we undertook some research to address a particular state problem involving Aroclor 1260. At our meeting with the state agency last year, I was describing the breadth of our search, admitting: "I know you're probably not terribly interested in this, but let me tell you the status of our aerobic work." The state official interrupted me and said, "Ron, we have other sites that aren't yours, that we're also interested in."

Martin Alexander said that EPA has no time and no money. Well, think about the money involved when companies are forced to truck waste. I know of one small company that trucked 150 loads of soil about 1,500 miles at a cost of $1.4 million. It was soil contaminated with Aroclor 1242, a PCB mixture that might have been biodegradable. The $1.4 million was spent not on research, but on a nonsolution. The soil is now sitting in another state. I think we should keep that in mind.

We want to address a PCB-specific issue, too. Unlike dioxins or PCDFs, all PCBs are considered toxic. Is that really a rational approach? As we look at waste problems involving PCBs, shouldn't we recognize that Aroclor 1331--which is mostly monochlorobiphenyls and includes a lot of biphenyl--is not the same as Aroclor 1260? It's not the same as 3,4- or 3',4'-tetrachlorobiphenyl. Dioxin is evaluated on the toxicity of the 2,3,7,8-tetrachloro compound and on the bioequivalence of other compounds against this standard. This is not done with PCBs. Steven Safe has done some very interesting work on what configuration really makes the PCB toxic in some of the animal studies where toxicity has been shown. Shouldn't we be evaluating sites on that basis? Shouldn't we evaluate a site not on what was dumped there but on what the contaminants are today? Polychlorinated biphenyls are not sitting there unchanged. We've demonstrated at many sites a transformation of PCBs which we believe is due to anaerobic dechlorination. Steve Hamilton has been talking to both federal and state agencies about this issue. Let's get a PCB equivalent, as we have a dioxin equivalent.

Finally, I would like to note that we don't just deal with old problems. Polychlorinated biphenyls are an old problem, in the sense that we have not discharged them into the environment for many years. General Electric is very actively pursuing biotechnology for our current waste problems. We have biological treatment of wastes from our chemical business such as phenol, acetone, and 1,4-butanediol. At one of our plants, we have an input of 1,000 or more pounds of phenol per day, and what leaves the plant is less than one pound, with concentration in the parts per billion range. (Of course, phenol is not a difficult compound.) At the R&D center we are looking for biological solutions also for waste streams from a variety of new chemical products. We are trying to develop processes that don't generate waste, for example, trying to change the chemistry of a polymer.

I could give you one other example where it costs a little more to do something today that in the long run is probably cheaper. At our R&D center in Schenectady, the staff of 200 generates a lot of hazardous wastes from the organic chemistry labs. We used to drum that waste and take it to landfills at roughly $100 per barrel; we're now incinerating those at $150 per barrel.

WEISHAAR: The major thrust of Monsanto's biotechnology business is not waste treatment. And that's by corporate design. If there is some technology fallout applicable for waste treatment, all the better. All our research in the waste treatment area has been with outside consulting firms and contractors. When the environmental and health issues of a major corporation or chemical company like Monsanto approach 15% of your cost of goods, it's a big part of doing business. We maintain a large staff to address those issues. The issues of the remedial projects and Superfund sites are accounted for out of a separate budget, through corporate reserves. So there's a significant amount of monies to address these issues. We are using outside companies to utilize biotechnologies in some of our reclamation projects.

OMENN: We have identified the remaining panel members as vendors, providers of engineering and microbiological services.

BRUBAKER: We are in an emerging industry that's discovering new kinds of technology. I think the three days of this conference have convinced us all more than ever that it's a partnership. On the panel today are four major companies which have joined in the partnership. Drs. McCarty and Alexander especially highlighted practical day-to-day problems that we need to solve. There's no way that providers of these services can invest the money and development effort to solve some of these problems without the shared risk-taking of the industry partners. Each project that we go into has a potential financial benefit. If the solution works, it will save $1 million or $10 million per project. So the investment of say, $100,000 or $50,000 on a bench-, laboratory-, or pilot-scale test is very reasonable and gets us to the next level of problem-solving, and to more reasonable solutions that can benefit all of us.

SIKES: Companies are being forced or asked of their own volition to spend very large sums of money to solve problems. As an emerging growth industry, we who represent the new, hopefully most cost-effective solutions must demonstrate that what we offer and what we propose have a sound scientific basis and are legitimate practical approaches. As this conference has demonstrated, the use of the biological sciences in hazardous waste management has finally gotten people's attention and legitimate scientific endeavors have been performed. We no longer see the individuals who, up until a very few years ago, were purporting to have "a dioxin bug in my back pocket." I think such hucksterism slowed the acceptance of the biological sciences as one of the baseline technologies of hazardous waste remediation, because a number of big companies have been burned by people promising an instant panacea. I know that we are committed to making sure that what we charge for is based on very good, sound science.

PFLUG: We face an environment today in which the people who are attempting to make the decisions to move from containment and redisposal strategies into treatment strategies are considered pioneers. Will I be the first one to try something and fail or make some mistake? We need to have courage, based on sound scientific and engineering principles, to make decisions that we are prepared to defend and then do the things that need to be done. For example, it's very difficult today to do treatability studies in a laboratory or in a pilot plant facility without full-scale permitting.

Let me cite more specifically the issues associated with permitting the mobile treatment units. Right now, there are draft federal regulations pertaining to the permitting on a state-wide basis of mobile treatment units. Yet, throughout this country, we have a wide disparity of attitudes about the rights of states to have more stringent requirements for treatment than the federal guidelines. In many cases, states have issued specific bans on the use of specific disposal technologies. We run the risk, thereby, of having federal regulations for permitting of mobile treatment units to be used on site, but whose implementation is not permitted by the state.

Finally, there is the whole issue of listed wastes and how they are to be treated. New draft regulations require a delisting process that involves Federal Register notice, complicating the selection of the treatment alternative.

GRUBBS: One of the difficulties, of course, in getting from the test tube to doing the field work is public awareness and public acceptance. Companies that have had successful field experience may find that client corporations are very reluctant to allow any release of the information. They do not want the public to know that they even had a problem which had to be solved. How do you advance technology with the legal liabilities we face from airing laundry in public?

BAYLEY: All corporations have within them people who deal with legal problems and with what we call public relations or corporate affairs, and they intersect in this area of environmental liability. The successes that companies have in dealing with their problems, getting an understanding of them, and figuring out what they can say outside about them relate to how those functions work together. There are many companies, I'm sure, where the senior management will say, "I don't want to know about that. Let's just wait until EPA comes along. If we have to do something, we will." I think that's a very shortsighted approach. I would urge technical people within companies that have that approach to try to educate the senior management, because the shortsightedness will eventually result in much greater expense.

Another problem for companies is that people responsible for management, legal work, and communication with government agencies often know absolutely nothing about the scientific aspects of these things. I'm certainly in that category. So, whether you're dealing with outside vendors or you have your own people who understand the chemistry or biology, there has to be an internal communication effort so that the process is demystified. Maybe then corporate officials will be less afraid to discuss such matters, if they can say with some scientific validity that they and you have solved a problem. The best way to deal with these things is straightforwardly. This involves increasing the communication between scientists and nonscientists, so that the process is dealt with in a less fearful way by those who are making public announcements.

DAVIS: To providers of these new technical services, let me advise that you avoid planting or encouraging claims in a neighborhood or some other area that would tend to make the public think the thing's a lot worse than it is. What most of us are trying to do is protect the corporation in terms of keeping the problem at the size that it really is, rather than have the people's perception of the problem be bigger than it is. I've approved a number of presentations by technical groups, but I wanted to run through the text of the document to make sure it was accurate. In any case, I think the regulatory environment is changing considerably. I think you can get approvals for good bioreclamation projects.

SIKES: I think that there has been an attitude that some kind of a stigma is applied to having had an accidental spill or release. It wasn't looked on as just another engineering problem that you needed to address and fix, or as a business problem. This perception is changing. Until a few years ago, the management of hazardous waste was not a conversation for the boardroom. For all major companies that have potential exposure today, environmental commitment is now at the very top level. One has to avoid, as Dick just mentioned, increasing a perceived problem because of the media interest and coverage of toxic compounds. In the last few years, instead of parts per million, everything is now defined in parts per billion. A thousand parts per billion is much worse sounding than one part per million. Unfortunately, this kind of hysteria is common. As responsible scientists, engineers, and business managers, we need to be cautious about how we communicate, so that the folks who earn their money by the law do not have any more advantage than they already do in this litigious world!

UNTERMAN: I can see a public relations problem if we would clean up a site or a process and then publicize the cleanup. Is there any public benefit to saying that you stopped beating your wife?

COX: Are insurance difficulties holding back the development of the business and of the technology?

SIKES: Yes. The fact that one cannot get useful environmental impairment insurance obviously limits risk-taking by organizations and forces complex, time-consuming contracts to be established between people buying and people supplying these kinds of services. The management of hazardous materials needs to be taken for what it really is, just another engineering, scientific discipline. One can go on the road behind a tanker full of gasoline, that's an everyday event; but following a truck with a thousand parts per million of gasoline-contaminated soil, that's a hazardous waste which has to be feared like a great leviathan. This is the wrong attitude, and we obviously need to do something about it. With people being sued successfully for huge awards because of underground tank spills where there was a $500 per year liability insurance premium, it's no surprise that we have an insurance crisis. I think we see some abating, however.

PFLUG: An insurance company expert recently said the difference between today and six months ago is that today you have the opportunity to pay for coverage but you still can't get it. Policies today are so filled with loopholes that you cannot get the necessary liability protection. Also, it's not unusual to see a bonding requirement for the performance of jobs that runs anywhere from 100% to 250% of the expected value of the whole project. It doesn't take long for a technology-based company to run out of assets to pledge against those specific liabilities.

BRUBAKER: The insurance problem in the United States also is limiting the number of companies that choose to participate in this industry. Many companies that have biotechnology expertise will not participate in this industry because they have found no way to assess and manage the risks of participating in a Superfund project if they're a "deep pockets" company. This was certainly one of the reasons why FMC--a $3 billion company--decided that having a small piece of their company involved in hazardous waste was not a strategic commitment and, therefore, didn't make sense.

ALLEN: Have you, either in cases of specific remediation or in terms of general company programs, found innovative ways to meet with, talk to, and explain to the public what it is you are doing?

SIKES: Yes. We've worked on a project in one of the southern states that involves a Superfund site with a number of industrial companies acting as the potentially responsible parties. This industry group has taken over the development of the remedial investigation feasibility study for cleanup of the site. A major commitment of this group was to put together a public relations effort, just as they had a technical committee, and they hired a local public relations agent. Because we were discussing a number of technologies that were quite unfamiliar to the average homeowner living near the site, we put on what was basically a science fair. It was a public meeting, but rather than get up and do our typical technical talk that only other people who speak our lingo understand, we were told to design the talks as if it was a junior high school science fair. We did that. We made very straightforward presentations, cutting out all lingo. It was extremely well received by parents, children, and the homeowners who lived around the site. I think that, of the many projects in which I've been involved, that was the most productive and creative way to talk to people about what was going on at the site.

WEISHAAR: At Monsanto it is a way of life to keep the public informed as to what we're doing, not only on issues associated with Superfund sites and on-site remediation, but also on issues of routine air permits and water permits. Permitting and regulations have become a big cost of doing business. Even if we think it's going to be a routine renewal of a permit, we hold community meetings. We let everybody know what is coming up, in addition to the public notices that are required by law to be published in the newspapers. We'll talk corporate-wide or at the plant site; we'll talk to anybody who's willing to listen. Some years back, when OSHA required material safety data sheets to be made available, I was preparing the material safety data. Our CEO said that when these were made available to OSHA, we should also make them available to anybody who wants them in the public. Sometimes when you sit behind the plant gates, your perception of the world is different from how the world perceives you. We printed 10,000 copies of material safety data sheets. We were waiting for an onslaught from the people at the firehouse, from the League of Women Voters, to ask for copies--anybody who wanted one could have one. We didn't get any requests.

PFLUG: One of the most effective techniques that I've seen is the establishment of a community advisory panel, on the basic premise that you trust your friends and those who live around you. The principal responsible parties and the regulatory agency identify within the local community key players whom they request to participate as an advisory group to the active operators of the site. These are generally government officials, community leaders, the Chamber of Commerce, the League of Women Voters. The panel meets on a regular basis to identify the issues associated with the activities and to establish a program for effective communications, with effective feedback and independent input.

DAVIS: A study was published, available from the Chemical Manufacturers Association, of a very effective program that was carried out in Memphis, Tennessee, relative to emergency preparedness.

PUBLIC PARTICIPATION

PUBLIC PARTICIPATION

Gilbert S. Omenn (Chairman), University of Washington
Caroline Allen, North Carolina League of Women Voters
John Dohrmann, Puget Sound Water Quality Authority
Tim Douglas, Mayor of Bellingham
Susan Hall, Hall and Associates
Gordon Orians, University of Washington

DOHRMANN: The Puget Sound Water Quality Authority is involved in water quality planning for Puget Sound; our concern is that the regulatory framework and structure be responsive to perceived and possible threats to the environment. We don't generally get involved in specific technologies. We work a lot with the public, in terms of public perceptions of environmental problems and of potential, future environmental problems. Instead of wearing my Authority hat, I'll try to project what I see as some of the probable public concerns with biotechnologies. I've been listening to these talks, looking at the posters, and reading through the abstracts; frankly, some of this stuff scares me to death.

One of the problems is the biotechnology jargon, because you may use biotechnology to develop nonliving products. But I have this tremendous fear about releasing viable organisms, either natural organisms that you've transported from their natural habitat or organisms that you've selected or designed. To me, transporting a bacterium that you found eating PCBs in Michigan and then releasing it in Southern California has some of the similarities of taking a bird or a rat from Europe and bringing it to the United States.

The ecological problems of relocating natural organisms out of their ecosystem are extremely hard to predict. I haven't got a clue about what the bacteria do in the dirt. And if I were on a citizen participation committee dealing with this issue, you'd have to spend a lot of time educating me. When we get to modified designer organisms, that's even worse. My primary concern is not that the organism that's supposed to prevent frost on the strawberry bushes is going to give me the plague. There are a lot of "ifs" involved in that. What I'm seeing here is that in the case of waste remediation, if you design an organism, the major waste we talk about contains useful products that have been thrown away. The fear then that I have is that when we start developing and using superbugs aimed at petroleum hydrocarbons, solvents, and chlorinated hydrocarbons, we're going to be in a situation where I will buy my gasoline at Texaco because it's pasteurized and contains antibiotic additives. Or, that I'll have to buy my paint thinner before the expiration date and keep it in the refrigerator after I open it. Our whole economy depends on the fact that certain products are not considered putrescible. If we start using organisms that have unique abilities to degrade these substances, we may find out that we're hurting ourselves.

On the other hand, I'll throw in some good news. There was an incident in Seattle a couple of years ago in which a breeding population of gypsy moth was discovered in a residential neighborhood just north of the university. The initial regulatory or protective agency response was to bring in the helicopters with the chemical pesticide. The community formed committees, had experts come in, talked about it, and the result was that the community advocated using a bacteria-based insect-killing technique (Bacillus thuringiensis) rather than a chemical-based one (Carbaryl). In this case, people had the experience of becoming comfortable with releasing selected bacteria into the environment.

DOUGLAS: I think John has probably captured some of the spirit that we all find now or will certainly be finding in our communities when we come to grips with the possibilities that your work offers us in solving some very important problems. I think the challenge to the people in local government is to understand biotechnology to the degree that we will be comfortable enough to go ahead and use it, because we are really where the rubber hits the road. We are at the final point where one has to come to grips with and implement solutions to these problems.

Certainly what I've heard today is encouraging to me, because I think that there are some real developments out there that may become very useful. Part of the good news is that you can expect from politics that we will usually come to closure on issues when there is a high sense of urgency. I think we're getting ever closer to that sense of urgency on a lot of the big problems facing us. Over the last few years we seem to be artfully eliminating every reasonable alternative to deal with the wastes that we are producing in this country. This is partly because as a society we are waking up to the fact that we live in a closed system. So what we do over here can't just be looked at in isolation. I think the urgency is there, and it's going to force a lot of us in government to more action than I think we've been accustomed to before. We've had to get on a quick learning curve.

About five years ago, as we left for a three-week summer drive back to the East coast, I was agonizing about how our incineration system in the city of Bellingham, which was about ten years old at that point, was wearing out. The landfill sites that the county owned were all being closed up because they were obvious disaster areas. So I was wondering, "What are we going to do?" We were having real political difficulties coming up with solutions. One evening we were driving through Wheeling, West Virginia, and on the radio news they said, "Last night the Wheeling City Council met and were informed that their last remaining landfill had two weeks of life left and there was no identifiable alternative." I felt a lot better when I heard that. I knew that we weren't alone.

I would like to suggest some things we could all be doing together that will improve the public climate. After all, building the proper atmosphere for good decision-making is what it's all about in politics--making sure the broader public is aware of what's happening and is as open as possible to the decisions that have to be made. Industry and research groups need to promote their own achievements in a way that the public understands. I recall the director of the Western Interstate Commission on Higher Education, which had an excellent newsletter, telling me, "Tim, the first thing I did when I was hired for the job was I went out and found the best writer that I could possibly find, who could take the most complex thing and get it across in simple language." Such a person is fundamental to the success of reaching the public.

We need to interpret concepts into clearer, but not simplistic, language that people can understand. We're living in a snapshot society. We're informed by watching 10- to 30-second shots on TV or seeing the most captivating photo or the most attention-grabbing headline in the newspaper and reading at most ten column inches about it. I think the public is more intelligent than that, but our convenience store lifestyle is reflected. I think we can be awakened to greater understanding if things are properly presented.

You can help us by acknowledging the risks and concerns that exist. Industries often come in and try to be very consoling and reassuring. You need to say, "There could be some problems here. There are some risks. There are some dangers. And here's what they are and here's how we deal with them." Without that balanced presentation, I think there's an immediate credibility gap that can never be overcome.

The suggestion of a community task force, if properly structured so that they understand their roles and responsibilities, can be very effective, and it certainly has been for us. I think that there's a high level of literacy in our society now on economic and environmental matters that we didn't have five to ten years ago; not only literacy, but interest and intrigue. So I think there is a receptive public.

We had an incident thrust upon our community. We became the first city in the Pacific Northwest that was going to have the honor of an incineration plant for burning and disposal of hazardous material under the new federal regulations. That's because we happen to have a cement plant on the edge of town. People were not involved ahead of time in the corporate decision to use that plant. In fact, it was only by accident that the public found out that for more than two and a half months the plant had been burning hazardous materials on an experimental basis. As those of us who were in government began finding out more about it, we discovered that very few data had been developed. It hadn't been done on standard protocols. So it was extremely hard to justify the activity that had gone on, much less try to reassure the public not to worry about that kind of an operation in the future when it was so close to residential areas, schools, and businesses. In the process, the public obtained records from the Washington State Department of Ecology on these companies, and discovered some public information that had not been released prior to the decision that showed a long history of some really serious problems. The combination of not having known ahead of time and discovering some bad news about a house that wasn't in order was just deadly to that project. Those of you from this area probably know the history. The governor stepped in, and the project was killed. There are other efforts going on right now to find a different location for the incineration site.

I'd like to underscore previous statements that demonstrating a commitment to the objectives that have been established, federally and state-wide, is very important. Those objectives are to recycle as much as possible these valuable products that happen to become waste. We need to work toward an economy that tries to recycle, to reuse, to minimize the amount of things that require disposal. This is going to take a very strong cooperative attitude on the part of government and industry, because it's also going to lead to lifestyle changes in this society which are unlike any we've experienced. Our whole consumption pattern is going to have to change.

I know that governments will want to be in league with people like you. I know that as a mayor I would hope not to have to tackle another hazardous waste-burning issue. I'd certainly rather use the kinds of technologies you people are talking about. I think that these would apply to a number of environmental problems that local governments face now.

ALLEN: Helen Childs Boyden, wife of the headmaster of the Deerfield Academy and teacher of chemistry at that school for more than 60 years, once wrote on her blackboard:

> There are few concepts so difficult that they do not yield to the repeated attack of the ordinary mind.

The key to successful translation of complex information to the public is, for me, Helen Boyden's phrase, "repeated attack." Call it the Chinese water torture method if you like. We know that there is not one public; there are many publics. These publics will reach conclusions based on information gathered through different channels. To inform broadly, <u>all</u> of these channels must be used.

Further, I would suggest that understanding begins at home. A useful device to further understanding and assistance in a wide range of issue matters is a standing dialogue or discussion group made up of representatives of industry, government, environmental groups, and concerned citizens. Such a group should be under the umbrella of a neutral group, perhaps the League of Women Voters or some other civic organization.

In my home county the League established a Hazardous Waste Task Force eight years ago. The dialogue established there among the different segments of our society has carried us through the siting and permitting of a hazardous waste treatment facility, the permitting under RCRA of several existing hazardous waste generators, and the efforts of one established waste treatment facility (solvent recovery) with a bad environmental history to clean up its act. The task force has worked on emergency response through the CAER (Community Awareness and Emergency Response) system proposed two years ago by the Chemical Manufacturers Association. We have worked closely with an organization called Clean Sites to facilitate the cleanup of two warehouses filled with several million pounds of PCB-laced oils. That situation involved the bankruptcy and disappearance of the officers of a corporation established to treat PCBs. One hundred eighty-one principal responsible parties were involved. Clean Sites gives the task force some credit; these warehouses were cleaned up (literally clean enough to eat off the floor!) in approximately nine months from the time it was realized that the wastes had been abandoned.

In the larger context, whatever your religious or philosophical persuasion, I would wager that the belief system of your childhood home is under attack. We no longer agree on definitions of parenthood or murder or even of what is "dead" and what is "live." Robotics in fifty years may yield a race of brilliant computers that tolerate as a relic species a small population of Homo sapiens. In this context, manipulation of genetic material is simply another enormously significant change and challenge which must get in line with all the other uncertainties with which we live today. This is tied to the notion of rate of change. In sociology, which is my discipline, it is noted that the only difference between "evolution" and "revolution" is the speed at which it happens. This is a truism, but "the faster things change, the harder it is to keep up." Put another way, total flux in our social framework makes it more difficult to come to terms with specific changes (such as recombinant DNA research).

Parenthetically, let me mention the obvious: for the microbiologist or geneticist in the laboratory, the very fact that you pursue your professional specialties means that for you these pursuits are OK. These activities meet your specifications for things that are good and of benefit to society. It should not surprise you that others do not always agree--that others may be fearful of what, to you, is a series of commonplace procedures.

Levels of awareness among the public bear some resemblance to Swiss cheese: on the surface almost no awareness (smooth, unruffled). But there are holes, some deep and some shallow. I would liken the deep holes to sitting issues. Within the affected area you have broad-based awareness of the issue and deep commitment to having it resolved favorably. (These are "health and safety" or "peace and tranquility" issues.) By contrast, in shallow holes people care about an issue--but not much. Though I can't document this, this may be where issues related to recombinant DNA applications fall. A few individuals are deeply committed to object to change. But most of our citizenry are not aware and do not care. It is far easier to "engage and enrage" to stop the killing of seal pups than it is to stop a new arrangement of the DNA code!

Rapidity of awareness makes a difference in the nature of public response. With "media events"--public hearings or a chemical spill or a fight over site location for a treatment facility--many people become aware of something in a hurry. Ongoing research occurring world-wide with little happening of dramatic impact is not newsworthy. The public, by and large, is not aware of these undertakings. Those with some slight awareness accept the notion that scientists have a right to "do their thing." Again, take note of Helen Boyden's "repeated attack." Awareness and understanding of the significant but undramatic is slow. One might argue that when, belatedly, the "sleeping giant" of public awareness is aroused, the cries of anguish and protest will be exponentially louder than if recognition of change has been cultivated on a regular basis.

Risk assessment is a matter which several very skilled practitioners of environmental public policy have carefully described, foremost among them being William Ruckelshaus (see above). I would simply call to your attention the excellent description developed by Peter Sandman of Rutgers, whose current occupation is attempting to enlighten journalists on the subject of reporting chemical hazards. Sandman points out that risks are not evaluated by a simple mathematical formula. Most of us differentiate between those risks we choose to take and those over which we have no control--whether or not the former are, in fact, statistically more likely to kill us. The latter, whether it is the siting of a hazardous waste treatment facility or exposure to airborne pollutants from nearby manufacturing plants, become an outrage. The outrage produces a vehement, negative response whether or not the risk is particularly great. It is likely that the longer the public remains uninformed on biotechnology, the more likely those changes that it produces will fall into the category of "outrage."

Credibility has received its share of discussion here. Without further elaboration, let me agree that neither government nor industry has much. This suggests the advisability of what I suggested earlier--that the ongoing dialogue group be sponsored by a "neutral party." Efforts to provide the public with information, whether it comes from industry or government, appear self-serving. To achieve a moderate amount of acceptance, information needs to come from a source that does not appear to have an immediate stake in its acceptance.

There is an additional component of credibility which should be explored. Very little has been said about this to my knowledge. "Pure science" is in the public's collective "mind" undertaken because of the desire to know, to extend the frontier of human understanding, whatever its consequences. Historically, we have assumed that those extraordinary practitioners of science who could, in fact, extend the warp and woof of understanding did not expect or even want payment (beyond their ordinary salaries or whatever sustained them) for what they did. Knowledge was for knowledge's sake. Reward lay in the investigations themselves. Biotechnology, as we are experiencing it today, presents a striking contrast to this apparently archaic notion. Explorers on the frontiers of knowledge shall apparently become fantastically wealthy as a consequence of their endeavors. Patents are being granted for new life forms, and commercial applications of these esoteric creations promise millions of dollars to the persons behind the test tubes. Science blends imperceptibly into technology, and the role of the university researcher fades into that of the commercial investigator whose objective is to further his own financial interests and those of the company. "Why not?" we say. Entrepreneurs in our society are richly rewarded for success.

There is no simple answer to this question. If genes are restructured, it is for a purpose: to do something better or make something happen according to the scientists' (or the company's) view of what is better and what <u>should</u> happen. There is probably little consensus as to what it is that is "better." When the boundaries between pure science and its applications are blurred, however, the confusion about the intent, public consciousness, and accountability of the investigator becomes enormous in the public's mind. Credibility again becomes a massive factor in public acceptance.

Finally, I would suggest that firms whose business it is to apply the findings of microbiology and related fields bear an extraordinary charge to conduct their business in responsible and public-spirited ways. Rightly or wrongly, they inherit the bad name which still adheres to hazardous waste management firms. In addition, the public's general lack of knowledge and its sense that fundamental values are at stake in the manipulation of life forms which occurs through rDNA research create the climate for rejection of specific applications of biotechnology. These concerns are legitimate. If the providers of biotechnology--and the creators as well--are not paying attention to citizens' concerns and are not prepared to deal with them in a forthright and thoroughgoing way, these providers deserve whatever strident and vociferous objections may be forthcoming when the sleeping public giant awakens.

One way to begin the education process is to provide ways for more members of the public to attend scientific meetings. Few people will dedicate three days to something of this sort, but quite a few would probably give three hours to it.

Professional meetings might include a session--perhaps several running concomitantly, depending on the conference city and ease or difficulty in recruiting "citizen" attendees--which would feature one biotechnology application presented at a very simple level, with plenty of time allowed for questions and give and take between the presenter, other specialists "on call," and members of the audience. Involvement of humanities students and social science students, trying to bridge C.P. Snow's "two cultures," would be desirable. The size of these groups should be kept small (I would guess no more than forty; twenty is better) so that those in attendance are not inhibited by numbers. Even the arrangement of seating should be controlled so that the presenter and his/her supporting experts are part of the group in a round setting, not separated from it. This type of program could be set up independently of professional meetings. For example, the Central North Carolina Section of the American Chemical Society has run workshops for journalists and public officials on hazardous waste management.

Another audience course lies in the "community club circuit," via a speakers' bureau of professional societies or univesity faculty. A corollary would be a series of articles for use by the various membership magazines of the national service and special interest organizations--the Rotary, Kiwanis, Junior League type publications.

Undoubtedly, creative ways can be found to extend science understanding through television. North Carolina's Agency for Public Telecommunications has used an "Open/Net" cable system for programming which includes call-in from audience members to program presenters.

ORIANS: Like everybody in this room, I wear a number of hats. I'm now putting on my public hat. I think it's important to know that each one of us becomes a member of the public the minute we step outside of our own area of expertise and show our concern with or express an opinion about something else. When I step outside of ecology to become involved in issues of nuclear power, aid to the Contras, or whatever, I become a member of the public. I have a certain sense of unease about some of the thrust of today's discussion. The sense of unease comes from the feeling that public participation is taken to mean informing the public so that the public can understand. The notion seems to be that the flow of useful information is unidirectional. I would like to develop a concept of the value of public information and public participation that views the process as multidirectional. The full process derives as much from the input and ideas that come from the public as it does from informing the public about complex issues in terms that can be understood.

I think this whole issue of the release of genetically altered organisms into the environment or the release of a nongenetically altered organism in a new place raises very significant public policy issues. There are many uncertainties involved in such endeavors. The professionals in any field are bound to view this matter narrowly. This is the classic case, as in most public policy issues, that the costs and benefits are not evenly distributed. If I'm a researcher in genetic engineering, I stand to gain from research contracts, professional advancement, and various other rewards. If I'm in a company that's in the business of marketing them, there will be financial benefits. Coming from such a perspective, you tend to emphasize certain things and deemphasize others--this is perfectly human, it's not at all sinister.

One of the major benefits of public participation and public involvement in these issues is that it gets more issues on the table. In almost every important environmental issue in the last decade or so--and our Institute for Environmental Studies here at the University of Washington, I might add, has been very much involved with the science/policy interface--one of the major roles that the broad public has played is to enlarge the agenda of items that are considered. We just had an interesting example. Suppose we're developing all sorts of superbugs that degrade products that heretofore have been undegradable; what are the consequences? It is very unlikely, in my opinion, that that issue will be raised by one of the technicians or technologists who are close to the issue. How valid a concern it is, I don't know; but such ideas need to be surfaced early on.

In addition, public participation should broaden the research agenda. My perception of many of the problems that have been associated with the implementation of new technologies in our society is that the full consequences and the full cycle of these technologies weren't thought out and investigated before they were employed. My perception is that perhaps three-fourths of the problems that the nuclear power industry faces were caused by a failure to view the whole cycle and start research in a timely manner to anticipate some of the problems that were going to arise. Now the whole technology is hung up on problems that should have been solved, or at least dealt with, long ago. If they weren't solvable, we should have made that judgment earlier. Had research been done properly, perhaps the incineration plant in Bellingham would be operating now.

With respect to releasing genetically altered organisms, there are several ecological issues that I think demand research at this point. We heard research results on how you track these organisms in the environment to know if they're surviving and what they're doing. But suppose they seem to be getting out of control--what would you do about it? If we do encounter problems, we surely would want to have some technologies available to cope and recoup. My perception of the perceived benefits of the release of these organisms and of the problems with sites that need to be cleaned up is that we are very likely to employ releases on a fairly wide scale before a lot of the necessary research is done.

My final comment concerns the notions of "timely" and "early." I think timing is very much of the essence with respect to the creative use of public participation. So often public participation efforts begin when we know what we want to do, and we're going to go out and convince the public. The public rightly perceives that someone is trying to talk us into something, to convince us, rather than to invite our opinions. Automatically, an adversarial procedure or suspicious attitude is set up. In contrast, if involvement is initiated early, before those petitions are hardened, before an enormous amount of emotional, intellectual, and physical capital has been invested, I think there is a better probability of avoiding adversarial relationships and of getting important ideas on the table and discussed.

HALL: Earlier, Mr. Brubaker talked about a partnership between the generators and vendors, which is definitely important. I think it is also important that there is a partnership between the generators, the vendors, and the public--we, representing ourselves now, and future generations. Put simply, in order to proceed with your new technologies, you're going to need the public's confidence. If the public believes that you are not acting in their best interest, they're going to stop you. They're going to pass laws to stop you. They're going to make sure you don't have the funds to do your research. On matters of health and welfare, the public insists on being consulted. One problem is that we fear that we don't understand. Because these programs are so highly technical, you're damned if you do and damned if you don't. You sometimes scare us more when you tell us the truth than if you didn't try at all. You've got to communicate with the public effectively if you want to proceed. But there are barriers.

Recently, I came across a review of the risk communication literature by Cavellow, Von Winterfeld and Slovic, highly respected researchers in the field. They listed four barriers in risk communication that probably will sound familiar to all of you. First, there's the message problem. The message is highly scientific, it's complex, and the uncertainties drive us crazy. When you scientists get up there and give all your qualifiers, it makes us think that you're either hiding something, making it up, or trying to make it look better. That's a problem that you and I, as a translator, deal with constantly. Second, there's the source problem. There's a lack of institutional trust and credibility. In other words, we see the experts disagreeing and we begin to believe that we can't believe anyone. Third, there's the channel problem. We get most of our information through the media. The media tend to focus on the sensational and the snapshot of the problem. They don't give us the background that we need. Lastly, there is the receiver problem. Individual citizens have inaccurate perceptions of risk. There's lack of interest, unless it hits them in their pocketbook or their backyard. There is also an overconfidence in our ability to avoid risk. If you've ever talked to a teenager about driving, you'll know what I'm talking about. Also, the public is very unrealistic about demands for scientific reliability.

So what can we do about these barriers? I have four recommendations. First of all, communicate early and often. Don't wait for all the facts in order to start engaging the public in a dialogue. Talk about the risks and benefits. But start discussing their values, what they're afraid of; then tell them about your methodologies, and about the history of the problem. Start talking with them about "What if we don't do this, what if we don't solve this problem; what's going to happen then?"

Second, don't expect people to speak your language. Acknowledge that not everybody in the world thinks the way that scientists do. Just because people don't think the way that a lot of scientists do does not mean that they have not yet evolved to your level of capability. It means that they think differently, that they approach things in a different way. Unfortunately, oftentimes many people get the impression from scientists that scientists look down upon us. It needs to come across that you acknowledge that some of us may be just as bright, but that we solve problems in a different way and that we understand things differently. Don't be your own worst enemy by coming across as arrogant or inflexible.

Third, use a translator. Include a communications expert on your team to help you from the very beginning. Find people who know how to analyze what kind of information needs to be transmitted at what time. Use that information; beg, borrow, or steal it, but get a communications person involved early on.

Finally, acknowledge that the public's fear is real. Just because it may not seem real to you doesn't mean that they aren't feeling it. I remember when I was a little girl there was a tiger under my bed. My father came in and said, "Don't be silly, there's no tiger under your bed. Go to sleep." It didn't make me believe or feel any less that there was a tiger under my bed. What it took was turning on the light, actually getting a flashlight, and getting down on the floor with me and lifting up the dust ruffle. "Look, there's nothing there but dust balls." I think that scientists need to do the same thing. You need to open the doors, turn the light on what you're doing, and say, "Look, this is what it is." If there is more there than dust balls, then we need to talk about what it is and what we are going to do with it.

FOCHT: I found this to be a very candid exchange. I think it is important for us to understand what it is we have to do. What we're doing wrong. What we're not conveying to you. I can understand Mr. Dohrmann's comments regarding being frightened about the release of nongenetically engineered organisms. My brother and sister asked me the same thing. It's important that we know when we seem condescending, if we're cutting things out, if we're short-changing you. If we do get into a dialogue, you should feel free to say, "I think you've left something out here."

I think there is an inherent fear of bacteria. We've been fighting against disease for centuries. Yet--and I don't mean to sound the least bit arrogant--the dispersal of microorganisms from one environment to another is very common, in the form of dust, particles on automobile tires that come from Michigan to California, etc.

HALL: I can also recommend that you make sure you're answering the questions that they have. Do a lot of listening. You're aware of so many things that the public maybe should be concerned about. Maybe you're trying to tell them more than they want to know. It's like the answer to, "Where did I come from?" Either you go through the whole thing about how you were conceived and born, or you came from New York. Which answer do they want?

KINGSBURY: Just two brief commments from someone who has been thrust into the interface between the public and the scientists over the last two years. First, I recall the comment that everybody is accepted as an expert among the scientists. It's a very serious issue. As the staff here from the House of Representatives can attest, it's very difficult when you get a group of scientists together and they get into an argument. Ironically, those of us who are scientists live with this behavior and, in fact, enjoy it. It's an intrinsic part of our environment. Many of us feel that the most exciting meetings that we go to are those where you can get up and have a very strenous, serious argument with your colleagues about a particular scientific issue. This is really what pushes knowledge ahead. When you do that in public, however, people get seriously confused. I think we've all suffered as a result.

My second point involves "revealing the risks." This is a frustration that I have felt especially when I've testified before the Congress about the pros and cons of biotechnology. We now have a little over a decade of experience with the commercial side of biotechnology. If you look at the safety record of the industry or at the safety record of the research labs, there's no problem. You open the closet and it's empty. What we have are wild, sometimes realistic, sometimes not realistic, speculations about what could happen under a given scenario. But we haven't seen any of it. We're living under perceived problems and not real ones. We have taken, in this sense, a preemptive strike with stringent regulation to respond to public speculations. If I want to open a biotechnical company in Bellingham, you say, "O.K., Kingsbury, come down to the City Council meeting and we're going to talk about giving you a license." And you ask, "What are the dangers?" What do I tell you? So far we don't have a track record of any dangers. All we have are people's discussions about what could become dangerous. This has made the discussion very difficult.

ORIANS: It's intrinsic to any new technology that all you have are possibilities when you start out. When you built your first dam and operated it for ten years, you were not likely to have seen the hundred-year flood. You won't have any useful statistics. At first you cannot have a track record. One doesn't know which of these speculations are closest to the mark and which are wild. How do we try to develop a process by which we can look at those possibilities and find out what is the most productive research or other activity that could help narrow these speculations? The broad public should be involved in this process from the beginning.

OMENN: I think environmental biotechnology is one topical area in which it should be possible to engage many interested parties and the public with less than the usual speculation about what might be the hazards of an inadvertent release of organisms to the environment. Many people are aware that we are dealing with real problems in terms of the chemical targets of these technologies. There are only so many alternatives that have been conceived. Each of the other alternatives has serious limitations. So there's an opportunity to discuss what's known, what's postulated, and what can be learned as more experience is gained with new technology, in a context in which the public is likely to feel that there is some advantage to finding out, rather than just satisfying the curiosity of scientists. It does meet an urgent need, as Mayor Douglas described. It's one of the reasons that some years ago I began to focus on this particular area, which was at first neglected, of the many aspects of biotechnology growing out of recombinant DNA work. It seemed to me that this was one area where the questions about what might be potential hazards of the new biotechnology could be stacked up against real hazards. The tradeoffs can be investigated by all parties concerned.

DOUGLAS: I think one of the problems that those of us in elected office are facing is the zero-risk objective. To me, this is the most troublesome thing of all. I don't believe that zero risk is achievable, so I don't buy into that concept. But the people in our community who are most interested in this topic are the people who are more oriented to the zero-risk solution. So it's not simply the risk of these organisms escaping and doing some awful things. If we use organisms, will they in fact do the entire job? Is there something still there that hasn't been measured, whose impact we don't yet know, that will be accumulating in the environment, and that will be potentially damaging?

I think what you've just said, Gil, is absolutely correct. We in office are going to have to work with the public through a reciprocal interchange to come to the conclusion that we've got some big problems out there that are going to have to be solved sooner than a zero-risk solution is available. Our job as elected officials is to reach the lowest-risk solution that we reasonably can. If we get to that point of understanding, which is quite a challenge, then I think the whole range of what you people do and a lot of other things can be fairly evaluated. I think the biggest challenge to elected people right now is coming to the realization that zero risk, or returning our environment to absolute pristine quality as if man had never been there, is just never going to happen.

LEVIN: The EPA studiously tries to keep a member of the public on all of its panels to help make these types of recommendations and decisions. We always have someone, for example, from the Environmental Defense Fund (EDF), the Audubon Society, or the League of Women Voters. Yet, from what I gather you're saying here today, that really doesn't convince the public. Is it not the responsibility of the public's representatives to go from that meeting to their membership, or is it still our responsibility at EPA to go one step further? It seems to me that we've done our job when we've communicated with your leadership.

ALLEN: You may need to lean on the representatives of public organizations to do more by way of communicating from that meeting to their membership and their members' friends.

HALL: I would suggest that in addition to leaning on them, perhaps they need some help. They often may not have the financial resources to be able to communicate back to their constituency, in the way that they would like to.

ORIANS: I happen to be a member of the National Audubon Society, and I contribute to organizations such as the EDF, but those are rather small membership organizations. That's a useful step, but it's only a very small part of what has to be done. Not that the EPA can do it all. But to expect to have a representative from the Audubon Society on a panel, and to think that goes any particular distance to solving the dialogue problem, seems to me to be sending out a platoon when you need a battalion.

OMENN: I would argue that the person who serves on that panel is no more an agent of the Audubon Society, the EDF, or the League of Women Voters than the scientist from, say, Cornell University who serves on the Science Advisory Board is an agent of and communicator to his university or the general scientific community. He acts as an individual--giving his best advice, participating as fully as possible in the discussions, raising questions, just as we're asking the public member to raise questions. It's still EPA's duty to find out how to put all that together, make the best decisions and proposals possible, elicit public comments, and explain what it's all about.

SUPERFUND STRATEGIES AND TECHNOLOGIES:

A ROLE FOR BIOTECHNOLOGY*

Joel S. Hirschhorn

Office of Technology Assessment
United States Congress
Washington, D.C. 20510

INTRODUCTION

One of the largest potential markets for biotechnology in the environmental area is the cleanup of chemically contaminated sites. Some site cleanups take place within the Federal Superfund program; but many others, supposedly less severe situations, occur within state or federal agency programs or are done by companies. The Office of Technology Assessment (OTA) has estimated that within the next 50 years as many as 10,000 sites may require cleanup within the Superfund program at a total cost of perhaps $300 billion (4). About a thousand sites are now on the National Priorities List and slated for a remedial cleanup.

OTA has also testified before the Congress that problems persist in using newer, improved cleanup technologies within the Superfund program (3). Some of the most significant testing and demonstration of newer cleanup technologies has been occurring outside of the Superfund program, although a cleanup technology demonstration program was established by Congress in the second authorization of the Superfund program. Congress also explicitly established the policy that permanently effective cleanup technologies should be used, instead of containment approaches that leave wastes in the ground, or the moving of wastes and contaminated materials to landfills (7).

The opportunity for biotechnology in cleanups is substantial. However, there continue to be major developments and advances in all generic cleanup technologies. Much progress is being made in commercializing newer forms of thermal destruction technologies, in chemical treatments such as fixation, and in physical approaches such as vacuum extraction of volatile organic chemicals from the ground. More-

---

*The views expressed herein are those of the author and not necessarily those of the Office of Technology Assessment.

over, in every technological area considerable emphasis is given to decreasing the cost of cleanup and to using the technology for onsite cleanup. During the past year or so, costs have dropped substantially in some areas; for example, incineration can now cost less than $300 per ton, instead of $600 to $800 per ton. There is significant movement toward using onsite cleanups because transportation poses risks and high costs. Offsite treatment facilities may not be available or cost-effective. As a result, mobile cleanup equipment has been developed by a number of companies.

Some of the newer cleanup technologies are also in situ in nature. This is very attractive because many sites involve large volumes of contaminated groundwater and soil; excavation is also expensive and sometimes poses potential environmental releases and risks to workers. In some cases, there is no reasonable prospect of removing contaminated materials for treatment. Pumping and treating enormous volumes of contaminated groundwater, although feasible, is time-consuming and expensive.

CURRENT STATUS OF APPLYING BIOTECHNOLOGY

Compared to the first conference four years ago, the prospects for using biotechnology in site cleanups appear to be more positive. Scientific research remains important, but it is less critical. However, current limits to scientific understanding and data place important limits on early use of biotechnology. For example, a recent study for the Environmental Protection Agency (EPA) said: "Not enough information is presently available to permit extrapolation from one environment to another or to utilize data on a similar compound to assess the biodegradability potential of a given substrate" (6). It is not possible, therefore, to make valid generalizations or extrapolations despite positive results at one or more cleanup sites. It has often been said that every cleanup site is unique. The truth of that statement may be greatest when it comes to using biotechnology.

Now it seems that the rate-controlling process is not producing more research results, but is overcoming obstacles on the application side--obstacles that exist even when there are scientific data showing cleanup effectiveness. The important implication is that public policies, regulations, public perceptions, and economics, as well as competition from alternative cleanup technologies, will determine the future use of biotechnology for cleanups.

People interested in advancing the _use_ of biotechnology need to understand, therefore, that cleanup of contaminated sites depends on public policy initiatives that, in the hazardous waste area, come from Congressional reactions to public concerns. Regardless of the technical merits of biotechnology, use of it will be hampered because much of the public lacks confidence in government cleanup programs. Many, many studies and data catalog the problems of the first five years of the Superfund program. This affects the testing, demonstration, and use of newer cleanup technologies. This also means that there is little latitude for technological and human failures. Cleanup technologies that are quick, unequivocally effective, low-cost, and free of environmental problems and uncertainties have the advantage.

Clearly, certain biological organisms can destroy some toxic chemicals of importance at contaminated sites. Biotechnology _will_ be used to a significant degree to clean up some toxic waste sites. _But_ this positive view needs to be tempered by other factors. Available data, which will not be reviewed here, on the effectiveness of biotechnology lead to the following observations:

(a) The number of toxic chemicals that are treatable with naturally occurring organisms is so large that there is no apparent, immediate need to put a large effort into genetic engineering to develop uniquely different organisms for destruction of toxic chemicals.

(b) The considerable success with on-site closed engineered systems, such as bioreactors, is very important and critical to the quick, practical use of proven biological approaches; this was an important outcome of the 1983 conference (5).

(c) Although some of the worst cleanup problems could use in situ biotechnology, there are many technical problems confronting this approach, and it should probably be delayed until more experience and success are gained from closed systems.

(d) There is a need for more awareness of regulatory requirements in the biotechnology community because there is evidence that many people are ignoring them.

(e) There is a critical need for more detailed technical and economic information from laboratory and field work to demonstrate, objectively, the technical feasibility and cost-effectiveness of specific biological treatments; general statements that biotechnology has broad technical feasibility and is more economical than alternative cleanup technologies are currently unsupportable; even in specific cases, it is not clear whether competitive cleanup technologies have been examined and compared with biotreatment which initially may be offered at very low cost in order to gain entry into the market.

(f) Advocates, proponents, and developers--including the large number of academic researchers in biotechnology and the financial community that provides venture capital to new firms--need to pay more attention to the limits of biotechnology, the competition from alternative cleanup technologies, and environmental concerns such as toxic by-products and mutations of microorganisms (even if they appear unrealistic to biological scientists).

The most important thing to do in the near term is to apply biological cleanups to simple problems in ways that build public confidence and a reliable database. This means biological application chiefly on sites where: there are only one or two chemical contaminants (e.g., spills, leaking underground storage tanks, and simple manufacturing sites); materials are easily transferred to closed treatment systems; naturally occurring microorganisms from the site are used; and complete information is made public on exactly what living organisms and other materials are used and, possibly, produced. It seems inappropriate at this time to attempt to apply biotechnology to chemically complex sites such as landfills and open dumps where there are also many physical obstacles to effective biotreatment.

SOME ISSUES AND PROBLEMS IN THE SUPERFUND SYSTEM

A more detailed discussion on the kinds of problems created when biotechnology enters the Superfund system is useful. After sites are discovered, identified, and assessed in a preliminary way to decide if a cleanup is necessary, the major step is to perform a Remedial Investigation (RI). An RI is expensive, often costing several hundred thousand dollars or much more, and taking many months or even years to complete. There has also been a lot of criticism of RIs, which leads to repeated studies and increased costs.

The traditional view has been that the RI defines the full scope of the problem at a site. However, it is becoming clearer that the RI must also provide site-specific information that can help rule out some cleanup approaches, evaluate the technical and economic feasibility of remaining cleanup technologies, and support the selection of one of them. In other words, as more generic kinds of cleanup technologies enter the picture, getting more information from an RI is necessary. Using permanently effective treatment technologies that destroy, detoxify, or otherwise render materials nonhazardous, rather than relying on containment and land disposal, places greater burdens on the RI.

This situation is particularly important for biotechnology, because there are uniquely different types of information necessary to rule out or to justify using biological treatment. This is more so for in situ treatment than for treatment of materials in a bioreactor. For example, information is needed on naturally occurring organisms and nutrients. Information on geophysical and climatic conditions is more critical, as is information on the presence of organisms and chemicals which might impede biological treatment and influence the final, residual level of contamination reached and, therefore, the ability of the technology to satisfy cleanup goals. Adsorption of contaminants on soil surfaces and porosity in soil particles is usually much more important for biological treatment than for incineration. The toxicity of the contaminants may also be important with regard to the treatment effectiveness of certain organisms, and not merely with regard to human health.

The diversity of chemical contamination at a site is now important, but it probably takes on much more importance if biological treatment is an option, because organisms are very specific in what they chemically alter. Using a short list of indicator contaminants at complex sites has become common practice, on the assumption that, if a cleanup technology works on them, it will work effectively on all the contaminants. This can be a valid argument for some cleanup technologies; but this shortcut is inappropriate if biological treatment is considered. Moreover, at the worst sites there can be substantial variation in contamination and site morphology from one location to another. This could be a much greater problem for biological treatment than for some other technologies, which means that more care may have to be taken in getting descriptive data on the three-dimensional nature of the site.

In the second phase, or the Feasibility Study (FS), laboratory and field tests are becoming more and more common to evaluate specific cleanup technologies or combinations of them. An FS can cost many hundreds of thousands of dollars, take a long time, and be very contentious. As more companies enter the market with variations of a generic technology, it becomes more difficult to evaluate them fairly by only considering one or two of them. Offsetting this problem is an accumulation of data and experience gained from other sites that help inform conclusions for a specific site. But, for the most part, this is still in the future, especially for biotechnology.

In the FS phase, the issue arises whether any toxic by-products are produced during biological treatment. This requires detailed theoretical and, sometimes, experimental work. Detailed mass balances that account for everything introduced during the biological treatment, including additives and nutrients, are probably also necessary to perform. In some cases, particularly for toxic metals, biotreatment may change the toxic material to a different form that may enter the environment, such as through volatilization, or may concentrate it. Some cleanup technologies have faced problems from proprietary compounds; this also may be a problem for biological treatment. Information on the kinetics of biotreatment is critical in estimating the time for cleanup (a factor of considerable importance to the public). Here, there may be problems related to scale-up of biotreatments with very limited field experience and with sudden changes in contaminant concentration that may cause the system to become less effective than anticipated.

Lastly, generating reliable cost data is necessary during the FS, because cost is a factor in selecting a cleanup plan. In this regard, the possibility of using combinations of different technologies for a complex site is important to consider in the FS. While it is true that biological treatment can and probably will be combined at times with other technologies, there may be a more cost-effective approach based on one technology. It has been suggested that biotreatment can be used first to degrade much of the toxic material present and then other techniques used to finish the job. This will be economical in some cases but not in others. Use of thermal destruction or vacuum extraction, for example, would make little sense after biotreatment; but air stripping or carbon adsorption, which use simple, low capital cost equipment and little energy, make more economic sense.

Clearly, making Superfund receptive to using biotreatment means more extensive site studies. For example, in rough but realistic terms, if more detailed site information is necessary in the RI to evaluate the applicability of biological treatment, and also more detailed data during the FS to evaluate specific biotreatments, then increases in the cost of these studies by $200,000 to $400,000 per site are possible. The increased cost for 2,000 sites over a period of five to ten years would be $400 million to $800 million. But biological treatment might be selected for only 10% of the sites on the basis of contaminant complexity alone. This means that for each of those sites where biological treatment is used, several million dollars must be spent to cover the increased study costs of all sites. Since the actual site cleanups might cost from $2 million to $10 million in most cases, there is a significant extra overhead cost from performing additional study.

The critical point is not to conclude that this extra study is inappropriate, but to realize how important it is to target biotechnology applications in order to reduce the general Superfund study effort as much as possible. Rules are needed to decide as early as possible, even at the first preliminary site assessment and investigation (before the RI), that biotreatment or any other generic cleanup technology is not technically feasible for a site. In the long run this will also benefit biotechnology, unless there is more interest in studying biotechnology (which may indeed prove to be lucrative to some people and some companies) than in using it.

In addition to the greater costs of getting more information, the studies are also likely to take more time and aggravate an already difficult situation. In this regard, initial actions at sites are becoming more important to recontrol and stabilize them in order to limit the spread of contamination into the environment. This is a key part of the strategy developed by OTA (4). In Superfund's reauthorization, Congress raised the amount of money that can be spent on such initial actions. Nevertheless, lengthy studies will remain a concern to communities that seek a permanent resolution and, often, the return of contaminated sites to productive use.

Another problem facing the emergence of biotechnology is that there probably are very few people now working within the Superfund system who have the necessary level of biological knowledge to conduct and evaluate site studies. Most of the civil, environmental, and chemical engineers and physical scientists working on RIs and FSs are not likely to have significant knowledge in microbiology, biochemistry, and molecular genetics. Almost all relevant expertise on biotreatment resides in the academic community and in those firms developing and commercializing biotechnology. However, Superfund is operated by a large number of engineering and consulting firms that are supervised by people in federal and state environmental agencies. These people have already had problems dealing with new cleanup technologies that are more closely related to their academic and work backgrounds than biotechnology is. Moreover, the actual field implementation of biotreatment requires considerable expertise which is probably not available now to a significant degree. Unless addressed effectively, this people problem could pose a major constraint to the use of biotechnology in the cleanup market.

Lastly, an important part of the Superfund reauthorization was the establishment of a grants program to communities to assist them in independently evaluating site studies. There have been problems and delays because of serious objections by communities to site studies and to cleanup decisions based on them. These grants will allow communities to hire their own consultants. Some people speculate that this will lead to more delays, while others see the possibility for faster studies and decisions. It is important to understand that in most cases community groups have been very receptive to newer, more permanently effective cleanup technologies. Indeed, grassroots public concerns were the prime force behind the Congressional policy shift away from containment and land disposal.

However, communities do not endorse newer cleanup technologies easily. It is important, therefore, for the developers of biotechnology to realize that there is a need to educate the public broadly about biotreatment, including its advantages, disadvantages, and limitations. There is now an urgent need for simple but comprehensive descriptions in plain language of how biological cleanup works. Many people have already been affected by zealous cleanup technology developers who have made claims that were unsupported by detailed technical and economic information. Already, one site that the EPA said was cleaned up has returned to the National Priorities List. And note that the experts communities hire must also have sufficient expertise to fairly evaluate biotechnology.

CONCLUSIONS

The results of experiments, pilot tests, and some limited field work support the belief that biotechnology will be used to clean up some chemically contaminated sites. However, it is necessary to move cautiously. Every generic cleanup technology has limits. It is wise to target cleanup applications very carefully so that biotechnology is based on indigenous natural organisms and is not used in situations where its effectiveness is uncertain or where another technology could do the job better, faster, or cheaper.

It would be very unfortunate if, for example, biotreatment at a Superfund site was later found to have left treated materials with unacceptable residual levels of contamination, or if some toxic by-product was formed which was as hazardous as the original contaminants and whose mobility might even be enhanced by the treatment. It is critical for the developers of biotechnology to avoid sites with a substantial diversity of toxic chemicals, where contamination may vary greatly in concentration and may exist in different physical forms (e.g., pockets of undissolved, concentrated toxic chemicals).

Developers must also do whatever is necessary to assure the public that no toxic by-products escape treatment. In most published results, there is no explicit explanation of what theoretical or laboratory work has confirmed the absence of toxic by-products. This situation is analogous to that faced by developers of thermal destruction technologies to demonstrate the absence of hazardous products of incomplete combustion. Standard procedures will be necessary to develop to assure the public that no toxic by-products, perhaps quite different from site contaminants, are formed during biotreatment.

It is also probably necessary in the near term to avoid the temptation to perform major in situ cleanups on complex sites. Regardless of how effective living organisms may be in destroying chemical contaminants, in situ biotreatment can be undermined by three-dimensional site diversity (geophysical, geochemical, hydrogeological, contamination, and biological). Standard procedures are also needed to assure the public that effective site sampling and testing are done to confirm the distributional effectiveness of biotreatment.

Biotechnology developers need to have the same sensitivity to environmental concerns and possible ecological problems as some biotechnology researchers. These concerns have been particularly well expressed by Alexander (Ref. 1 and 2; Alexander, this Volume). Many questions may be raised that make little sense to experts, but that must be addressed in plain language to allay public concerns. With other generic cleanup technologies, this has meant doing more testing than originally planned.

Superfund now provides significant resources for research and development (R&D). Clearly, there are a number of areas for biotreatment R&D. Serious long-term attention could be given to anaerobic organisms, other types of organisms than bacteria, systems of organisms that may mean even more sensitivity to exact site conditions, genetically engineered organisms, development of engineered systems capable of dealing with varying contaminant concentrations, and techniques for in situ treatment.

Government agencies, elected officials, the courts, responsible parties, communities, environmental organizations, and the news media all play important roles in the Superfund system. To varying degrees, they are all bombarded with claims from developers of cleanup technologies. The cleanup market in the United States and abroad is so enormous that more technology developers enter it continuously. Competition is already fierce and will grow substantially. Although there remains a need to demonstrate the effectiveness of treatment compared to containment and land disposal, the competition will increasingly be between different types of treatment to render toxic materials non-hazardous.

Also, competition becomes tough and changes character when different firms compete with the same basic treatment technology, as has happened in recent times with thermal destruction. Just as firms have developed competitive types of furnaces (e.g., fuel-fired kilns, infrared, plasma arc), biotechnology companies will probably compete in terms of different types of microorganisms, such as aerobic, anaerobic, bacteria, or fungi. This inevitable development also suggests the need for biotechnology firms to focus on certain contaminants, a method of application (e.g., in situ versus bioreactor) or certain types of site problems (e.g., soil, lagoons, or groundwater), or some combination.

Technology developers have a right to have their technologies evaluated fairly for specific site cleanups. As competition grows, there is a greater burden on technology developers to provide very detailed technical and economic data. Proven field experience also is critical. Competition among different biotechnology firms will put more emphasis on quality of work, reliability, and cost--factors of considerable importance to the public. As competition grows, there is also a greater burden on the entire Superfund system to acquire and analyze more information in order to make decisions that can stand up to scrutiny and challenge.

To avoid delays and excessive study costs, it would be wise for developers, in their own and the public interest, not to try to get their technology considered for all or most sites, but instead to target Superfund sites as early as possible. Otherwise, the high level of competition itself contributes to gridlock, where endless studies and contested cleanup decisions delay effective cleanups, reduce environmental protection, and impede the introduction of newer cleanup technologies.

REFERENCES

1. Alexander, M. (1981) Biodegradation of chemicals of environmental concern. Science 211:132-138.
2. Alexander, M. (1984) Ecological constraints on genetic engineering. Genetic Control of Environmental Pollutants, G.S. Omenn and A. Hollaender, eds. Plenum Press, New York, pp. 151-168.
3. Hirschhorn, J.S., Testimony, Subcommittee on Superfund and Environmental Oversight, Committee on Environment and Public Works, United States Senate, April 14, 1987 and December 10, 1987.
4. Office of Technology Assessment (1985) Superfund Strategy, Government Printing Office, Washington, D.C.
5. Omenn, G.S., and A. Hollaender, eds. (1984) Genetic Control of Environmental Pollutants, Plenum Press, New York.

6. Rochkind, M.L., et al. (1986) Microbial Decomposition of Chlorinated Aromatic Compounds, EPA/600/2-86/090.
7. Superfund Amendments and Reauthorization Act of 1986, signed October 11, 1986; Public Law 99-499; extends Superfund program for five years.

QUESTIONS AND ANSWERS FOLLOWING HIRSCHHORN PRESENTATION

FOCHT: I was struck by your remark regarding the diversity of chemicals in dumpsites. Part of the problem is unmarked barrels of toxic waste. I think we all realize that nothing is going to happen with something as biodegradable as methanol, if it's got mercury in it.

HIRSCHHORN: I think for the first five or ten years of the application of biotechnology, people are going to work on very simple sites. When you get into complex Superfund sites, the diversity is incredible. It's not just buried drums, it's also chemical wastes that have been openly distributed and poured into the ground or that have seeped into the groundwater. In situ application of biotechnology, ultimately, is the most important application of biotechnology, but it runs counter to this other reality. Most of those situations where you really need an in situ technology are the same ones that involve chemical diversity. So, dealing with chemical diversity is a good research problem.

SIMON: I heard a piece of good news recently from a representative of a major pharmaceutical company about the grants that go to communities. The first time he heard that they were going to use Superfund money to let local communities hire experts, he fainted dead away at the thought that he was going to pay Superfund taxes for the other side to retain a hired gun. Previously, the company would go to the community and nobody would talk to them, much less listen to them. When a few communities hired their own scientific experts, the company found that the community would talk and listen to their own expert who was in turn talking to the industry, so a dialogue was established. The industry is going to get behind that idea that the community have its own independent scientific evaluation.

HIRSCHHORN: There were two sites, I believe, where grants were provided to communities, even before the Congressional action demanding it. I think the potential is very good, but there's still a lot of apprehension out there. I've been told by people in industry that they fear that the grant program is going to delay decisions for several years at every site. We have to wait and see how it plays out.

LEWIS: Besides providing a site manager for one of the SITE programs, the EPA pubishes a number of guides and manuals, trying to gather information from projects that we have funded and from information that other people have sent us. At present, we have available a revised handbook on remedial action techniques, underground storage tanks, and tank collection actions, and we will be preparing others. We try to get this information out to the regions, the states, and the user community. We're also cooperating very strongly with the NATO CCMS program, trying to gather similar information from many other countries.

HIRSCHHORN: I would call that system a very passive information transfer system. It depends a great deal on people wanting that information and going after it. I think we need a much more active form of information transfer. We review these feasibility studies, by the way, and the most glaring problem is that the people doing the studies don't even know about the latest technological advances and are using data that are several years old. We need a better way to spread information about technology. The other problem is that the information reported in the literature simply is not detailed enough.

LEWIS: We have been trying to reach people with technology transfer seminars. Often the actual site manager for an individual task in the Superfund has only one or two years of experience out of school--a general engineer. So we release documents showing our own R&D programs, identifying with staff technical expertise, hoping that these people will come back to us for help and advice.

HIRSCHHORN: It's not working. The decisions being made on selection of cleanup technologies are still very conservative decisions. That is, people are not selecting new forms of physical and chemical processes that have been well proven, that have been totally demonstrated in field tests. People who actually make the decisions want to use proven technology--which is land disposal. My first reaction is, "It's proven to fail." But, to the conservative engineers and the young engineers out there, land disposal is proven technology. Everything else--and I do mean, literally, everything else--is unproven in the minds of the people doing the feasibility studies and making the decisions.

BRUBAKER: I would like to respond to your comment about the problems with the diversity of Superfund sites. I didn't hear anyone today saying that biotechnology was the magic bullet that would solve all problems on Superfund sites. What I think the industry is trying to do is to marry biotechnology as a complementary tool with physical and chemical methods. I have participated in RI and FS studies on a variety of sites. We don't say, "You've got 47 chemicals and we can degrade all 47." Instead, we say, "We've got 47 chemicals and we can degrade 37 of them. Now we have 10 others to look at in terms of how to treat them in some other fashion."

HIRSCHHORN: I agree. One of the complications from a process perspective, though, is that as soon as you start getting into a train of technologies you've added enormous complexity to the evaluation and selection of technologies. It's going to cost much more money, and a lot of companies are going to make fortunes doing the studies. More money is spent on doing studies than on physically cleaning up sites. It would be simpler if we knew we could use one technology at a particular type of site, confirm it efficiently, and go for it. To evaluate all the possibilities at all sites will mean a lot more money and delay.

GARRISON: Regarding public participation, don't feel that you're putting us down. We want to know what you have to say. Keep us informed. Don't ever feel that we don't want to know what you're saying. If we feel that you're coming across in a way we don't understand, we'll let you know. As far as the EPA and Department of Ecology are concerned, I feel that you should always go one step further. We appreciate everything that you do to help keep us informed. Since you are our employees, you should be obligated to the citizens to always go that one step further.

Finally, the comment you made regarding having read all the feasibility studies. Have you read the Colbert Landfill Site Study?

HIRSCHHORN: I didn't mean all. But we have read a number of them.

GARRISON: The Colbert Site Study was not written to be reviewed by a scientific panel. I feel that I was able to understand most of that report. I feel it was written, at our insistence, so that we could understand it. We have a good working relationship with the consulting firm, the U.S. Department of Energy, and the representative from EPA that we have been working with for two years as a community. The Colbert area is basically a melting pot. We have people from all walks of life. I can't be more emphatic about keeping us informed. We're out there. It's our water; it's our backyard. You won't ever be sorry if you have good public relations.

HIRSCHHORN: In most sites that we have worked on, the local community groups have been very supportive of using newer innovative cleanup technologies. The public is not standing in the way of using new technology. But in order for them to support this, they demand to enter into serious dialogues with all the proponents of the alternative new technologies.

FINAL DISCUSSION

Gilbert S. Omenn (Chairman), University of Washington
Martin Alexander, Cornell University
Alan T. Bull, The University of Canterbury
Ananda M. Chakrabarty, The University of Illinois
Rita Colwell, The University of Maryland
Susan Hall, Hall and Associates
David Kingsbury, National Science Foundation
Hester Kobayashi, Standard Oil

COLWELL: If we don't develop technology, we'll be in the place where we would be if we hadn't done the basic research on tissue culture to develop the polio vaccine. If we had depended on the original technology, we would now have tremendous miniaturized iron lungs. Instead of healthy, vaccinated kids, we'd have kids running around with their own portable iron lungs. My point is that we've got to move ahead with the kinds of experiments that have been reported at this meeting. I don't think anybody has claimed to have all the answers or would corner the market on the Superfund money. What we are trying to do is to get a cost-effective, clean, efficient way of dealing with a complicated problem.

Somehow the impression has been conveyed that an organism that degrades hydrocarbons is going to be stopped in its tracks by heavy metals or other common toxins. In Baltimore Harbor, where you've got everything including the kitchen sink floating by, the organisms that can degrade oil--crankcase drippings usually--are also resistant to heavy metals at very high concentrations, because they concentrate in the oil. Multiple plasmids carry genes to degrade hydrocarbons as well as to detoxify the heavy metal. I think we'll find this situation arise regularly whether we isolate organisms by the enrichment techniques of the past one hundred years or by selective gene insertion techniques that we're developing through biotechnology.

Gil Omenn asked me to give an overview of the last few days. First of all, there have been tremendous advances since the last conference in 1983. At that conference, we were talking about what we ought to do. At this conference, we've been talking about what we've done and where we ought to go from where we are now. I would expect that at the next conference, we would have many further advances. We will be able to understand the kinds of organisms that Jim Tiedje talked about--these new creatures, having new characteristics, that we can't even name because they don't look like anything we've seen before or they have unexpected activities. We will have information about bioreactors--multipathway bioreactors--that will do a lot more than one single pathway step.

These are large steps to take, but once we use our minds and our money we will make advances. We do, in fact, have increased funding from the NSF, the EPA, and other agencies to look at microbial systematics, to do microbial ecology, and to do bioengineering (three areas that have been sadly neglected over the last two decades). With recent developments in ribosomal RNA and DNA hybridization and sequencing, we'll have an improved systematics and, therefore, a better understanding of the kinds of organisms that are in the environment with which we must become familiar. We will also have improved techniques for detection. But there are problems with regard to detection in the environment, particularly detecting organisms that are in a dormant state, that are viable but nonculturable. I think this will need to be addressed.

Finally, I think that the frustration we find as microbial ecologists and systematists is that we are expected to answer questions in a very short period of time, whereas there have been three decades of enormous amounts of money, brilliant minds, and ingenious techniques applied to unraveling the genetic code and developing applications of molecular biology and genetic engineering. You learn how to clone genes and then how to express them, and now you turn to the ecologists and say, "O.K., we want to be able to apply these new discoveries to the environment immediately. Tell us how. Tell us the answers to the problems," and so forth. We have to lean back and say, "Sure, we'll do that, but give us time; give us an opportunity to learn the basic things we need to know to answer your questions." Then the biotechnology will be better able to be applied.

KINGSBURY: First, for the sake of this mixed audience and from a personal point of view, I would like to take this opportunity to compliment the microbial ecologists as a community for the extraordinary level of achievement that we have seen in recent years. For those of us who have had experience with this field and who have watched its decline over the past couple of decades, the level that they're trying to achieve in a very short period of time is extraordinary. When we see the resources available to the microbial ecologists and the soil and wildlife microbiologists compared to those available to the molecular biologists, I think that the accomplishments made by the ecologists and microbiologists are amazing. In that sense, we should not be as impatient as it seems that we are. I think that we're infused with the impatience of the molecular biologists--of which I am one, and I admit to a great amount of impatience--because I think it's a mindset that one learns to get into in this field. The techniques are so powerful and the progress is so rapid that one very quickly wants everything to happen. Sometimes it take a little longer.

My other comment is that this is a public policy area whether we like it or not, as has been brought out so well at this conference. Those of us who are involved in this interface between science and the public have an enormous job. I think that we're not always doing it well. Somehow we have to learn how to bring this message out for people to better understand it. Some of the things you hear are very unsettling, especially concerning the dangers of turning microorganisms loose to munch up toxics. Sometimes my initial reaction is one of great frustration. On the other hand, as I reflect on it, it's simply a failure on many of our parts to bring forward the message as to what these organisms really are and what they can really do. I think that by the time that we're back here four years from now, hopefully, we will have figured out how to do this better.

KOBAYASHI: I, too, am rather impressed with the advances that have been made during the past three years or so, having been watching the changes that have been occurring since the last conference of this nature in 1983. There are a few areas, however, which I feel do require a lot more work. Yes, we do have a problem with complex waste and undefinable waste. I hope that by the time that we're back here four years from now, we will have figured out how to confront this problem.

Another area is toxicity. Some wastes are much too toxic to be dealt with directly. In our experience, this requires a combination of possible physical and chemical techniques along with biotechnology. There is still a place for biotechnology, whether it be in situ or in reactors. You can use them in combination, thereby cutting down on some of the costs associated with the physical and chemical treatment techniques. The problem we're dealing with is a moving target. As Dr. Hirschhorn mentioned before, many sites have to be cleaned up several times.

Of course, our requirements are changing. I work in an industry in which we're watching the requirements change while we are working in process development for environmental control, biological and physical. We're impressed with how these requirements are changing. For instance, we're now required to look at very low concentrations. At one time it was just BOD, COD, and pH, perhaps, and a few compounds that we were concerned with. It seems as though we're going to be soon looking at toxicity, not just chemical concentrations. This would mean looking beyond specific compounds to the interactions among these metabolites. Many microbiologists who today deal with individual organisms may be willing to work on interactions. Finally, we have to look at the total toxicity effect.

Another area is cost effectiveness. It's very important to look at the technology, whether it's biological, physical, or chemical. Each can be expensive, depending on how we use them. It's important to see how we might combine them, to make costs come down. We should not work individually as microbiologists, molecular biologists, engineers, or hydrologists; we have to work together. We may develop very useful organisms in the laboratory, but their application will depend on how well we can apply them in the field or in reactors. So, it's very important that we biologists communicate with the engineers and the hydrologists.

CHAKRABARTY: I have been listening, of course, to all the problems that have been enumerated. There are no simple solutions. I was struck by Mr. Hirschhorn's prediction that genetically engineered microorganisms will not play a role in hazardous waste cleanup, at least not in the near future. I disagree with him on this point, since I like to believe that genetic engineering has the potential to play an important role if it's given an opportunity.

I think that Mr. Hirschhorn ignores the fact that most of the applications of genetic engineering we are seeing today are in the medical areas, because for decades the NIH has invested tremendous amounts of money into getting the knowledge base that is necessary in order to develop any kind of a technology. In contrast, one of the most neglected areas in terms of funding from the U.S. Congress is the area of the environment, including microbial ecology and genetics applicable to ecological problems. If you had asked yourself or any scientists in the early- to mid-seventies what they felt was the potential of *E. coli* to make insulin or interferon, they would have laughed at you. Yet, it happened because there were scientific breakthroughs and ingenuities. I believe that there are people who can make some real breakthrough, who can come up with very interesting ideas, if they're supported.

The EPA is a major source of funding for environmental pollution research. However, the EPA supports mostly transient research that is geared to problem-solving and not to fundamental basic work. I submit to you that we cannot develop the new technology that we in the academic community have been asked to develop with the meager support that's been provided. The only other source of support is the industry. Yet, as mentioned by others, Monsanto and other companies have not supported development in this area because of potential political and legal problems. Why would they support research and technology development in an area which is so much engulfed with political controversies? So that's a problem to be addressed.

In the academic community, besides funding problems, there is another problem concerning recruitment of the bright minds. In the academic community, potential post-doctorates coming to work in the area of environmental pollution first will ask: "How long is it going to take? How many publications am I going to have in the next couple of years so that I can get a job? What are my possibilities?" From my experience, it has been difficult to attract good people, including students, because the areas seem too difficult. We are talking about complex problems. It's not easy at all. So what's the solution?

Well, one obvious solution, of course, is for Congress to use some of the Superfund money for innovative research. Congress is starting to do so. Congress has asked the National Institute of Environmental Health Sciences (NIEHS) to get some money from the EPA and spend it for innovative research. I was delighted. I went to NIEHS and talked with them. They said, "You know, we basically support toxicology research. So we got money from the Superfund program and we're going to do some more toxicological research to find out which of the chemicals are causing problems." I said, "But you already know quite a bit about at least some of the chemicals that are causing problems. Don't you think that you should do something to remove them?" But it's not in the tradition. Again, the NIEHS is basically a part of the NIH. They're a lot more interested in health-related research, even though that may involve them in environmental pollution, than they are in actually developing a technology to remove the toxic chemicals or hazardous wastes from the environment.

Perhaps a better approach is the NSF centers initiative. As I mentioned, it's difficult to attract the bright students to this academic research area because they don't see too much potential and because we're talking about handling toxic chemicals, which must be done in laboratories that are well equipped to protect the graduate students. Perhaps a university/industry collaborative program could specifically address this problem. We need to tell people that this is an important and nasty problem about which something ought to be done. We understand that there are a lot of challenges in developing the knowledge base that would lead to the appropriate technology; so we should find the optimal way to develop this kind of a program.

BULL: I congratulate Gil Omenn for putting Mr. Hirschhorn on as the final speaker. I actually agree with a good deal of what he was saying. However, he elicited in me one of the classic biological reactions--that of irritability. Rita Colwell has already addressed some of my points. Nevertheless, I feel pressed to make a modest defense of people like myself who call ourselves biotechnologists. We do recognize that there are other competing technologies. For example, this was probably the main reason why my colleagues and I prepared a report for the OECD.

I kept applauding those people that I heard this week talking about trying to grapple with in situ problems. For goodness sake, don't put off those people who are trying to make biotechnological approaches to those problems. I suspect most of them realize the complexity of the situations they are dealing with and also the possibility that they will make only modest advances.

I think we ought to be looking more closely at "upstream" developments in two particular areas. The first is changing the manufacturing process to reduce the pollutant load. For example, we could develop new lubricating oils that degrade very rapidly if they get into waterways. The other is the development of upon-the-line or in-line detoxification processes. This is much closer to my own experience and understanding; it's probably where the major value of genetically modified organisms will come. Dialogue with the manufacturing industry will be very important in bringing these developments about.

One of the many impressions I'll take away from Seattle was Stanley Cohen's comment the first evening: it's the status quo that is fraught with hazards. Again, I've been impressed with the innovation and also the confidence with which people have tried to apply new technology. I think this deserves encouragement, despite all the conjectural risks that one may encounter.

ALEXANDER: To me, the necessary actors are the public, government, industry, and academia. The public has to be involved. But there has to be some evaluation of the comparative risks of all the things that the public is exposed to. To make chemicals a major concern, when we have major problems with a variety of other public health stresses, is inappropriate. People who come to our hearings who are worried about genetically engineered organisms often are driving cars which are emitting air pollutants and are smoking at the meetings. We should examine the comparative risks and, without writing off other problems, seek a balance in our reactions to the problems.

I think there is a fantastic degree of insularity in industry and academia. The left hand does not know what the right hand is doing; we really do not have meaningful cooperation when it comes to solving real problems. Finally, as a member of the Science Advisory Board, I would focus on the "bureaucratization" of government. We have put too many regulatory layers in the way of new technologies. I think the greatest problem in Washington is not necessarily money, time, competence or honesty, but the bureaucratic framework.

With regard to innovative technologies, often there is no track record. There has to be an objective evaluation, without hiding behind the necessary proprietary cloak, of the utility and the potential hazards associated with the various technologies. On the positive side, government, industry, academia, and the public working together can move ahead in solving some very real problems.

HALL: I re-emphasize that you need advocates for your programs to be successful. You yourselves have to build a constituency that understands what you're doing and will join your parade; otherwise, they're going to stop you. The approach should be consultative. It should definitely not be unidirectional. It does take time.

I was struck by the comments both about how fast this technology is moving and about frustration that it's not being used sooner. You are light years ahead of most of the citizens of this city, country, and world. You are asking us, the public, to understand the risks and make a decision about what we're willing to live with. You're also affecting our value systems in the deepest way. You're asking us to look at our social structures and consider religious, economic, and political questions.

COLWELL: I feel compelled to take issue with Al Chakrabarty on one small point. I think EPA has done an incredibly good job in building in-house at Gulf Breeze and at Corvallis programs in molecular microbial ecology. They've also built an external funding program in this area. I think they deserve credit for what they've done in the last two or three years.

CONCLUDING REMARKS

Gilbert S. Omenn

School of Public Health and Community Medicine
University of Washington
Seattle, Washington 98195

My impressions of this meeting are quite positive. We have heard significant advances in microbial ecology, physiology, and genetics, including new organisms that were discovered, not created, and new concepts of how pathways might be tied together with new techniques. We've heard about major field applications that were considered unlikely speculations at the meeting four years ago. Many participants stated their surprise at how much progress in real applications had been achieved in this period.

Several speakers presented data on multiple target chemicals of different classes, and the use of consortia to cover the more resistant chemical species. A dominant theme is the use of combinations of physical, chemical, and biological methods, not necessarily simultaneously applied. In some settings, pretreatment with biological methods may make incineration acceptable; in other cases, chemical and physical engineering maneuvers will make biological methods more successful.

Finally, the comments and questions of many speakers have demonstrated sensitivity to the larger community. There are environmental problems for which these technologies should be mobilized. There is a need to understand and explain better what we mean by toxicity. Biology has a lot to offer in measuring overall toxicity of environmental samples. Rather than increasing the sensitivity in our chemical analyses down to parts per trillion and beyond, we might use fish kill, Daphnia test, or some other tests to indicate adequate safety.

There has been a lot of comment about another meeting in four years. However, quite a few other meetings are now addressing biodegradation. Sometime after the book is published and a lot more research has been accomplished, I will consider your post-meeting comments and rejuvenate our Steering Committee to assess the need for a bridging meeting such as this, to reach across disciplines and scientific and public policy aspects of this nationally important area of environmental biotechnology.

CONTENTS: POSTER ABSTRACTS (Alphabetical)

4-CHLOROBENZOATE METABOLISM BY ACINETOBACTER SP. 4-CB 1 AND ITS RELEVANCE TO BIODEGRADATION OF 4,4'-DICHLOROBIPHENYL

P. Adriaens and D.D. Focht

University of California
Riverside, California 92521

A microorganism was isolated from a PCB-contaminated soil by liquid culture enrichment on 4-chlorobenzoate (4-CB) as a sole source of carbon and energy. It was identified as an Acinetobacter sp. The pure culture was able to grow on benzoate, 4-hydroxybenzoate, 3-hydroxybenzoate, and mixtures of 3-chlorobenzoate + 3-hydroxybenzoate and 2-chlorobenzoate + 2-hydroxybenzoate. No growth was observed on 2-hydroxybenzoate, 2-chlorobenzoate, and 3-chlorobenzoate as the sole carbon and energy sources. Oxygen uptake measurements with resting cells, grown on benzoate or 4-chlorobenzoate, yielded low oxygen uptake rates with 4-hydroxybenzoate or protocatechuate, as compared to uptake rates of 4-hydroxybenzoate-grown cells (about 10%). This indicates that 4-CB degradation does not follow the protocatechuate pathway and that dechlorination of the aromatic ring is unlikely. The rate of oxygen uptake of resting cells grown on 4-CB was in the following order: 4-chlorobenzoate > 2-chlorobenzoate > 3-chlorobenzoate. Aerobic assays with cells grown on each of the previously mentioned substrates were performed over an incubation period of four hours to check substrate disappearance and chloride release. Acinetobacter P6 (Furukawa et al., 1979) was incubated with 0.5 mM 4,4'-dichlorobiphenyl (4,4'-DCB) and 2.0 mM biphenyl in liquid culture alone and in co-culture with Acinetobacter sp. 4-CB 1. Whereas 4-CB accumulated in the monoculture, no 4-CB was found in the co-culture.

ANALYSIS OF THE PROMOTER FOR THE PSEUDOMONAS PUTIDA catBC OPERON

Teri L. Aldrich and A.M. Chakrabarty

University of Illinois at Chicago
Chicago, Illinois 60612

Pseudomonas putida is capable of dissimilating benzoate to succinate and acetyl Co-A via the β-ketoadipate pathway. Two of the essential enzymes involved in this process are muconate lactonizing enzyme I, which catalyzes the conversion of cis,cis-muconate to muconolactone, and muconolactone isomerase, which catalyzes the conversion of muconolactone to β-ketoadipate enol-lactone. These enzymes are encoded by the catB and catC genes, respectively. These tightly linked genes have been cloned from the chromosome of P. putida strain RB1 and their nucleotide sequences determined. Based upon the observation that a single mutation will affect the expression of both genes, catB and catC are believed to comprise an operon. Through the use of deletion clones and a promoter probe vector, a single promoter region has been located 5' of catBC. Sequences located between -44 and -198 were determined to be essential for promoter activity. S1 nuclease and reverse transcriptase mapping were used to locate the transcription initiation site 63 base pairs upstream of the catB ATG initiation codon. Transcripts originating from this site were found only in RNA prepared from benzoate-grown cells. The nucleotide sequence of the catBC promoter was compared to the sequences of other prokaryotic promoters; however, a consensus sequence was not apparent.

These results indicate that the catB and catC genes form an inducible operon which is regulated at the transcriptional level and that nucleotide sequences upstream of -44 are required for promoter activity.

TWO JOINT OFFICE OF RESEARCH AND DEVELOPMENT/OFFICE OF SOLID WASTE AND EMERGENCY RESPONSE PROJECTS

Jodi Bakst

U.S. Environmental Protection Agency
Office of Solid Waste
Washington, D.C. 20460

The Office of Research and Development (ORD) and the Office of Solid Waste and Emergency Response (OSWER) are jointly working on two projects which may have significant implications for the development of biotechnology and chemical waste degradation. One project is the Biosystems Concept, a five-year research plan designed to develop and select both microorganisms and biological treatment systems which can reduce or eliminate the risks from hazardous waste. The second project is the Superfund Innovative Technology Evaluation (SITE) Program, where OSWER and ORD will fund and evaluate a minimum of ten pilot-scale or full-scale demonstrations of alternative technologies at Superfund sites per year through 1990. The poster paper presents an overview of these programs, specifically focusing on their goals and objectives, and the implications for the development/application of microbial degradation.

MODELING THE GROWTH KINETICS AND SUBSTRATE UTILIZATION OF A TCE-DEGRADING MICROORGANISM

L.W. Barnthouse,[1] A.V. Palumbo,[1] V.S. Tripath,[1] and C.D. Little[2]

[1]Environmental Sciences Division
Oak Ridge National Laboratory
Oak Ridge, Tennessee 37831*

[2]Institute of Molecular Biophysics
Florida State University
Tallahassee, Florida 32306

Development of practical methods for microbiological treatment of contaminated groundwater requires a quantitative understanding of the degradation process. We have developed a generalized model of microbial activity that couples growth kinetics to primary substrate utilization and contaminant disappearance. The model incorporates chemical limiting factors and accounts for the toxic effects of the contaminant itself. It has been coupled directly to an existing model of groundwater hydrogeochemistry and can be used to predict biological activity and contaminant degradation in a contaminated plume.

We applied the model to an aerobic, methane-utilizing microorganism isolated from a trichloroethylene (TCE)-contaminated well. This organism degrades TCE into water-soluble intermediates via a methane monooxygenase enzyme system. A factorial experiment was used to isolate effects of methane, copper, and TCE concentrations on microbial biomass, methane disappearance, and TCE degradation in closed cultures. Later experiments were used to refine the initial parameter estimates.

Current experiments are testing a modified version of the model that incorporates oxygen as a limiting factor. Column experiments have been initiated to test the integrated biological-hydrogeochemical model. The completed model can be used in field experiments to optimize system design and to assess potential environmental risks.

---

*Funded under the Oak Ridge National Laboratory Exploratory Studies Program. Operated by Martin Marietta Energy Systems, Inc., under Contract No. DE-AC05-OR21400 with the U.S. Department of Energy.

A SHUTTLE SYSTEM FOR THE RAPID INTRODUCTION OF TRACKING AND OTHER GENES INTO THE CHROMOSOMES OF SOIL BACTERIA

Gerald F. Barry

Plant Microbiology Group
Biological Sciences Department
Monsanto Company AA2G
St. Louis, Missouri 63198

The ability to easily and accurately track genetically engineered bacteria is an important aspect of their introduction into the environment. The genes introduced into these bacteria should also be inserted in such a way that their likelihood of loss or transfer to other bacteria is minimized. These two aims have been addressed for the fluorescent Pseudomonads by development of the Tn-derivative (Tn7-lac) gene insertion system that inserts the termini of the transposon but not those genes required for transposition. The genes (lacZY) that provide the phenotype used to select the recombinant derivatives also provide a phenotype that can be used very effectively in the tracking of these recombinant bacteria even in the presence of high numbers of similar indigenous bacteria. In the system above, the addition of other genes to the element (and then to the bacteria) was difficult, as was the introduction of plasmids of different incompatibilities into many target bacteria.

A more useful system has been developed as follows: a small high copy Escherichia coli replicon with few restriction sites was derived from a pUC plasmid, and a small fragment of the E. coli chromosome containing the Tn7 insertion site was cloned into it. A Tn7-lac element was then transposed into this plasmid with the aid of a transposition helper plasmid. This pUC::Tn7-lac derivative facilitated the manipulation of the Tn7-lac element and the introduction of new DNA segments. To transfer the new elements to the chromosomes of soil bacteria, the elements were first transposed to an IncQ plasmid containing the Tn7 transposition genes and the same E. coli chromosome fragment as above. This delivery plasmid was then transmobilized into the target bacteria and selection for lactose utilization was applied. The delivery plasmid was highly unstable and easily lost, thereby facilitating the rapid isolation of plasmid-free recombinant derivatives.

THE MOLECULAR MICROBIAL ECOLOGY OF A CATABOLIC GENOTYPE IN A BIOLOGICAL TREATMENT PROCESS

J.W. Blackburn,[1] T.L. Donaldson,[2] R.K. Jain,[1] and G.S. Sayler[1]

[1]The University of Tennessee
Knoxville, Tennessee

[2]Oak Ridge National Laboratory
Oak Ridge, Tennessee 37831*

Gene probe technology has been used in conjunction with conventional biochemical assays to characterize the in situ dynamic structure of a microbial ecosystem containing a critical subpopulation that degrades naphthalene. A continuous, lab-scale, activated sludge system fed with a real industrial waste and inoculated with real plant sludge was used. Naphthalene was only 0.2% to 0.06% of the dissolved organic carbon in the industrial feed stream, and the naphthalene-positive genotype (cells containing the NAH7 plasmid) ranged from 0.007% to 0.6% of the total cells (measured by acridine orange direct counts).

The use of gene probes showed that the critical subpopulation carrying the naphthalene genotype correlated directly with the mineralization rate of naphthalene and inversely with the naphthalene stripping rate in the system. Biotransformation and mineralization of naphthalene consistently occurred, and naphthalene stripping was consistently absent when the critical genotype exceeded ten million NAH7-positive cell-forming units (cfu)/mL, or 0.2% of the total cells. Perturbations in the reactor-operating conditions led to intermittent excursions below the ten million genotypic cfu/mL threshold and subsequent reductions in mineralization and increases in stripping.

---

*Operated by Martin Marietta Energy Systems, Inc., for the U.S. Department of Energy under Contract No. DE-AC05-84OR21400.

PILOT-SCALE BIOLOGICAL TREATMENT OF CONTAMINATED GROUNDWATER AT AN ABANDONED WOOD TREATMENT PLANT

Lena Blais

Remediation Technologies, Inc.
Kent, Washington 98032

A pilot-scale fixed-film biological reactor was designed and installed by Remediation Technologies, Inc. at an abandoned wood treating plant to degrade contaminants present in groundwater and in storm waters. Wood preservatives used at the plant were creosote, pentachlorophenol (PCP), and chromated copper arsenate (CCA).

The treatment system consists of a feed tank and a 55-gallon reactor with a bottom outlet. Water is fed from the feed tank to the reactor using a variable speed pump. The reactor is filled with 1-inch pall ring packing material which acts as a support medium for biological growth. Aeration for the reactor is supplied by an air pump which delivers air to the bottom of the reactor through air diffusers (aquarium air stones). Nutrients are added to the feed tank in the form of a nitrogen-phosphorus fertilizer.

The reactor was initially fed water at a rate of ten gal/day in order to produce a retention time in the reactor of five days. Upon acclimation of the system, the feed rate was increased so that the retention time was reduced to two and a half days.

Removal efficiencies in the first months of operation ranged from 70% to 85%. The major organic compound present, naphthalene, was immediately reduced by more than 99%. PCP, however, resisted biodegradation until the fifth month. PCP was reduced at that time from 2.8 mg/l to 0.2 mg/l, a greater than 90% reduction. Total BNA concentrations during the first five months averaged 19 mg/l in the system influent and 2.8 mg/l in the effluent, for an 85% removal. Plans for a full-scale system are in preparation.

CLONING AND SEQUENCING OF *PSEUDOMONAS* GENES INVOLVED IN THE DEMETHYLATION OF THE LIGNIN DEGRADATION DERIVATIVE, VANILLATE

F. Brunel, N. Chevalier, and J. Davison

Unit of Molecular Biology
Brussels, Belgium

Bacterial species are quite readily isolated which are capable of degrading aromatic hydrocarbons. With rare exceptions, they do not perform the opening of the benzene ring unless two hydroxyl groups have been introduced in cis into the nucleus. Thus, before ring fission, methyl-substituted derivatives undergo oxidative demethylation.

Our own interest has recently focused on the oxidative demethylation of the lignin degradation derivative, vanillate (3-methoxy-4-hydroxybenzoate).

We have isolated and characterized two genes from *Pseudomonas* sp. encoding the proteins necessary for demethylation. These genes are situated on a 2.6-kb *BamHI-HincII* fragment. Their transcription is initiated within the fragment and proceeds from *BamHI* to *HincII*.

The DNA sequence permits further predictions on the genes involved in demethylation. Two open reading frames are present on the DNA from bases 258-1246 and bases 1248-2190. They show a bias in G/C content of the third codon (A: 90%; B: 92%) which is characteristic of *Pseudomonas* chromosomal genes. They are preceded at 11 bp and 8 bp, respectively, by Shine and Dalgarno-like sequences (GGCAAG; GAGGA). Their translation would give rise to two proteins of 36,000 and 41,000 daltons.

Further work is now in progress to verify this prediction. New constructs are being made to locate the promoter and quantify its strength, and gene expression is being attempted in *Escherichia coli*.

This research was carried out under research contracts GBI-3-016-B of the Biomolecular Engineering Programme and BAP-0048-B of the Biotechnology Action Programme of the Commission of the European Communities.

A NOVEL METHOD TO DETECT GENETICALLY ALTERED MICROORGANISMS IN THE ENVIRONMENT

Deb K. Chatterjee*

Environmental Protection Agency
Sabine Island
Gulf Breeze, Florida 32561

Current methods for tracking genetically engineered microorganisms or engineered gene(s) contained in microorganisms are poorly developed. Novel approaches have been undertaken to detect such organisms or gene(s) which utilize selective and/or genetic probe methods. A model recombinant plasmid, called pDC91, was constructed which contains 3-chlorobenzoate (3-CB) degradative genes and a tetracycline-resistance gene (both serve as selective markers), as well as a portion of mouse B-globin gene (serves as a unique genetic marker). The plasmid was introduced into *Pseudomonas putida* for tracking experiments in simulated natural environments. Freshwater, saltwater, and their sediments were used. It was observed that *P. putida*/pDC91 survived for a few days in each experimental environment and then decreased rapidly to less than 50 per ml and failed to produce any detectable colonies on the 3-CB-containing plates. Upon addition of 3-CB to each environment, growth of *P. putida*/pDC91 resumed in the freshwater sediment and, to some extent, in the freshwater, but not in the saltwater or its sediment. pDC91 plasmid DNA was found to be present in colonies that grew from the bacteria in the freshwater and its sediment.

To test the feasibility of using a unique genetic marker as a tag for an engineered organism, total DNA was isolated from sediments and waters in which *P. putida*/pDC91 was inoculated and permitted to decrease to levels undetectable by colony screening methods. The DNA was spotted onto nitrocellulose paper and hybridized with a mouse globin gene. Positive signals were found in the freshwater and freshwater sediment, but not in the saltwater or its sediment. The results suggest that either a selective growth pressure or a unique gene probe can be used to detect recombinant organisms in the environment. Therefore, introduction of novel metabolic gene(s) and/or a unique gene sequence into organisms desired for environmental release should facilitate monitoring the survival of such organisms in the environment.

---

*Presently at Bethesda Research Laboratories, Life Technologies, Inc., Gaithersburg, Maryland 20877.

SUCCESSFUL MICROBIAL REMEDIATIONS OF LARGE-SCALE HAZARDOUS WASTE PROBLEMS, SOILS, AND WASTE SITE LEACHATE

James J. Duffy and Anthony C. Ying

TreaTek
Occidental Chemical Corporation
Grand Island, New York 14072

The successful application of microbial biotechnologies to the solution of hazardous waste site problems on a full industrial scale will be described. Results of laboratory and pilot studies leading to full-scale designs and operations will be examined.

Microbial populations may be developed which show metabolic activities on hazardous wastes. Numerous laboratory and small-scale studies have given promise of actual scale-up and full-site remediations. Some have questioned the application of microbially based technologies to diverse situations and the part they may play in total waste problem remediations.

We will describe the efforts for two successful projects where microbial biotechnology has led to full-scale remediations.

The first was a project for laboratory to actual full-scale soil remediation at an abandoned gas works site. Target organic contaminants were reduced to levels of acceptance by the property owner. Microbial technology, coupled with traditional remediation techniques, played the major role in this reduction.

The second was a detailed investigation of the remediation of a waste site leachate to acceptable levels. Again, the combination of microbial technology with traditional technologies led to success.

In both cases, an economic advantage was achieved over alternate technologies evaluated.

BIODEGRADATION OF THE HERBICIDE ALACHLOR IN LIQUID CULTURE AND A SOIL SLURRY

W.K. Gauger, N.R. Adrian, and D.P. Matthees

Microbiology and Chemistry Departments
South Dakota State University
Brookings, South Dakota 57007

Alachlor is a pre-emergence chloroacylanilide herbicide in widespread use, marketed as Lasso. In South Dakota, chemical spills amounting to nearly 9,000 gallons of pesticides and their solvents have been reported for the past two growing seasons. A large portion of this material has been alachlor; since many spills are not reported, its presence in the environment in high levels may be substantially greater than this.

Studies were undertaken to isolate and identify indigenous soil bacteria capable of degrading alachlor. A streptomycete, designated Isolate 5, was obtained from an agronomic soil. It degraded alachlor from 100 mcg/ml to undetectable levels (< 1 mcg/ml) in a period of seven days in complex and semi-defined liquid culture media, but not in a chemically defined culture medium. Specific degradation products are not yet known; however, thin layer chromatography analyses have revealed that aromatic ring cleavage did not occur suggesting that the organism degraded only the aliphatic portions of the molecule. To simulate a spill situation, a laboratory experiment was devised where a sterile soil slurry supplemented with glucose (0.3% w/v) and alachlor (570 mcg/g) was inoculated with Isolate 5. Residual levels of alachlor were monitored over a period equivalent to one growing season (four months) and were observed to drop to less than 5 mcg/g.

It was also determined that this organism lacks plasmid DNA. Therefore, the genome coding for alachlor degradation is located on the chromosome. Current studies are attempting to isolate degradative enzymes in an effort to facilitate spill cleanup. Degradation product analysis to elucidate biochemical pathway features is being performed using mass spectroscopic methods.

MICROBIAL VOLATILIZATION OF SELENIUM FROM SOILS

U. Karlson and W.T. Frankenberger, Jr.

Department of Soil and Environmental Sciences
University of California
Riverside, California 92521

Microbial volatilization of selenium (Se) was evaluated as a potential strategy for detoxifying Se-affected soils. The factors governing Se volatilization were investigated, employing a ^{75}Se labeling technique with activated carbon traps. Three fungi that volatilized selenite and selenate were isolated from seleniferous agricultural soils and identified as *Ulocladium tuberculatum*, *Penicillium citrinum*, and *Acremonium falciforme*.

In kinetic studies using uninoculated Los Banos, Panoche, and Panhill clay loam soils, volatilization rates were constant over a period of 30 days. Maximum volatilization was observed with the application of 300 μg/g Se (IV) for all three soils. Without carbon amendments, the volatilization potential in Los Banos was 7.3% in 13 days; amendments with 0.2% carbon (pectin) increased that potential to 21.7%. With other carbon sources the volatilization in 13 days was 16.5% (corn straw), 12.5% (starch), 9.2% (diary manure), 7.8% (cellulose), and 6.5% (sewage sludge). Inoculation with the isolated fungi did not significantly enhance Se volatilization. Amendment with 0.01% to 0.04% nitrate-N decreased volatilization rates up to 35%. Salinizing the soils with $CaCl_2$, NaCl, and Na_2SO_4, up to an EC_e of 20 dS/m, inhibited Se volatilization. The addition of molybdenum, mercury, chromium, and lead greatly inhibited Se volatilization, while arsenic and boron had little effect. The presence of cobalt, zinc, and nickel at relatively high levels (up to 250 μmol kg^{-1} soil) increased Se volatilization rates by up to 300% after a lag period of five to 14 days. It is speculated that these latter three elements serve as co-factors in biomethylation of Se.

TRANSFORMATION AND MINERALIZATION OF AROCLOR 1254

H.-P.E. Kohler, D. Kohler-Staub, and D.D. Focht

University of California
Riverside, California 92521

Aroclor 1254 was transformed by actively growing cells of Acinetobacter P6 to a greater extent than by resting cell suspensions. This was commensurate with the exponential loss of biphenyl-oxidizing activity with whole cells to 5% of the original after 28 hours. Acinetobacter P6 growing on biphenyl was able to transform 34% of Aroclor 1254 (10 ppm). P6 was also grown in co-culture together with three strains that show prospects of transforming metabolites produced by Acinetobacter P6 during co-metabolism of Aroclor. P6, strains HBP1, TCC1, and 4ClBz1 were grown on different combinations of the following substrates: biphenyl, 2,2'-dihydroxybiphenyl, trans-3-chlorocrotonic acid, 4-chlorobenzoic acid, and 10 ppm of Aroclor 1254 spiked with ^{14}C-Aroclor 1254. In the case of Acinetobacter P6 growing solely on biphenyl, 0.6% of the original radioactivity could be found as ^{14}C-CO_2. Although up to 45% of the radioactive Aroclor 1254 was transformed in co-cultures of Acinetobacter P6 together with the three additional strains, no more than 2.6% could be recovered as ^{14}C-CO_2. Degradation of metabolites formed by Acinetobacter P6, rather than the initial oxidation of Aroclor 1254, might be the rate-limiting step in the overall mineralization of these polychlorinated biphenyls.

MICROBIAL DEGRADATION OF CHLORINATED UNSATURATED FATTY ACIDS

D. Kohler-Staub and D.D. Focht

University of California
Riverside, California 92521

Incomplete degradation of multiple halogenated aromatic compounds [e.g., polychlorinated biphenyls (PCBs)] through the meta-fission pathway may be due to accumulation of halogenated unsaturated fatty acids. In order to study their further degradation, the enrichment culture technique was applied to isolate a microorganism capable of growing on the model compound chlorocrotonic acid (3-chlorobutenoic acid). The isolated strain, called CC1, grows on cis-chlorocrotonic acid, trans-chlorocrotonic acid, 3-chlorobutanoic acid, 2-chlorobutanoic acid, 2-chloropropionic acid, and chloracetic acid as sole sources of energy and carbon. Experiments with both resting cells and cell-free extract assays revealed that CC1 produces constitutively a 2-chloroalcanoic acid dehalogenase which acts on mono- and dichloroacetate, 2-chloropropionate, 2,2-dichloropropionate, and 2-chlorobutanate. The specific activities in terms of chloride release that were measured with cell-free extracts were highest for 2-chloropropionate and 2-chloroacetate--ranging from 0.6 to 1.6 μMol/min/mg protein. Dehalogenase activity on β-chloro-C4 acids was not observed from cell-free extracts of cells grown on chlorinated or nonchlorinated acids, whereas it was observed from whole cells.

ANAEROBIC BIODEGRADATION CAPABILITIES OF FRESHWATER LAKE SEDIMENTS: INFLUENCE OF TEMPERATURE AND SULFATE CONCENTRATION ON THE DEGRADATION OF 2,4-DICHLOROPHENOL

G.-W. Kohring and J. Wiegel

Department of Microbiology
University of Georgia
Athens, Georgia 30602

Freshwater lake sediment samples from the region around Athens, Georgia, were taken anaerobically after determination of the temperature, pH, and Eh. The slurries were homogenized and incubated in anaerobic crimp tubes (20 ml samples with 16 g wet weight in 20 mM phosphate buffer pH 7.0 containing 20 ppm of added 2,4-dichlorophenol) at 18 different temperatures in the range between 5°C and 72°C. To one set of samples 0.25% (w/v) sulfate was added. Three times autoclaved mud samples were used as controls, and all samples were done in duplicate. Chlorophenol concentrations were determined using HPLC after extraction of 0.2 ml samples with 50% acetonitrile. Methane formation was measured by gas chromatography. In the tested mud samples, 2,4-dichlorophenol was transformed to 4-chlorophenol which, after additional lag periods of several weeks, was then further degraded. Adaptation periods decreased between 5°C and 30°C and the dechlorination rates showed maxima at 30°C, 35°C, and 43°C with a pronounced minimum at 40°C, suggesting that at least two groups of dechlorinating organisms were present. No degradation occurred above 53°C, although there was biological activity up to 60°C, indicated by high methane production. Between 15°C and 30°C, the dechlorination rates and the methane formation rates increased exponentially (linear part of the Arrhenius plot) and the dechlorination rate peaks coincided with methane formation peaks. Addition of 0.25% sulfate to the slurries led to elongated lag periods and low dechlorination rates (about 10% of the values without sulfate added). Experiments with mud from a mangrove area, as an example of a bacterial community dominated by sulfate-reducing bacteria, showed no degradation during 110 days of incubation.

A NEW METHOD FOR BACTERIAL STRAIN CONSTRUCTION IN BIODEGRADATION OF XENOBIOTIC COMPOUNDS

L. Krockel and D.D. Focht

University of California
Riverside, California 92521

Separate continuous cultures of Pseudomonas putida R5-3 (benzene$^+$, toluene$^+$, xylene$^+$, benzoate$^+$), grown on toluene, and Pseudomonas alcaligenes C-O (benzoate$^+$, 3-chlorobenzoate$^+$), grown on benzoate, were concentrated and continuously amalgamated on a ceramic bead-column, which was subjected to a continuous stream of chlorobenzene vapors. A recombinant strain of P. putida CB1-9, able to grow on chlorobenzene, 1,4-dichlorobenzene (1,4-DCB), and 3-chlorobenzoate, was isolated in less than one month. Chloride was released in stoichiometric amounts when P. putida CB1-9 was grown on either chlorobenzene or 1,4-DCB. Chlorobenzene-utilizing strains were not found in continuous cultures run at the lowest dilution rate (0.05 hr^{-1}) or in the absence of the donor strain, P.

alcaligenes C-0. The recombinant strain was related to P. putida R5-3, phenotypically and genetically. The recombinant, however, lost its ability to grow on p- and m-xylenes and the corresponding methylbenzoates. P. putida R5-3 carries a nonconjugative XYL-like plasmid with induction patterns similar to the TOL plasmid pWWO. P. putida CB1-9 carries a 33-kb plasmid which is derived from the 57-kb plasmid of P. putida R5-3, probably by deletion of genes coding for key enzymes in xylene and toluate catabolism.

THE ORGANOFLUOROPHOSPHATE HYDROLASES OF TETRAHYMENA THERMOPHILA AND RANGIA CUNEATA

Wayne G. Landis, Nancy A. Chester, Mark V. Haley, Dennis W. Johnson, and Renee M. Tauber

Environmental Toxicology Branch
Toxicology Division
Chemical Research, Development and
Engineering Center
Aberdeen Proving Grounds, Maryland 21010-5423

Organofluorophosphate hydrolases (DFPases) are a diverse and widespread group of enzymes that are able to hydrolyze the potent acetylcholinesterase inhibitors 0,0-diisopropylphosphofluoridate (DFP) and 0-1,2,2-trimethylpropylmethylphosphonofluoridate (soman). Two broad categories of DFPases are currently recognized. Mazur-type DFPase has a molecular weight of 67,000 to 90,000 D, generally hydrolyzes soman faster than DFP, and is stimulated by Mn^{2+}. Squid-type DFPase is much smaller, 22,000-26,000 D, hydrolyzes DFP faster than soman, and Mn^{2+} has no effect or is slightly inhibitory.

Five enzymatic activities have been identified in the protozoan, Tetrahymena thermophila. In some cases the activities share characteristics of both the squid-type and Mazur-type DFPases. A chromogenic substrate related to paraxon, 4-nitrophenylethylphenyl(phenyl)phosphinate (NEPP), is hydrolyzed by Tetrahymena DFPase-2 and DFPase-3 and by the Mazur-type DFPase from hog kidney. The close DFP analog N,N'-diisopropyldiamidofluoridate (Mipafox) is a potent inhibitor of the Tetrahymena DFPases. DFPase activity has also been found in the clam, Rangia cuneata. Two major groups of activity can be identified at molecular weights of 20,000 to 30,000 D and 70,000 to 100,000 D. The higher molecular weight activity resembles a Mazur-type DFPase in that it hydrolyzes soman faster than DFP, is Mn^{2+} stimulated, and hydrolyzes NEPP. The lower molecular weight activity is in the range of the squid-type DFPase. An enzymatic activity that hydrolyzes Mipafox is apparent at a molecular weight of 130,000 D.

A variety of organophosphate hydrolyzing enzymes have been identified. A potential exists for the use of these materials in the detoxification of hazardous wastes and effluents. Pure enzymes could be usefully incorporated into bioreactors for processing waste streams. Engineered biological communities incorporating a variety of trophic levels and interactions may be the way to reduce toxicant concentrations in important habitats.

FLUIDITY OF PLASMID DNA FRAGMENTS INVOLVED IN BIODEGRADATION OF AROMATIC HYDROCARBONS BY *PSEUDOMONAS PUTIDA* STRAIN R5-3

J.V. Leary, B. Carney, and D.D. Focht

University of California
Riverside, California 92521

Pseudomonas putida strain R5-3 isolated from sewage sludge contained two plasmids 100 and 80 kb in size after growth on media containing 4-methylbenzoate (4-MB) as sole carbon source. When transferred to toluene (Tol) or 3-xylene (3-Xyl) as the sole carbon source, the plasmid profile changed with loss of the 80-kb plasmid and the appearance of two plasmids 50 and 60 kb in size. When R5-3 was transferred back to 4-MB from either Tol or 3-Xyl, the original 100- and 80-kb plasmids were present and the 50- and 60-kb plasmids were lost. Restriction enzyme analysis of total plasmid DNA indicated rearrangements and loss of DNA restriction fragments in the strains maintained on Tol and 3-Xyl when compared to the original R5-3 maintained on 4-MB. In the isolate which "retrieved" the original plasmid profile, the restriction enzyme fragment pattern was identical to the original R5-3, in that the fragments which were missing after prolonged growth on Tol or 3-Xyl were again present. The possibility that selected fragments are sequestered in the chromosome is presently under investigation.

BIOLOGICAL TREATMENT OF PETROLEUM HYDROCARBONS AND POLYNUCLEAR AROMATIC HYDROCARBONS IN WASTE LAGOONS

M. Tony Lieberman and Jason A. Caplan

Biorestoration Division
O.H. Materials Corporation
Findlay, Ohio

Four lagoons containing approximately two million gallons of mixed chemical washings, principally petroleum hydrocarbons (PHCs) from a railcar repair service facility in Texas, were treated biologically in a timely and cost-effective manner. Floating sludge was removed from the largest of the four impoundments and it was converted into an in situ bioreactor. Analytical results obtained from water column samples collected weekly showed excellent removals of oil and grease, phenolics, and polynuclear aromatic hydrocarbons (PAHs). Concentrations of anthracene, benzo(a)pyrene, fluoranthene, and phenanthrene showed ten-fold reductions within one week. In addition, PHCs were reduced quickly and significantly, with successful biodegradation indicated by decreases in the nC_{17}/pristane and nC_{18}/phytane ratios. These ratios are concentration-independent indicators of PHC biodegradation. Periodic, measured introduction of additional contaminated sludge and water to the bioactive lagoon resulted in predicted increases in contaminant levels followed by significant reductions in the following two- to three-week period. Bacterial numbers remained high throughout the winter months. It is anticipated that continued transfer of material from the adjacent three lagoons will result in complete bioremediation of almost two million gallons of water and sludge in preparation for closure. The cost of these activities is projected at less than $.25 per gallon, and duration of cleanup is projected at one and a half years.

ENHANCED BIOLOGICAL REMOVAL OF TOLUENE FROM AQUEOUS SYSTEMS

Jon Lindstrom, Conrad Vernon, and Ed Brown

Water Research Center
University of Alaska-Fairbanks
Fairbanks, Alaska 99775-1760

Steady-state cultures of aquatic microorganisms have been shown to respond to the presence of hydrocarbons in a variety of ways. Dichlorobiphenyl in μg/l concentrations stopped phosphate transport in the green alga Selenastrum capricornutum, while several mg/l of toluene and more than one g/l of methanol were required to achieve the same effect. Other organisms (e.g., Pseudomonas putida F1) have been shown to tolerate and/or metabolize toluene when it is present as the sole carbon source in mg/l concentrations.

To biologically remove waste organic solvents, such as toluene, from aqueous systems, it is desirable to enhance biodegradation in a manner which is both cost-effective and efficient. It is usually not practical to enhance biodegradation solely by increasing levels of toxic substrates, since growth inhibition normally occurs at relatively low concentrations of the toxin. However, in a carbon-limited system, biodegradation of toxic organic substrates is proportional to microbial biomass. Large biomasses of cells can be grown, processed, and then directly added to a contaminated ecosystem. Alternatively, enhancement of biodegradation by biomass increase may be achieved continuously in the contaminated system if a second nontoxic carbon/energy source is provided for cell growth. P. putida cells immobilized in alginate and exposed to a continuously flowing nutrient supply will simultaneously metabolize valine and toluene as long as the cells remain carbon limited. Such a system can be designed to enhance the removal rate of any biodegradable toxic compound in a waste effluent or from any contaminated aqueous phase.

ON-SITE BIOLOGICAL TREATMENT OF CREOSOTE-CONTAMINATED SOILS

R. Linkenheil

Remediation Technologies
Kent, Washington 98032

On-site biological treatment was selected as the preferred remedial alternative for creosote sludges and contaminated soils resulting from the closure of RCRA and CERCLA impoundments at a former wood-treating plant in Minnesota. This option was selected after completing a one-year pilot-scale study. The three-acre treatment facility was constructed in the fall of 1985 and has been in operation since the spring of 1986. The pilot-scale studies consisted of six box plots, each 50 feet^{-2}, loaded with waste/soil mixtures. The studies evaluated the degradation of the creosote constituents [primarily polynuclear aromatic hydrocarbons (PAHs) and phenolic compounds] under three waste-loading rates. The use of nutrients and commercially available bacteria were evaluated. The optimum treatment method consisted of a batch-loading rate at 5% benzene extractables in the soil and the use of nutrients and irrigation. Seeding the treatment area with commercially available bacteria did not have any measurable advantage.

The full-scale facility was constructed within the confines of one of the former surface impoundments after removing the sludges and contaminated soils. The treatment facility is three acres in size and is lined with a 100 ml synthetic liner. A leachate collection system, consisting of two feet of sand and pea gravel with drain pipes on 100-foot centers, overlays the liner. Two feet of native fine sands were placed above the leachate collection system and serve as the mixing and treatment zone. A center pivot irrigation system recirculates the leachate collected from the system.

Prior to the initial waste application, 120 tons of manure were applied to the treatment area to provide nutrients and increase the organic carbon content of the soil. The first waste application (in May, 1986) involved 1,200 cubic yards of sludges and contaminated soils. A second waste application occurred in April, 1987. Ultimately, this facility will treat 12,000 cubic yards of waste over a five-year period.

The facility is tilled on a biweekly basis and irrigated as needed, based on soil moisture content. Samples of the treatment soil and the leachate are collected on a periodic basis and analyzed for PAH, phenols, and benzene extractables. Results from the first year indicate that over 85% of the PAHs have been degraded. There were no detectable levels of most constituents in the leachate. Results of the full-scale studies confirm the performance predictions from the pilot-scale studies.

Total capital and operating costs are estimated to be less than $\$75/\text{yard}^3$. Other alternatives (on-site incineration or off-site disposal) were estimated to be greater than $\$200/\text{yard}^3$.

RISK ASSESSMENT OF APPLICATION OF BIODEGRADATIVE MICROORGANISMS FOR HAZARDOUS WASTE DESTRUCTION

Vlasta Molak

Environmental Criteria and Assessment Office
U.S. Environmental Protection Agency

Problems with hazardous waste sites have led to the limitation of land disposal practices for certain classes of chemicals. EPA's National Priority List of hazardous waste sites includes 888 sites (Superfund sites) that have to be cleaned up in the near future because of their potential for human health hazards. The Office of Emergency and Remedial Response (OERR), within the Office of Solid Waste and Emergency Response (OSWER), recently issued guidelines to increase the use of alternative technologies at Superfund sites. Also, in response to the Superfund Amendments and Reauthorization Act of 1986 (SARA), the Office of Research and Development (ORD) and OSWER have established a formal program to accelerate the development, demonstration, and use of new or innovative technologies.

One of the possibly leading innovative technologies is an in situ biodegradation of various hazardous substances using natural microorganisms and/or genetically engineered microorganisms (GEMs). Since any new technology used on a large scale may introduce new risks, it is important to evaluate in advance the potential risks posed by the full-scale use of this technology. This presentation will describe a model system for biodegradation and risks posed by microbial growth and transport in the environment. It will include natural and genetically engineered microorganisms. The model system could then be applied to a real-life situation where microorganisms are to be used for remedial action at the specific hazardous waste sites.

ANAEROBIC TREATMENT OF WASTE STREAMS FROM THE ORGANIC CHEMICALS INDUSTRY

A.S. Ng, M.F. Torpy, and C.M. Rose

Argonne National Laboratory

Results of laboratory experimentation will be presented that describe the anaerobic treatability of certain organic compounds and of a range of process waste streams generated by the organic chemicals industry.

Experimentation was conducted using Biochemical Methane Potential assays and various bench-scale reactors, including fixed-film filters, two-phase reactor systems, and mixed reactors operated in differing modes. At the bench-scale testing level, several wastes were found to be readily amenable to anaerobic digestion, and most of the wastes tested only required simple pretreatment (i.e., neutralization of alkalinity supplementation) to ensure high-rate treatability and process stability. Other results indicate the attributes of specific pretreatments to toxicity and of cotreatments to avoid/reduce toxicity or stimulate acclimatization for successful mineralization of the organic substrate.

The results also provide an understanding of appropriate acclimatization periods and of the benefits for identifying reasonable sources of acclimatized organisms. Data are presented from assay testing of more than 40 organic compounds, illustrating the potential applicability of anaerobic digestion for treatment of a wide array of industrial wastes.

(Work supported by the U.S. Department of Energy, Assistant Secretary for the Office of Industrial Programs, under contract W-31-109-Eng-38.)

BIOLOGICAL TREATMENT FOR SOILS CONTAMINATED WITH POLYCHLORINATED BIPHENYLS

P. Nix[1] and T. Kakida[2]

[1]E.V.S. Consultants, Inc.
Seattle, Washington 98102

[2]Seattle City Light
Seattle, Washington

An analysis of soils at certain substations operated by Seattle City Light has indicated a number of areas contaminated with polychlorinated biphenyls (PCBs). PCB levels are generally less than 50 ppm in surficial soils. In situ biological treatment technologies have been proven for the degradation of certain waste types (i.e., petrochemical wastes), and published literature has indicated that certain PCB congeners are biodegradable. Our conceptual approach, therefore, will enhance the biomass of indigenous PCB-degrading microbes.

Laboratory experimental trials will involve:

landfarming, enhancing the ability of indigenous soil microbes by augmentation of the soil with selective compounds;

composting, mixing contaminated soil in varying ratios with amendments such as alfalfa hay and nutrients to promote thermophilic microbial degradation in a phase 1 and to promote fungal degradation in a phase 2.

Both concepts will be tested using modified soil perfusion columns with recycling of a liquid medium containing amendments (i.e., biphenyl, sodium lignin sulfonate, sodium acetate, nutrients) to promote PCB degradation. This approach is realistic, since our concept for field work will incorporate the use of greenhouses in which the soil/compost can be treated on site with regulation of such factors as temperature, leachate recycling, and air monitoring.

DEGRADATION OF SUBSTITUTED PHENOLS BY IMMOBILIZED BACTERIA

Kirk O'Reilly and Ronald Crawford

Department of Bacteriology and Biochemistry
University of Idaho
Moscow, Idaho 83843

Bacteria can degrade a wide range of toxic chemicals. Our goal is to harness this ability to remove chemical toxins from polluted waters. A promising method for water treatment involves the use of immobilized microbial cells. For example, immobilized bacteria have been used to remove pollutants such as phenol and benzene from water. Here we report results of studies of the degradation of substituted phenols by immobilized bacterial cells.

A pentachlorophenol-degrading Flavobacterium (ATCC 39723) and a p-cresol-degrading bacterium thought to be a Pseudomonas sp. were used. After growth in the presence of the appropriate substrate to induce catabolic enzymes, the cells were immobilized in Ca-alginate beads (mean diameter, 3 mm; mean wet weight, 15 mg). The degradative activities of the immobilized bacteria were tested at various bead and substrate concentrations, and a variety of bioreactor designs were examined.

Immobilized Flavobacterium cells were effective degraders of Pentachlorophenol (PCP). Complete degradation of up to 200 ppm of PCP could be obtained in batch reactors. The reaction was shown to be zero-order with respect to both PCP and oxygen concentration. Immobilized Pseudomonas cells completely degraded p-cresol in reactors with initial concentrations of 250 ppm or less. Partial degradation was detected in reactors with p-cresol concentrations as high as 1,000 ppm. The reaction was zero-order with respect to p-cresol concentration over the range tested. The degradation rate was dependent on the method of aeration, with the highest rate detected in oxygen gas air-lift reactors.

Immobilized-cell bioreactors should be applicable to the decontamination of other compounds. Many xenobiotic molecules are susceptible to biodegradation. It should be possible in many instances to customize bioreactors for degradation of specific mixtures of compounds by using the proper immobilized bacteria.

METABOLIC CONVERSION OF LINDANE TO CHLOROPHENOLS, HYDROPHOBIC ACIDS, AND MORE HYDROPHILIC PRODUCTS BY STREPTOMYCES PSI

B. Pogell, M. Speedie, M. MacDonald,
R. Kline, and Y. Huang

School of Pharmacy
University of Maryland
Baltimore, Maryland 21201

Several streptomycetes known to metabolize organochlorides or lignocellulose were tested for their ability to break the recalcitrant carbon-chlorine bond using the insecticide lindane (hexachlorocyclohexane) as a model substrate. Streptomyces PSI rapidly converted [U-^{14}C] lindane to a mixture of water-soluble metabolites when grown on tryptone-yeast extract at 28°C for one to five days. Chemical analysis of the products showed extensive aromatization with formation of a trichlorophenol, as well as extensive conversion to water-soluble acids, other more hydrophilic products, and Co_2. Several pigment-negative mutants were isolated which still converted lindane to the same metabolites, but the wild-type organism would not grow on lindane as its sole carbon source. With higher concentrations of lindane (0.1 mg/ml), both the wild-type and a pig-mutant strain converted 20% to water solubility in six days.

RESPIROMETRIC STUDY OF MICROBIAL DEGRADATION OF PHENOL

P.A. Pryfogle, D.F. Suciu, P.M. Wikoff, and P.L. Wichlacz

National Engineering Laboratory
Idaho Falls, Idaho 83415

Microbial oxidation of phenol is a recognized, cost-effective method of treating relatively low concentration (<500 ppm) phenol-containing waste streams. However, little information exists on naturally occurring bacteria or their ability to degrade phenol at relatively high concentrations. The present study characterized the physiological response of a self-sustaining bacterial population harvested from a fixed-film biological reactor used for the degradation of concentrated phenolic (>1,000 ppm) paint-stripping solutions containing cadmium and chromium. Microbial activity was monitored by the biological oxygen uptake rate, phenol concentration, ATP concentration, and dry solid weights. The effects of temperature, pH, phenol concentration, $(NH_4)_2HPO_4$ concentrations, Cr ($K_2Cr_2O_7$), and Cd ($CdSO_4$) metals on the respiration rate were determined, and the ability to degrade various phenolic compounds was evaluated. The bacteria were aerobic, gram-negative rods, predominantly pseudomonads. A temperature of 20°C and pH near-neutrality produced optimal respiration rates. The highest level of ATP activity was observed for a phenol concentration of 2,000 ppm, and above or below this phenol concentration the ATP levels were lower. Little or no effect on respiration rate was observed for Cr concentrations up to 120 ppm; but above this level it decreased, and at 320 ppm Cr it ceased. Cadmium, at all concentrations tested, exhibited an effect on the respiration rate. At 120 ppm Cd, the rate had decreased to one-half of the endogenous rate, and no further changes were detected by additions of Cd up to 325 ppm.

FIELD DEMONSTRATION OF THE EFFECTIVENESS OF LAND APPLICATION FOR THE TREATMENT OF CREOSOTE-CONTAMINATED WASTES

June A. Reineman

Remediation Technologies
Kent, Washington 98032

A land treatment demonstration was conducted at a former wood-treatment plant in Montana, as part of the RCRA closure process, for disposal of 16,000 cubic yards of creosote-contaminated sludges and soils. The primary hazardous constituents in the waste are polycyclic aromatic hydrocarbons (PAHs).

The site is well suited for land treatment, consisting of soils with moderate permeability, neutral pH, and moderate CEC and organic carbon content. Rainfall in the area is 23 inches per year. Soil tensiometer data indicate recirculating soil water patterns which would increase the hydraulic residence time and reduce the potential for hazardous constituent migration through the system.

Three types of lysimeters were placed at the site: trench, pan, and vacuum. Although checked on a regular basis, no water was ever accumulated in either pan or trench units. The vacuum lysimeter produced enough water for sampling only once. Analytical results on the lysimeter water showed all PAHs to be below a detectable limit of 5 μg per liter.

As the objective of a land treatment demonstration is to show degradation, immobilization, and/or transformation of the hazardous constituents in the treatment zone (usually considered the upper five-foot soil column), zone of incorporation (8 inches) and soil core samples were periodically taken for PAH analysis. First-order kinetics were used to describe the loss of PAHs from the soil. Half-lives of 3- and 4-ring PAHs were less than 50 days. The larger compounds degraded at a slower rate, yet were still observed to decrease with time.

Following completion of the land treatment demonstration, a trench was dug into the test plot area in order to obtain soil samples at specific depths and locations below the zone of incorporation. Results from this sampling event indicate no contaminant migration at or below 1.5 feet.

The hazardous constituents were found to be degraded, transformed, or immobilized in the upper 1.5-foot soil zone. Results from the land treatment demonstration and "worst case" values were used as input parameters into a land treatment model to predict contaminant movement and concentrations after ten years. The results further confirmed the appropriateness of land application for the treatment and ultimate disposal of the creosote-contaminated waste at this site.

THE LAND TREATABILITY OF PETROLEUM REFINERY WASTE CONSTITUENTS

John R. Ryan

Remediation Technologies
Kent, Washington 98032

The degradation and immobilization of refinery waste constituents within land treatment systems was evaluated through a series of laboratory and modeling studies. The study was completed for the American Petroleum Institute (API) and evaluated two soil types, three waste-loading rates, and two rainfall rates. The waste used was a composite of API separator sludge, DAF float, and emulsion solid slop oil. The principal constituents of interest were the volatile aromatics and polyaromatic hydrocarbons (PAH).

Degradation rates were estimated from chemical loss measured over an eight-month time period using laboratory microcosms. Degradation rates were corrected for volatilization losses in these experiments. There was no apparent effect of soil type or loading rate on the degradation of the PAH under the set of conditions evaluated.

The soil column studies confirmed this hypothesis. Soil columns were constructed using five-foot high, four-inch wide glass columns packed with the two soil types and loaded with the three waste loading rates. Two recharge amounts were applied over the course of a six-month time period equivalent to two pore volumes and four pore volumes. Leachate was collected on a monthly basis, and the soil cores were destructively analyzed at the end of the study. The leachate was consistently below detection limits and there was no constituent migration measured in the soil beneath the zone of incorporation.

Results from these studies were integrated into a land treatment model developed by the U.S. Environmental Protection Agency. The modeling results compared quite well to values measured in the soil column studies. The modeling effort indicates that the model can be used to reasonably predict treatment performance of a land treatment system.

The study has shown that the relative rates of degradation and mobility will vary among refinery waste constituents in a land treatment system. Although a variety of site and operational factors can affect these rates, it has been shown that the soil is an extremely flexible and effective treatment medium. The soils used in this study varied significantly in terms of important physical and chemical characteristics and received a wide range of waste loading rates; yet they produced the same results: degradation, immobilization, and transformation of the constituents within the waste application zone.

USE OF CATABOLIC MUTANTS OF PSEUDOMONAS CEPACIA AC1100 TO ISOLATE GENETIC ELEMENTS RESPONSIBLE FOR 2,4,5-TRICHLOROPHENOXYACETIC ACID DEGRADATION

U.M.X. Sangodkar,[1] P.J. Chapman,[2]
R.A. Haugland,[3] and A.M. Chakrabarty[4]

[1]U.S. Environmental Protection Agency
Gulf Breeze, Florida

[2]University of Minnesota
Minneapolis, Minnesota

[3]U.S. Environmental Protection Agency
Cincinnati, Ohio

[4]University of Illinois at Chicago
Chicago, Illinois

Pseudomonas cepacia AC1100, constructed by plasmid-assisted molecular breeding, is a unique microorganism capable of complete mineralization of the herbicide 2,4,5-trichlorophenoxyacetic acid (2,4,5-T). 2,4,5-T is a major component of the defoliant Agent Orange. Due to its ability to remove over 95% of the pollutant from contaminated soils, AC1100 is considered a candidate for deliberate release into the environment. We have isolated a number of 2,4,5-T minus mutants which appear spontaneously after a few generations of growth on nonselective media. Some of the blocked mutants accumulated colored metabolites of 2,4,5-T. This genetic instability may be a reflection of the gene rearrangements involved in initial evolution of AC1100 in the chemostat. To isolate the genetic elements contributing to 2,4,5-T degradation, we mobilized a cosmid-gene bank of AC1100 into several Tn5-induced blocked mutants. The partial diploids, resulting from complementation of insertion mutations, restored the 2,4,5-T degradative ability. A hybrid cosmid pUS1, carrying an

AC1100 DNA fragment, complemented the Tn5-induced mutant PT88 and was stably maintained along with the indigenous plasmids in this strain. pUS1 complemented two other Tn5-induced mutants and also restored 2,4,5-T^{+} phenotype of SM9A, one of the spontaneous blocked mutants of AC1100. A 25-kb insert from pUS1 was mapped for its restriction endonuclease sites and was used for subcloning smaller fragments on a broad host-range plasmid.

BIOLOGICAL TREATMENT OF ORGANIC CONTAMINANTS USING SUBMERGED FIXED-FILM REACTORS

George J. Skladany[1] and Kevin M. Sullivan[2]

[1]DETOX, Inc.
Newport Beach, California 92660

[2]DETOX, Inc.
Ithaca, New York 14852

Biological treatment of contaminated groundwater and process waters is an efficient and cost-effective remediation technology. Fixed-film biological processes have demonstrated excellent removal efficiencies and stability both in the laboratory and in the field. This poster will present laboratory and on-site results of submerged fixed-film biological treatment systems treating both high- and low-influent organic concentration waters.

Much of the theoretical and real-world experience to date has centered on the treatment of high (>50 ppm) concentrations of organics. While this condition is usually met in industrial pretreatment waste streams, groundwater contaminant concentrations are frequently much lower. Academic research, using submerged fixed-film systems, has demonstrated that biofilms have limited ability to survive when exposed to carbon levels below those needed for active growth. Expanding on this work, biological reactors can be designed to capitalize on the maintenance and decay of organisms present in a biofilm, rather than classical microbial growth processes. This technology has been used in the laboratory to treat influent streams containing 1-5 ppm of benzene and/or methyl ethyl ketone to <5 ppb each. Pilot-scale work in the field has reduced influent benzene concentrations of 500-600 ppb to <2-3 ppb. Data from full-scale operations will also be presented.

More commonly, influent organic concentrations are >50 ppm. Under these conditions, submerged fixed-film bioreactors are able to reduce influent concentrations to the low ppm or high ppb range. These designs have been used to treat waters contaminated with phenols and various organic solvents. Data from the operation of several of these installations will be presented.

MONOCLONAL ANTIBODIES FOR DETECTION OF ENVIRONMENTAL CONTAMINATION: AN IMMUNOASSAY FOR DIOXIN

Larry H. Stanker, Bruce Watkins, and Martin Vanderlaan

Biomedial Sciences Division
Lawrence Livermore National Laboratory
Livermore, California 94550

Environmental monitoring for chemical pollutants has traditionally been done using a variety of analytical techniques including gas chromatography and mass spectroscopy. These techniques, while very precise, are usually time-consuming, and require sophisticated equipment, dedicated laboratories, and highly trained personnel. All result in a costly analysis that is not amenable to automation. Alternative assays that detect specific chemicals and are less expensive, faster, and adaptable to automated systems are desirable. They also should be simple enough to be performed in mobile field laboratories by relatively untrained personnel.

Immunoassays meet these criteria and have been extensively used to detect infectious agents and compounds of medical importance. Albro and his colleagues (Toxicol. Appl. Pharmacol., 1979) early on recognized the utility of immunoassays for environmental monitoring. However, immunoassays for environmental monitoring are only now beginning to be developed. This is due in part to the problems associated with using polyclonal antisera as the source of antibody. The degree of specificity required in many cases cannot be attained with polyclonal antiserum. Also, the quality of immune serum varies between individual animals and between subsequent bleedings of the same animal. The recent development of monoclonal antibody technology, however, makes it feasible to produce antibodies that are highly specific and uniform.

We are in the process of developing immunoassays capable of detecting a number of toxic chemicals including dioxin, heptachlor, and heptachlor epoxide, as well as assays of various pesticides and veterinary drugs. Our findings clearly indicate that highly specific immunoassays are possible. The antidioxin antibodies that we have isolated recognize the most toxic tetra- and pentachloro dioxins, including 2,3,7,8-tetrachlorodibenzodioxin (TCDD). We have incorporated these antibodies into a competition enzyme-linked immunosorbent assay that easily is able to detect 2,3,7,8-TCDD at the part-per-billion level in soil extracts (the Environmental Protection Agency action limit). We have characterized the reactivity of these mabs to a large panel of dioxin and dibenzofuran isomers, as well as their reactivity to different polychlorobiphenyl isomers and a large number of chlorinated organics. The ability of these antidioxin antibodies to discriminate between different isomers raises the possibility of engineering congener-specific assays. They also serve as an example of the exquisite specificity of monoclonal antibodies and demonstrate the advantage of using monoclonal versus polyclonal antibodies when developing immunoassays to small organic molecules.

(This work supported by the U.S. Environmental Protection Agency and the U.S. Department of Agriculture through interagency agreement DW89931433-01-0 and 12-34-5-043, respectively, under the auspices of the U.S. Department of Energy, by Lawrence Livermore National Laboratory under contract No. W-7405-ENG-48.)

MODELING THE METHANE AND TOXIC-DEGRADATION KINETICS OF METHANE-OXIDIZING BIOFILM REACTORS

S.E. Strand, R.M. Seamons, and H.D. Stensel

College of Forest Resources
University of Washington
Seattle, Washington 98195

The kinetics of methane-oxidizing bioreactors for the degradation of toxic organics are modeled. Calculations of the fluxes of methane and chlorinated hydrocarbons such as 1,1,1-trichloroethane (TCA) were made using biofilm model. The model simulates the effects of competition by toxics and methane on their enzymatic oxidation by the methane monooxygenase. Dual-competitive-substrate/diffusion kinetics are used to model biofilm cometabolism, integrating the following equations:

$$\frac{d^2S_1}{dx^2} = \frac{-r_1S_1}{1 + \frac{S_1}{K_1} + \frac{S_2}{K_2}}$$

$$\frac{d^2S_2}{dx^2} = \frac{-r_2S_2}{1 + \frac{S_1}{K_1} + \frac{S_2}{K_2}}$$

where S_1 and S_2 are the local concentrations of methane and TCA, respectively, and r and K are the maximum uptake rates and Monod coefficients, and x is the distance into the biofilm.

Submerged aerated biofilm reactors and unsaturated filters were considered with mass transfer rates of methane and oxygen dependent on the gas flow rate. The equations were solved using finite difference methods to give profiles through the reactor and the biofilm. The results of model calculations will be used to assist in the design of laboratory-scale reactors for experimentally determining the rates of methane and toxic chemical uptake by methane-oxidizing bacteria in biofilm reactors.

TWO-STAGE DEGRADATION OF PARATHION BY *FLAVOBACTERIUM* SP. AND *PSEUDOMONAS* SP.

Ronen Tchelet,[1] Dan Levanon,[1] and Yigal Henis[2]

[1]Migal-Galilee Technological Center
Kiryat Shmona, 10-200 Israel

[2]Faculty of Agriculture
Hebrew University
Rehovot, Israel

Degradation of parathion by strains of *Flavobacterium* sp. and *Pseudomonas* sp. is a two-step process. In the first stage, parathion is degraded constitutively to para-nitrophenol. In the second step, the presence of para-nitrophenol induces its degradation after a lag period.

Parathion degradation as sole source of carbon by *Flavobacterium* sp. was enhanced by the presence of Cu^{++} ions (0.25 ppm) in the medium, accompanied by a slightly higher growth rate of the bacterium.

Para-nitrophenol degradation is influenced by its concentration in the medium. An increase of para-nitrophenol concentration in the medium from 5.0 to 10.0 ppm resulted in a lag period of 30 hours. A higher concentration (20.0 ppm) caused a decrease in the degradation rate, and complete inhibition was achieved with a concentration of 50.0 ppm. *Flavobacterium* growth rate is also influenced by para-nitrophenol concentration. The 50 ppm of para-nitrophenol was toxic and resulted in a decline in the bacterial population.

The influence of cations on parathion degradation by *Flavobacterium* and *Pseudomonas* and by crude extracts of their enzymes was observed. Cu^{++} ions were most effective, causing inhibition with *Pseudomonas* and enhancement with *Flavobacterium*. Both bacterial strains experienced the same effect on degradation by living bacteria cells and their crude enzymes. From these observations it is suggested that Cu^{++} ions have a direct effect on the bacterial enzymes active in parathion degradation. The capability of *Flavobacterium* to degrade parathion is correlated with the age of the bacterial culture. There is an increase in degradation ability with the age of the culture.

BACTERIAL DEGRADATION OF CHLORINATED BIPHENYLS

S. Walia, R. Tewari, G. Brieger, V. Ellis,
V. Thimm, M. Villareal, A. Kaiser,
and T. MacGuire

Oakland University
Rochester, Michigan 48063

Chlorinated biphenyls are widely used in industry and agriculture. These compounds are known to cause teratogenic effects and are widely distributed in the environment. Bacterial strains degrading chlorinated biphenyls were isolated by enrichment and direct screening procedures from activated sewage and polychlorinated biphenyl- (PCB) contaminated soils. The PCB-degrading bacteria were tentatively identified as *Pseudomonas*, *Aeromonas*, *Achromobacter*, *Alcaligenes*, *Enterobacter*, *Bacillus*, and gram-positive pleomorphic rods. These PCB-degrading bacterial strains were resistant to several antibiotics including newer β-lactam antibiotics (imipenem, moxalactam, cephoxitin, cefizoxine, and aztreonam). The electrophoretic pattern of plasmid DNAs from these strains showed presence of plasmids (molecular weight ranging 2 to >40 Mdal). Thin layer chromatographic analysis of the water-soluble metabolites of 4-chlorobiphenyl by resting cell suspension of *Pseudomonas putida* (strain 83) and gram-positive bacteria (strain OBS23) showed different metabolites of Rf value: 0.71, 0.78, 0.83, 0.14, 0.18, 0.37, 0.53, 0.60, and 0.69, respectively. The gas chromatographic analysis and ultraviolet absorption spectrum of water-soluble metabolites of 2,3- and 4-chlorinated biphenyls showed accumulation of benzoic acid and 2,3- and 4-chlorobenzoic acid, respectively. Our results demonstrate that PCB can be degraded by pure culture of bacteria and further suggest a preliminary evidence of reductive dehalogenation of chlorobiphenyls with P. putida strain 83.

THE YEAST CUP1 GENE: A MODEL FOR BIOLOGICAL DETOXIFICATION OF HEAVY METAL EFFLUENTS

J. Welch and S. Fogel

Department of Genetics
University of California
Berkeley, California 94720

In Saccharomyces cerevisiae, CUP1 codes for a small copper-binding, cysteine-rich protein, copper chelatin. We cloned and sequenced the CUP1 gene and find that the protein encoded is similar but not identical to mammalian metallothionein. Strains sensitive to copper contain only a single gene copy, while resistant strains harbor many tandemly organized copies. Resistant laboratory strains contain 12, 13, or 14 copies while industrial strains have 2-16 repeats. Moderate selection for enhanced resistance yields strains with up to 28 copies. These strains retain their high copy numbers in the absence of continued selection. Copper chelatin is produced by cells challenged with copper in direct proportion to the gene copy number. CUP1 genes are transcribed; mRNA levels being induced approximately 20-fold above the basal level by Cu^{++} ions. A strain's basal level is directly related to its gene number. Copper uptake is extremely rapid; 50% of the maximal level is achieved in 15 seconds, and Weser (1979) showed that the metal ion is bound covalently.

We demonstrated that strains could be systematically selected for high CUP1 copy numbers by selection for elevated copper resistance. Another scheme for deriving increased copy numbers utilizes our multi-copy CUP1 clone which can be introduced at nonresident sites on the yeast genome. Selection for increased copy numbers starting with multiple copies at multiple sites would undoubtedly generate strains harboring extremely high copy numbers. Utilizing industrial yeast strains offers a major advantage since they represent an untapped reservoir of variability relative to metal-binding capacity. Different metal-binding systems--such as those for copper, cadmium, mercury, etc.--may be combined into a single elite strain via conventional breeding and/or recombinant DNA strategies. Such strains could be harnessed to immobilize or remove heavy metals from industrial effluents.

DEGRADATION OF NITROGUANIDINE WASTEWATER COMPONENTS

R.T. Williams, A.R. MacGillivray, and Wayne E. Sisk

Roy F. Weston, Inc.
West Chester, Pennsylvania 19380

Nitroguanidine (NQ) is a water-soluble nitroamino compound used as an ingredient in propellants. NQ wastewaters contain NQ (0.4-2,5000 mg/L), guanidine nitrate (1.0-200 mg/L), ammonia (1.0-200 mg/L), nitrate

(1.0-11,800 mg/L), and sulfate (0-5,500 mg/L). Wastewater is being stored in lagoons while a treatment method is developed.

Weston, Inc. is investigating enhanced biodegradation for a land treatment system. Continuous flow and perfusion soil columns were utilized, and the influent and effluent were analyzed for wastewater components and transformation products (nitrosoguanidine, quandine, cyanamide, melamine, and cyanoguanidine). Simulated NQ wastewater was applied. Whey, molasses, and glucose were tested as carbon supplements. In addition, mineralization rate and soil mobility experiments were conducted using ^{14}C-NQ and ^{14}C-GN.

Data from soil column experiments indicate that NQ would be poorly removed in a land treatment system. Batch mineralization experiments support these findings. A multiple-component biological system receiving pretreated wastewater may be a viable method for reducing the environmental risk posed by these wastewaters.

APPENDIX: LISTS OF PRIORITY POLLUTANTS, TARGETS OF OPPORTUNITY

Michael F. Torpy,[1*] Jodi Bakst,[2]
and Gilbert S. Omenn[3]

[1]ECOVA Corporation
Redmond, Washington 98052

[2]Office of Solid Waste
U.S. Environmental Protection Agency
Washington, D.C. 20460

[3]University of Washington
Seattle, Washington 98195

---

* AICHE representative on Steering Committee.

THE FOLLOWING MATERIAL WAS EXCERPTED FROM THE FOLLOWING SOURCES:

Appendix 1. State of California
"Safe Drinking Water and Toxics Enforcement Act of 1986 (Proposition 65): Chemicals Known to Cause Cancer or Reproductive Toxicity (October 1, 1987)" and "Safe Drinking Water and Toxics Enforcement Act of 1986: Scientific Advisory Panel, University of California, Davis, Oct. 30, 1987," Health and Welfare Agency, State of California, Oct. 31, 1987.

Appendix 2. U.S. Federal Clean Water Act
"Table 116.4A: List of Hazardous Substances," 40 CFR 116, Chapt. 1, pp. 34-43, July 1, 1986.

Appendix 3. U.S. Groundwater Monitoring List
U.S. Federal Register 52(131):25947-25952, July 9, 1987.

Appendix 4. U.S. Clean Air Act
"National Emission Standards for Hazardous Air Pollutants," Sect. 112 of the Clean Air Act, as of July 15, 1986.

Appendix 5. U.S. Federal Safe Drinking Water Act
"Fact Sheet: Drinking Water Regulations under 1986 Amendments to SDWA," U.S. Environmental Protection Agency, Office of Drinking Water, Washington, D.C., Dec. 4, 1986.

Appendix 6. Hazardous Constituents as Described by the Resource Conservation and Recovery Act (RCRA)
"Appendix VIII, Hazardous Constituents," 40 CFR 261, pp. 395-400, July 1, 1986.

Appendix 7. Superfund
First Priority List of Hazardous Substances That Will Be the Subject of Toxicological Profiles. U.S. Federal Register 52(74):12866-12874, April 17, 1987.

Appendix 1. State of California Safe Drinking Water Enforcement Act of 1986 (Proposition 65) as of October 30, 1987.

<u>Carcinogens</u>

2-Acetylaminofluorene
Acrylonitrile
Adriamycin
AF-2[2-furyl)-3-(5-nitro-2-furyl)]acrylamide
ortho-Aminoazotoluene
4-Aminobiphenyl(4-Aminodiphenyl)
2 Amino-5-(5-nitro-2-furyl)-1,3,4-thiadizole
Amitrole
ortho-Anisidine and ortho-Anisidien hydrochloride
Analgesic mixturs containing phenacetin
Aramite
Arsenic (inorganic arsenic compounds)
Asbestos
Auramine
Azaserine
Azathioprine

Benz(a)aanthracene
Benzene
Benzidine (and its salts)
Benzo(b)fluoranthene
Benzo(j)fluoranthene
Benzo(k)fluoranthene
Benzo(a)pyrene
Benzotrichloride
Benzyl violte 4B
Beryllium and beryllium compounds
N,N-Bis(2-chloroethyl)-2-napththylamine napazine)
Bischloroethyl nitrosourea (BCNU)
Bis(chloromethyl)ether
1,4-Butanediol dimethanesufonate (Myleran)
beta-Butyrolactone

Cadmium and cadmium compounds
Carbon tetrachloride
Certain combined chemotherapy for lymphomas
Chlorambucil
Chloroform
Chloromethyl methyl ether (technical grade)
Chromium (hexavalent compounds
Coke oven emissions
Conjugated estrogens
Cyclophosphamide
DDT (1,1,1-Trichloro-2,2-bis(p-chlorophenyl)-ethane
1,2-Dibromo-3-chloropropane (DBCP)
3-3'-Dichlorobenzidine
Diethylstilbestrol

Epichlorohydrin
Ethylene dibromide
1,2-Dichloroethane (Ethylene dichloride)
Ethylene oxide

Hexachlorobenzene
Hexachlorocyclohexane (technical grade)
Melphalan
Methoxsalen with ultraviolet A therapy (PUVA)
4,4'-Methylene-bis(2-chloroaniline)
Mustard Gas

2-Naphthylamine
Nickel refinery dust form the pyrometallurgical process
Nickel carbonyl
Nickel subsulfide
N-nitrosodi-n-butylamine
N-Nitrosodiethylamine
N-nitrosodimethylamine
N-nitroso-N-ethylurea
N-nitrosopyrrolidine

Soots, tars and lubricant base oils and derived products, specifically vacuum distillates, acid treated oils, aromatic oils, used engine oils, and mineral oils, when used in occupations such as mulespinning, metal machining and jute processing.

Thorium dioxide
Treosulfan

Vinyl chloride

<u>Reproductive Toxicants</u>

Aminopterin
Chlorcyclizine hydrochloride
1,2-Dibromo-3-chloropropane (DBCP)
Diathylstilbestrol (DES)
Diphenylhydantoin
EThyl alcohol in alcoholic beverages
Ethylene oxide
Etretinate
Isotretinoin
Lead
Methyl mercury
Thalidomide
Valproate
Warfarin

<u>Consideration of Chemicals as Known to the State to Cause Cancer</u>

Polychlorinated biphenyls
2,3,7,8-Tetrachlorodibenzo-para-dioxin (TCDD)
Toxaphene (polychlorinated camphenes)
2,4,6-Trichlorophenol

<u>Candidates</u>

Aflatoxins
Carrageenan (degraded)
Chordecone (Kepone)
1-(2-Chloroethyl)-3-cyclohexyl-1-nitrosourea (CCNU)
4-Chloro-ortho-phenylene-diamine
Citrus Red No. 2
para-Cresidine
Cupferron
Cycasin

Dacarbazine
Daunomycin
N,N'-Diacetylbenzidine
3,4-Diaminoanisole sulfate
4,4'-Daminodiphenyl ether
2,4'-Diaminotoluene
Dibenz(a,h)acridine
Dibenz(a,j)acridine
Dibenz(a,h)anthracene
7H-Dibenzo(c,g)carbazole
Dibenzo(a,e)pyrene
Dibenzo(a,h)pyrene
Dibenzo(a,i)pyrene
Dibenzo(a,l)pyrene
3,3'-Dichloro-4,4'-diaminodiphenyl ether
Diepoxybutane
Di(2-ethylhexyl)phthalate
1,2-Diethylhydrazine
Diethyl sulfate
Dihydrosafrole
Dihydroxymethylfuratizine (Panfuran S)
3-3'-Dimethoxybenzidine (ortho-Dianiasidine)
4-Dimethylaminoazobenzene
trans-2-[(Dimethylamino) methylimino]-5-[2-5-nitro-2-furyl)vinyl]-1,3,4-oxadiazole
3,3'Dimethylbenzidine (ortho-Tolidine)
Dimethylcarbamoyl chloride
1,1 Dimethylhydrazine
1,2 Dimethylhydrazine
Dimethyl sulfate
1,4-Dioxane
Direct Black 38 (technical grade)
Direct Blue 6 (technical grade)

Estradiol 17 B
Estrone
Ethinylestradiol
Ethyleneimine
Ethylene thiourea
Ethyl methanesulfonate

Formaldehyde (gas)
Formylhydrazino-4-(5-nitro-2-furyl)thiazole

Glycialdehyde
Gyromitrin (acetaldehyde methylformylhydrazone)
Hexamethylphosphoramide
Hydrazine and hydrazine sulfate
Hydrazobenzene

Appendix 2. U.S. Federal Clean Water Act.

<u>Designated Hazardous defined under Section 311(b)(2)A</u>

Acetaldehyde
Acetic acid
Acetic anhydride
Acetone cyanohydrin
Acetyl bromide
Acetyl chloride
Acrolein
Acrylonitrile
Adipic acid
Aldrin
Allyl Alcohol
Allyl Chloride
Aluminum sulfate
Ammonia
Ammonium acetate
Ammonium benzoate
Ammonium bicarbonate
Ammonium bichromate
Ammonium bifluoride
Ammonium bisulfite
Ammonium carbamate
Ammonium carbonate
Ammonium chloride
Ammonium chromate
Ammonium citrate dibasic
Ammonium fluoborate
Ammonium fluoride
Ammonium hydroxide
Ammonium oxalate
Ammonium silicolfluoride
Ammonium sulfamate
Ammonium sulfide
Ammonium sulfite
Ammonium tartrate
Ammonium thiocyanate
Ammonium thiosulfate
Amyl acetate
Aniline
Antimony pentachloride
Antimony potassium tartrate
Antimony tribromide
Antimony trichloride
Anitmony trifluoride
Anitmony trioxide
Arsenic disulfide
Arsenic pentoxide
Arsenic trichloride
Arsenic trioxide
Arsenic trisulfide
Barium cyanide
Benzene
Benzoic acid
Benzonitrile
Benzoyl chloride
Benzyl chloride
Berylium chloride
Berylium fluoride
Berylium nitrate
Butyl acetate
Butylamine
n-Butyl phthalate
Butyric acid
Cadmium acetate
Cadmium bromide
Cadmium chloride
Calcium arsenate
Calcium arsenite
Calcium carbide
Calcium chromate
Calcium cyanide
Calcium dodecylbenzenesulfonate
Calcium hypochlorite
Captan
Carbaryl
Carbofuran
Carbon disulfide
Carbon tetrachloride
Chlordane
Chlorine
Chlorobenzene
Chloroform
Chlorpyrifos
Chlorosulfonic acid
Chromic acetate
Chromic acid
Chromic sulfate
Chromous chloride
Cobaltous bromide
Cobaltous formate
Cobaltous sulfamate
Coumaphos
Cresol
Crotonaldehyde
Cupric acetate
Cupric acetoarsenite
Cupric chloride
Cupric nitrate
Cupric oxalate
Cupric sulfate
Cupric sulfate, ammoniated
Cupric tartrate
Cyanogen chloride
Cyclohexane
2,4-D acid
2,4-D ester
DDT
Diazinon
Dicamba
Dichlobenil
Dichlone
Dichlorobenzene
Dichloropropane
Dichloropropene-dichloro
propane (mixture)
2,2-Dichloropropionic acid
Dichlorvos
Dieldrin
Diethylamine
Dimehtylamine
Dinitrobenzene (mixed)
Dinitrophenol
Dinitrotoluene
Diquat
Disulfoton
Diuron
Dodecylbenzenesulfonic acid
Endosulfan
Endrin
Epichlorohydrin
Ethion
Ethylbenzene
Ehtylenediamine
Ethylenediamine-tetra acetic acid (EDTA)
Ethylene dibromide
Ethylene dichloride
Ferric ammonium citrate
Ferric ammonium oxalate
Ferric chloride
Ferric fluoride
Ferric nitrate
Ferric sulfate
Ferrous ammonium sulfate
Ferrous chloride
Ferrous sulfate
Formaldehyde
Formic acid
Fumaric acid
Furfural
Guthion
Heptachlor
Hexachlorocyclopentadiene
Hydrochloric acid
Hydrofluoric acid
Hydrogen cyanide
Hydrogen sulfide
Isoprene
Isopropanolamine dodecylbenzene sulfonate
Kelthane
Kepone
Lead acetate
Lead arsenate
Lead chloride
Lead fluoborate
Lead fluoride
Lead iodide
Lead nitrate
Lead stearate
Lead sulfate
Lead sulfide
Lead thiocyanate
Lindane
Lithium chromate
Malathion
Maleic acid
Maleic anhydride
Mercaptodimethur
Mercuric cyanide
Mercuric nitrate
Mercuric sulfate
Mercuric thiocyanate
Mercurous nitrate
Methoxychlor
Methyl mercaptan
Methyl methacrylate
Methyl parathion

Mevinphos
Mexacarbate
Monoethylamine
Monomethylamine
Naled
Naphthalene
Naphthenic acid
Nickel ammonium sulfate
Nickel chloride
Nickel hydroxide
Nickel nitrate
Nickel sulfate
Nitric acid
Nitrobenzene
Nitrogen dioxide
Nitrophenol (mixed)
Nitrotoluene
Paragormaldehyde
Parathion
Pentachlorophenol
Phenol
Phosgene
Phosphoric acid
Phosphorus
Phosphorus oxychloride
Phosphorous pentasulfide
Phosphorus trichloride
Polychlorinated biphenyls
Potassium arsenate
Potassium arsenite
Potassium bichromate
Potassium chromate
Potassium cyanide
Potassium hydroxide
Potassium permanganate
Propargite
Propionic acid
Propionic anhydride
Propylene oxide
Pyrethrins
Quinoline
Resocinol
Selenium oxide
Silver nitrate
Sodium
Sodium arsenate
Sodium arsenite
Sodium bichromate
Sodium bifluoride
Sodium bisulfite
Sodium chromate
Sodium cyanide
Sodium dodecylbenzene-sulfonate
Sodium fluoride
Sodium hydrosulfide
Sodium hydroxide
Sodium hypochlorite
Sodium methylate
Sodium nitrite
Sodium phosphate,dibasic
Sodium phosphate, tribasic
Sodium selenite
Strontium chromate
Strychnine
Styrene
Sulfuric acid
Sulfur monochloride
2,4,5-T acid
2,4,5-T amines
2,4,5-T esters
2,4,5-T salts
TDE
2,4,5-TP acid
2,4,5-TP esters
Tetraethyl lead
Tetraethyl pyrophosphate
Thallium sulfate
Toluene
Toxaphene
Trichlorfon
Tichlorethylene
Trichlorophenol
Triethanolamine dodecyl benzene-sulfonate
Triethylamine
Trimethylamine
Uranyl acetate
Uranyl nitrate
Vanadium pentoxide
Vanadyl sulfate
Vinyl acetate
Vinylidene chloride
Xylene (mixed)
Xylenol
Zinc acetate
Zinc ammonium chloride
Zinc borate
Zinc bromide
Zinc carbonate
Zinc chloride
Zinc cyanide
Zinc fluoride
Zinc formate
Zinc hydrosulfite
Zinc nitrate
Zinc phenolsulfonate
Zinc phosphide
Zinc silicofluoride
Zinc sulfate
Zirconium nitrate
Zirconium potassium fluoride
Zirconium sulfate
Zirconium tetrachloride

<u>Consent Decree Pollutants Under Section 304</u>

Acenaphthene
Acrolein
Acrylonitrile
Aldrin/Dieldrin
Antimony
Arsenic
Asbestos
Benzene
Benzidine
Beryllium
Cadmium
Carbon tetrachloride
Chordane
Chlorinated benzenes
Chlorinated ethanes
Chloralkyl ethers
Chlorinated naphthalene
Chlorinated phenols
Chloroform
2-chlorophenol
Chormium
Copper
Cyanides
DDT
Dichlorobenzenes
Dichlorobenzidine
Dichloroethylenes
2,4-dichlorophenol
Dichloropropanes/propenes
2,4-dimethylphenol
Dinitrotoluene
Diphenylhydrazine
Endosulfan
Endrin
Ethylbenzene
Fluoranthene
Haloethers
Halomethanes
Heptachlor
Hexachlorobutadiene
Hexachlorocyclohexane
Hexachlorocyclopentadiene
Isophorone
Lead
Mercury
Naphthalene
Nickel
Nitrobenzene
Nitrophenols
Nitrosamines
Pentachlorophenol
Phenol
Phthalate esters
Polychlorinated biphenyls (PCBs)
Polynuclear aromatic hydrocarbons
Selenium
Silver
2,3,7,8-tetrachlorodibenzo-p-dioxin (TCDD)
Tetrachloroethylene
Thallium
Toluene
Toxaphene
Trichloroethylene
Vinyl chloride
Zinc

Appendix 3. U.S. Groundwater Monitoring List.

Acenaphthene
Acenaphthylene
Acetone
Acetophenone
Acetonitile
2-Acetylaminofluorene
Acrolein
Acrylonitrile
Aldrin
Allyl Chloride
4-Aminobiphenyl
Aniline
Anthracene
Antimony
Aramite
Arsenic
Barium
Benzene
Benzo(a)anthracene
Benzo(b)fluoranthene
Benzo(k)fluoranthene
Benzo(ghi)perylene
Benzo(a)pyrene
Benzyl alcohol
Beryllium
alpha-BHC
beta-BHC
delta-BHC
gamma-BHC
Bis(2-chloroethoxy)-methane
Bis(2-chloroethyl)ether
Bis(2-chloro-1methyl-ethyl) ether
Bis(2-ethylhexyl) phthalate
Bromodichloromethane
Bromoform
4-Bromophenyl phenyl ether
Butyl benzyl phthalate
Cadmium
Carbon disulfide
Carbon tetrachloride
Chlordane
p-Chloroaniline
Chlorobenzene
Chlorobenzilate
p-Chloro-m-cresol
Chloroethane
Chloroform
2-Chloronaphthalene
2-Chlorophenol
4-Chlorophenyl phenyl ether
Chloroprene
Chromium
Chrysene
Cobalt
Copper
m-Cresol
o-Cresol
p-Cresol
Cyanide
2,4-D
4,4'-DDD
4,4'-DDE
4,4'-DDT
Diallate
Debenz(a,h)anthracene
Dibenzofuran
Dibromochloromethane
1,2-Dibromo-3-chloropropane
1,2-Dibromoethane
Di-n-butyl phthalate
o-Dichlorobenzene
m-Dichlorobenzene
p-Dichlorobenzene
3,3-Dichlorobenzidine
trans-1,4-Dichloro-2-butene
Dichlorodifluoromethane
1,1-Dichloroethane
1,2-Dichloroethane
1,1-Dichloroethylene
trans-1,2-Dichloro-ethylene
2,4-Dichlorophenol
2,6-Dichlorophenol
1,2-Dichloropropane
cis-1,3-Dichloropropene
trans-1,3-Dichloropropene
Dieldrin
Diethyl phthalate
Thionazin
Dimethoate
p-(Dimehtylamino)azobenzene
7,12-Dimethylbenz(a)-anthracene
3.3'-Dimethylbenzidine
alpha, alpha-Dimethyl-amine
2,4-Dimethylphenol
Dimethyl phthalate
m-Dinitrobenzene
4,6-Dinitro-o-cresol
2,4-Dinitrophenol
2,4-Dinitrotoluene
2,6-Dinitrotoluene
Dinoseb
Di-n-octyl phthalate
1,4-Dioxane
Diphenylamine
Disulfoton
Endosulfan I
Endosulfan II
Endosulfan sulfate
Endrin
Endrin aldehyde
Ethylbenzene
Ethyl methacrylate
Ethyl methanesulfonate
Famphur
Fluoranthene
Fluorene
Heptachlor
Heptachlor epoxide
Hexachlorobenzene
Hexachlorobutadiene
Hexachlorocyclo pentadiene
Hexachloroethane
Hexachlorophene
Hexachloropropene
2-Hexanone
Indeno(1,2,3-cd)pyrene
Isobutyl alcohol
Isodrin
Isophorone
Isosalfrole
Kepone
Lead
Mercury
Methacrylonitrile
Methapyrilene
Methoxychlor
Methyl bromide
Methyl chloride
3-Methylcholanthrene
Methylene bromide
Methylene chloride
Methyl ethyl ketone
Methyl iodide
Methyl methacrylate
Methyl methanesulfonate
2-Methylnaphthalene
Methyl parathion
4-Methyl-2-pentanone
Naphthalene
1,4-Naphthoquinone
1-Naphthylamine
2-Naphthylamine
Nickel
o-Nitroaniline
m-Nitroaniline
p-Nitroaniline
Nitrobenzene
o-Nitrophenol
p-Nitrophenol
4-Nitroquinoline 1-oxide
N-Nitrosodi-n-butylamine
N-Nitrosodiethylamine
N-Nitrosodimethylamine
N-Nitrosodiphenylamine
N-Nitrosodipropylamine
N-Nitrosomethylethyl amine
N-Nitrosomorpholine
N-Nitrosopiperidine
N-Nitrosopyrolidine
5-Nitro-o-toluidine
Parathon
Polychlorinated biphenyls
Polychlorinated dibenzo-p-dioxins
Polychlorinated dibenzofurans
Pentachlorobenzene
Pentachloroethane
Pentachloronitrobenzene
Pentachlorophenol
Phenacetin
Phenanthrene
Phenol
p-Phenylenediamine
Phorate
2-Picoline
Pronamide
Propionitrile
Pyrene
Pyridine
Salrole
Selenium
Silver
Silvex
Styrene
Sulfide
2,4,5-T
2,3,7,8-TCDD
1,2,4,5-Tetrachlorobenzene
1,1,1,2-Tetrachloro-ethane
1,1,2,2 Tetrachloro ethane
Tetrachloroethylene
2,3,4,6-Tetrachloro-phenol
Tetraethyl dithiopyro-phosphate
Thallium
Tin
Toluene
o-Toluidine
Toxaphene
1,2,4-Trichlorobenzene
1,1,1-Trichloroethane
1,1,2-Trichloroethane
Trichloroethylene
Trhichlorofluoromethane
2,4,5-Trichlorophenol
2,4,6-Trichlorophenol
1,2,3-Trichloropropane
0,0,0-Triethyl phos phorothioate
sym-Trinitrobenzene
Vanadium
Vinyl acetate
Vinyl chloride
Xylene (total)
Zinc

Appendix 4. U.S. Clean Air Act.

Chemicals addressed Through the Toxic Air Pollutant Evaluation and Control Program Under the National Emission Standards for Hazardous Air Pollutants (NESHAP)

Preliminary Health and Source Screening

Propylene
Chlorine and HCI
Ammonia
Zinc and zinc oxide
Styrene
Naphthalene
Toluene diisocyanate
Xylenes
Ethyl chloride
Methyl methacrylate
Maleic anhydride
Phthalic anhydride
Methyl isocyanate
Mineral fibers
Hydrogen fluoride
Phosphorous
Sodium hydroxide
Hydrocyanic acid
Selenium and cpds
Mercuric chloride
Bromine & inorganic cpds
Dimethyl amine
Methanol

Detailed Assessment
Phosgene
Acetaldehyde
Acrolein
Hydrogen sulfide
Municipal waste cumbustor emissions

Board Review
Dibenzofurans
Beryllium

Regulatory Options Analysis
Nickel
Gasoline vapors
Asbestos
Dioxins
Formaldehyde
Copper
Propylene oxide

Regulatory Decisions
Asbestos
Beryllium
Vinyl chloride
Coke oven emissions
Benzene
Arsenic
Radionuclides
Mercury
Toluene*
POM*
CFC-113*
Methyl chloroform*
Epichlorohydrin*
Manganese*
Chlorobenzenes*
Vinylidene chloride*
Hexachlorocyclopenta diene*
Chloroprene*
Phenol*
Acrylonitrile
Chromium
Carbon tetrachloride
Ethylene oxide
Chloroform
Ethylene dichloride
Cadmium
1,3-Butadiene
Methylene chloride
Perchloroethylene
Trichloroethylene

Proposed
Benzene

Promulgated
Mercury
Beryllium
Asbestos
Vinyl chloride
Benzene
Radionuclides
Arsenic

*Decision not to Regulate

Appendix 5. U.S. Federal Safe Drinking Water Act

Contaminants Required to be Regulated

Volatile Organic Chemicals
Trichloroethylene
Tetrachloroethylene
Carbon tetrachloride
1,1,1-Trichloroethane
1,2-Dichloroethane
Vinyl chloride
Methylene chloride
Benzene
Chlorobenzene
Dichlorobenzene
1,1-Dichloroethylene
cis-1,2-Dichloroethylene

Inorganics
Arsenic
Barium
Cadmium
Chromium
Lead
Mercury
Nitrate
Selenium
Silver
Fluoride
Aluminum
Antimony
Molybdenum
Asbestos
Sulfate
Copper
Vanadium
Sodium
Nickel
Zinc
Thallium
Beryllium
Cyanide

Organics
Endrin
Lindane
Methoxychlor
Toxaphene
2,4-D
2,4,5-TP
Aldicarb
Chlordane
Dalapon
Diquat
Endothall
Glyphosate
Carbofuran
Alachlor
Epichlorohydrin
Toluene
Adipates
2,3,7,8-TCDD (Dioxin)
1,1,2-Trichloroethane
Vydate
Simazine
PAHs
PCBs
Atrazine
Phthalates
Acrylamide
Dibromochloropropane (DBCP)
1,2-dichloropropane
Pentachlorophenol
Pichloram
Dinoseb
Ethylene dibromide (EDB)
Dibromomethane
Xylene
Hexachlorocyclo-pentadiene

Radionuclides
Radium 226 and 228
Beta particle and photon radioactivity
Uranium
Gross alpha particle activity
Radon

Microbiology and Turbidity
Total Coliforms
Turbidity
Giardia lamblia
Viruses
Standard plate count
Legionella

Chemicals Affected by Proposed Maximum Contaminant Level Goals (MCLGs) for Synthetic Organic Chemicals (SOCs)

Acrylamide
Alachlor
Aldicarb
Aldicarb sulfoxide
Aldicarb sulfonone
Carbofuran
Chlordane
cis-1,2-Dichloroethylene
DBCP
1,2-Dichloropropane
o-Dichlorobenzene
2,4-D
EDB
Epichlorohydrin
Ethylbenzene
Heptachlor
Heptachlor epoxide
Lindane
Methoxychlor
Monochlorobenzene
PCBs
Pentachlorophenol
Styrene
Toluene
2,4,5-TP
Toxaphene
trans-1,2-Dichloroethylene
Xylene

Appendix 6. Hazardous constituents as described by the Resources Conservation and Recovery Act (40 CFR 261, Appendix VIII).

Acetonitrile (Ethanenitrile)
Acetophenone (Ethanone, 1-phenyl)
3-(alpha-Acetonylbenzyl)-4-hydroxycoumarin and salts (Warfarin)
2-Acetylaminofluorene (Acetamide, N-(9H-fluoren-2-yl)-)
Acetyl chloride (Ethanoyl chloride)
1-Acetyl-2-thiourea (Acetamide, N-(aminothioxomethyl)-)
Acrolein (2-Propenal)
Acrylamide (2-Propenamide)
Acrylonitrile (2-Propenenitrile)
Aflatoxins
Aldrin (1,2,3,4,10,10-Hexachloro-1,4,4a,5,8,8a,8b-hexahydro-endo,exo-1,4:5,8-Dimethanonaphthalene)
Allyl alcohol (2-Propen-1-ol)
Aluminum phosphide
4-Aminobiphenyl ([1,1′-Biphenyl]-4-amine)
6-Amino-1,1a,2,8,8a,8b-hexahydro-8-(hydroxymethyl)-8a-methoxy-5-methylcarbamate azirino[2′,3′:3,4]pyrrolo[1,2-a]indole-4,7-dione, (ester) (Mitomycin C) (Azirino[2′3′:3,4]pyrrolo(1,2-a)indole-4,7-dione, 6-amino-8-[((aminocarbonyl)oxy)methyl]-1,1a,2,8,8a,8b-hexahydro-8amethoxy-5-methy-)
5-(Aminomethyl)-3-isoxazolol (3(2H)-Isoxazolone, 5-(aminomethyl)-) 4-Aminopyridine (4-Pyridinamine)
Amitrole (1H-1,2,4-Triazol-3-amine)
Aniline (Benzenamine)
Antimony and compounds, N.O.S.*
Aramite (Sulfurous acid, 2-chloroethyl-, 2-[4-(1,1-dimethylethyl)phenoxy]-1-methylethyl ester)
Arsenic and compounds, N.O.S.*
Arsenic acid (Orthoarsenic acid)
Arsenic pentoxide (Arsenic (V) oxide)
Arsenic trioxide (Arsenic (III) oxide)
Auramine (Benzenamine, 4,4′-carbonimidoylbis[N,N-Dimethyl-, monohydrochloride)
Azaserine (L-Serine, diazoacetate (ester))
Barium and compounds, N.O.S.*
Barium cyanide
Benz[c]acridine (3,4-Benzacridine)
Benz[a]anthracene (1,2-Benzanthracene)
Benzene (Cyclohexatriene)
Benzene, 2-amino-1-methyl (o-Toluidine)
Benzene, 4-amino-1-methyl (p-Toluidine)
Benzene, dichloromethyl- (Benzal chloride)
Benzenearsonic acid (Arsonic acid, phenyl-)
Benzenethiol (Thiophenol)
Benzidine ([1,1′-Biphenyl]-4,4′diamine)
Benzo[b]fluoranthene (2,3-Benzofluoranthene)
Benzo[j]fluoranthene (7,8-Benzofluoranthene)
Benzo[a]pyrene (3,4-Benzopyrene)
p-Benzoquinone (1,4-Cyclohexadienedione)
Benzotrichloride (Benzene, trichloromethyl-)
Benzyl chloride (Benzene, (chloromethyl)-)
Beryllium and compounds, N.O.S.*
Bis(2-chloroethoxy)methane (Ethane, 1,1′-[methylenebis(oxy)]bis[2-chloro-])
Bis(2-chloroethyl) ether (Ethane, 1,1′-oxybis[2-chloro-])
N,N-Bis(2-chloroethyl)-2-naphthylamine (Chlornaphazine)
Bis(2-chloroisopropyl) ether (Propane, 2,2′-oxybis[2-chloro-])
Bis(chloromethyl) ether (Methane, oxybis[chloro-])
Bis(2-ethylhexyl) phthalate (1,2-Benzenedicarboxylic acid, bis(2-ethylhexyl) ester)
Bromoacetone (2-Propanone, 1-bromo-)
Bromomethane (Methyl bromide)
4-Bromophenyl phenyl ether (Benzene, 1-bromo-4-phenoxy-)
Brucine (Strychnidin-10-one, 2,3-dimethoxy-)
2-Butanone peroxide (Methyl ethyl ketone, peroxide)
Butyl benzyl phthalate (1,2-Benzenedicarboxylic acid, butyl phenylmethyl ester)
2-sec-Butyl-4,6-dinitrophenol (DNBP) (Phenol, 2,4-dinitro-6-(1-methylpropyl)-)
Cadmium and compounds, N.O.S.*
Calcium chromate (Chromic acid, calcium salt)
Calcium cyanide
Carbon disulfide (Carbon bisulfide)
Carbon oxyfluoride (Carbonyl fluoride)
Chloral (Acetaldehyde, trichloro-)
Chlorambucil (Butanoic acid, 4-[bis(2-chloroethyl)amino]benzene-)
Chlordane (alpha and gamma isomers) (4,7-Methanoindan, 1,2,4,5,6,7,8,8-octachloro-3,4,7,7a-tetrahydro-) (alpha and gamma isomers)
Chlorinated benzenes, N.O.S.*
Chlorinated ethane, N.O.S.*
Chlorinated fluorocarbons, N.O.S.*
Chlorinated naphthalene, N.O.S.*
Chlorinated phenol, N.O.S.*
Chloroacetaldehyde (Acetaldehyde, chloro-)
Chloroalkyl ethers, N.O.S.*
p-Chloroaniline (Benzenamine, 4-chloro-)
Chlorobenzene (Benzene, chloro-)
Chlorobenzilate (Benzeneacetic acid, 4-chloro-alpha-(4-chlorophenyl)-alpha-hydroxy-, ethyl ester)
2-Chloro-1, 3-butadiene (chloroprene)
p-Chloro-m-cresol (Phenol, 4-chloro-3-methyl)
1-Chloro-2,3-epoxypropane (Oxirane, 2-(chloromethyl)-)
2-Chloroethyl vinyl ether (Ethene, (2-chloroethoxy)-)
Chloroform (Methane, trichloro-)
Chloromethane (Methyl chloride)
Chloromethyl methyl ether (Methane, chloromethoxy-)
2-Chloronaphthalene (Naphthalene, beta-chloro-)
2-Chlorophenol (Phenol, o-chloro-)
1-(o-Chlorophenyl)thiourea (Thiourea, (2-chlorophenyl)-)
3-Chloropropene (allyl chloride)
3-Chloropropionitrile (Propanenitrile, 3-chloro-)
Chromium and compounds, N.O.S.*
Chrysene (1,2-Benzphenanthrene)
Citrus red No. 2 (2-Naphthol, 1-[(2,5-dimethoxyphenyl)azo]-)
Coal tars
Copper cyanide
Creosote (Creosote, wood)
Cresols (Cresylic acid) (Phenol, methyl-)
Crotonaldehyde (2-Butenal)
Cyanides (soluble salts and complexes), N.O.S.*
Cyanogen (Ethanedinitrile)
Cyanogen bromide (Bromine cyanide)
Cyanogen chloride (Chlorine cyanide)

---

* The abbreviation N.O.S. (not otherwise specified) signifies those members of the general class not specifically listed by name in this appendix.

Cycasin (beta-D-Glucopyranoside, (methyl-ONN-azoxy)methyl-)
2-Cyclohexyl-4,6-dinitrophenol (Phenol, 2-cyclohexyl-4,6-dinitro-)
Cyclophosphamide (2H-1,3,2,-Oxazaphosphorine, [bis(2-chloroethyl)amino]-tetrahydro-, 2-oxide)
Daunomycin (5,12-Naphthacenedione, (8S-cis)-8-acetyl-10-[(3-amino-2,3,6-trideoxy)-alpha-L-lyxo-hexopyranosyl)oxy]-7,8,9,10-tetrahydro-6,8,11-trihydroxy-1-methoxy-)
DDD (Dichlorodiphenyldichloroethane) (Ethane, 1,1-dichloro-2,2-bis(p-chlorophenyl)-)
DDE (Ethylene, 1,1-dichloro-2,2-bis(4-chlorophenyl)-)
DDT (Dichlorodiphenyltrichloroethane) (Ethane, 1,1,1-trichloro-2,2-bis(p-chlorophenyl)-)
Diallate (S-(2,3-dichloroallyl) diisopropylthiocarbamate)
Dibenz[a,h]acridine (1,2,5,6-Dibenzacridine)
Dibenz[a,j]acridine (1,2,7,8-Dibenzacridine)
Dibenz[a,h]anthracene (1,2,5,6-Dibenzanthracene)
7H-Dibenzo[c,g]carbazole (3,4,5,6-Dibenzcarbazole)
Dibenzo[a,e]pyrene (1,2,4,5-Dibenzpyrene)
Dibenzo[a,h]pyrene (1,2,5,6-Dibenzpyrene)
Dibenzo[a,i]pyrene (1,2,7,8-Dibenzpyrene)
1,2-Dibromo-3-chloropropane (Propane, 1,2-dibromo-3-chloro-)
1,2-Dibromoethane (Ethylene dibromide)
Dibromomethane (Methylene bromide)
Di-n-butyl phthalate (1,2-Benzenedicarboxylic acid, dibutyl ester)
o-Dichlorobenzene (Benzene, 1,2-dichloro-)
m-Dichlorobenzene (Benzene, 1,3-dichloro-)
p-Dichlorobenzene (Benzene, 1,4-dichloro-)
Dichlorobenzene, N.O.S.* (Benzene, dichloro-, N.O.S.*)
3,3'-Dichlorobenzidine ([1,1'-Biphenyl]-4,4'-diamine, 3,3'-dichloro-)
1,4-Dichloro-2-butene (2-Butene, 1,4-dichloro-)
Dichlorodifluoromethane (Methane, dichlorodifluoro-)
1,1-Dichloroethane (Ethylidene dichloride)
1,2-Dichloroethane (Ethylene dichloride)
trans-1,2-Dichloroethene (1,2-Dichloroethylene)
Dichloroethylene, N.O.S.* (Ethene, dichloro-, N.O.S.*)
1,1-Dichloroethylene (Ethene, 1,1-dichloro-)
Dichloromethane (Methylene chloride)
2,4-Dichlorophenol (Phenol, 2,4-dichloro-)
2,6-Dichlorophenol (Phenol, 2,6-dichloro-)
2,4-Dichlorophenoxyacetic acid (2,4-D), salts and esters (Acetic acid, 2,4-dichlorophenoxy-, salts and esters)
Dichlorophenylarsine (Phenyl dichloroarsine)
Dichloropropane, N.O.S.* (Propane, dichloro-, N.O.S.*)
1,2-Dichloropropane (Propylene dichloride)
Dichloropropanol, N.O.S.* (Propanol, dichloro-, N.O.S.*)
Dichloropropene, N.O.S.* (Propene, dichloro-, N.O.S.*)
1,3-Dichloropropene (1-Propene, 1,3-dichloro-)
Dieldrin (1,2,3,4,10.10-hexachloro-6,7-epoxy-1,4,4a,5,6,7,8,8a-octa-hydro-endo,exo-1,4:5,8-Dimethanonaphthalene)
1,2:3,4-Diepoxybutane (2,2'-Bioxirane)
Diethylarsine (Arsine, diethyl-)
N,N-Diethylhydrazine (Hydrazine, 1,2-diethyl)
O,O-Diethyl S-methyl ester of phosphorodithioic acid (Phosphorodithioic acid, O,O-diethyl S-methyl ester
O,O-Diethylphosphoric acid, O-p-nitrophenyl ester (Phosphoric acid, diethyl p-nitrophenyl ester)
Diethyl phthalate (1,2-Benzenedicarboxylic acid, diethyl ester)
O,O-Diethyl O-2-pyrazinyl phosphorothioate (Phosphorothioic acid, O,O-diethyl O-pyrazinyl ester
Diethylstilbesterol (4,4'-Stilbenediol, alpha,alpha-diethyl, bis(dihydrogen phosphate, (E)-)
Dihydrosafrole (Benzene, 1,2-methylenedioxy-4-propyl-)
3,4-Dihydroxy-alpha-(methylamino)methyl benzyl alcohol (1,2-Benzenediol, 4-[1-hydroxy-2-(methylamino)ethyl]-)
Diisopropylfluorophosphate (DFP) (Phosphorofluoridic acid, bis(1-methylethyl) ester)
Dimethoate (Phosphorodithioic acid, O,O-dimethyl S-[2-(methylamino)-2-oxoethyl] ester
3,3'-Dimethoxybenzidine ([1,1'-Biphenyl]-4,4'diamine, 3-3'-dimethoxy-)
p-Dimethylaminoazobenzene (Benzenamine, N,N-dimethyl-4-(phenylazo)-)
7,12-Dimethylbenz[a]anthracene (1,2-Benzanthracene, 7,12-dimethyl-)
3,3'-Dimethylbenzidine ([1,1'-Biphenyl]-4,4'-diamine, 3,3'-dimethyl-)
Dimethylcarbamoyl chloride (Carbamoyl chloride, dimethyl-)
1,1-Dimethylhydrazine (Hydrazine, 1,1-dimethyl-)
1,2-Dimethylhydrazine (Hydrazine, 1,2-dimethyl-)
3,3-Dimethyl-1-(methylthio)-2-butanone, O-[(methylamino) carbonyl]oxime (Thiofanox)
alpha,alpha-Dimethylphenethylamine (Ethanamine, 1,1-dimethyl-2-phenyl-)
2,4-Dimethylphenol (Phenol, 2,4-dimethyl-)
Dimethyl phthalate (1,2-Benzenedicarboxylic acid, dimethyl ester)
Dimethyl sulfate (Sulfuric acid, dimethyl ester)
Dinitrobenzene, N.O.S.* (Benzene, dinitro-, N.O.S.*)
4,6-Dinitro-o-cresol and salts (Phenol, 2,4-dinitro-6-methyl-, and salts)
2,4-Dinitrophenol (Phenol, 2,4-dinitro-)
2,4-Dinitrotoluene (Benzene, 1-methyl-2,4-dinitro-)
2,6-Dinitrotoluene (Benzene, 1-methyl-2,6-dinitro-)
Di-n-octyl phthalate (1,2-Benzenedicarboxylic acid, dioctyl ester)
1,4-Dioxane (1,4-Diethylene oxide)
Diphenylamine (Benzenamine, N-phenyl-)
1,2-Diphenylhydrazine (Hydrazine, 1,2-diphenyl-)
Di-n-propylnitrosamine (N-Nitroso-di-n-propylamine)
Disulfoton (O,O-diethyl S-[2-(ethylthio)ethyl] phosphorodithioate)
2,4-Dithiobiuret (Thioimidodicarbonic diamide)
Endosulfan (5-Norbornene, 2,3-dimethanol, 1,4,5,6,7,7-hexachloro-, cyclic sulfite)
Endrin and metabolites (1,2,3,4,10,10-hexachloro-6,7-epoxy-1,4,4a,5,6,7,8,8a-octahydro-endo,endo-1,4:5,8-dimethanonaphthalene, and metabolites)
Ethyl carbamate (Urethan) (Carbamic acid, ethyl ester)
Ethyl cyanide (propanenitrile)
Ethylene glycol monoethyl ether (Ethanol, 2-ethoxy)

Ethylene oxide (Oxirane)
Ethylenebisdithiocarbamic acid, salts and esters (1,2-Ethanediylbiscarbamodithioic acid, salts and esters
Ethyleneimine (Aziridine)
Ethylenethiourea (2-Imidazolidinethione)
Ethyl methacrylate (2-Propenoic acid, 2-methyl-, ethyl ester)
Ethyl methanesulfonate (Methanesulfonic acid, ethyl ester)
Fluoranthene (Benzo[j,k]fluorene)
Fluorine
2-Fluoroacetamide (Acetamide, 2-fluoro-)
Fluoroacetic acid, sodium salt (Acetic acid, fluoro-, sodium salt)
Formaldehyde (Methylene oxide)
Formic acid (Methanoic acid)
Glycidylaldehyde (1-Propanol-2,3-epoxy)
Halomethane, N.O.S.*
Heptachlor (4,7-Methano-1H-indene, 1,4,5,6,7,8,8-heptachloro-3a,4,7,7a-tetrahydro-)
Heptachlor epoxide (alpha, beta, and gamma isomers) (4,7-Methano-1H-indene, 1,4,5,6,7,8,8-heptachloro-2,3-epoxy-3a,4,7,7-tetrahydro-, alpha, beta, and gamma isomers)
Hexachlorobenzene (Benzene, hexachloro-)
Hexachlorobutadiene (1,3-Butadiene, 1,1,2,3,4,4-hexachloro-)
Hexachlorocyclohexane (all isomers) (Lindane and isomers)
Hexachlorocyclopentadiene (1,3-Cyclopentadiene, 1,2,3,4,5,5-hexachloro-)
Hexachlorodibenzo-p-dioxins
Hexachlorodibenzofurans
Hexachloroethane (Ethane, 1,1,1,2,2,2-hexachloro-)
1,2,3,4,10,10-Hexachloro-1,4,4a,5,8,8a-hexahydro-1,4:5,8-endo,endo-dimethanonaphthalene (Hexachlorohexahydro-endo,endo-dimethanonaphthalene)
Hexachlorophene (2,2′-Methylenebis(3,4,6-trichlorophenol))
Hexachloropropene (1-Propene, 1,1,2,3,3,3-hexachloro-)
Hexaethyl tetraphosphate (Tetraphosphoric acid, hexaethyl ester)
Hydrazine (Diamine)
Hydrocyanic acid (Hydrogen cyanide)
Hydrofluoric acid (Hydrogen fluoride)
Hydrogen sulfide (Sulfur hydride)
Hydroxydimethylarsine oxide (Cacodylic acid)
Indeno(1,2,3-cd)pyrene (1,10-(1,2-phenylene)pyrene)
Iodomethane (Methyl iodide)
Iron dextran (Ferric dextran)
Isocyanic acid, methyl ester (Methyl isocyanate)
Isobutyl alcohol (1-Propanol, 2-methyl-)
Isosafrole (Benzene, 1,2-methylenedioxy-4-allyl-)
Kepone (Decachlorooctahydro-1,3,4-Methano-2H-cyclobuta[cd]pentalen-2-one)
Lasiocarpine (2-Butenoic acid, 2-methyl-, 7-[(2,3-dihydroxy-2-(1-methoxyethyl)-3-methyl-1-oxobutoxy)methyl]-2,3,5,7a-tetrahydro-1H-pyrrolizin-1-yl ester)
Lead and compounds, N.O.S.*
Lead acetate (Acetic acid, lead salt)
Lead phosphate (Phosphoric acid, lead salt)
Lead subacetate (Lead, bis(acetato-O)tetrahydroxytri-)
Maleic anhydride (2,5-Furandione)
Maleic hydrazide (1,2-Dihydro-3,6-pyridazinedione)
Malononitrile (Propanedinitrile)
Melphalan (Alanine, 3-[p-bis(2-chloroethyl)amino]phenyl-, L-)
Mercury fulminate (Fulminic acid, mercury salt)
Mercury and compounds, N.O.S.*
Methacrylonitrile (2-Propenenitrile, 2-methyl-)
Methanethiol (Thiomethanol)
Methapyrilene (Pyridine, 2-[(2-dimethylamino)ethyl]-2-thenylamino-)
Metholmyl (Acetimidic acid, N-[(methylcarbamoyl)oxy]thio-, methyl ester
Methoxychlor (Ethane, 1,1,1-trichloro-2,2′-bis(p-methoxyphenyl)-)
2-Methylaziridine (1,2-Propylenimine)
3-Methylcholanthrene (Benz[j]aceanthrylene, 1,2-dihydro-3-methyl-)
Methyl chlorocarbonate (Carbonochloridic acid, methyl ester)
4,4′-Methylenebis(2-chloroaniline) (Benzenamine, 4,4′-methylenebis-(2-chloro-)
Methyl ethyl ketone (MEK) (2-Butanone)
Methyl hydrazine (Hydrazine, methyl-)
2-Methyllactonitrile (Propanenitrile, 2-hydroxy-2-methyl-)
Methyl methacrylate (2-Propenoic acid, 2-methyl-, methyl ester)
Methyl methanesulfonate (Methanesulfonic acid, methyl ester)
2-Methyl-2-(methylthio)propionaldehyde-o-(methylcarbonyl) oxime (Propanal, 2-methyl-2-(methylthio)-, O-[(methylamino)carbonyl]oxime)
N-Methyl-N′-nitro-N-nitrosoguanidine (Guanidine, N-nitroso-N-methyl-N′-nitro-)
Methyl parathion (O,O-dimethyl O-(4-nitrophenyl) phosphorothioate)
Methylthiouracil (4-1H-Pyrimidinone, 2,3-dihydro-6-methyl-2-thioxo-)
Mustard gas (Sulfide, bis(2-chloroethyl)-)
Naphthalene
1,4-Naphthoquinone (1,4-Naphthalenedione)
1-Naphthylamine (alpha-Naphthylamine)
2-Naphthylamine (beta-Naphthylamine)
1-Naphthyl-2-thiourea (Thiourea, 1-naphthalenyl-)
Nickel and compounds, N.O.S.*
Nickel carbonyl (Nickel tetracarbonyl)
Nickel cyanide (Nickel (II) cyanide)
Nicotine and salts (Pyridine, (S)-3-(1-methyl-2-pyrrolidinyl)-, and salts)
Nitric oxide (Nitrogen (II) oxide)
Nitrobenzine (Benzene, nitro-)
Nitrogen dioxide (Nitrogen (IV) oxide)
Nitrogen mustard and hydrochloride salt (Ethanamine, 2-chloro-, N-(2-chloroethyl)-N-methyl-, and hydrochloride salt)
Nitrogen mustard N-Oxide and hydrochloride salt (Ethanamine, 2-chloro-, N-(2-chloroethyl)-N-methyl-, and hydrochloride salt)
Nitroglycerine (1,2,3-Propanetriol, trinitrate)
Nitrosamine, N.O.S.*
p-Nitroaniline (Benzenamine, 4-nitro-)
4-Nitrophenol (Phenol, 4-nitro-)
2-Nitropropane (Propane, 2-nitro)
4-Nitroquinoline-1-oxide (Quinoline, 4-nitro-1-oxide-)
N-Nitrosodi-n-butylamine (1-Butanamine, N-butyl-N-nitroso-)
N-Nitrosodiethanolamine (Ethanol, 2,2′-(nitrosoimino)bis-)
N-Nitrosodiethylamine (Ethanamine, N-ethyl-N-nitroso-)
N-Nitrosodimethylamine (Dimethylnitrosamine)
N-Nitroso-N-ethylurea (Carbamide, N-ethyl-N-nitroso-)

N-Nitrosomethylethylamine (Ethanamine, N-methyl-N-nitroso-)
N-Nitroso-N-methylurea (Carbamide, N-methyl-N-nitroso-)
N-Nitroso-N-methylurethane (Carbamic acid, methylnitroso-, ethyl ester)
N-Nitrosomethylvinylamine (Ethenamine, N-methyl-N-nitroso-)
N-Nitrosomorpholine (Morpholine, N-nitroso-)
N-Nitrosonornicotine (Nornicotine, N-nitroso-)
N-Nitrosopiperidine (Pyridine, hexahydro-, N-nitroso-)
Nitrosopyrrolidine (Pyrrole, tetrahydro-, N-nitroso-)
N-Nitrososarcosine (Sarcosine, N-nitroso-)
5-Nitro-o-toluidine (Benzenamine, 2-methyl-5-nitro-)
Octamethylpyrophosphoramide (Diphosphoramide, octamethyl-)
Osmium tetroxide (Osmium (VIII) oxide)
7-Oxabicyclo[2.2.1]heptane-2,3-dicarboxylic acid (Endothal)
Paraldehyde (1,3,5-Trioxane, 2,4,6-trimethyl-)
Parathion (Phosphorothioic acid, O,O-diethyl O-(p-nitrophenyl) ester
Pentachlorobenzene (Benzene, pentachloro-)
Pentachlorodibenzo-*p*-dioxins
Pentachlorodibenzofurans
Pentachloroethane (Ethane, pentachloro-)
Pentachloronitrobenzene (PCNB) (Benzene, pentachloronitro-)
Pentachlorophenol (Phenol, pentachloro-)
Phenacetin (Acetamide, N-(4-ethoxyphenyl)-)
Phenol (Benzene, hydroxy-)
Phenylenediamine (Benzenediamine)
Phenylmercury acetate (Mercury, acetatophenyl-)
N-Phenylthiourea (Thiourea, phenyl-)
Phosgene (Carbonyl chloride)
Phosphine (Hydrogen phosphide)
Phosphorodithioic acid, O,O-diethyl S-[(ethylthio)methyl] ester (Phorate)
Phosphorothioic acid, O,O-dimethyl O-[p-((dimethylamino)sulfonyl)phenyl] ester (Famphur)
Phthalic acid esters, N.O.S.* (Benzene, 1,2-dicarboxylic acid, esters, N.O.S.*)
Phthalic anhydride (1,2-Benzenedicarboxylic acid anhydride)
2-Picoline (Pyridine, 2-methyl-)
Polychlorinated biphenyl, N.O.S.*
Potassium cyanide
Potassium silver cyanide (Argentate(1-), dicyano-, potassium)
Pronamide (3,5-Dichloro-N-(1,1-dimethyl-2-propynyl)benzamide)
1,3-Propane sultone (1,2-Oxathiolane, 2,2-dioxide)
n-Propylamine (1-Propanamine)
Propylthiouracil (Undecamethylenediamine, N,N'-bis(2-chlorobenzyl)-, dihydrochloride)
2-Propyn-1-ol (Propargyl alcohol)
Pyridine
Reserpine (Yohimban-16-carboxylic acid, 11,17-dimethoxy-18-[(3,4,5-trimethoxybenzoyl)oxy]-, methyl ester)
Resorcinol (1,3-Benzenediol)
Saccharin and salts (1,2-Benzoisothiazolin-3-one, 1,1-dioxide, and salts)
Safrole (Benzene, 1,2-methylenedioxy-4-allyl-)
Selenious acid (Selenium dioxide)
Selenium and compounds, N.O.S.*
Selenium sulfide (Sulfur selenide)
Selenourea (Carbamimidoselenoic acid)
Silver and compounds, N.O.S.*
Silver cyanide
Sodium cyanide
Streptozotocin (D-Glucopyranose, 2-deoxy-2-(3-methyl-3-nitrosoureido)-)
Strontium sulfide
Strychnine and salts (Strychnidin-10-one, and salts)
1,2,4,5-Tetrachlorobenzene (Benzene, 1,2,4,5-tetrachloro-)
2,3,7,8-Tetrachlorodibenzo-p-dioxin (TCDD) (Dibenzo-p-dioxin, 2,3,7,8-tetrachloro-)
Tetrachlorodibenzo-*p*-dioxins
Tetrachlorodibenzofurans
Tetrachloroethane, N.O.S.* (Ethane, tetrachloro-, N.O.S.*)
1,1,1,2-Tetrachlorethane (Ethane, 1,1,1,2-tetrachloro-)
1,1,2,2-Tetrachlorethane (Ethane, 1,1,2,2-tetrachloro-)
Tetrachloroethane (Ethene, 1,1,2,2-tetrachloro-)
Tetrachloromethane (Carbon tetrachloride)
2,3,4,6,-Tetrachlorophenol (Phenol, 2,3,4,6-tetrachloro-)
Tetraethyldithiopyrophosphate (Dithiopyrophosphoric acid, tetraethyl-ester)
Tetraethyl lead (Plumbane, tetraethyl-)
Tetraethylpyrophosphate (Pyrophosphoric acide, tetraethyl ester)
Tetranitromethane (Methane, tetranitro-)
Thallium and compounds, N.O.S.*
Thallic oxide (Thallium (III) oxide)
Thallium (I) acetate (Acetic acid, thallium (I) salt)
Thallium (I) carbonate (Carbonic acid, dithallium (I) salt)
Thallium (I) chloride
Thallium (I) nitrate (Nitric acid, thallium (I) salt)
Thallium selenite
Thallium (I) sulfate (Sulfuric acid, thallium (I) salt)
Thioacetamide (Ethanethioamide)
Thiosemicarbazide (Hydrazinecarbothioamide)
Thiourea (Carbamide thio-)
Thiuram (Bis(dimethylthiocarbamoyl) disulfide)
Toluene (Benzene, methyl-)
Toluenediamine, N.O.S* (Diaminotoluene, N.O.S*)
2,4-Toluenediamine
2,6-Toluenediamine
3,4-Toluenediamine
o-Toluidine hydrochloride (Benzenamine, 2-methyl-, hydrochloride)
Tolylene diisocyanate (Benzene, 1,3-diisocyanatomethyl-)
Toxaphene (Camphene, octachloro-)
Tribromomethane (Bromoform)
1,2,4-Trichlorobenzene (Benzene, 1,2,4-trichloro-)
1,1,1-Trichloroethane (Methyl chloroform)
1,1,2-Trichloroethane (Ethane, 1,1,2-trichloro-)
Trichloroethene (Trichloroethylene)
Trichloromethanethiol (Methanethiol, trichloro-)
Trichloromonofluoromethane (Methane, trichlorofluoro-)
2,4,5-Trichlorophenol (Phenol, 2,4,5-trichloro-)
2,4,6-Trichlorophenol (Phenol, 2,4,6-trichloro-)
2,4,5-Trichlorophenoxyacetic acid (2,4,5-T) (Acetic acid, 2,4,5-trichlorophenoxy-)
2,4,5-Trichlorophenoxypropionic acid (2,4,5-TP) (Silvex) (Propionoic acid, 2-(2,4,5-trichlorophenoxy)-)
Trichloropropane, N.O.S.* (Propane, trichloro-, N.O.S.*)
1,2,3-Trichloropropane (Propane, 1,2,3-trichloro-)
O,O,O-Triethyl phosphorothioate (Phosphorothioic acid, O,O,O-triethyl ester)
sym-Trinitrobenzene (Benzene, 1,3,5-trinitro-)
Tris(1-azridinyl) phosphine sulfide (Phosphine sulfide, tris(1-aziridinyl-)
Tris(2,3-dibromopropyl) phosphate (1-Propanol, 2,3-dibromo-, phosphate)
Trypan blue (2,7-Naphthalenedisulfonic acid, 3,3'-[(3,3'-dimethyl(1,1'-biphenyl)-4,4'-diyl)bis(azo)]bis(5-amino-4-hydroxy-, tetrasodium salt)
Uracil mustard (Uracil 5-[bis(2-chloroethyl)amino]-)
Vanadic acid, ammonium salt (ammonium vanadate)
Vanadium pentoxide (Vanadium (V) oxide)
Vinyl chloride (Ethene, chloro-)
Zinc cyanide
Zinc phosphide

Appendix 7. Priority List of Substances for Toxicological Profiles by ASTDR (Agency for Toxic Substances and Disease Registry) under Superfund Amendments (SAPA) of 1986 (as of 17 April, 1987).*

Priority Group 1		Priority Group 2		Priority Group 3		Priority Group 4	
50328	Benzo(a)pyrene	56235	Carbon tetrachloride	71556	1,1,1-Trichloroethane	51285	2,4-Diitrophenol
53703	Dibenzo(a),h)anthracene	57749	Chlordane	74873	Chloromethane	59507	P-Chloro-m-cresol
56553	Benzo(a)anthracene	62759	N-nitrosodimethylamine	75218	Oxirane	62533	Aniline
57125	Cyanide	72559	4,4'-DDE, DDT, DDD	75252	Bromoform	65850	Benzoic acid
60571	Dieldrin/aldrin	75003	Chloroethane	75343	1,1-Dichloroethane	67721	Hexachloroethane
67663	Chloroform	75274	Bromodichloromethane	84742	Di-N-butyl phthalate	74839	Bromomethane
71432	Benzene	75354	1,1-Dichloroethene	88062	2,4,6-Trichlorophenol	75150	Carbondisulfide
75014	Vinyl chloride	78591	Isophorone	91203	Napthalene	75694	Fluorotrichloromethane
75092	Methylene chloride	78875	1,2-Dichloropropane	98953	Nitrobenzene	75718	Dichlorodifluoromethane
76448	Heptachlor/heptachlor epoxide	79005	1,1,2-Trichloroethane	100414	Ethylbenzene	78933	2-Butanone
79016	Trichloroethene	79435	1,1,2,2-Tetrachloroethane	107028	Acrolein	84662	Diethyl phthalate
86306	N-nitrosodiphenylamine	87865	Pentachlorophenol	107131	Acrylonitrile	85018	Phenanthrene
106467	1,4-Dichlorobenzene	91941	3,3'-Dichlorobenzidine	108907	Chlorobenzene	87683	Hexachlorobutadiene
117817	Bis(2-ethylhexyl)phthalate	92875	Benzidine	118741	Hexachlorobenzene	95487	Phenol,2-methyl
127184	Tetrachloroethene	107062	1,2-Dichloroethane	122667	1,2-Diphenylhydrazine	95501	1,2-Dichlorobenzene
205992	Benzo(b)fluoranthene	108883	Toluene	124481	Chlorodibromomethane	105679	2,4-Dimethylphenol
218019	Chrysene	108952	Phenol	156606	1,2-Trans-dichloroethene	108101	2-Pentanone, 4-Methyl
1745016	P-Dioxin	111444	Bis(2-chloroethyl)ether	193395	Indeno(1,2,3-cd)pyrene	120821	1,2,4-Trichlorobenzene
7439921	Lead	121142	2,4-Dinitrotoluene	606202	2,6,Dinitrotoluene	120832	2,4-Dichlorophenol
7440020	Nickel	319846	BHC-alpha, gamma, beta, delta	1330207	Total xylenes	123911	1,4-Dioxane
7440382	Arsenic	542881	Bis(chloromethyl)ether	7221934	Endrin aldehyde/endrin	131113	Dimethyl phthalate
7440417	Beryllium	621647	N-nitrosodi-n-propylamine	7440224	Silver	206440	Fluoranthene
7440439	Cadmium	7439976	Mercury	7440508	Copper	534521	4,6-Dinitro-2-methylphenol
7440473	Chromium	7440666	Zinc	7664417	Ammonia	541731	1,3-Dichlorobenzene
11196625	PCB-1260,54,48,42,32,21,1016	7782492	Selenium	8001352	Toxaphene	7440280	Thallium

*Order within priority group is by CAS No.

ROSTER OF SPEAKERS, CHAIRMEN, AND PARTICIPANTS

Adams, Randy	ECOVA Corporation, Redmond, WA 98052
Adams, William J.	Monsanto Company, St. Louis, MO 63167
Adriaens, Peter	University of California, Riverside, CA 92521
Adrian, Neal	South Dakota State University, Brookings, SD 57006
Aggerholm, David	Port of Seattle, Seattle, WA 98111
Aldrich, Teri	University of Illinois at Chicago, Chicago, IL 60680
Alexander, Martin	Cornell University, Ithaca, NY 14853
Allen, Caroline	North Carolina League of Women Voters, Greensboro, NC 27401
Anderson, Dan	Cyanotech Corporation, Woodinville, WA 98072
Angell, Philip	United States Congress, Washington, DC 20515
Ashcom, David W.	ECOVA Corporation, Redmond, WA 98052
Atlas, Ronald M.	University of Louisville, Louisville, KY 40292
Attaway, Hubert	Manville Corporation, Denver, CO 80217
Augerot, Xan	Washington Seagrant, Seattle, WA 98105
Augino, Alan	University of Washington, Seattle, WA 98195
Ayyagari, L. Rao	Lindenwood College, St. Charles, MO 63301
Bagdasarian, Michael	Michigan Biotechnology Institute, Lansing, MI 48909
Bagdasarian, Mira	Michigan Biotechnology Institute, Lansing, MI 48909
Bakst, Jodi	U.S. Environmental Protection Agency, Washington, DC 20460
Ball, Christopher	Panlabs, Inc., Seattle, WA 98115
Ballard, Ernesta	Pacific Celebration '89, Seattle, WA 98154-1101
Banks, Robert Darryl	New York State Department of Environmental Conservation, Albany, NY 12233-1013
Barnthouse, Lawrence W.	Oak Ridge National Laboratory, Oak Ridge, TN 37831-6036
Barry, Gerard F.	Monsanto Company, AA2G, St. Louis, MO 63198
Baugh, Kent	IT Corporation, 4585 Pacheco Blvd., Martinez, CA 94553
Bauman, Jeff	METRO, Seattle, WA 98104
Bayley, Christopher	Burlington Northern Railroad, Seattle, WA 98104
Bayley, Emery P.	ECOVA Corporation, Redmond, WA 98052

Bhuta, Prakash H.	Eastern Washington University, Cheney, WA 99004
Bissonnette, David A.	PACCAR, Inc., Bellevue, WA 98009
Bjelland, Mark	Department of Environmental Health, Seattle, WA 98195
Blair, James E.	Lancy International, Inc., Warrendale, PA 15086
Boatman, Edwin S.	University of Washington, Seattle, WA 98195
Bolton, George	University of Washington, Seattle, WA 98195
Bolton, Jr., Harvey	Battelle Northwest, Richland, WA 99352
Bonifield, Thomas R.	U.S. Public Health Service, Seattle, WA 98174
Bouma, Dirk	Levine-Fricke, Oakland, CA 94611-4567
Bourquin, Al	ECOVA Corporation, Redmond, WA 98052
Boyce, Sheila	University of Washington, Seattle, WA 98195
Breysse, Peter A.	University of Washington, Seattle, WA 98195
Britton, Larry	Texas Research Institute, Inc., Austin, TX 78733
Brockman, Fred J.	Battelle, Pacific Northwest Laboratories Richland, WA 99352
Brooks, Patricia A.	ECOVA Corporation, Redmond, WA 98052
Brown, Mark P.	New York State Department of Environmental Conservation, Albany, NY 12233
Brubaker, Gaylen R.	International Technology Corporation, Torrance, CA 90505
Bull, Alan T.	The University of Canterbury, Kent CT2 7NJ, ENGLAND
Burbidge, Richard G.	NUS Corporation, Houston, TX 77054
Cable, Thomas J.	ECOVA Corporation, Bellevue, WA 98004
Campbell, Charlotte R.	Zymo Genetics, Seattle, WA 98011
Campion, Jack	Camp Dresser & McKee, Inc., Boston, MA 02108
Caplan, Jason	O.H. Materials Corp., Findlay, OH
Chadsey, Matt	Seattle Water Department, Everett, WA 98201
Chakrabarty, A. M.	The University of Illinois, Chicago, IL 60680
Chapman, Peter J.	U.S. Environmental Protection Agency, Gulf Breeze, FL 32561-3999
Chee, Joanne	ECOVA Corporation, Redmond, WA 98052
Chen, Jeffrey J.	Lancy International, Inc., Warrendale, PA 15086
Cichowiicz, Judith	Univar Corporation, Seattle, WA 98104
Clark, Elizabeth	University of Washington, Seattle, WA 98195
Cohen, Lori	U.S. Environmental Protection Agency, Seattle, WA 98101
Cohen, Stanley N.	Stanford University School of Medicine, Stanford, CA 94305
Coleman, Bob	Argonne National Laboratory, Argonne, IL 60439
Colwell, Rita	The University of Maryland, Adelphi, MD 20783
Cone, Nelson	University of Washington, Seattle, WA 98195
Cornish, Blake M.	United States Congress, Washington, DC 20510-8025
Costa, Lucio G.	University of Washington, Seattle, WA 98195
Cotton, Douglas W.	ECOVA Corporation, Redmond, WA 98052

Cox, Johnson	Interbio, Inc., Naperville, IL 60540
Cox, Jonson	Land Restoration Systems Ltd., London W12 8AW, ENGLAND
Cozad, John	Celgene, Warren, NJ 07060
Crawford, Ronald L.	University of Idaho, Moscow, ID 83843
Davis, Dick A.	Univar Corporation, Seattle, WA 98104
Davis, Lois	Sybron Chemicals, Inc., Salem, VA 24153
Davison, John	ICP Unit of Molecular Biology, Brussels B1200, BELGIUM
Davoli, Dana J.	U.S. Environmental Protection Agency, Seattle, WA 98101
Day, Paul	U.S. Environmental Protection Agency, Seattle, WA 98101
DelVecchio, Vito G.	Navy Ocean Systems Center, San Diego, CA 92152
Demain, Arnold	Massachusetts Institute of Technology, Cambridge, MA 02139
de Meyer, Rita	University of California, Riverside, CA 92521
Denovan, Barbara	Battelle Northwest, Richland, WA 99352
Deobald, Lee A.	University of Idaho, Moscow, ID 83843
DeRoos, Roger	University of Washington, Seattle, WA 98195
DeWalle, Foppe	University of Washington, Seattle, WA 98195
Dibble, J.T.	Texaco Research, Port Arthur, TX 77641
Dohrmann, John	Puget Sound Water Quality Authority, Seattle, WA 98101
Donaldson, Terrence L.	Oak Ridge National Laboratory/Martin Marietta Energy Systems, Oak Ridge, TN 37831-6227
Doolittle, James A.	Converse GES, Seattle, WA 98121
Douglas, Tim	Mayor of Bellingham, Bellingham, WA 98225
Drury, David D.	Groundwater Technology, Inc., Concord, CA 94520
Duffy, James J.	Occidental Chemical Corporation, Grand Island, NY 14072
Dunnette, David	Portland State University, Portland, OR 97207
Eaton, David L.	University of Washington, Seattle, WA 98195
Eckhoff, Joe	University of Washington, Seattle, WA 98195
Ely, Roger	University of Washington, Seattle, WA 98195
Englert, Karl	ECOVA Corporation, Redmond, WA 98052
Ensley, Burt D.	AMGEN, Thousand Oaks, CA 91360
Evans, Susan	ERM Northwest, Redmond, WA 98052
Falkow, Stanley	Stanford University, Stanford, CA 94305
Farley, Frederick E.	Microlife Technics, Sarasota, FL 34230
Federle, Tom	Procter & Gamble Company, Cincinnati, OH 45217
Feilen, Jean	ECOVA Corporation, Redmond, WA 98052
Feole, Cathy M.	ECOVA Corporation, Redmond, WA 98052
Ferguson, John	University of Washington, Seattle, WA 98195
Fernyhough, Brent	Cyanotech Corporation, Woodinville, WA 98072
Fisher, Willa A.	Bremerton-Kitsap County Health Department, Bremerton, WA 98312

Flanagan, Patrick	National Science Foundation, Washington, DC 20550
Fletcher, Linnea	University of Texas, Austin, TX 78712
Focht, Dennis D.	University of California, Riverside, CA 92521
Fogel, Seymour	University of California, Berkeley, CA 94720
Foght, Julia	University of Alberta, Edmonton, Alberta T6G 2E9, CANADA
Formoso, Carl	University of Washington, Seattle, WA 98195
Francis, Jacqueline R.	Naval Energy & Environmental Support Activity, Port Hueneme, CA 93043-5014
Frankenberger, Jr., William T.	University of California, Riverside, CA 92521
Fuentes, Rene	U.S. Environmental Protection Agency, Seattle, WA 98101
Furlong, Clement	University of Washington, Seattle, WA 98195
Gakstatter, Jack	U.S. Environmental Protection Agency, Seattle, WA 98101
Gartland, William	National Institutes of Health, Rockville, MD 20852
Gauger, W. Kennedy	South Dakota State University, Brookings, SD 57007
Gersberg, Richard M.	San Diego State University, San Diego, CA 92182
Glaser, John	U.S. Environmental Protection Agency, Cincinnati, OH 45268
Glowinski, Irene B.	United States Congress, Washington, DC 20515
Goldstein, Robert	Electric Power Research Institute, Palo Alto, CA 94303
Graves, Nathan A.	Kennedy, Jenks, Chilton, Federal Way, WA 98003
Grubbs, R.B. "Jones"	Solmar Corporation, Orange, CA 92667
Guarraia, Leonard J.	Monsanto Company, St. Louis, MO 63167
Gunderson-Harrison, Lisen	Cyanotech Corporation, Woodinville, WA 98072
Gutzmer, Michael P.	Texas Water Commission, Austin, TX 78711-3087
Hadley, Jane	Seattle Post-Intelligencer, Seattle, WA 98119
Hagedorn, Scott	Celgene, Summit, NJ 07901
Hall, Susan	Hall and Associates, Seattle, WA 98121
Hamel, Denise M.	University of Washington, Seattle, WA 98195
Hamilton, Steve	General Electric Company, Schenectady, NY 12301
Hanks, Thrift G.	University of California at Irvine, Los Angeles, CA 90027
Hanson, Thomas	University of Washington, Seattle, WA 98195
Harding, Gwen Lee	University of Texas at Austin, Austin, TX 78705
Hassett, Christopher	University of Washington, Seattle, WA 98195
Hastings, Janis	U.S. Environmental Protection Agency, Seattle, WA 98101
Hatch, Kenneth L.	ECOVA Corporation, Seattle, WA 98111-7077
Haugland, Richard A.	University of Illinois at Chicago, Cincinnati, OH 45268

Hayes, Paul	U.S. Environmental Protection Agency, Washington, DC 20460
Hedine, Karen L.	ECOVA Corporation, Redmond, WA 98052
Hertel, Greg	University of Idaho, Moscow, ID 83843
Heyward, Andrew A.	METRO, Seattle, WA 98119
Hickman, Douglas O.	ECOVA Corporation, Baltimore, MD 21202
Higgins, Cynthia	Cyanotech Corporation, Woodinville, WA 98072
Hill, Kathy	ECOVA Corporation, Redmond, WA 98052
Hill, Richard	ECOVA Corporation, Redmond, WA 98052
Hirschhorn, Joel S.	United States Congress, Washington, DC 20510
Holm, Harvey W.	Montana State University, Bozeman, MT 59717
Hsieh, Chu-Chin	University of Washington, Seattle, WA 98195
Huang, Chi-Min	University of California, Riverside, CA 92521
Huber, Karen J.	Work Environment Safety Services, Everett, WA 98208
Hudson, Rick	University of Washington, Seattle, WA 98195
Hufford, David	City of Tacoma - Sewer Utility, Tacoma, WA 98421
Igbene, Akali	Levine-Fricke, Inc., Oakland, CA 94611-4567
Isbister, Jenefir D.	Atlantic Research Corporation, Alexandria, VA 22313
Johnson, Lamar	Idaho National Engineering Laboratory, Idaho Falls, ID 83415
Johnston, Robert F.	Johnston Associates, Inc., Princeton, NJ 08540
Jones, Gwyn	Fresh Water Biological Association, Ambleside, Cumbria LA220LP, ENGLAND
Jowise, Peter	Ecology and Environment, Inc., Seattle, WA 98104
Kadakia, Rakesh K.	University of Idaho, Moscow, ID 83843
Kakida, Terry J.	Seattle City Light, Seattle, WA 98104
Karlson, Ulrich	University of California, Riverside, CA 92521
Kaufman, A. Keith	BIOTA, Inc., Los Angeles, CA 90048
Keller, Michael	Chemical Processors, Seattle, WA 98108
Kenis, Paul	Naval Ocean Systems Center, San Diego, CA 92152-5000
Kepiro, Joseph L.	University of Washington, Seattle, WA 98195
Kievit, Bob	U.S. Environmental Protection Agency, Seattle, WA 98101
Kimerle, Richard A.	Monsanto Company, St. Louis, MO 63167
Kingsbury, David	National Science Foundation, Washington, DC 20550
Kirschner, Larry	General Dynamics/Fort Worth Division, Ft. Worth, TX 76101
Kobayashi, Hester	Standard Oil, Cleveland, OH 44128
Koch, Richard	Department of Ecology, Redmond, WA 98052
Koegler, Syd S.	Battelle Northwest Laboratory, Richland, WA 99352
Kohler, Hans-Peter	University of California, Riverside, CA 92521

Kohler-Staub, Doris	University of California, Riverside, CA 92521
Kohring, Gert-Wieland	University of Georgia, Athens, GA 30602
Komisar, Sim	University of Washington, Seattle, WA 98195
Kopecky, Anne L.	Myriad Environmental Services, Seattle, WA 98107
Kotturi, Gopaul	University of Waterloo, Victoria, B.C. V8N 4B6, CANADA
Kreitinger, Joseph	ECOVA Corporation, Redmond, WA 98052
Krockel, Lothar	University of California, Riverside, CA 92521
Kulik, Conrad	Electric Power Research Institute, Palo Alto, CA 94303
Kuny, Gregory	Council for Research Planning in Biological Sciences, Washington, DC 20036-2077
Kunz, Daniel	North Texas State University, Denton, TX 76203
Labib, Foroozan	University of Washington, Seattle, WA 98195
Landis, Wayne G.	Chemical Research, Development, and Engineering Center, Aberdeen Proving Grounds, MD 21010-5423
Landon-Arnold, Susan	Naval Civil Engineering Laboratory, Port Hueneme, CA 93043
Leary, John V.	University of California, Riverside, CA 92521
Leder, John	Landau Associates, Edmonds, WA 98020
Levanon, Dan	MIGAUL-Galilee Technological Center, Kiryat Shmona 10-200, ISRAEL
Levin, Morris	U.S. Environmental Protection Agency, Washington, DC 20460
Lewis, Cindy	University of Washington, Seattle, WA 98195
Lewis, Ronald F.	U.S. Environmental Protection Agency, Cincinnati, OH 45268
Lieberman, Tony	O.H. Materials Corporation, Findlay, OH
Lightfoot, Don	Eastern Washington University, Cheney, WA 99004
Lindstrom, Jon	University of Alaska, Fairbanks, AK 99775-1760
Linkenheil, Ron	Remediation Technologies, Inc., Fort Collins, CO 80524
List, Mark	University of Washington, Seattle, WA 98195
Litchfield, Carol D.	E.I. DuPont DeNemours & Company, Wilmington, DE 19898
Little, C. Deane	Oak Ridge National Laboratory, Oak Ridge, TN 37831
Lloyd, Michael	University of Puget Sound, Mercer Island, WA 98101
Loeppke, Ron	University of Washington, Seattle, WA 98125
Long, Jennifer	Seattle, WA 98103
Long, Phil	U.S. Environmental Protection Agency, Seattle, WA 98101
Loo, Walter W.	ECOVA Corporation, Redmond, WA 98052
Lucas, Herbert L.	ECOVA Corporation, Los Angeles, CA 90049
Lyle, Lori K.	ECOVA Corporation, Redmond, WA 98052
Lyon, Stephen R.	San Diego State University, San Diego, CA 92105

Maier, William	University of Washington, Seattle, WA 98195
Manning, Jr., John F.	University of Alabama, Birmingham, AL 35294
McCarty, Perry L.	Stanford University, Stanford, CA 94305
McCrary, Brian F.	University of Arizona, Tucson, AZ 85724
McFarland, Beverly L.	Chevron Research Company, Richmond, CA 94802-0621
McLearn, Mary	Electric Power Research Institute, Palo Alto, CA 94303
Meador, James	University of Washington, Seattle, WA 98195
Mechalas, Byron	Southern California Edison, Rosemead, CA 91770
Meyer, Roger J.	University of Washington, Tacoma, WA 98403
Meyers-Schulte, Kathy	Computer Sciences Corporation, San Diego, CA 92110
Miles, Don	Bremerton-Kitsap County Health Department, Bremerton, WA 98312
Miller, Mel	Chemical Processors, Seattle, WA 98108
Molak, Vlasta	U.S. Environmental Protection Agency, Washington, DC 20460
Molnaa, Barry	Solmar Corporation, Orange, CA 92667
Moreland, Sue A.	The Markland Group, Washington, DC 20003
Morgan, Michael S.	University of Washington, Seattle, WA 98195
Morrison, James L.	ECOVA Corporation, Redmond, WA 98052
Murakami, Yoshitaka	Cornell University, Ithaca, NY 14853
Murphy, Sheldon D.	University of Washington, Seattle, WA 98195
Nadim, Louise M.	University of Texas, Austin, TX 78712
Nelson, Michael J.K.	U.S. Environmental Protection Agency, Gulf Breeze, FL 32561
Ninteman, Jerry	University of Washington, Bellevue, WA 98007
Nix, Peter	E.V.S. Consultants, Inc., Seattle, WA 98102
Njagi, G.D.E.	Kenyatta University, Nairobi, KENYA
Nolan, Claire	ECOVA Corporation, Redmond, WA 98052
O'Brien, Maura S.	ECOVA Corporation, Redmond, WA 98052
Omenn, Gilbert S.	University of Washington, Seattle, WA 98195
Omiecinski, Curt	University of Washington, Seattle, WA 98195
Ongerth, Jerry	University of Washington, Seattle, WA 98195
O'Reilly, Kirk	University of Idaho, Moscow, ID 83843
Orians, Gordon	University of Washington, Seattle, WA 98195
Owen, Derek	ECOVA Corporation, Redmond, WA 98052
Owen, Elizabeth	ECOVA Corporation, Redmond, WA 98052
Parker, Jean C.	U.S. Environmental Protection Agency, Denver, CO 80202-2405
Parker, John C.	Microbiological Associates, Inc., Rockville, MD 20850
Payne, Gregory	Macmillan Publishing Company, New York, NY 10022
Pedersen, Jens	National Food Agency, DK-2860 Soborg, DENMARK
Peeler, Maria V.	St. Martin's College, Olympia, WA 98504-6811
Perkins, Ed	Washington State University, Pullman, WA 99164
Perla, Donna	U.S. Environmental Protection Agency, Washington, DC 20460

Peyton, Thomas	AmTech Consultants, Lafayette, IN 47902
Pflug, Dale	BioTrol, Inc., Chaska, MN 55318
Pickwell, George V.	Naval Ocean Systems Center, San Diego, CA 92152-5000
Pogell, Burton M.	University of Maryland, Baltimore, MD 21201
Pollock, Thomas J.	Syntro Corporation, San Diego, CA 92121
Pritchard, P.H.	U.S. Environmental Protection Agency, Gulf Breeze, FL 32561
Pryfogle, Peter	Idaho National Engineering Laboratory, Idaho Falls, ID 83415
Rams, Jacqueline M.	CDM Federal Programs Corporation, Fairfax, VA 22033
Roberts, Leslie	Science Magazine, Washington, DC 20005
Roberts, Todd	University of Washington, Seattle, WA 98195
Robertson, Tom	U.S. Environmental Protection Agency, Portland, OR 97204
Rose, Candace M.	Argonne National Laboratory, Argonne, IL 60439
Rosenstock, Paul D.	Rohm and Haas Delaware Valley, Inc., Bristol, PA 19007
Ross, Bill	U.S. Environmental Protection Agency, Seattle, WA 98101
Ross, Derek	ECOVA Corporation, Redmond, WA 98052
Ruckelshaus, William D.	William D. Ruckelshaus Associates, Seattle, WA 98101
Ruddick, John	SRH Associates, Portland, OR 97232
Rushin, Carol	U.S. Environmental Protection Agency, Seattle, WA 98101
Ryan, John	Remediation Technologies, Inc., Kent, WA 98032
Sambataro, Joseph	ECOVA Corporation, Redmond, WA 98052
Sanford, Robert	ECOVA Corporation, Redmond, WA 98052
Sangodkar, Udaykumar	U.S. Environmental Protection Agency, University of Illinois, Gulf Breeze, FL 32561
Sather, Norm	Argonne National Laboratory, Argonne, IL 60439
Schafer, Carl W.	ECOVA Corporation, Princeton, NJ 08542
Schroeder, Alan	U.S. Department of Energy, Washington, DC 20585
Seamons, Robert	University of Washington, Seattle, WA 98195
Serell, Marco	Tektronix, Inc., Beaverton, OR 97077
Sferra, P.R.	U.S. Environmental Protection Agency, Cincinnati, OH 45268
Shanks, Patricia L.	McCutchen, Doyle, Brown & Enersen, San Francisco, CA 94111
Shellooe, Jim	ECOVA Corporation, Redmond, WA 98052
Shelton, Daniel R.	U.S. Department of Agriculture, Beltsville, MD 20705
Shepherd, Meggen	University of Washington, Seattle, WA 98195
Shivji, Mahmood S.	University of Washington, Seattle, WA 98195
Sikes, David	ECOVA Corporation, Redmond, WA 98052
Silver, Marc A.	Johnston Associates, Inc., Princeton, NJ 08540
Simon, Greg	United States Congress, Washington, DC 20515
Sims, Ronald C.	Utah State University, Logan, UT 84332

Singh, Sadhana C.	University of Texas at Austin, Austin, TX 78712
Skladany, George J.	Detox, Inc., Newport Beach, CA 92660
Small, Margaret	U.S. Environmental Protection Agency, Seattle, WA 98101
Smallbeck, Donald	Harding Lawson Associates, Novato, CA 94947
Smith, George B.	AWD Technologies, South San Francisco, CA 94080
Smith, Peter	University College, Galway, IRELAND
Spain, James	United States Air Force, Tyndall Air Force Base, FL 32403-6001
Speece, R.E.	Drexel University, Philadelphia, PA 19104
Spiker, Jennie	University of Idaho, Moscow, ID 83843
St. John, William	ECOVA Corporation, Redmond, WA 98052
Stamnes, Robert L.	U.S. Environmental Protection Agency, Seattle, WA 98101
Stanker, Larry	Lawrence Livermore National Laboratory, Livermore, CA 94550
Stefan, Fran	Department of Ecology, Southeast Regional Office, Olympia, WA 98504-0811
Steinert, Scott A.	Computer Sciences Corporation, San Diego, CA 92110
Stensel, David	University of Washington, Seattle, WA 98195
Stevenson, Ronald	Sacramento Army Depot, Sacramento, CA 95813
Stewart, Bernice C.	Prince Georges Community College, Largo, MD 20772
Stoddard, Charles L.	Wyckoff Company, Seattle, WA 98126
Storm, Patricia C.	U.S. Environmental Protection Agency, Seattle, WA 98101
Strauss, W.M.	Monsanto Corporation, St. Louis, MO 63167
Strehler, Andrew	ECOVA Corporation, Redmond, WA 98052
Stroo, Hans	ECOVA Corporation, Redmond, WA 98052
Stroud, John A.	Professional Corporation, Sacramento, CA 95819
Sugino, Alan	University of Washington, Seattle, WA 98195
Suidan, Makram T.	University of Illinois, Urbana, IL 61801
Suk, William A.	National Institute of Environmental Health Sciences, Research Triangle Park, NC 27709
Summers, Anne O.	Massachusetts Institute of Technology, Cambridge, MA 02139
Swanson, Robin	KeyTronic, Inc., Spokane, WA 99214
Tabb, Eric	Department of Labor & Industries, Seattle, WA 98119
Tacoronti, Rudolph V.	University of Washington, Seattle, WA 98125
Taylor, Patricia A.	Hazardous Waste Department, CH2M HILL, Bellevue, WA 98009
Taub, Frieda B.	University of Washington, Seattle, WA 98195
Tchelet, Ronen	Industrial Area, Kiryat-Shmona 10-200, ISRAEL
Teeter, David	U.S. Environmental Protection Agency, Seattle, WA 98101
Thomas, Siobhan	BioTech Northwest, Woodinville, WA 98072
Thompson, Neil	U.S. Environmental Protection Agency, Seattle, WA 98101

Tiedje, James	Michigan State University, East Lansing, MI 48824
Timmis, Kenneth N.	University of Geneva, 1211 Geneva 4, SWITZERLAND
Topp, Ed	University of Minnesota, Navarre, MN 55392
Torpy, Michael	ECOVA Corporation, Redmond, WA 98052
Trautman, Walter	ECOVA Corporation, Redmond, WA 98052
Treser, Chuck	University of Washington, Seattle, WA 98195
Unterman, Ronald	General Electric Company, Schenectady, NY 12301
Van Leuven, Leah J.	ECOVA Corporation, Redmond, WA 98052
Vasconcelos, Jay	U.S. Environmental Protection Agency, Seattle, WA 98101
Venosa, Albert D.	U.S. Environmental Protection Agency, Cincinnati, OH 45268
Ventullo, Roy M.	University of Dayton, Dayton, OH 45469
Verhey, Steve	University of Washington, Seattle, WA 98195
Vernon, Conrad	University of Alaska, Fairbanks, AK 99775-1760
Vestal, J. Robie	University of Cincinnati, Cincinnati, OH 45221-0006
Vonifield, Thomas	Department of Health and Human Services, Seattle, WA 98121
Walia, Satish K.	Oakland University, Rochester, MI 48063
Wallace, Blair	B.C. Research, Vancouver, B.C. V6S 2L2, CANADA
Walsh, John	University of Massachusetts, Dedham, MA 02026
Wang, David	Department of Labor & Industries, Vancouver, WA 98666
Wann, David	U.S. Environmental Protection Agency, Denver, CO 80202-2405
Watt, Joe C.	Stearns Catalytic Division of United Engineers, Philadelphia, PA 19101
Weathersby, Jeff	Tacoma News-Tribune, Tacoma, WA 98411
Weishaar, Michael	Monsanto Chemical Company, St. Louis, MO 63167
Weist, Robert D.	Amgen Incorporated, Thousand Oaks, CA 91320
Welch, Juliet W.	University of California, Berkeley, CA 94720
West, Hillary	New York State Legislative Commission on Science and Technology Assembly, Albany, NY 12248
Westlake, D.	University of Alberta, Edmonton, Alberta T6G 2E9, CANADA
White, Megan	Washington State Department of Ecology, Olympia, WA 98504
Wicks, Patrick	ERM Northwest, Redmond, WA 98052
Wiegel, Juergen	University of Georgia, Athens, GA 30602
Wildung, R. E.	Battelle Northwest, Richland, WA 99352
Williams, Barbara R.	OTA/National Institutes of Health, Washington, DC 20510-8025
Williams, Marcia E.	U.S. Environmental Protection Agency, Washington, DC 20460
Williams, Richard	Roy F. Weston, West Chester, PA 19317

Williamson, Kenneth	Oregon State University, Corvallis, OR 97330
Wilson, Claire M.	Council for Research Planning in Biological Sciences, Washington, DC 20036-2077
Wilson, Neil M.	University of Washington, Seattle, WA 98195
Winter, Robert	Amgen, Inc., Thousand Oaks, CA 91320
Wojinski, Stanley F.	ECOVA Corporation, Redmond, WA 98052
Wolleson, Ward	J.R. Simplot, Pocatello, ID 83204
Wong, Philip	U.S. Environmental Protection Agency, Seattle, WA 98101
Woods, James S.	University of Washington, Seattle, WA 98195
Woodward, Dick	ERT, Inc., Houston, TX 77036
Yamamoto, Deb	U.S. Environmental Protection Agency, Seattle, WA 98101
Yerkes, Valerie	Department of Public Services, Seattle, WA 98105
Ying, Anthony	Occidental Chemicals/Treatek, Niagara Falls, NY 14302
Zylstra, Gerben	University of Texas at Austin, Austin, TX 78712-4941

INDEX